Applications of Laser-Driven Particle Acceleration

Applications of Laser-Driven Particle Acceleration

Edited by
Paul R. Bolton
Katia Parodi
Jörg Schreiber

CRC Press
Taylor & Francis Group
Boca Raton London New York

CRC Press is an imprint of the
Taylor & Francis Group, an **informa** business

MATLAB® is a trademark of The MathWorks, Inc. and is used with permission. The MathWorks does not warrant the accuracy of the text or exercises in this book. This book's use or discussion of MATLAB® software or related products does not constitute endorsement or sponsorship by The MathWorks of a particular pedagogical approach or particular use of the MATLAB® software.

CRC Press
Taylor & Francis Group
6000 Broken Sound Parkway NW, Suite 300
Boca Raton, FL 33487-2742

First issued in paperback 2020

© 2018 by Taylor & Francis Group, LLC
CRC Press is an imprint of Taylor & Francis Group, an Informa business

No claim to original U.S. Government works

ISBN-13: 978-0-367-57151-1 (pbk)
ISBN-13: 978-1-4987-6641-8 (hbk)

Library of Congress Cataloging-in-Publication Data

Names: Bolton, Paul R., editor. | Parodi, Katia, 1975- editor. | Schreiber, Jorg, 1976- editor.
Title: Applications of laser-driven particle acceleration / edited by Paul R. Bolton
(Ludwig-Maximilians-Universitat Munchen, Germany), Katia Parodi
(Ludwig-Maximilians-Universitat Munchen, Germany), Jorg Schreiber
(Ludwig-Maximilians-Universitat Munchen, Germany).
Description: Boca Raton, FL : CRC Press, Taylor & Francis Group, [2018] | Includes bibliographical references and index.
Identifiers: LCCN 2018000497 | ISBN 9781498766418 (hardback ; alk. paper) | ISBN 1498766412
(hardback ; alk. paper) | ISBN 9781498766425 (ebook) | ISBN 1498766420 (ebook)
Subjects: LCSH: Particle accelerators. | Lasers.
Classification: LCC QC787.P3 A72 2018 | DDC 539.7/3--dc23
LC record available at https://lccn.loc.gov/2018000497

Visit the Taylor & Francis Web site at
http://www.taylorandfrancis.com

and the CRC Press Web site at
http://www.crcpress.com

Contents

PART I   Laser-Driven Particle Acceleration and Associated Energetic Photon and Neutron Generation: Current Understanding and Basic Capabilities

PART II   Applications of Laser-Driven Particle Acceleration

Foreword

It is a great pleasure for me to introduce this timely work which highlights a wide variety of possible applications for laser-driven particle acceleration. The editors of the ALPA book are experts in medical and laser–plasma physics research with interests that include applications to biomedicine. Their book with twenty-four chapters represents a concerted effort on the part of the many authors to present data addressing relevant scientific developments and selected applications. These results will promote and guide further progress in a research field with a high potential for the benefit of modern society.

As a basis, Part I focusses on a general understanding of acceleration mechanisms and demonstrated capabilities of new technologies under unique laser-driven acceleration conditions. While the achieved maximum kinetic energies reported to date for ions remain below 100 MeV/nucleon, scaling what we know for reaching much higher energies mandates higher focused laser intensities and higher power laser drivers in general. This further motivates many ongoing experimental programmes which can ultimately determine impactful limitations to the energy yield with such laser scaling.

Part II is devoted to potential applications of laser-driven generation of energetic photon and particle beams. Expert authors provide a broad selection of applied investigations for which required sophistication can vary significantly. For example, some material and device irradiation can be simply implemented by direct source emission, while charged particle radiography and experimental radiation biology can require controlled beams of the more sophisticated 'accelerator system', ultimately paving the way for a new kind of laser-driven clinical cancer treatment.

The value and necessity of long-term strategies and commensurate funding that enables the phased progression of laser-driven particle accelerator capabilities for multiple uses is made clear in this assortment of candidate application topics. This is particularly the case for the more distant uses with significantly more challenging and stringent beam requirements, such as laser-driven radiotherapy. Challenging cases like this will especially benefit from combining the long-term view with relevant precursory application achievements in the laboratory. We then recognize that demonstrated versatile utility in the near-term for laser-driven particle acceleration can help realize the extended support that is essential for moving this research forward on the grander scale. It is clear that the near-term and long-term viewpoints can therefore be cooperative and coordinative. Contributors to the ALPA book have collectively presented this healthy, broader view and a compelling case to the reader for the first time. The ALPA book is important and necessary.

Dr. Michael Molls
Professor Emeritus, Radiation Oncology
Spokesperson, Technical University of Munich Emeriti of Excellence
Munich, Germany

Editors' Preface

The laser has historically been a paragon for innovative science and expansive technical development. Our purpose in compiling the ALPA book has been to emphasize this encouraging feature as it relates to particle acceleration and accelerators. For decades, lasers have been a basis for innovative contribution to particle accelerator advancement. Appropriately, the context of the book at this embryonic stage, is formative accelerator development with laser-drivers and the potentially broad utility of laser-driven particle acceleration. We have presented this in two parts. Part I reviews our current understanding, capabilities and state-of-the-art demonstration of laser-driven particle acceleration as well as associated energetic neutron and photon generation. Part II is a broader encounter in which experts describe a varied sampling of potential applications that are addressed to applied scientific and engineering researchers. Particularly for Part II, one need not be a laser or laser–plasma expert to benefit from this material. (Although figures in the ALPA book are printed using grayscale, those figures for which correct or effective interpretation rely on colour presentation have been selected from each chapter and are separately available on the CRC Press website.)

Consistent with the title, we adopt an application-centric view. This view must also be intrinsically broad, realizing from the history of laser development and laser interactions that such exploration can readily lead pioneers to new fields of scientific enquiry and technical capability, significantly impacting methodologies and strategies. The presented application examples highlight a vast potential and its diversity. Befitting it is that this published result is a cumulative effort by many experts providing a broad scope of research subject matter. Because of the intrinsic coupling with high-power laser technology, applications of laser-driven particle acceleration can be one of the more notable and valuable outcomes of the petawatt laser era. By informing readers representing multiple research communities about application possibilities for laser-driven particle acceleration, we hope that the material in this book will inspire new thinking to envision and develop more applications that can uniquely benefit from the laser-driven case. We enthusiastically pursued this project to present meaningful prospective opportunities that can be realistic and even exciting.

Paul R. Bolton
Katia Parodi
Jörg Schreiber
Garching, Germany

MATLAB® is a registered trademark of The MathWorks, Inc. For product information, please contact:

The MathWorks, Inc.
3 Apple Hill Drive
Natick, MA 01760-2098 USA
Tel: (508) 647-7000
Fax: (508) 647-7001
E-mail: info@mathworks.com
Web: http://www.mathworks.com

About the Editors

Prof. Dr. Paul R. Bolton is a visiting professor in the physics faculty at the Ludwig-Maximilians-Universität München (LMU) in Germany where he explores laser-driven ion accelerator sources and system development for novel applications. His early graduate studies of laser spectroscopy were conducted at the University of Toronto. As part of his PhD research at Yale University, he studied accelerator-generated exotic atoms. Paul Bolton began his professional career in 1982 with industrial research and development in the San Francisco Bay area where he also held an adjunct professorship in the physics department of San Jose State University. He later joined the Lawrence Livermore National Laboratory and, as a senior scientist, investigated atomic and plasma physics with high-power laser and high-power microwave sources. In 2000, he began exploring the ubiquitous role of lasers in accelerator advancement at the SLAC National Accelerator Laboratory with RF photoinjector development for its X-ray free-electron laser. In 2008 this interest led him to the Kansai Photon Science Institute in Kyoto where he became the Deputy Director General responsible for research that included laser-driven particle acceleration and associated generation of energetic photon sources. Paul Bolton joined the Chair of Medical Physics at LMU in 2014. He maintains keen interest in broadband, high-power laser systems and their specialized application to ultra-fast laser–atomic physics, extreme laser–plasma physics, plasma optics and particle acceleration.

Prof. Dr. Katia Parodi received her PhD in Physics from the University of Dresden, Germany in 2004. She then worked as a postdoctoral fellow at the Massachusetts General Hospital and Harvard Medical School in Boston, USA. In 2006, she returned to Germany as tenured scientist and group leader at the Heidelberg Ion Therapy Center, obtaining in 2009 her habilitation from Heidelberg University. Since 2012 she is a full professor and Chair of Medical Physics in the physics faculty of Ludwig-Maximilians-Universität München (LMU Munich), where she initiated a dedicated specialization curriculum in medical physics. Her main research interests and ongoing projects are high-precision, image-guided radiotherapy with special focus on ion beams from both established conventional sources and novel laser-driven systems.

Prof. Dr. Jörg Schreiber received his PhD degree in physics at the Ludwig-Maximilians-Universität München (LMU) in 2006. He was awarded a postdoctoral fellowship from the German Academic Exchange Council to study in depth laser–plasma interactions at Imperial College in London. On his return to Germany in 2010, he accepted a position at the Max Planck Institute for Quantum Optics, where he led a team within the Munich Centre for Advanced Photonics to demonstrate applicability of laser-ion acceleration to biological studies. Since 2011, Jörg Schreiber is an associate professor in the Chair of Medical Physics in the physics faculty at LMU. Developing laser-particle acceleration within the Centre for Advanced Laser Applications into a mature, versatile and applicable tool represents his prime research interest.

Contributors

Devesh K. Avasthi
Amity Institute of Nanotechnology
Directorate of Engineering and Technology
Amity University
Uttar Pradesh, India

Paul R. Bolton
Department of Medical Physics
Faculty of Physics
Ludwig-Maximilians-Universität München
München, Germany

Peter Böni
Physik-Department E21
Lehrstuhl für Neutronenstreuung
Technische Universität München
Garching, Germany

David Denis-Petit
Université de Bordeaux, CNRS-IN2P3
Centre Etudes Nucléaires de Bordeaux
    Gradignan
Gradignan, France

Flyura Djurabekova
Helsinki Institute of Physics and Department
    of Physics
University of Helsinki, PB 43
Helsinki, Finland

Brendan Dromey
Centre for Plasma Physics
School of Mathematics and Physics
Queens University Belfast
Belfast, United Kingdom

Wolfgang Enghardt
OncoRay – National Center for Radiation
    Research in Oncology
Faculty of Medicine
University Hospital Carl Gustav Carus
Technische Universität Dresden
and
Department of Radiotherapy and Radiation
    Oncology
University Hospital Carl Gustav Carus
Technische Universität Dresden
and
Institute of Radiooncology – OncoRay
Helmholtz-Zentrum Dresden - Rossendorf
Dresden, Germany

R. Joel England
SLAC National Accelerator Laboratory
Menlo Park, California

Anna A. Friedl
Department of Radiation Oncology
University Hospital
Ludwig-Maximilians-Universität München
München, Germany

Kazuhisa Fujita
The Graduate School for the Creation of New
    Photonics Industries
Shuzuoka, Japan

Christine E. Hellweg
German Aerospace Center
Institute of Aerospace Medicine
Radiation Biology
Cologne, Germany

Peter Hommelhoff
Physics Department
Friedrich-Alexander-Universität
    Erlangen-Nürnberg
Erlangen, Germany

Mitsuru Imaizumi
Research and Development Directorate
Japan Aerospace Exploration Agency (JAXA)
Tsukuba, Ibaraki, Japan

Shunsukei Inoue
Institute for Chemistry
Kyoto University
Kyoto, Japan

Stefan Karsch
Faculty of Physics
Ludwig-Maximilians-Universität München
München, Germany

Yoshiaki Kato
The Graduate School for the Creation of New
    Photonics Industries
Shuzuoka, Japan

Konstantin Khrennikov
Faculty of Physics
Ludwig-Maximilians-Universität München
München, Germany

Ken W.D. Ledingham
University of Strathclyde
and
Scottish Universities Physics Alliance (SUPA)
University of Glasgow
Glasgow, United Kingdom

Robert J. Ledoux
Passport Systems, Inc.
North Billerica, Massachusetts

Jun Ma
Laboratoire de Chimie Physique CNRS
Université Paris-Sud
Orsay, France

Andrea Macchi
Consiglio Nazionale delle Ricerche
Istituto Nazionale di Ottica (CNR/INO)
Adriano Gozzini Laboratory
Pisa, Italy

Victor Malka
Laboratoire d'Optique Appliquee – UMR7639,
    ENSTA-Paris Tech
CNRS, Ecole Polytechnique
Université Paris-Saclay
Palaiseau, France

and

Department of Physics and Complex Systems
Weizmann Institute of Science
Rehovot, Israel

David Mascali
INFN-Laboratori Nazionali del Sud
Catania, Italy

Paul McKenna
University of Strathclyde
and
Scottish Universities Physics Alliance (SUPA)
University of Glasgow
Glasgow, United Kingdom

Kunioki Mima
The Graduate School for the Creation of New
    Photonics Industries
Shuzuoka, Japan

Kengo Moribayashi
Kansai Photon Science Institute
National Institutes for Quantum and
    Radiological Science and Technology
Kyoto, Japan

Mehran Mostafavi
Laboratoire de Chimie Physique CNRS
Université Paris-Sud
Orsay, France

Kai Nordlund
PB 43 (Pietari Kalmin katu 2)
Helsinki Institute of Physics and Department
    of Physics
University of Helsinki
Helsinki, Finland

Takeshi Ohshima
Takasaki Advanced Radiation Research
    Institute
Quantum Beam Science Research Directorate
National Institutes for Quantum and
    Radiological Science and Technology (QST)
Takasaki, Japan

Yoshie Otake
Neutron Beam Technology Team
RIKEN Center for Advanced Photonics
Saitama, Japan

Katia Parodi
Department of Medical Physics
Faculty of Physics
Ludwig-Maximilians-Universität München
München, Germany

Jörg Pawelke
OncoRay – National Center for Radiation
    Research in Oncology
Faculty of Medicine
University Hospital Carl Gustav Carus
Technische Universität Dresden
and
Institute of Radiooncology – OncoRay
Helmholtz-Zentrum Dresden - Rossendorf
Dresden, Germany

Winfried Petry
Technische Universität München
Physik Department &
    Forschungsneutronenquelle Heinz
    Maier-Leibnitz (FRM II)
Garching, Germany

Günther Reitz
German Aerospace Center
Institute of Aerospace Medicine, Radiation
    Biology
Cologne, Germany

and

Nuclear Physics Institute of Czech Academy
    of Sciences
Department of Radiation Dosimetry
Prague, Czech Republic

Luis Roso
Centro de Láseres Pulsados, CLPU
Salamanca, Spain

Markus Roth
Institut für Kernphysik
Technische Universität Darmstadt
Darmstadt, Germany

Shuji Sakabe
Institute for Chemistry
Kyoto University
Kyoto, Japan

Thomas E. Schmid
Institute of Innovative Radiotherapy (iRT)
Helmholtz Zentrum München
Oberschleißheim, Germany

and

Department of Radiation Oncology
Klinikum rechts der Isar
Technische Universität München
München, Germany

Uli Schmidhammer
Laboratoire de Chimie Physique CNRS
Université Paris-Sud
Orsay, France

Jörg Schreiber
Department of Medical Physics
Faculty of Physics
Ludwig-Maximilians-Universität München
München, Germany

Reinhard Schulte
Division of Biomedical Sciences
Department of Basic Sciences
Loma Linda University
Loma Linda, California

Naoya Shikazono
Kansai Photon Science Institute
National Institutes for Quantum and
    Radiological Science and Technology
Kyoto, Japan

Klaus M. Spohr
University of Strathclyde
Glasgow, United Kingdom
and
University of the West of Scotland Paisley
    Campus
Paisley, United Kingdom

Sanjeev K. Srivastava
Department of Physics
Indian Institute of Technology Kharagpur
West Bengal, India

Medhi Tarisien
Université de Bordeaux, CNRS-IN2P3
Centre Etudes Nucléaires de Bordeaux
    Gradignan
Gradignan, France

Peter G. Thirolf
Department of Medical Physics
Faculty of Physics
Ludwig-Maximilians-Universität München
München, Germany

Salvatore Tudisco
INFN-Laboratori Nazionali del Sud
Catania, Italy

Jan J. Wilkens
Department of Radiation Oncology
Klinikum rechts der Isar
Technische Universität München
München, Germany

and

Physics Department
Technische Universität München
Garching, Germany

1

ALPA Introduction

Paul R. Bolton

Katia Parodi

Jörg Schreiber

For the past few decades researchers have continued to explore the extent to which extreme laser–plasma environments, abruptly created at a target site by pulsed laser irradiation of very high intensity, can generate energetic particle and associated photon emission. In the case of ion acceleration, this has been especially true since 2000 [Snavely 2000]. For electrons, laser systems with peak powers of hundreds of terawatts (TW) can accelerate bunches to GeV energy levels with charges of 0.1 nanoCoulombs or more and for an intrinsically ultrashort 'at-source' duration. For laser-driven ion acceleration, higher laser powers are typically needed, but the highest kinetic energies to date remain subrelativistic. For example, hundreds of terawatts of peak laser power can accelerate proton bunches to tens of MeV energies (the recently published maximum energies approach 100 MeV [Wagner 2016, Higginson 2018]) with nano-Coulomb charge levels and with an intrinsically ultrashort 'at-source' duration. These proton energies are therefore adequate for also generating energetic neutrons in nuclear reactions. In general, we can regard laser-driven particle acceleration as an innovative optical component of accelerator advancement. This ALPA book presents details of laser-driven particle acceleration in two parts; where Part I describes our current state of understanding, development and capability for sources of ions, neutrons, electrons and associated energetic photons. The case of direct electron acceleration by laser fields is also included in this first part. Addressing the core theme of this compiled work, Part II presents an assortment of potential applications for laser-driven energetic sources. This we believe to be a subset of application possibilities, where those that can notably benefit from and even exploit the unique features of laser-driven energetic sources are of special interest in the near term. Experimental demonstrations that reveal improved technical capabilities can engender a growing optimism for the laser-driven case as a source for meaningful applications and inquiry in various research fields.

1.1 General Aims of This Book

Written for the general scientific and technical researcher, our primary aim is to inform experts and non-experts about laser-driven particle acceleration and the wealth of research opportunities afforded by its novel source capability. Purposeful scientific and technical development of laser-driven acceleration requires meaningful applications. Guiding and otherwise advancing scientific and technical development to higher maturity levels requires that applications be ultimately doable (especially in the near term). As incremental milestone accomplishments, applications can serve to measure stepwise progress with laser-driven acceleration. In parallel, the unique energetic particle and photon yields generated by a high-power laser can be the basis of useful test beds that enable further development of sophisticated targetry, novel instrumentation (for diagnostics and control of laser pulses, plasmas, and particle or photon beams) and beam line components.

Therefore, we hope this compiled work will ultimately inspire and engage a pioneering community of visionary users from a broad spectrum of scientific and technical inquiry. Healthy engagement of this sort can promote diverse research agendas and organized scientific exchange (typically in meetings and published documentation) that can highlight, promote and steer advancement of the laser-driven case. This naturally includes the evolution of common concepts and terminology as a basis for workable connections and improved understanding between multiple communities. Further, it ushers a heightened awareness of trends in evolving capabilities. As a new technical path, it remains essential to continually assess laser-driven capabilities with increasing sophistication and this must be done in a meaningful context (for example, in the context of current accelerator development and its history). In so doing, progressive and comprehensive strategic visions for the future can emerge. It is clear that these assessments are dynamic and must therefore be ongoing; particularly given the continued rapid progress of conventional accelerator development. Further, the pursuit of applications in this active environment brings a practical added value to novel schemes for particle acceleration by lasers.

On account of the typically high-power requirements, studies of laser-driven ion acceleration are closely coupled with high power (~ petawatt) laser development. We anticipate that the current petawatt (PW) era for laser systems will necessarily feature the codevelopment of necessary companion technologies that can enable and enhance our understanding of ultrafast extreme field phenomena in extreme laser–plasma interactions. They can also be essential for finessing laser-driven ion and electron acceleration processes. Reciprocally, laser-driven particle acceleration and photon generation in this extreme setting can also provide key diagnostics for high-power (and high intensity/field) lasers and studies of laser–plasma interactions at these levels. We therefore aim to also heighten the reader's awareness of this symbiosis between high-power laser technology development in the petawatt era, extreme laser–plasma interactions and laser-driven particle acceleration.

Nonetheless, we are still in an embryonic stage of its development. This is especially true if one considers full accelerator systems (of which more is written later in this introduction and in the conclusion) where the 'unanticipated' is a key environmental feature in exploratory research. It will be necessary to accommodate new ideas and favour applications that become increasingly feasible as the laser-driven capability is extended and inevitably becomes more sophisticated. In the end, we hope that this work will also motivate the reader to adopt a practical comprehensive view of achieved progress and directions for the laser-driven case and to consider expansive future prospects.

1.2 Lasers in Accelerator Development: Some Important Concepts and Milestones

The history of laser-based contribution to accelerator advancement exposes a naturally evolving path of guided development with notable achievements in sources, diagnostics, particle optics and acceleration mechanisms. Addressing this important history is not the purpose of this book. However, the following four examples are worthy of mention: the laser-irradiated electron gun as the photoemission source for the 'RF photoinjector' which was first reported in 1985 [Fraser 1985], ultrafast (broadband) electro-optic detection of ultrashort electron bunches [Jamison 2003], direct acceleration of electrons by longitudinal laser fields in microstructured dielectric accelerator sections (the so-called dielectric laser accelerator, DLA [England 2014] which is described in Part I) and laser-induced localized plasmas used as either accelerator sections or optics (lenses) for high energy electrons and positrons [Ng 2001] (please note that it can be useful to more specifically categorize plasma optics applications as 'plasma photonics' for handling photons, 'plasma electronics' for handling electrons and positions and 'plasma ionics' for handling ions). In comparison with the electron gun (i.e. the well-established RF photoinjector), we can regard the laser-driven source as a 'laser-plasma photoinjector' in both electron and ion cases. Because the plasma at the target is the endpoint for the driving laser pulse and the site for converting laser pulse energy to particle kinetic energy, it is also primarily a source of secondary particles and photons (the laser-driver being the primary one).

For ions, the laser-plasma photoinjector typically emits from an irradiated target region a divergent 'spray' of particles which can feature a significant energy spread (being much greater than the 0.1% level typical of conventional (not laser-driven) accelerators). Some simpler applications might take advantage of these intrinsic at-source features. The source comprises the laser, the target and the laser-driven plasma which includes all plasma fields. We do not regard this source as a complete accelerator system. Because more challenging applications can feature more sophisticated and stringent particle and photon requirements, the development of particle and photon beams that are repetition-rated, stable, well-controlled and therefore well-diagnosed is mandated in these more difficult cases. Ideally this is provided by the integrated laser-driven particle accelerator system (for example, the 'integrated laser-driven ion accelerator system' ILDIAS for ions [Bolton 2016, Schreiber 2016] or the 'integrated laser-driven electron accelerator system' ILDEAS for electrons). Figure 1.1 illustrates the ion example (ILDIAS) and highlights five key components: laser system(s) and photon beam delivery, (relativistic) laser–plasma engineering, targetry, instrumentation for diagnostics and control of charged particle (and neutron) bunches and energetic photon pulses, and beam line optics and design as the transport subsystem for bunch delivery to applications. Applications should be based on unique capabilities; some of which are indicated in the sixth category at the far right in the figure. For a given laser system it is the development of targetry, instrumentation and transport optics for bunch delivery that completes ILDIAS as an ion accelerator. 'Delivery' is then the function that marks the separation between ILDIAS as the machine and applications. The bracketing terms, 'integrated' and 'systems' in ILDIAS functionally highlight the important extension beyond the laser-driven plasma source. It is then important to appreciate the distinction between candidate source development (the subject of many single shot laser–plasma studies to date) and the full integrated laser-driven particle accelerator system. A few additional comments on this issue can also be found in the ALPA Conclusion (Chapter 24).

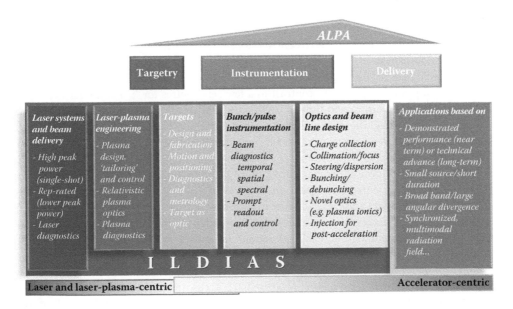

FIGURE 1.1 The ILDIAS concept for ions and its key components. The sixth category on the far right is not part of ILDIAS and lists examples of important features to consider for novel applications.

It is therefore clear that our use of the term, particle 'accelerator' does not refer simply to that which accelerates particles (one might use the term with slightly altered spelling, 'accelerater' for this simpler logical definition which would then include many more devices). We take the common and comprehensive view that an 'accelerator' is a well-controlled integrated multidisciplinary machine of high sophistication delivering high-quality and high directivity beams that are capable of both high and low kinetic energy spread, typically with high precision, stability, reproducibility and reliability. The previous sentence attests the impressive history and technological sophistication of current conventional accelerators. Electric fields are generally used to accelerate charged particles to higher kinetic energies and magnetic fields typically steer and focus them. So, the history of conventional charged particle accelerator development is closely aligned with the history of high-powered electrostatic (DC) and RF/microwave source technology needed to provide adequate accelerating fields and acceleration gradients in beamline structures. Although understandably delayed relative to electrostatic and RF/microwave driven acceleration (which benefitted from a naturally close association with early electron and ion accelerators), particle acceleration driven by lasers can similarly invoke coupled histories of high power laser system development, extreme laser–plasma science and associated particle acceleration.

It is important to understand ongoing accelerator development (laser-driven or otherwise) in the context of its rich history. Initially proposed by Ising for electrons almost a century ago [Ising 1924], the world's first demonstrated accelerator was an RF linac for sodium and potassium ions developed by R. Wideroe [Wideroe 1928]. In 1932 the first cyclotron (which was surprisingly compact) was reported by Lawrence and Livingston [Lawrence 1932]. The next two decades also witnessed new developments of high voltage electrostatic and RF/microwave sources that readily became the drivers for early relativistic electron and subrelativistic proton accelerators as research 'machines'. Notable milestones include the first high voltage generator [van de Graaff 1931, van de Graaff 1933], the klystron tube [Varian 1937, Varian 1939], the first synchrotron proposals [Veksler 1944, McMillan 1945], the first proton linac in 1946 [Alvarez 1955] and the 6 MeV electron linac [Ginzton 1948]. By the 1960s, the basic science and technology of high energy electron linacs was becoming relatively well-established, but the laser was still quite new; being first demonstrated in 1960 [Maiman 1960]. The long linac at the SLAC National Accelerator Laboratory at Stanford University (formerly the Stanford Linear Accelerator Center or SLAC) is a culminating example from that time where accelerating fields of inline RF cavities operating at 2856 MHz reached field amplitudes of tens of MV/m to obtain tens of MeV/m acceleration gradients for electrons and positrons [Chodorow 1955]. It is also the site where impressive laser-based contributions to accelerator technology and even accelerator-based contributions to laser technology (i.e. the LCLS XFEL [Emma 2010]) have been developed.

Typically, laser emission wavelengths are significantly shorter than microwave wavelengths (about 1 micron for high power ir systems in contrast to a few centimetres for 'S' band microwaves). So, we can expect tighter (near) diffraction-limited focusing to field amplitudes and more localized acceleration gradients that are significantly higher than those obtained with microwaves which are guided inside waveguide structures. Seven years after the first demonstration of the maser principle in 1953, the first laser (optical maser) was demonstrated. As a result, extending direct acceleration (of the Wideroe type) of electrons into the optical regime (with much higher frequencies than those of the RF or microwave regime) has been explored for some time. For brevity, we make reference here only to the more recent versions with ir lasers and compact dielectric microstructures. In this direct scheme (dielectric laser acceleration or DLA) there is no mediating plasma and localized longitudinal field amplitudes of 3 GV/m that generate about 250 MeV/m acceleration gradients have been reported with modest laser pulse energies below a millijoule [Peralta 2013].

But even before the arrival of the laser, a 'new principle' of charged particle acceleration, termed 'coherent acceleration', was proposed by Veksler [Veksler 1957]. According to this new thesis, the accelerating field was no longer externally applied and independently controlled, but instead the result of superimposed fields of the charged particles themselves. This prescient work featured, for example, acceleration of highly localized clusters of charged particles in relativistic projectile–plasma target configurations capable of high energy transfer (in one of his suggested configurations Veksler had estimated capability for proton acceleration to TeV levels) with intrinsically perfect synchronism. His visionary proposal clearly foreshadowed laser-driven particle acceleration in plasmas with high power lasers. We might then consider the initial plasma formation and subsequent radiation pressure exerted on it by impinging intense light (i.e. radiation pressure acceleration, RPA) to be a modern realization of Veksler's coherent acceleration principle. For Velsker type schemes, high power laser systems would be needed.

A key enabling optical technique for developing broadband laser systems with high peak (single pulse) powers adequate for exploring particle acceleration in laser-generated plasmas is chirped pulse amplification (CPA). It is interesting that the first report of CPA demonstration [Strickland 1985] was made in the same year as the reported RF photoinjector demonstration. CPA quickly enabled single pulse peak powers of broadband systems to reach terawatt levels by efficiently compressing pulse durations into the femtosecond regime. Laser pulses compressed in this way to about 10 TW levels were used in pivotal reported demonstrations of electron acceleration (generating few femtosecond bunches of significantly improved quality) by laser–plasma wakefields in 2004 [Faure 2004, Geddes 2004, Mangles 2004]. Featuring reduced divergence (to the 10 mrad level), enhanced efficiency (up to 10%), greatly reduced energy spread (as low as 2%) and central energies up to 170 MeV they were collectively referred to as 'Dream beam'. Indeed, the laser–plasma is a Veksler type setting and the accelerating field is attributed to the charge density distribution but such wakefield acceleration of electrons can also be regarded as a localized plasma realization of Wideroe type acceleration; particularly in cases where seed electrons are injected with appropriate timing into a wakefield region.

In 2000, laser–plasma-driven proton acceleration to maximum energies near 58 MeV using the first petawatt ir laser was reported [Snavely 2000]. Laser–plasma schemes investigated for ion acceleration are Veksler-like. Since then, laser-driven proton sources have routinely demonstrated capability for tens of MeV energies (reaching maximum energies that approach 100 MeV [Wagner 2016, Higginson 2018]) driven by tens of TV/m (peak) laser field amplitudes and establishing record level TeV/m acceleration gradients in plasmas. A near century of particle accelerator history (90 years) and near half century (58 years) of laser history (where the last 33 years include CPA) have brought us to the present assessment of laser-driven particle acceleration capabilities and their potential uses.

1.3 Uniqueness in the Laser Case

Much research to date (as described in Part I) has focused on characterizations of energetic particle emission as artefacts of extreme ultrafast laser–plasma interactions which has mainly included laser and targetry development. Typically, this has been done at the single shot (laser pulse) level with keen interest in extreme laser powers and intensities aimed at reaching highest possible particle kinetic energies with high energy conversion efficiencies. Some of these demonstrated or simulated sources can be considered as candidates for the front end of the more sophisticated integrated laser-driven particle accelerator system.

We have already made clear that laser-driven source studies for ions and electrons reveal unique emergent particle bunch characteristics. The energy spread is typically much greater than the 0.1% typical of conventional accelerators which is especially the case for ion sources where it can actually exceed 100%. Electron and ion bunches can be of ultrashort duration (for example, femtoseconds for electrons and tens of femtoseconds for protons) at the source so they can debunch rapidly in transport due to the large energy spread. Notable for ions is also the very large angular divergence which can be tens of degrees. Although appearing unfavourable to conventional acceleration, this parameter space characterizes plasma-based laser-driven particle acceleration. The source descriptions in the chapters of Part I make evident another notable laser-driven source feature which is the promise for synchronous (and tunable) multiple beam capability. This can include synchronous combinations of various particle and photon beams (including also part of the ir pulse energy or its harmonics) as well as different energy beams for a given particle type (for example, two proton beams at different energies where the higher energy one might be used as a probe for prompt diagnostics including imaging). This potential feature can have direct implications for parallel multi-target development.

Because we can anticipate future machine requirements for applications to vary significantly, we envision the laser-driven 'machine' to usher a variety of innovative accelerator concepts which can enable niche-like alternatives that can augment conventional accelerator machine capability. In many cases new diagnostic techniques, multiple laser target and source configurations, new beam optics (and other inline components) and novel beam architectures can emerge; especially for efficient charge collection and bunch transport near the source. It is clear in general that a critical laser-driven accelerator challenge is to mitigate obstacles and exploit uniqueness.

1.4 Organization of This Book

Each chapter is written by expert authors for readers at the scientific and engineering research level. Because we aim to reach a diverse readership with an applications theme as core, we have attempted to use, where possible, consistent definitions and nomenclature throughout. Specific chapter contents and alphabetically ordered reference lists are included in each chapter. Part I follows this introduction and addresses the state-of-the-art of our understanding, development and capability of laser-driven particle acceleration (dealing with ions, neutrons, electrons and energetic photon sources associated with electron acceleration). The demonstrated capability presented in this part is the scientific and technical basis for the selected applications described in Part II. Laser and laser–plasma scientists and engineers might focus initially on the relevant Part I chapters prior to their perusal of Part II applications. More details can be found in the Preamble to Part I.

Part II, as the primary subject matter of this book, highlights an assortment of potential applications of laser-driven particle acceleration. Both near and far term possibilities are included for which the contributors' visionary mindset has been encouraged. We expect Part II to be particularly relevant to readers with specific application priorities in mind. Scientists and engineers associated with fields other than laser and laser-plasma physics might prefer to first consult the Part II chapters that are aligned with their interests. This information can be subsequently augmented by referring to the relevant laser-driven source capability in Part I. Immediately preceding Part II, the Preamble to Part II provides more specific detail and briefly categorizes the included applications. In Chapter 24, readers will find the editors' summary assessment with future prospects for the potential application of laser-driven particle acceleration and associated energetic photon and neutron sources.

References for Chapter 1

Alvarez, L.W. et al., 1955. "Berkeley Proton Linear Accelerator" *Rev. Sci. Inst.* **26**: 111–133.

Bolton, P.R. 2016. "The Integrated Laser-driven Ion Accelerator System and the Laser-driven Ion Beam Radiotherapy Challenge" *Nucl. Inst. and Meth. A* **809**: 149.

Chodorow, M. et al., 1955. "Stanford High-Energy Linear Electron Accelerator (Mark III)" *Rev. Sci. Inst.* **26**: 131–203.

Emma, P. et al., 2010. "First Lasing and Operation of an Angstrom-Wavelength Free-Electron Laser" *Nature Photonics* **4**: 641–647.

England, R.J. et al., 2014. "Dielectric laser accelerators" *Rev. Mod. Phys.* **86**: 1337.

Faure, J. et al., 2004. "A Laser-plasma Accelerator Producing Monoenergetic Electron Beams" *Nature* **431**: 541–544.

Fraser, J.S., Sheffield, R.L. and Gray, E.R. 1985. "High-Brightness Photoemitter Development for Electron Accelerator Injectors" in "Laser Acceleration of Particles" eds. Joshi, C. and Katsouleas, T. *AIP Conference Proceedings* **130**: 598.

Geddes, C.G.R. et al., 2004. "High-quality Electron Beams from a Laser Wakefield Accelerator Using Plasma-channel Guiding" *Nature* **431**: 538–541.

Ginzton, E.L., Hansen, W.W. and Kennedy, W.R. 1948. "A Linear Electron Accelerator" *Rev. Sci. Inst.* **19**: 89.

Higginson, A. et al., 2018. "Near-100 MeV protons via a laser-driven transparency-enhanced hybrid acceleration scheme" *Nature Comm.* **9**:724.

Ising, G. 1924. *"Prinzip einer Methode zur Herstellung von Kanalstrahlen hoher Voltzahl"* Arkiv för Matematik, Astronomi och Fysik 18: 1–4.

Jamison, S.P. et al., 2003. "High-Temporal-Resolution, Single-Shot Characterization of Terahertz Pulses" *Opt. Lett.* **28**: 1710–1712.

Lawrence, E.O. and Livingston, M.S. 1932. "The Production of High Speed Light Ions Without the Use of High Voltages" *Phys. Rev.* **40**: 19.

Maiman, T.H. 1960. "Stimulated Optical Radiation in Ruby" *Nature* 187: 493–494.

Mangles, S.P.D. et al., 2004. "Monoenergetic Beams of Relativistic Electrons from Intense Laser–Plasma Interactions" *Nature* **431**: 535–538.

McMillan, E.M. 1945. "A Proposed High Energy Particle Accelerator" *Phys. Rev.* **68**: 143–144.

Ng, J.S.T. et al., 2001. "Observation of Plasma Focusing of a 28.5 GeV Positron Beam", *Phys. Rev. Lett.* **87**: 244801.

Peralta, E. A. et al., 2013. "Demonstration of Electron Acceleration in a Laser-Driven Dielectric Microstructure" *Nature* **503**: 91–94.

Schreiber, J., Bolton, P.R. and Parodi, K. 2016. "Hands-on Laser-driven Ion Acceleration: A Primer for Laser-Driven Source Development and Potential Applications" *Rev. Sci. Inst.* **87**: 071101.

Snavely, R.A. et al., 2000. "Intense High-Energy Proton Beams from Petawatt-Laser Irradiation of Solids", *Phys. Rev. Lett.* **85**: 2945–2948.

Strickland, D. and Mourou, G. 1985. "Compression of Amplified Chirped Optical Pulses" *Opt. Comm.* **56**: 219–221.

Van de Graaff, R.J. 1931. *Phys. Rev.* **38**: 1919.

Van de Graaff, R.J. et al., 1933. "The Electrostatic Production of High Voltage for Nuclear Investigations", *Phys. Rev.* **43**: 149.

Varian, R.H. 1937. 'from May – July 1937 entries of notebook', *National Archives*, Washington, D.C.

Varian, R.H. 1939. 'U.S. Patent No. 2242275' (applied for in 1937).

Veksler, V.I. 1944. "A New Method of Accelerating Relativistic Particles" *Comptes Rendus (Doklady) de l'Academie des Sciences de l'URSS* 43: 346–348.

Veksler, V.I. 1957. "The Principle of Coherent Acceleration of Charged Particles" *The Soviet Journal of Atomic Energy* 2: 525–528.

Wagner, F. et al., 2016. "Maximum Proton Energy Above 85 MeV from the Relativistic Interaction of Laser Pulses with Micrometer Thick CH_2 Targets", *Phys. Rev. Lett.* **116**: 205002.

Wideroe, R. 1928. *"Über ein neues Prinzip zur Herstellung hoher Spannungen"* Archiv für Elektrotechnik und Übertragungstechnik 21: 387 (in German).

I

Laser-Driven Particle Acceleration and Associated Energetic Photon and Neutron Generation: Current Understanding and Basic Capabilities

Preamble to Part I

Significant experimental achievements and distinctive features of laser-driven particle acceleration are presented in Part I. Organized according to particle type, this part can be viewed in two groups; group one (Chapters 2, 3 and 4) pertains to electrons and associated energetic photon generation and group two (Chapters 5 and 6) pertains to proton (and other ion) acceleration and related energetic neutron production. The two groups can be read in any order. In the first group, Chapters 2 and 3 focus on two kinds of laser-driven electron acceleration; direct acceleration by the longitudinal component of the laser field using microstructured dielectric material and laser wakefield acceleration in laser-induced plasmas, respectively. Chapter 4 presents a broad assessment of laser-accelerated electrons as X-ray and gamma ray sources by various mechanisms. Chapters 2, 3 and 4 also include brief final comments about some applications. It can be helpful to the reader to read Chapter 3 before Chapter 4.

In the second group, Chapter 5 comprehensively addresses our current understanding of ion acceleration in laser plasmas and its unique capabilities, while Chapter 6 describes energetic (nuclear) neutron generation techniques (and a few select applications to which reference is also made in the Preamble to Part II). Such 'hot' neutron generation relies on nuclear processes that follow laser-driven acceleration of proton (and deuteron) projectiles. It is recommended that Chapter 5 be read prior to Chapter 6. Part I then provides for the reader a scientific and technical basis for applying laser-driven particle acceleration and associated energetic photon and neutron generation to meaningful scientific inquiry. Highlighting unique laser-driven capabilities can usher novel applications and facilitate critical assessment of viability and designs for the future.

2

Laser Wakefield
Acceleration of Electrons

Victor Malka

2.1 Introduction

By reaching peak intensities of 10^{18} W/cm², laser pulses interacting with matter are producing, in the relativistic regime, bunches of energetic particles and radiation [Esarey 2009, Malka 2012, Joshi 2007, Malka 2008]. With the recent demonstration of new approaches for producing energetic particle bunches, this field of research is impressively growing. Extremely high electric field amplitudes that can exceed hundreds of the GV/m levels in the laser-created plasma have been shown to be of relevance for particle acceleration. Because of the extreme values of the accelerating field, about 10,000 times larger than those produced in conventional radio-frequency (microwave) cavities, plasma accelerators appear to be very promising for the development of compact accelerators. The fine manipulation of the electron motion in the plasma medium allows an excellent control of electron injection and acceleration. The bunches delivered by laser plasma accelerators have a number of interesting properties (such as short bunch duration, high peak brightness and spatial quality) from which applications in many fields, including medicine, radiobiology, chemistry, physics and material science, security (material inspection) and of course accelerator science, can benefit.

2.2 Principle

In a plasma medium, the accelerating electric field results from the collective oscillation of electrons around ions that are, during the timescale of interest, assumed to be motionless. The accelerating gradient results from the rapid electron plasma oscillation that follows the initial electronic perturbation. In an initially uniform and non-collisional plasma, the displacement of a slab of electrons from their equilibrium position produces a restoring force that drives them back towards the initial equilibrium position. The typical frequency of electron oscillations around the equilibrium position is called the electron plasma frequency ω_{pe}, with $\omega_{pe} = (n_e e^2/m_e \varepsilon_0)^{1/2}$, where n_e is the unperturbed electron density, e is the magnitude of electron charge, m_e is the electron rest mass and ε_0 is the vacuum permittivity.

If $\omega_{pe} < \omega_L$ (where ω_L is the laser frequency) then the characteristic time scale of plasma electron motion is greater than the optical period of the incoming radiation and the medium cannot stop the propagation of the electromagnetic wave (i.e. the laser waveform or pulse). The medium is then transparent and is called 'under-dense'. In the opposite case, when $\omega_{pe} > \omega_L$, then the characteristic time scale of plasma electron motion is fast enough to adapt to the incoming wave and to reflect totally or partially the laser radiation. In this case the medium is called 'over-dense'. These two domains are separated by the frequency ω_L, which therefore corresponds to the critical density, $n_c = \left(\omega_L^2 m_e \varepsilon_0 / e^2 \right)^{1/2}$. For a wavelength $\lambda_L = 1$ µm, one obtains $n_c = 1.1 \times 10^{21}$ cm^{-3}. The typical working range of plasma densities, with current laser technology, is within 10^{17} cm^{-3} to 10^{19} cm^{-3}.

One can easily estimate the value of the electric field, $\mathbf{E}(z, t)$ in the plasma that results from collective oscillation of electrons. In the simplest case, for a periodic sinusoidal perturbation of the electron density in a uniform ion layer, the density perturbation $\delta n = \delta n_e \sin(k_p z - \omega_{pe} t)$, where ω_{pe} and k_p are the angular frequency and the wave number of the plasma wave, respectively. The associated electric field $\delta \mathbf{E}$ is simply derived from the Poisson equation, $\nabla . \delta \mathbf{E} = -\delta n_e e / \varepsilon_0$, with the following expression: $\mathbf{E}(z, t)/E_0 = (\delta n_e / n_e) \cos(k_p z - \omega_{pe} t) \mathbf{e_z}$, where $E_0 = m_e c \omega_{pe} / e$, c is the speed of light and $\mathbf{e_z}$ is the unit vector in the z direction.

The collective oscillation of electrons is, in the case of laser wakefield, excited by the ponderomotive force of the laser. Within a single oscillation period the laser pulse exerts two successive pushes in opposite directions on the plasma electrons. The laser wakefield is a resonant process for which excitation of the electron plasma wave is maximum when the laser pulse duration is of the order of $1/\omega_{pe}$. The peak amplitude of the plasma wave, in the linear regime $a_0 \ll 1$, scales with laser intensity which is proportional to a_0^2 [Gorbunov 1987]:

$$\mathbf{E}(z, t) = E_0 \left(a_0^2 \sqrt{\pi}/4 \right) k_p L \exp\left(-k_p^2 L^2 / 4 \right) \cos(k_L z - \omega_L t) \mathbf{e_z} \qquad (2.1)$$

in which $L = c\tau_L$ is the length of the laser pulse, k_L is the laser wavevector, a_0 is the laser normalized vector potential that scales for a linear polarization as $a_0 = 0.86 \left(I_{18} \lambda_{\mu m}^2 \right)^{1/2}$, I_{18} being the laser peak intensity in 10^{18} W/cm^2 units and $\lambda_{\mu m}$ is the laser wavelength expressed in micrometres. Equation 2.1 explicitly shows the dependence of the amplitude of the wave on the length of the exciting pulse. As shown on Figure 2.1, the maximal value of the plasma wave amplitude is obtained for a laser pulse length $L = \sqrt{2}/k_p$. One can note that in the linear regime, the electric field, as the gradient of a potential, has a sinusoidal variation and can reach maximal amplitudes of a few GV/m.

To accelerate electrons in the electric potential of the plasma wave that moves with a phase velocity close to the speed of light, one has to inject relativistic electrons. This is comparable to the sea wave case, where a surfer has to crawl to gain enough velocity to catch the wave. In the simplest case of such

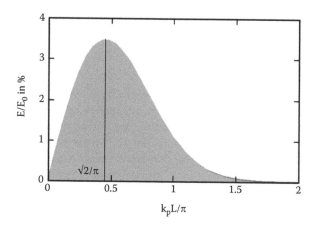

FIGURE 2.1 Variation of the plasma wave amplitude as a function of $k_p L$ for a laser pulse with a normalized vector potential of $a_0 = 0.3$.

very fast plasma waves, the velocity normalized to c, β_p is close to 1, when the electron density is much lower than the critical density $n_e \ll n_c$. In this case, the associated Lorentz's factor $\gamma_p = \left(1 - \beta_p^2\right)^{-1/2}$ is greater than 1, and the maximum energy that an electron can gain is simply given by

$$W_{\max} = 4\gamma_p^2 (\delta n_e / n_e) m_e c^2 \tag{2.2}$$

This energy is reached when the electrons stay in the whole accelerating part of the electric potential of the plasma wave, which is of length $\lambda_p/2$, in the plasma wave frame. This length, which is called the dephasing length L_{deph}, corresponds to a $\lambda_p/2$ rotation in the phase space, is given in the laboratory frame by $L_{\mathrm{deph}} \sim \gamma_p^2 \lambda_p$.

In these formulas, we have considered a unique test electron, which has no influence on the plasma wave. In reality, in the case of efficient trapping of the particles, the high peak current of the injected electrons can modify the electric fields and distort the plasma wave. This is called the space-charge effect or more exactly the beam loading effect (which results from the Coulomb repulsion force between electrons).

This linear theory is difficult to apply to nonlinear regimes, which are explored experimentally and only 3D PIC simulations can explain in detail the experimental results.

Figures 2.2 and 2.3 show the spatial distribution of electron density (top), the accelerating electric field and the radial electric field for two laser and plasma conditions. In the first case linear case (Figure 2.2), the peak normalized laser amplitude is $a_0 = 0.2$, the laser pulse duration is 40 fs and the laser waist is 9 μm. The homogeneous plasma (before laser pulse arrival) has an electron density of $7.7 \times 10^{18}\,\mathrm{cm}^{-3}$.

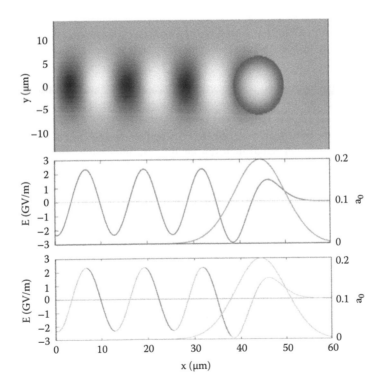

FIGURE 2.2 Linear laser wakefield. Top: isodensity in false colour with the laser pulse in orange, bright regions indicate high electron density. Middle: on-axis longitudinal field in violet (negative values of the electric field means that electrons are accelerated toward right) and laser normalized vector potential in green. Bottom: on-axis radial field in green (red) for focusing (defocusing) and laser normalized vector potential in yellow.

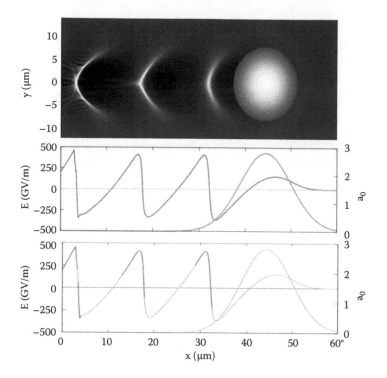

FIGURE 2.3   Nonlinear laser wakefield. Top: isodensity in false colour with the laser pulse in orange. Middle: on-axis longitudinal field in violet and laser normalised vector potential in green. Bottom: on-axis radial field in green (red) for focusing (defocusing) and laser normalized vector potential in yellow.

For this relatively small value of a_0, the plasma responds almost linearly to the perturbation generated by the laser pulse [Tajima 1979]. In this regime, as shown on Figure 2.2 the wakes are smooth with sinusoidal variation for the amplitudes of both components of the electric field. One can also note that the focusing fields have the same lengths as the defocusing fields and that the maxima of longitudinal electric field separates regions of radial focusing fields from defocusing regions. Therefore, bunches centred at these maxima are not well trapped [Malka 2006]. In addition, because the accelerating field is limited to the few GV/m level, it requires tens of centimetres of acceleration to reach the GeV energy level. To achieve such an energy gain, one has to face the problem of guiding the intense laser pulse over distances of tens of centimetres.

Trapping around the maximum of electric field requires bent wakes. Bent wakes are generated in inhomogeneous plasmas (for example, a parabolic density profile) or by working in the nonlinear regime using higher laser intensity. In this case, the bending is due to nonlinear effects. As a_0 grows, wakes become steeper and the wave front becomes more curved due to the increased relativistic shifting of the plasma frequency. As shown in Figure 2.3, the region of focusing field is much larger than in the linear homogeneous case, and interestingly, the focusing field is the highest where the accelerating field is the highest. More importantly, in this nonlinear regime [Malka 2002], since a few hundred GV/m is easily reachable, the GeV energy level can be obtained over a few centimetres distance without any external guiding structure.

2.3 The Different Injection Schemes

In the first experiments, electrons were externally injected into relativistic plasma waves driven by one or two laser pulses. Injected at a few MeV, few electrons were indeed accelerated in the GV/m longitudinal field gaining 1.5 MeV [Amiranoff 1998] and even up to tens of MeV energies [Clayton 1993, Everett 1994]. The accelerated electron bunch was of poor quality and a broad energy distribution because the bunch length of the injected electrons was longer than the plasma wavelength. In 1994, an important step was achieved with the direct production of an electron beam from a gas target irradiated by an intense laser pulse in the so-called self-modulated laser wakefield regime [Sprangle 1992, Antonsen 1992, Andreev 1992]. The high amplitudes reached by the plasma waves together with the efficient heating of electrons allow the breaking of the wakefield (wave-breaking), i.e. the trapping of a copious number of hot electrons that consequently can be accelerated from 0 to tens of MeV within a millimetre length scale [Modena 1995]. With a 1 J, 30 fs laser operating at a 10 Hz repetition rate, the benefit of near-resonant density in the forced laser wakefield regime was first demonstrated. Since then the electron beam quality has been improved towards a narrower divergence angle and energy distribution. In 2004, the acceleration scheme was improved, using the so-called "bubble regime" [Pukhov 2002] with first time acceleration of an electron bunch with a quasi-monoenergetic distribution [Geddes 2004, Faure 2004, Mangles 2004]. However, at that time, the reproducibility of the beam parameters was still not excellent. Several ideas have been proposed since then to improve the quality and the stability of the accelerated electron bunches.

In 2006, stable and energy-tunable electron bunches were observed by using two counter-propagating laser pulses in the colliding scheme [Faure 2006]. The use of two laser pulses enables separation of the injection and acceleration processes. The first laser pulse, the pump pulse, was used to excite the wakefield while the second pulse, the injection pulse, was used to heat electrons during the collision with the pump pulse. After this collision, electrons that gain enough longitudinal momentum (to reach the trapping energy) were trapped and further accelerated in the wakefield, as shown in Figure 2.4. To trap electrons in a regime where self-trapping does not occur, one has to either inject electrons with energies greater than the trapping energy or de-phase electrons with respect to the plasma wave. As mentioned earlier, providing high-quality electron bunches with a low energy spread (i.e. monochromatic bunches), requires injection of electrons within a very short time (or over a very short distance). This can be achieved using additional ultra-short laser pulses whose only purpose is to trigger electron injection [Esarey 1997]. In this scheme, the main pulse (pump pulse) creates a high amplitude plasma wave and collides with a secondary pulse of lower intensity. The interference of the two pulses creates a beat-wave pattern, with zero phase velocity, that heats some electrons from the plasma background. The force associated with this ponderomotive beat-wave is proportional to the laser frequency, ω_L. It is therefore many times greater than the ponderomotive force associated with the pump laser that is inversely proportional to the pulse duration at resonance and therefore proportional to the plasma frequency ω_p. As a result, the mechanism is still efficient even for modest laser intensities. Upon interacting with this field pattern, some background electrons gain enough momentum to be trapped in the main plasma wave and accelerated to high energies. As the overlap time of the lasers is short, the electrons are injected over

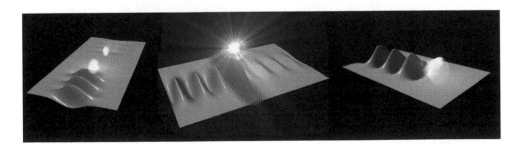

FIGURE 2.4 Principle of the colliding laser pulse scheme. Left: the pump laser pulse creates a wakefield; centre: the two laser pulses collide; right: the injected electrons are accelerated in the wake field created by the pump laser pulse.

a very short distance and can be accelerated to very well-defined energies. This concept was validated in an experiment [Faure 2006], using two counter-propagating pulses. It was shown that the collision of the two lasers leads to narrow-energy spread electron bunches. Moreover, their central energy could be controllably tuned by changing the location of the collision region in the plasma.

High-quality and stable electrons can also be obtained with a single laser pulse using the density gradient injection scheme. This scheme [Bulanov 1998] involves a downward density ramp and relies on the slowing down of the plasma wave velocity at the density ramp. This decrease of the plasma wave phase velocity lowers the threshold for trapping the plasma background electrons and causes wave-breaking of the wakefield in the density ramp.

Injection of electrons in a density transition is based on the idea of controlling the bubble expansion. The bubble size scales as $n_e^{-1/2}$, thus the cavity expands as the laser travels through a longitudinally decreasing plasma density. Before the density decrease, the wake phase velocity is too large for electrons inside the cavity to be trapped: they slip across the bubble without being injected. When the laser enters the down-ramp, the back of the bubble experiences a larger plasma density than the front, and is thus locally decelerated, leading to an increase of the bubble size during the propagation. Electrons inside the cavity are then able to stay a longer time in the accelerating gradients, gaining more energy up to the point where they are eventually trapped in the wake. Long down-ramp injection of electrons presents some drawbacks. First, electrons may enter the bubble and be trapped continuously in the density gradient, leading to rather high charges but also large energy spread. Second, the cavity slows down during the electron bunch acceleration, and the bunch quickly reaches the centre of the cavity where the accelerating field reverses sign.

With a sharper density transition, the injection mechanism is slightly different, as seen in Figure 2.5. When crossing the down-ramp, the sudden cavity half-size increase causes electrons just behind the bubble to be instantly loaded in the accelerating phase of the wakefield. The phase velocity of the bubble is frozen in the sharp transition; thus, the injection is caused by the sudden plasma wavelength increase rather than the reduced phase velocity in this case. Electron injection in a sharp density transition formed by a shock in a supersonic gas flow using a razor blade [Schmid 2010, Buck 2013] has been experimentally demonstrated in the past few years and resulted in high-quality electron bunches. This scheme has been improved further by using a gas mixture consisted of low- and high-Z gases. Electrons ionized from internal atomic levels of the high-Z gas, where the laser crosses the shock front, spend more time in the accelerating field because the cavity expands. They can thus be injected below the threshold for regular ionization injection, leading to localized trapping and low energy spreads [Thaury 2015].

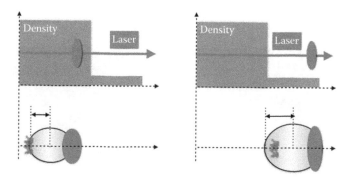

FIGURE 2.5 Schematic representation of a down-ramp electron injection in a sharp density transition.

TABLE 2.1 Electron Beam Parameters for Three Different Laser Energy Ranges: (1–10 mJ), (1–2 J) and (15–30 J)

Table Parameter	Category 1	Category 2	Category 3
Energy (MeV)	0.001–0.01	300–600	1500–4500
Relative energy spread dE/E	0.1	0.01	<0.1
Normalized emittance ε (in πmm.mrad)	1–3	1	1
Transverse size σ_x in μm	1	1	1
Bunch length σ_z in fs	< few fs	few fs	few fs
Charge Q in pC	0.1	10–100	10–500
Divergence in mrad	5–10	1–3	1–3

2.4 Electron Bunch Parameters

Up to now, electron beams with a quasi-monoenergetic distribution have been obtained in the nonlinear regimes by using the injection schemes reported in the different injection schemes. In the many published articles on laser plasma accelerators, the electron beam parameters are reported. Because of the complexity of such measurements, one cannot find the measurement of the electron beam parameters in the literature performed all together. Table 2.1 lists some typical bunch parameter values for three different laser energy ranges: Category 1 (1–10 mJ), Category 2 (1–2 J) and Category 3 (15–30 J) with respectively different repetition rates kHz, 10 Hz and 0.1 Hz, and costs 0.2–0.4 M€, 1–2 M€ and 5–10 M€, respectively.

2.5 Conclusion

The important progress that has been made in laser plasma acceleration, from the first acceleration of externally injected electrons in a GV/m laser wakefield and self-injection in a 100 GV/m laser wakefield with first a broad spectrum, to a series of experiments resulting in narrow energy spread electron bunches with compact 10 Hz laser systems, have boosted this field of research in which tens of laboratories/teams are currently playing important roles. The evolution of short-pulse laser technology with diode pump lasers or fibre lasers, a field experiencing rapid progress, will eventually contribute to their societal applications; in material science, for example, for high resolution gamma radiography [Glinec 2005, Ben-Ismail 2011], in medicine for cancer treatment [Glinec 2006, Fuchs 2009], in chemistry [Brozek-Pluska 2005, Gauduel 2010] and in radiobiology [Malka 2010, Rigaud 2010]. In the near future, the development of compact free electron lasers could open the way for production of intense X-ray pulses, by coupling the electron bunch with an undulator. Due to the very high peak current of a few kiloamperes [Lundh 2011], which is comparable to that used at LCLS (SLAC's XFEL), laser plasma accelerators for free electron lasers, the so-called fifth generation light source, are clearly identified by the scientific community as a major development. Alternative schemes to produce ultra-short X-ray pulses, using Compton, betatron or Bremsstrahlung X-ray generation mechanisms, have also been considered. Tremendous progress has been made in the study of betatron radiation in a laser plasma accelerator. Since its first observation in 2004 [Rousse 2004], and the first monitoring of electron betatron motion in 2008 [Glinec 2008], a number of articles have reported this new source type in more detail, including measures of a sub ps duration [Ta Phuoc 2007] and of transverse size in the micron range [Ta Phuoc 2006]. Betatron radiation was used recently to perform high spatial resolution (about 10 microns) X-ray phase contrast imaging in a single-shot operation mode [Fourmaux 2011, Kneip 2011]. A new idea using plasma undulators has been proposed in which is indicated potential production of tunable quasi-monoenergetic X-rays in the hundreds of keV [Andriyash 2015] regions. A detailed review of X-ray generation using laser plasma accelerators can be found in Corde [2013]. Also in chapter 4 of his book, the reader will find a detailed discussion of laser-accelerated electrons as X-ray and gamma-ray sources by Khrennikov and Karsch.

References for Chapter 2

Amiranoff, F. et al. 1998. "Observation of Laser Wakefield Acceleration of Electrons" *Phys. Rev. Lett.* **81**: 995.

Andreev, N.E., Gorbunov, L.M., Kirsanov, V.I., Pogosova, A.A., and Ramazashvili, R.R. 1992. "Resonant excitation of wake fields by a laser pulse in a plasma" *JETP Lett.* **55**: 571.

Andriyash, I.A., Lehe, R., Lifschitz, A., Thaury, C., Rax, J.-M., Krushelnick, K., and Malka, V. 2015. "Nano-structured plasma wiggler as a source of high-brightness femtosecond X-ray" *Nature Communications* **5**: 4736. (DOI: 10.1038/ncomms5736)

Antonsen Jr., T.M., and Mora, P. 1992. "Self-focusing and Raman scattering of laser pulses in tenuous plasmas" *Phys. Rev. Lett.* **69** (15): 2204.

Ben-Ismail, A. et al. 2011. "Compact and high-quality gamma-ray source applied to 10 μm-range resolution radiography" *Appl. Phys. Lett.* **98**(26): 264101.

Brozek-Pluska, B., Gliger, D., Hallou, A., Malka, V., and Gauduel, Y.A., 2005. "Direct observation of elementary radical events: Low- and high-energy radiation femtochemistry in solutions" *Radiation Chem.* **72**(2–3): 149.

Buck, A. et al. 2013. "Shock-Front Injector for High-Quality Laser-Plasma Acceleration" *Phys. Rev. Lett.* **110**: 185006.

Bulanov, S., Naumova, N., Pegoraro, F., and Sakai, J. 1998. "Particle injection into the wave acceleration phase due to nonlinear wake wave breaking" *Phys. Rev. E* **58**(5): R5257(R).

Clayton, C.E. et al. 1993. "Ultrahigh-gradient acceleration of injected electrons by laser-excited relativistic electron plasma waves" *Phys. Rev. Lett.* **70**: 37.

Corde, S., Ta Phuoc, K., Lambert, G., Fitour, R., Malka, V., Rousse, A., Beck, A., and Lefebvre, E. 2013. "Femtosecond x rays from laser-plasma accelerators" *Rev. of Modern Physics* **85**: 1.

Esarey, E., Hubbard, R.F., Leemans, W.P., Ting, A., and Sprangle, P. 1997. "Electron Injection into Plasma Wakefields by Colliding Laser Pulses" *Phys. Rev. Lett.* **79**: 2682.

Esarey, E., Schroeder, C.B., and Leemans, W.P. 2009. "Physics of laser-driven plasma-based electron accelerators" *Rev. Mod. Phys.* **81**(3): 1229–1285.

Everett, M.J. et al. 1994. "Trapped electron acceleration by a laser-driven relativistic plasma wave" *Nature (London)* **368**: 527.

Faure, J. et al. 2004. "A laser plasma accelerator producing monenergetic electron beams" *Nature* **431**: 541.

Faure, J., Rechatin, C., Norlin, A., Lifschitz, A., Glinec, Y., and Malka V. 2006. "Controlled injection and acceleration of electrons in plasma wakefields by colliding laser pulses" *Nature* **444**: 737.

Fourmaux, S. et al. 2011. "Single shot phase contrast imaging using laser-produced Betatron x-ray beams" *Opt. Lett.* **36**(13): 2426.

Fuchs, T., Szymanowski, H., Oelfke, U., Glinec, Y., Rechatin, C., Faure, J., and Malka, V. 2009. "Treatment planning for laser-accelerated very-high energy electrons" *Phys. Med. Biol.* **54**(11): 3315.

Gauduel, Y., Glinec, Y., Rousseau, J.-P., Burgy, F., and Malka, V. 2010, "High energy radiation femtochemistry of water molecules: Early electron-radical pairs processes" *Eur. Phys. J. D* **60**(1): 121.

Geddes, C.G.R. et al. 2004. "High-quality electron beams from laser wakefield accelerator using plasma-channel guiding" *Nature* **431**: 538.

Glinec, Y. et al. 2005. "High-Resolution γ-Ray Radiography Produced by a Laser-Plasma Driven Electron Source" *Phys. Rev. Lett.* **94**(2): 025003.

Glinec, Y., Faure, J., Lifschitz, A., Vieira, J.M., Fonseca, R.A., Silva, L.O., and Malka, V. 2008. "Direct observation of betatron oscillations in a laser-plasma electron accelerator" *Eur. Phys. Lett.* **81**(6): 64001.

Glinec, Y., Faure, J., Malka, V., Fuchs, T., Szymanowski, H., and Oelfke, U. 2006. "Radiotherapy with laser-plasma accelerators: Monte Carlo simulation of dose deposited by an experimental quasimonoenergetic electron beam" *Med. Phys.* **33**(1): 155.

Gorbunov, L.M. and Kirsanov, V.I. 1987. "Excitation of plasma waves by an electromagnetic wave packet" *Sov. Phys. JETP* **66**: 290.

Joshi, C. 2007. "The development of laser-and beam-driven plasma accelerators as an experimental field *Phys. Plasmas* **14**(5): 055501.

Kneip, S. et al. 2011. "X-ray phase contrast imaging of biological specimens with femtosecond pulses of betatron radiation from a compact laser plasma wakefield accelerator" *Appl. Phys. Lett.* **99**(9): 093701.

Lundh, O. et al. 2011. "Few-femtosecond, few kiloAmperes laser-accelerated electron bunches" *Nature Phys.* **7**: 219.

Malka, V. et al. 2002. "Electron Acceleration by a Wake Field Forced by an Intense Ultrashort Laser Pulse" *Science* **22**: 298.

Malka, V. 2012. "Laser plasma accelerators" *Phys. Plasmas* **19**(5): 055501.

Malka, V., Faure, J., Gauduel, Y.A., Lefebvre, E., Rousse, A., and Ta Phuoc, K. 2008. "Principles and applications of compact laser-plasma accelerators" *Nature Phys.* **4**: 447–453.

Malka, V., Faure, J., and Gauduel, Y.A. 2010. "Ultra-short electron beams based spatio-temporal radiation biology and radiotherapy" *Mutation Research* **704**(1–3): 142.

Malka, V., Lifschitz, A., Faure, J., and Glinec, Y. 2006. "Staged concept of laser-plasma acceleration toward multi-GeV electron beams" *Phys. Rev. ST Accel. Beams* **9**: 091301.

Mangles, S.P.D. et al. 2004. "Monenergetic beams of relativistic electrons from intense laser-plasma interactions" *Nature* **431**: 535.

Modena, A. et al. 1995. "Electron acceleration from the breaking of relativistic plasma waves" *Nature* **377**: 606.

Pukhov, A., and Meyer-Ter-Vehn, J. 2002. "Laser wake field acceleration: The highly non-linear broken-wave regime" *Appl. Phys. B* **74** (4–5): 355.

Rigaud, O. et al. 2010. "Exploring ultrashort high-energy electron-induced damage in human carcinoma cells" *Cell Death Disease* **1** (2010) e73.

Rousse, A. et al. 2004. "Production of a keV X-ray beam from synchroton radiation in relativistic laser plasma interaction" *Phys. Rev. Lett.* **93**: 135005.

Schmid, K. et al. 2010. "Density-transition based electron injector for laser driven wakefield accelerators" *Phys. Rev. ST Accel. Beams* **13**(9): 091301.

Sprangle, P., Esarey, E., Krall, J., and Joyce, G. 1992. "Propagation and guiding of intense laser pulses in plasmas" *Phys. Rev. Lett.* **69**(15): 2200.

Tajima, T. and Dawson, J. 1979. "Laser electron accelerator" *Phys. Rev. Lett.* **43**: 267.

Ta Phuoc, K. et al. 2006. "Imaging electron trajectories in a laser-wakefield cavity using betatron x-ray radiation" *Phys. Rev. Lett.* **97**(22): 225002.

Ta Phuoc, K. et al. 2007. "Demonstration of the ultrafast nature of laser produced betatron radiation" *Phys. Plasmas* **14**(8): 080701.

Thaury, C. et al. 2015. "Shock assisted ionization injection in laser-plasma accelerators" *Scientific Reports* **5**:16310, DOI: 10.1038/srep16310.

3

Dielectric Laser Acceleration of Electrons

3.1 Introduction and Historical Background

Particle accelerators are a critical scientific and industrial tool in many fields, from probing the structure of matter at its fundamental levels, to industrial radioisotope production, medical radiation therapy and security scanners. In the past few years, the first demonstrations of a revolutionary new approach have been conducted, termed 'dielectric laser-driven acceleration' (DLA), that combines the fabrication methods of the microchip industry with modern pulsed lasers to greatly shrink the size and cost of particle accelerators. As experimental firsts, these results are reminiscent of demonstrations of the first radio-frequency linear accelerator conducted at Stanford in 1948, upon which the current state-of-the-art particle accelerator (the microwave linac) is based [Ginzton 1948]. That early device was one metre in length, and required several people to carry it. By comparison, the first laser-powered dielectric structures developed at Stanford for particle acceleration were one millimetre long (see Figure 3.1), and have demonstrated accelerating fields over an order of magnitude higher than typical operating conditions for conventional particle accelerators. This approach, which has been colloquially referred to as an "accelerator on a chip," opens the door to creating a future generation of miniaturized devices with applications in science, industry and security.

The use of lasers to accelerate charged particles in material structures has been a topic of considerable interest since the optical laser was invented in the early 1960s. Early concepts proposed using lasers to accelerate particles by operating known radiative processes in reverse, including the inverse Cherenkov accelerator [Shimoda 1961] and the inverse Smith-Purcell accelerator [Lohmann 1962, Takeda 1968, Palmer 1980]. The latter effect has been demonstrated experimentally with nonrelativistic electrons using a metallic grating at submillimetre wavelengths, albeit with small acceleration gradients in the keV/m range [Mizuno 1987]. Energy modulation of both relativistic and subrelativistic electrons has also been observed in a laser field truncated by a thin downstream metallic film [Plettner 2005, Kirchner 2014]. More recently, several experiments have reached acceleration gradients on par with and exceeding those of classical RF accelerators [Breuer 2013, Leedle 2015a].

These and related initial experiments, which we will discuss later, were important in laying the groundwork for efficient phased laser acceleration. However, the interaction mechanism used to accelerate the particles in the experiments of references, [Plettner 2005, Sears 2008, Feist 2015] (inverse transition radiation at a metallic interface) exhibits a relatively weak effect, and the required laser fluences

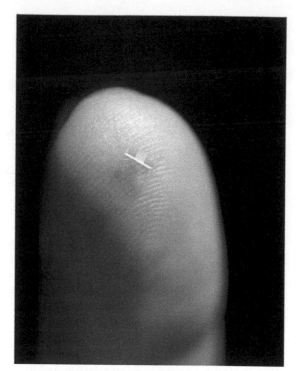

Microwave linac, invented 1948

Microchip accelerator, invented 2013

(a)   (b)

FIGURE 3.1 Micro-accelerator technology opens a path to revolutionize particle accelerators: (a) the first microwave linac was "only" 3 ft long with a gradient of 2 MV/m (Courtesy Stanford Special Collections and University Archive); (b) by comparison the first microchip accelerators developed at Stanford were 1 mm long with a gradient of 300 MV/m (photo credit is SLAC National Accelerator Laboratory and reproduced here with its permission).

can readily exceed the damage limit of the metal surface. This points to the need to use materials with high damage limits combined with acceleration mechanisms that are more efficient. Due to these considerations, combined with the fact that metallic surfaces suffer high ohmic losses and low damage threshold limits at optical wavelengths, a shift in focus has occurred towards photonic structures made of transparent dielectric materials and incorporating new technologies, such as photonic crystals and metasurfaces [Rosing 1990, Lin 2001, Cowan 2003, Mizrahi 2004, Schachter 2004, Naranjo 2012, Bar-Lev 2014]. The proposed structures have in common with each other the use of periodic variations in dielectric constant via either layers of different materials or interspacing materials with the vacuum, to generate a uniform phase-synchronous accelerating wave inside of a narrow vacuum channel where the charged particle beam can travel. These approaches can be classified based upon the dimensionality of the periodic variation employed (1D, 2D, or 3D). These three general categories have different relative benefits. The slab-symmetric or planar 1D type of geometry is simpler to fabricate and its wide aspect ratio helps in improving charge transmission. This has led to structures of this type as the first to be successfully fabricated and undergo demonstration experiments [Peralta 2013, Breuer 2013]. In addition, both resonant (non-zero quality factor) [Yoder 2005] and non-resonant (zero quality factor) [Plettner 2006, Chang 2014] variants of this concept have been proposed. The 2D (fibre) type geometry is similar conceptually to more conventional radio frequency accelerators in that the confined accelerating mode is almost azimuthally symmetric, mimicking the transverse magnetic TM_{01} mode of a conventional accelerating cavity [Cowan 2003]. However, the preferred fabrication technique (telecom fibre drawing of so-called photonic crystal fibres [Russel 2003]) is less amenable to a fully on-chip approach.

Fully 3D photonic crystal structures offer complete mode confinement in all three dimensions, making them highly amenable to laser coupling using integrated on-chip waveguiding, but due to their complex geometry are challenging to fabricate [Staude 2012, Wu 2014, Wu 2015].

Laser damage studies have been performed on a variety of optical materials in order to evaluate their suitability as base materials for fabrication of high-gradient DLA accelerators. These measurements employed a pump-probe configuration where the decrease in specular reflection of a HeNe probe laser coincident with a pulsed IR laser on the sample provides a metric for the onset of damage to the material. A variety of common and exotic materials, including silicon, silica, and oxides of aluminum, zirconium, yttrium and hafnium have been studied [Stuart 1995, Soong 2012]. The damage fluence of patterned silica gratings, corresponding to those of Figures 3.1b and 3.3, was found to be 0.85 J/cm² at a wavelength $\lambda = 800$ nm (half the damage fluence for bulk silica). When focused to micron-scale spot sizes, this corresponds to peak electric fields in the multi-GV/m range at the surface of the accelerating structure. Assuming a ratio of 1/2 for the field the electrons experience to the surface field in a silica accelerating structure at this wavelength gives an approximate upper limit of 1 GeV/m on achievable gradient with 1 ps pulse duration. By using a combination of higher damage threshold materials, such as sapphire and calcium fluoride, and sub-ps pulse durations, this limit could possibly be improved by another order of magnitude, reaching up to 10 GeV/m accelerating gradients. By comparison, the 2-mile linear accelerator at SLAC operates with an accelerating gradient of 20 to 30 MeV/m, and research cavities operating with microwave frequencies in the range of 10–35 GHz have reached gradients of slightly above 100 MeV/m [Grudiev 2009].

3.2 Conceptual Outline for an Accelerator on a Chip

The basic conceptual layout for an integrated accelerator based entirely upon DLA and compatible optical elements on the scale of a single six-inch wafer can be divided into three main components: (1) the injector, including the particle source; (2) a multi-staged DLA accelerator; and (3) compatible on-chip mechanisms for radiation generation. These basic components are schematically illustrated in Figure 3.2. The power distribution scheme is envisioned as a coupler that brings a pulse from an external fibre laser onto the integrated chip, distributes it between multiple structures via guided-wave power splitters, and then recombines the spent laser pulse and extracts it from the chip via a mirror-image output coupler, after which the power is either dumped, or for optimal efficiency, recycled [Colby 2011].

The electron beam parameters required for laser-driven structures are quite different from those in conventional radio-frequency accelerators. Assuming that the guiding channel's transverse dimensions are of the order of the drive laser wavelength (i.e. 1 to 10 microns) the power coupling efficiency to the particle bunches can, in principle, be tens of percent, with optimal efficiency at bunch charges of 1 to 20 fC (femtocoulombs) corresponding to 10^4 to 10^5 electrons [Siemann 2004]. In order for successive bunches to coincide with the accelerating phase of the wave, the desired bunch durations are on the attosecond scale. For example, a bunch that occupies 1% of the laser period would have a duration of 67 attoseconds for a 2 μm drive wavelength. And since each driving laser pulse may be many optical cycles in duration, a train of these ultrashort electron bunches could be loaded into each laser pulse, with the spacing between bunches equal to the laser wavelength (or an integer multiple thereof). By comparison, the electron bunches in the conventional linear accelerators at SLAC typically contain a few pC to several nC of charge, and have durations of several hundred femtoseconds. A technique for generating the requisite optically microbunched attosecond scale beams needed for DLA has been demonstrated at SLAC [Sears 2008]. In this approach, a laser beam interacts with electrons traveling through a periodic magnetic undulator, or inverse free-electron laser (IFEL). The electrons take oscillatory trajectories through the undulator, thereby introducing a transverse velocity component. At the proper resonance condition, the electrons stay synchronized in phase with the laser electromagnetic wave and can continuously acquire energy from the laser field, producing a sinusoidal

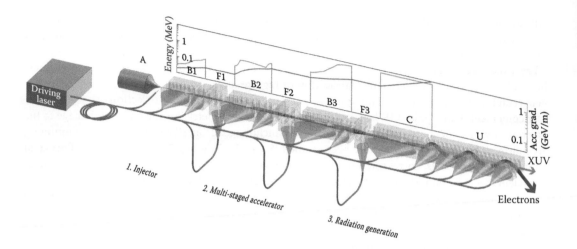

FIGURE 3.2 Schematic showing the primary components that must be developed to make a laser-driven electron source (A) and low-beta acceleration structure (B) (the combination of A and B we call injector), accelerator (C) and undulator (U) for production of relativistic electrons and generation of XUV and X-ray radiation. Focusing and deflection elements (F) are also represented as a necessary component of a fully operational system. The plot shows the influence of the fields as the electrons propagate through the system. Note that the plot is not to scale and that fibre laser amplifiers are not shown. The photonic coupling elements required to couple the laser power into the DLA elements are also not shown.

energy modulation along the bunch with the same period as the laser. After the IFEL, the electrons pass through a magnetic compressor to transform the energy modulation into a longitudinal density modulation. This technique has been used to produce microbunches with FWHM durations of 410 attoseconds at 60 MeV particle energy. For this example, 50% of the total charge is contained within the FWHM of the microbunches. Ultimately, compatible bunching schemes are needed that operate at subrelativistic energies and that can be integrated with on-chip devices, an example of which is proposed in [Schachter 2016]. By microbunching the beam, the coupling efficiency of the axial accelerating field to the particles in a DLA can in principle be as high as 60% [Siemann 2004, Na 2005]. Combined with recent advances in power efficiencies of solid state lasers, which now exceed 30% [Moulton 2009] and designs for near 100% power coupling of laser power into a DLA structure [Wu 2015], the estimates of a wall-plug power efficiency for a DLA based system are in the range of 10–12%, which is comparable to more conventional approaches [England 2014]. Note, however, that high power efficiency is only mandatory for large, science-driven accelerators with MW-scale beam powers, and is a less critical concern for low and medium energy beams in medical and security settings.

Recent work in field emission needle-tip emitters and emitter arrays indicates that electron beams with the requisite charge and emittance requirements are within reach [Hommelhoff 2006a, Borman 2010, Kanungo 2015, Graves 2012, Breuer 2014, McNeur 2016a]. Electron emission triggered by femtosecond class lasers in nanotip emitters provides a pathway to produce ultra-low (nanometer scale) normalized emittance beams of electrons ideally suited for acceleration in optical-scale devices [Jensen 2010, Spence 2013, Hommelhoff 2006a, Hommelhoff 2006b], with the recent demonstration of a miniaturized 30 keV electron gun and the observation of up to 2000 electron-per-pulse emission from a tip-based source [Hoffrogge 2014]. Similar to conventional accelerators, the source properties determine the initial beam properties. Hence, also in analogy to modern accelerators and free-electron laser facilities, much source work will be required to provide the unique beam quality demanded by the small DLA structures.

Due to the four to five orders of magnitude reduction in operating wavelength in going from RF to laser-driven accelerators, the corresponding normalized accelerating field per interaction period λ (assuming an accelerating field of $E_0 = 1$ GV/m and a wavelength λ of 2 μm) is of the order $\alpha = e\lambda E_0/2\pi mc^2 = 0.001$. This is 3 orders of magnitude reduced from the case of a conventional radio-frequency photoinjector, where typically $\alpha \sim 1$. Consequently, in order to capture subrelativistic particles with starting energies in the 10 to 40 keV range in the accelerating field, we must vary the wave vector $k(z)$ as a function of particle travel distance z. This could be done in practice by a longitudinal variation of the DLA accelerator structure in order to maintain local phase synchronicity with the particles. Experiments to demonstrate this technique by using chirped gratings have recently been conducted [McNeur 2016b].

3.3 Recent Experimental Results

An experiment conducted jointly at SLAC and Stanford University in 2013 demonstrated accelerating gradients exceeding 300 MeV/m in a laser-driven fused silica structure at 800 nm wavelength, representing the first observation of particle acceleration within an enclosed laser-driven dielectric structure. With shorter laser pulses, and hence higher sustainable electric field intensity in the dielectric material, the subsequent record high gradients of 690 MeV/m have been observed [Wootton 2016]. The DLA prototype structure, shown in Figure 3.3a and its inset, is based upon the periodic phase reset structure proposed by Plettner, Lu and Byer [Plettner 2006]. In this scheme, electrons were accelerated in the vacuum gap between two parallel gratings with a channel gap of 400 nm.

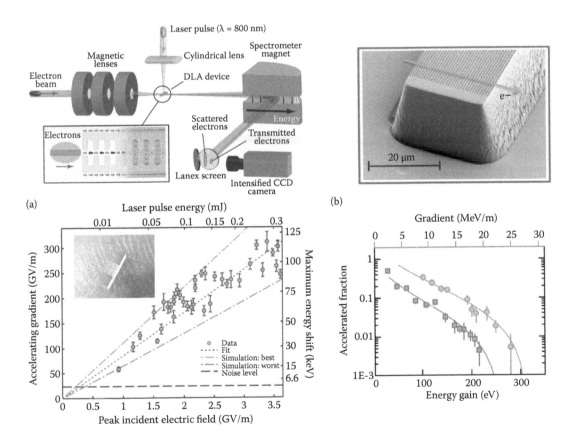

FIGURE 3.3 The first DLA acceleration demonstrations conducted in 2013 (a) with 60 MeV electrons at SLAC (top) with accelerating gradients (bottom) up to 300 MeV/m and (b) at MPQ/FAU with 28 keV electrons on an open grating accelerator (top) and gradients up to 25 MeV/m (bottom). (Adapted from A. Peralta et al. 2013, *Nature* 503: 91–94. With permission; and J. Breuer and P. Hommelhoff 2013, *Phys. Rev. Lett.* 111: 134803).

TABLE 3.1   Several Recent DLA Experimental Results

Parameter	SLAC [Peralta 2013]	MPQ/FAU [Breuer 2013]	Stanford [Leedle 2015a]	FAU/Stanford [Kozak 2016]	SLAC [Wootton 2016]
Material	Fused Silica	Fused Silica	Silicon	Silicon	Fused Silica
Incident Beam Energy	60 MeV	28 keV	96.3 keV	28 keV	60 MeV
Particle Velocity (v/c)	0.9996	0.32	0.54	0.32	0.9996
Laser Wavelength	800 nm	790 nm	907 nm	1800 nm	800 nm
Laser Repetition Rate	10 Hz	2.7 MHz	76 MHz	100 kHz	10 Hz
Laser Pulse Energy	330 μJ	160 nJ	5.2 nJ	50 nJ	128 μJ
Pulse Duration	1.24 ps	110 fs	130 fs	20 fs	64 fs
Interaction Length	450 μm	11 μm	5.6 μm	1.8 μm	16.3 μm
Max. Energy Gain	120 keV	275 eV	2.05 keV	0.3 keV	24 keV
Max.Gradient	310 MeV/m	25 MeV/m	370 MeV/m	210 MeV/m	690 MeV/m

The gradient was determined from the width of the resultant energy modulation of the transmitted population as a function of laser pulse energy, yielding a maximum gradient in excess of 300 MeV/m, as shown in Figure 3.3a (bottom). Concurrent with the relativistic acceleration experiments of Figure 3.3a, researchers at Max Planck Institute for Quantum Optics (MPQ) and Friedrich-Alexander University Erlangen-Nuremberg (FAU), simultaneously demonstrated acceleration of 28 keV electrons in a fused silica "mesa" structure, as shown in the top image of Figure 3.3b, using a 110 fs Ti:Sapphire laser (800 nm), producing gradients up to 25 MeV/m, as shown in the bottom image of Figure 3.3b.

Subsequent experiments at higher subrelativistic energies (near 100 keV) have been conducted at Stanford University, using silicon-based structures [Leedle 2015a, Leedle 2015b]. The more recently demonstrated silicon structures are based on a dual-pillar geometry, with maximal gradients up to 370 MeV/m. Comparable gradients of 200 MeV/m have also been reached with an electron energy of less than 30 keV by driving a silicon structure with laser pulses with a duration of only two cycles [Kozak 2016]. In Table 3.1 we compare basic experimental parameters of several recent demonstration experiments. We note that the subrelativistic experiments described in the references [Breuer 2013, Leedle 2015b, Kozak 2016] utilize open (single-grating) geometries where the accelerating field falls off exponentially as a function of distance from the grating surface, whereas those of references [Peralta 2013, Leedle 2015a] used enclosed or dual-sided structures.

In these direct optical-scale interactions, an electron bunch longer than the operating wavelength of the accelerator will experience an energy modulation, as some electrons are accelerated and others decelerated depending upon the local phase of the optical wave at their respective positions within the bunch. In order to instead produce a net acceleration of the electrons using the DLA concept, the bunch must be microbunched with a periodicity equal to the laser wavelength. Techniques for accomplishing this at the optical period of a laser have been previously demonstrated at wavelengths of 10 μm and 800 nm [Sears 2008, Kimura 2001], as discussed in Section 3.2. Additionally, in the various DLA acceleration experiments described above, pre-accelerated electron beams were provided by either conventional radio frequency or by DC electron guns similar to electron microscope columns. In order to make truly integrated on-chip accelerators, it is desirable to also develop miniaturized particle sources and subrelativistic laser acceleration schemes to accelerate particles to MeV level energies for injection into speed-of-light structures.

3.4 Future Prospects and Conclusions

In the last four years, much progress has been made since the proof-of-concept demonstrations of dielectric laser acceleration with individual structures. High gradient acceleration approaching 1 GeV/m has been shown [Wootton 2015]; structures for particle deflection and focusing, and compatible optical-scale beam position monitors have been demonstrated [Leedle 2015a, Leedle 2015b, Soong 2014]. Two

DLA elements have been concatenated, representing a first step towards a more complex, multi-element DLA accelerator [McNeur 2016b]. These are encouraging initial steps to a future DLA-based 'accelerator on a chip', but much more is required, including the development of suitable diagnostics and additional beam manipulation techniques. Integration of these subcomponents into an optical-scale accelerator that can be mass-produced using available nanofabrication techniques requires that they be constructed using compatible fabrication methods, making the possibility of using the same structures for both the accelerator and for these complementary beam manipulation devices highly attractive. Due to the high damage fluence of dielectric materials at IR wavelengths, the resultant field strengths can exceed those of more traditional RF and magnetostatic devices by 1 to 2 orders of magnitude.

Obtaining wall-plug efficiencies suitable for applications with a high beam power, such as a free-electron laser or a linear collider, will additionally require the development of integrated couplers with high efficiency, fed by a network of waveguides that split the laser power from a common feed among various accelerator components, with phase-stable fibre laser power amplifiers in between the common feed laser and the accelerator components. Initial results in simulating such couplers using silicon-on-insulator (SOI) waveguides indicate coupling efficiencies from the input waveguide to the accelerating mode close to 100%. Maintaining phase synchronicity of the laser pulse and the accelerated electrons between many separately fed structures (as indicated in Figure 3.2) would be accomplished by fabricating the requisite phase delays into the lengths of the waveguide feeds.

Applications for compact accelerators with target energies in the MeV range, such as medical radiation oncology, could provide compelling near term uses for a DLA based system. Radiation doses ranging from 1 to 10 Gray (1 Gray = Joule/kilogram radiation energy deposition) per treatment are used to avoid the damage of healthy tissue as the beams pass through the body. Direct electron irradiation has been used only for skin cancer or shallow treatment in the past. An encapsulated DLA built onto the end of a fibre-optic catheter could be placed directly adjacent to a tumour site through existing intracavitary, interstitial and intraluminal brachytherapy placement methods, allowing controlled internal delivery of megavoltage electron radiation beams. Electrons have an energy loss rate of about 2 MeV/cm in water, allowing for irradiation volumes to be tightly controlled, and a 10–20 MeV electron DLA can potentially provide improved dosimetry relative to existing brachytherapy sources that primarily utilize photons from radioactive seeds.

We show a set of estimated baseline electron beam parameters for a future DLA based accelerator at 5- and 10- to 20-year time scales in Table 3.2. The nearer-term (5-year) parameters are relevant for lower-energy and lower beam power applications such as security, medicine and imaging. Even with the modest trapped bunch charge of 2000 electrons per laser pulse and a laser repetition rate of 2 MHz, medically interesting dose rates of 1.4 Gray per second appear feasible. The higher overall efficiency

TABLE 3.2 Projected DLA Beam Parameter Ranges for Applications

Parameter	Units	DLA (5-Year)	DLA (10–20 Year)
Beam Energy	MeV	10	250,000
Gradient	GeV/m	1	1
Microbunch Charge	# electrons	8	38000
Microbunch Duration	attosecond	400	10
Microbunches per Train	#	250	159
Bunch Train Rep Rate	MHz	2	60
Normalized Emittance	nm rad	4	2
Spot Size	nm	1000	2
Beam Power	W	0.007	7.3×10^6
Wall Plug Efficiency	%	2.9	9.5
Wall Plug Power	W	< 100	76.2×10^6
Active Linac Length	m	0.01	300

needed for higher energy and higher beam power applications, represented by the '10–20 Year' column, are achievable in principle and can be realized given new developments on a longer-term time scale. Given that the first DLA demonstration experiments are only a few years old, it is hard to predict where the field will stand in five years, let alone in 10 to 20 years. However, in spite of the conceptual and technical challenges that remain to be addressed, recent encouraging and rapid progress, combined with the ability to leverage well-developed nanofabrication methods and power-efficient fibre laser technology, point to a bright future ahead for the DLA approach.

References for Chapter 3

Bar-Lev, D. and Scheuer, J. 2014. "Plasmonic metasurface for efficient ultrashort pulse laser-driven particle acceleration" *Phys. Rev. ST Accel. Beams* **17**: 121302.

Bormann, R. et al. 2010. "Tip-Enhanced Strong-Field Photoemission" *Phys. Rev. Lett.* **105**: 147601.

Breuer, J. and Hommelhoff, P. 2013. "Laser-based acceleration of nonrelativistic electrons at a dielectric structure" *Phys. Rev. Lett.* **111**: 134803.

Breuer, J., McNeur, J. and Hommelhoff, P. 2014. "Dielectric laser acceleration of electrons in the vicinity of single and double grating structures – theory and simulations" *Journal of Physics B: Atomic, Molecular and Optical Physics* **47**: 234004.

Chang, C.M. and Solgaard, O. 2014. "Silicon buried gratings for dielectric laser electron accelerators" *Applied Physics Letters* **104**: 184102.

Colby, E.R., England, R.J. and Noble, R.J. 2011. "A Laser-Driven Linear Collider: Sample Machine Parameters and Configuration" *2011 Particle Accelerator Conference Proceedings*, New York, 262.

Cowan, B.M. 2003. "Two-dimensional photonic crystal accelerator structures" *Phys. Rev. ST Accel. Beams* **6**: 101301.

England, R.J. et al. 2014. "Dielectric laser accelerators" *Rev. Mod. Phys.* **86**: 1337.

Feist, A. et al. 2015. "Quantum coherent optical phase modulation in an ultrafast transmission electron microscope" *Nature* **521**: 200–203.

Ginzton, E.L., Hansen, W.W. and Kennedy, W.R. 1948. "A Linear Electron Accelerator" *Rev. Sci. Instrum.* **19**: 89–108.

Graves, W.S., Kaertner, F.X., Moncton, D.E. and Piot, P. 2012. "Intense Superradiant X Rays from a Compact Source Using a Nanocathode Array and Emittance Exchange" *Phys. Rev. Lett.* **108**: 263904.

Grudiev, A., Calatroni, S. and Wuensch, W. 2009. "New local field quantity describing the high gradient limit of accelerating structures" *Phys. Rev. Accel. Beams* **12**: 102001.

Hoffrogge, J., Stein, J.P., Kruger, M., Forster, M., Hammer, J., Ehberger, D., Baum, P. and Hommelhoff, P. 2014. "Tip-based source of femtosecond electron pulses at 30 keV" *J. Appl. Phys.* **115**: 094506.

Hommelhoff, P., Sortais, Y., Aghajani-Talesh, A. and Kasevich M.A. 2006a. "Field emission tip as a nanometer source of free-electron femtosecond pulses" *Phys. Rev. Lett.* **96**: 077401.

Hommelhoff, P., Kealhofer, C., and Kasevich, M.A. 2006b. "Ultrafast Electron Pulses from a Tungsten Tip Triggered by Low-Power Femtosecond Laser Pulses" *Phys. Rev. Lett.* **97**: 247402.

Jensen, K.L., O'Shea, P.G., Feldman, D.W. and Shaw, J.L. 2010. "Emittance of a field emission electron source" *J. Appl. Phys.* **107**: 014903.

Kanungo, P.D. et al. 2015. "Electron beam collimation from an all-metal double-gate 40000 nanotip array: Improved emission current and beam uniformity upon neon gas conditioning" *J. Vac. Sci. Technol. B* **33**: 03C113.

Kirchner, F.O., Gliserin, A., Krausz, F. and Baum, P. 2014. "Laser streaking of free electrons at 25 keV" *Nature Photonics* **8**: 52–57.

Kozak, M., Förster, M., McNeur, J., Schönenberger, N., Leedle, K., Deng, H., Harris, J.S., Byer, R.L and Hommelhoff, P. 2016. "Dielectric laser acceleration of sub-relativistic electrons by few-cycle laser pulses" *Proceedings of High Brightness Beams Workshop, Havanna, Cuba.*

Kimura, W.D. et al. 2001. "First Staging of Two Laser Accelerators" *Phys. Rev. Lett.* **86**: 4041.

Leedle, K. et al. 2015a. "Dielectric laser acceleration of sub-100 keV electrons with silicon dual-pillar grating structures" *Optics Letters* **40**: 4344.

Leedle, K.J., Pease, R.F., Byer, R.L. and Harris, J.S. 2015b. "Laser acceleration and deflection of 96.3 kev electrons with a silicon dielectric structure" *Optica* **2**: 158–161.

Lin, X.E. 2001. "Photonic band gap fiber accelerator" *Phys. Rev. ST Accel. Beams* **4**: 051301.

Lohmann, A. 1962. "Electron Acceleration by Light Waves" IBM Technical Note TN-5.

McNeur, J. et al. 2016a. "A miniaturized electron source based on dielectric laser accelerator operation at higher spatial harmonics and a nanotip photoemitter" *Journal of Physics B: Atomic, Molecular and Optical Physics* **49**: 034006.

McNeur, J. et al. 2016b. "Elements of a dielectric laser accelerator" arXiv:1604.07684.

Mizrahi, A. and Schachter, L. 2004. "Optical Bragg accelerators" *Phys. Rev. E* **70**: 016505.

Mizuno, K., Pae, J., Nozokido, T. and Furuya, K. 1987. "Experimental evidence of the inverse smith-purcell effect" *Nature* **328**: 45–47.

Moulton, P.F., Rines, G.A., Slobodtchikov, E., Wall, K.F., Frith, G., Sampson, B. and Carter, A. 2009. "Tm-doped fiber lasers: fundamentals and power scaling," *IEEE J. Selected topics in Quantum Elect* **15**: 85.

Na, Y.C.N., Siemann, R. and Byer, R.L. 2005. "Energy efficiency of an intracavity coupled, laser-driven linear accelerator pumped by an external laser" *Phys. Rev. ST Accel. Beams* **8**: 031301.

Naranjo, B., Valloni, A., Putterman, S. and Rosenzweig, J.B. 2012. "Stable Charged-Particle Acceleration and Focusing in a Laser Accelerator Using Spatial Harmonics" *Phys. Rev. Lett.* **109**: 164803.

Palmer, R.B. 1980. "A laser-driven grating linac" *Particle Accelerators* **11**: 81.

Peralta, E.A, Soong, K., England, R.J., Colby, E.R., Wu, Z., Montazeri, B., McGuinness, C., McNeur, J., Leedle, K.J., Walz, D., Sozer, E.B., Cowan, B., Schwartz, B., Travish, G. and Byer, R.L. 2013. "Demonstration of electron acceleration in a laser-driven dielectric microstructure" *Nature* **503**: 91–94.

Plettner, T., Byer, R.L., Colby, E.R., Sears, C.M.S., Spencer, J. and Siemann, R.H. 2005. "Visible-laser acceleration of relativistic electrons in a semi-infinite vacuum" *Phys. Rev. Lett.,* **95**: 134801.

Plettner, T., Lu, P.P. and Byer, R.L. 2006." Proposed few-optical cycle laser-driven particle accelerator structure" *Phys. Rev. ST Accel. Beams* **9**: 111301.

Rosing, M. and Gai, W. 1990. "Longitudinal and transverse wake-field effects in dielectric structures" *Phys. Rev. D* **42**: 1829.

Russel, Ph. 2003. "Photonic Crystal Fibers" *Science* **299**: 358–362.

Schachter, L. 2004. "Energy recovery in an optical linear collider" *Phys. Rev. E* **70**: 016504.

Schachter, L., Kimura, W.D. and Ben-Zvi, I. 2016. "Ultrashort microbunch electron source" *Proc. of the 16th Advanced Accelerator Concepts Workshop*, AIP Conf. Proc. **1777**: 080013.

Sears, C.M.S., Colby, Ischebeck, E.R., McGuinness, C., Nelson, J., Noble, R., Siemann, R.H., Spencer, J., Walz, D., Plettner, T. and Byer, R.L. 2008. "Production and characterization of attosecond electron bunch trains" *Phys. Rev. ST Accel. Beams,* **11**: 061301.

Shimoda, K. 1961. "Proposal for an electron accelerator using an optical maser" Applied Optics 1: 33.

Siemann, R. 2004. "Energy efficiency of laser driven, structure based accelerators" *Phys. Rev. ST Accel. Beams* **7**: 061303.

Soong, K., Byer, R.L., Colby, E.R., England, R.J. and Peralta, E.A. 2012. "Laser damage threshold measurements of optical materials for direct laser accelerators" *Proc. of the 2012 Advanced Accelerator Concepts Workshop. A.I.P. Conf. Proc,* **1507**: 511.

Soong, K. et al. 2014. "Electron beam position monitor for a dielectric microaccelerator" *Opt. Lett.* **39**: 4747.

Spence, J. 2013. Chapters 4, 9 and 10 in "High Resolution Electron Microscopy. Monographs on the Physics and Chemistry of Materials. Fourth Edition". Oxford Univ. Press.

Staude, I. et al. 2012. "Waveguides in three-dimensional photonic bandgap materials for particle-accelerator on a chip architectures" *Optics Express* **20**: 5607–5609.

Stuart, B. C., Feit, M.D., Rubenchick, A.M., Shore, B.W. and Perry, M.D. 1995. "Laser-Induced Damage in Dielectrics with Nanosecond to Subpicosecond Pulses", *Phys. Rev. Lett.* **74**: 2248–2251.

Takeda, Y. and Matsui, I. 1968. "Laser linac with grating" *Nuclear Instruments and Methods* **62**: 306–310.

Wootton, K. et al. 2016. "Demonstration of acceleration of electrons at a dielectric microstructure using femtosecond laser pulses" *Optics Letters* **41** (12): 2672.

Wu, Z., England, R.J., Ng, C.-K., Cowan, B., McGuinness, C., Lee, C., Qi, M. and Tantawi, S. 2014. "Coupling power into accelerating mode of a three-dimensional silicon woodpile photonics band-gap waveguide," *Phys. Rev. ST-AB*, **17**: 081301.

Wu, Z., Lee, C., Wootton, K.P., Ng, C.-K., Qi, M. and England, R.J. 2015. "A Traveling-Wave Forward Coupler Design for a New Accelerating Mode in a Silicon Woodpile Accelerator" *IEEE Journal of Selected Topics in Quantum Electronics* **22**: 4400909.

Yoder, R.B. and Rosenzweig, J.B. 2005. "Side-coupled slab-symmetric structure for high-gradient acceleration using terahertz power" *Phys. Rev. ST AB* **8**: 111301.

4

Laser-Accelerated Electrons as X-Ray/γ-Ray Sources

Stefan Karsch

Konstantin
Khrennikov

4.1 Introduction

The strong space charge forces in laser-wakefield accelerators (LWFAs) naturally yield ultra-dense and ultra-short electron (e⁻) bunches without complicated cooling, making them very attractive for the high-brilliance X-ray generation. Table 4.1 lists typical LWFA parameters:

Assuming Gaussian bunches, this corresponds to a low normalized transverse emittance of $\leq 1\pi$ mm mrad, and 10 kA peak currents. The femtosecond laser pulses that drive an LWFA define the phase of the accelerating wave, and therefore synchronize the electron bunches to the laser with few-fs precision. Since these properties are intrinsic to LWFAs, their electron bunches are promising drivers for compact, brilliant X-ray sources. If an electron bunch is wiggled, its emittance governs the minimum emittance of the generated X-rays.

The brilliance of an X-ray source is defined as follows Figure 4.1:

$$\text{brilliance} = \frac{\text{photons}}{(\text{mm}^2 \cdot \text{mrad}^2) \cdot (\text{s} \cdot 0.1\% \text{ bandwidth})} \tag{4.1}$$

Here, the left bracket in the denominator is the product of the emittances in both transverse directions, while the right bracket gives the longitudinal emittance. Low emittance is synonymous with a high phase-space density, meaning that all electrons (and consequently photons) are confined to a small volume in space and time and have a high directionality and low energy spread. This underlines the paramount importance of using an accelerator that provides a small electron bunch emittance. Note that for pulsed sources, the peak and average brilliance differ by orders of magnitude according to the duty factor.

TABLE 4.1 Observed LWFA Electron Bunch Parameters

Parameter/Item	Parameter Range/Comment
Energy	50–500 MeV
FWHM energy spread	10%
Charge	10–100 pC
FWHM source diameter	2 μm
FWHM pulse duration	5 fs
FWHM beam divergence	2 mrad

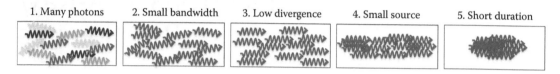

1. Many photons   2. Small bandwidth   3. Low divergence   4. Small source   5. Short duration

FIGURE 4.1 Ingredients of high brilliance in a photon beam.

However, there are some important differences between LWFA bunches and large-scale RF machines for driving, e.g. an X-ray free-electron laser: Since the longitudinal accelerating gradient of an LWFA is four to five orders of magnitude larger, the acceleration process is much quicker. This implies that the time of trapping in the accelerating field for each electron has a much higher influence on its final energy. Therefore, it is much harder to create monochromatic electron beams in an LWFA, but monochromatic (or highly energy-time correlated) beams are a prerequisite for achieving self-amplified spontaneous emission (SASE) free-electron lasing (FEL). Moreover, plasma-based accelerators only recently became stable enough for meaningful parameter studies; typical parameter fluctuations are on the 10% level. Finally, the limited efficiency and repetition rate of current high-intensity lasers prevents high average electron currents and hence photon flux. However, for many applications, the short duration, high peak current and intrinsic synchronization are reason enough to pursue such all-optical, free-electron-based photon sources as enabling tools. In the following we will cover three different free-electron sources, as depicted in Figure 4.2.

They all have the same emission principle in common, namely Larmor radiation of an accelerated charge. This text aims at giving the reader a very basic introduction into the physics of these radiation

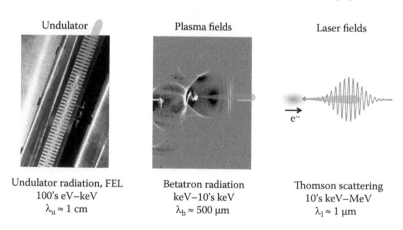

Undulator   Plasma fields   Laser fields

Undulator radiation, FEL   Betatron radiation   Thomson scattering
100's eV–keV   keV–10's keV   10's keV–MeV
$\lambda_u \approx 1$ cm $\lambda_b \approx 500$ μm $\lambda_l \approx 1$ μm

FIGURE 4.2 Wiggler electron sources. Top: source of wiggling field; bottom: characteristic photon energy for 0.1–1 GeV electron beams and typical wiggling wavelengths. (Undulator cartoon courtesy of Thorsten Naeser and Christian Hackenberger. Reproduced with permission.)

processes. Starting from the simple case of an electron on a circular orbit, we will work out the basic resonance criterion for undulator emission and the details of an undulator spectrum, before applying the same principles to the related Thomson scattering and betatron emission cases, including a survey of recent experimental findings. In the Appendix, we provide a step-by-step guide for calculating emission spectra from electron trajectories in arbitrary fields using MATLAB®.

4.2 Radiation by Accelerated Free Electrons

The radiation of a free accelerated electron will be treated by closely following the excellent description by Wille [39]. Starting with the well-known equations for Larmor radiation of an accelerated charge [17]

$$P_r = \frac{e^2}{6\pi\epsilon_0 m_0^2 c^3}\left(\frac{d\vec{p}}{dt}\right)^2 \quad \text{and} \quad \frac{dP_r}{d\Omega} = \frac{e^2}{16\pi^2\epsilon_0 m_0^2 c^3}\left(\frac{d\vec{p}}{dt}\right)^2 \sin^2\Psi \tag{4.2}$$

for the total and angularly resolved radiated power, respectively, we obtain a relativistic invariant form of the total power by transforming the coordinate time derivative d/dt of the three-momentum by the proper time derivative $d/d\tau$ of the four-momentum P_μ:

$$dt \rightarrow d\tau = \frac{1}{\gamma}dt \quad \text{and} \quad \left(\frac{d\vec{p}}{dt}\right)^2 \rightarrow \left(\frac{dP\mu}{d\tau}\right)^2 = \left(\frac{d\vec{p}}{d\tau}\right)^2 - \frac{1}{c^2}\left(\frac{E}{d\tau}\right)^2 \tag{4.3}$$

Here $\vec{p}$ denotes the three-momentum, E the energy, γ the relativistic factor, e and m_0 the charge and rest mass of the electron, respectively, c is the speed of light, ϵ_0 is the permittivity of vacuum and Ω is the solid angle of detection around an emission angle Ψ from the electron trajectory. Plugging this into Equation 4.2 (left) yields

$$P_r = \frac{e^2 c}{6\pi\epsilon_0 (m_0 c^2)^2}\left[\left(\frac{d\vec{p}}{d\tau}\right)^2 - \frac{1}{c^2}\left(\frac{E}{d\tau}\right)^2\right] \tag{4.4}$$

4.2.1 Circular Trajectory: Synchrotron Radiation

If we assume deflection in a purely magnetic field, $dE/dt = 0$ and 4.4 simplifies to

$$P_r = \frac{e^2 c}{6\pi\epsilon_0 (m_0 c^2)^2}\left(\frac{d\vec{p}}{d\tau}\right)^2 = \frac{e^2 c\gamma^2}{6\pi\epsilon_0 (m_0 c^2)^2}\left(\frac{d\vec{p}}{dt}\right)^2 \tag{4.5}$$

in the internal and laboratory frames, respectively. With the identities $dp/dt = p\omega = p \cdot v/R$ for a charge with velocity v circulating in the field with frequency ω, $E = pc$ for the relativistic energy-momentum correlation and $\gamma \approx E/m_0 c^2$, we obtain the well-known result:

$$P_r = \frac{e^2 c}{6\pi\epsilon_0 (m_0 c^2)^4}\frac{E^4}{R^2} \tag{4.6}$$

The radiated power increases with the particle's energy to the 4th power and quadratically with the inverse deflection radius R. The left part of Figure 4.3 depicts the Hertzian dipole pattern given by

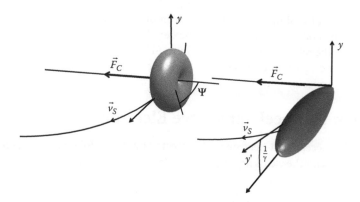

FIGURE 4.3 Instantaneous lab-frame radiation distributions of an electron on a circular orbit: (left) nonrelativistic, (right) $\gamma = 3$. The majority of the radiation is emitted in a $1/\gamma$-cone.

Equation 4.2 of a nonrelativistic electron at each point on a circular orbit. The radiation distribution for a relativistic ($\gamma = 10$) electron is shown in the right panel, which is a Lorentz-boosted version of Hertzian dipole. The holes and the forward-pointing half-torus are sharply bent into a $1/\gamma$ cone, while the radiation frequency is boosted in forward direction as well. An observer in the laboratory frame sees this cone of emitted photons sweep across his field of view as the electron follows its circular path (see Figure 4.4). At points A and B on the electron trajectory, the cone enters and leaves the observer's field of view, so light is only detected between these points.

The distance from A to B measured along the arc is $2R/\gamma$, where R is its radius. The first light (emitted at A) reaches the observer on a straight line through B at $t_\gamma = \overline{AB}/c = 2R\sin(1/\gamma)/c$, while the electrons reach point B at $t_e = 2R/c\beta\gamma$ and emit the last radiation. Therefore the duration of the light flash is

$$\Delta t = t_e - t_\gamma = \frac{2R}{c\beta\gamma} - \frac{2R}{c}\sin\frac{1}{\gamma} \approx \frac{4R}{3c\gamma^3} \tag{4.7}$$

Here, $\beta = v/c$ denotes the usual normalized velocity. Since $\Delta t \propto 1/c\gamma^3$ is short, the emitted spectrum is a broad continuum, cantered around the so-called 'critical frequency'

$$\omega_{cr} = \frac{2}{\Delta t} = \frac{3c\gamma^3}{2R} \tag{4.8}$$

It divides the emitted spectrum into two parts with equal energy content. The spectral photon emission rate $\dot{N}$ follows a universal synchrotron distribution (see Figure 4.5):

$$\frac{d\dot{N}}{d\epsilon/\epsilon} = \frac{3\sqrt{3}e\gamma^4 I_b}{8\pi\epsilon_0 R\omega_{cr}\hbar}\xi\int_\xi^\infty K_{\frac{5}{3}}(\xi)d\xi; \quad \xi = \frac{\omega}{\omega_{cr}} \tag{4.9}$$

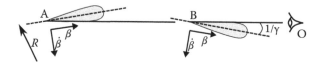

FIGURE 4.4 Electron on a circular orbit emitting light.

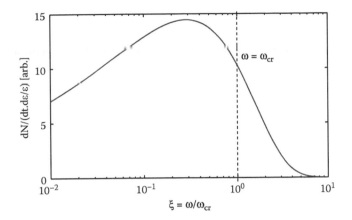

FIGURE 4.5   Universal synchrotron spectrum integrated over all directions.

Here, ϵ is the photon energy, I_b is the beam current, ω the radiated frequency, $\hbar$ the reduced Planck constant and $K_{5/3}$ the modified Bessel function of the second kind.

4.2.2 Periodic Deflection: The Undulator

The previous section demonstrated that the radiation of a relativistic electron on a circular orbit covers a broad continuum. However, if the electron is deflected periodically on a sinusoidal trajectory, the radiation is coherently ($\propto N_p^2$) enhanced at the wiggling frequency, when N_p refers to the number of periods, as shown in Figure 4.6. The derivation of the emission frequency is straightforward and will be briefly described here, since it holds for all free-electron X-ray sources. In the laboratory frame and in the limit of zero transverse deflection (weak field), the wiggling frequency Ω_w of the electron in the undulator is:

$$\Omega_w = \frac{2\pi}{T} = \frac{2\pi v}{\lambda_u} = k_u \beta c \qquad (4.10)$$

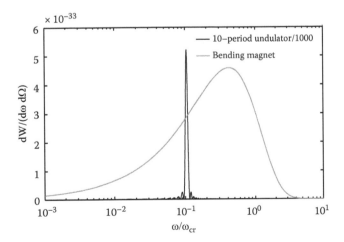

FIGURE 4.6   Spectral energy flux density $dW/(d\omega d\Omega)$ of a 10 period undulator (suppressed by a factor of 1000) compared bending magnet. Emission calculated along axis.

Here, T ist the wiggling period duration, v the electron's velocity, λ_u the undulator period and k_u the undulator wavenumber. In the co-moving (i.e. proper) frame of the electron the wiggling (and consequently radiating) frequency is transformed to $\omega' = \gamma \Omega_W$. Back in the laboratory frame, an observer sees ω' Doppler-upshifted:

$$\omega_{X-ray,undulator} = \omega' \sqrt{\frac{c+v}{c-v}} = \omega' \sqrt{\frac{1+\beta}{1-\beta}} = \omega' \sqrt{\frac{(1+\beta)^2}{1-\beta^2}} \approx 2\gamma\omega' = 2\gamma^2\Omega_w \tag{4.11}$$

This result holds for a stationary wiggling field in the laboratory frame, such as an undulator. If the field is moving, as the laser in the case of Thomson backscattering, the electron sees the frequency of the laser ω_L Doppler-shifted in its proper frame. For the scattered frequency in the lab frame, the Doppler shift has to be applied twice, resulting in (for head-on collision):

$$\omega_{X-ray,Thomson} = \omega' \sqrt{\frac{c+v}{c-v}} = \omega' \sqrt{\frac{1+\beta}{1-\beta}} = \omega' \sqrt{\frac{(1+\beta)^2}{1-\beta^2}} = \omega_L \frac{(1+\beta)^2}{1-\beta^2} \approx 4\gamma^2\omega_L \tag{4.12}$$

These expressions were derived in the limit of a weak wiggling field. Stronger wiggling fields lead to a reduction of the emitted frequency, since the transverse excursion causes a period lengthening since the maximum electron velocity is c. Therefore an electron on a sinusoidal trajectory will always have a lower on-axis velocity v_z. Consequently, we will now consider an electron that has forward (p_z) and transverse (p_x) momentum. The photon, observed at an angle Θ from the axis has an energy $E = \hbar\omega$ and a momentum $p = \hbar\omega / c$. In the proper frame, the photon's four-momentum is then

$$P'_\mu = \begin{pmatrix} \dfrac{E'}{c} \\ P'_x \\ P'_y \\ P'_z \end{pmatrix} = \begin{pmatrix} \gamma & 0 & 0 & -\beta\gamma \\ 0 & 0 & 0 & 0 \\ 0 & 0 & 0 & 0 \\ -\beta\gamma & 0 & 0 & \gamma \end{pmatrix} \begin{pmatrix} \dfrac{E}{c} \\ p\sin(\Theta) \\ 0 \\ p\cos(\Theta) \end{pmatrix} \tag{4.13}$$

and the energy of the emitted photon has now acquired an angular dependence:

$$\frac{E'}{c} = \gamma\frac{E}{c} - \beta\gamma p\cos(\Theta) = \gamma\frac{\hbar\omega_{X-ray}}{c}(1-\beta\cos(\Theta)) \tag{4.14}$$

Consequently the frequency/wavelength reads:

$$\omega_{X-ray} = \frac{\omega'}{\gamma(1-\beta\cos(\Theta))} = \frac{\Omega_u}{(1-\beta\cos(\Theta))} \tag{4.15}$$

$$\lambda_{X-ray} = \lambda_u(1-\beta\cos(\Theta)) \tag{4.16}$$

With the approximation $\cos(\Theta) \approx 1 - \Theta^2/2$ (because the maximum angle $\Theta \approx 1/\gamma \ll 1$ for a non-vanishing radiation signal) and replacing β by $1 - (1 + K^2/2)/2\gamma^2$, we get:

$$
\begin{aligned}
\lambda_u(1-\beta\cos(\Theta)) &= \lambda_u\left[1-\left(1-\frac{1+\frac{K^2}{2}}{2\gamma^2}\right)\left(1-\frac{\Theta^2}{2}\right)\right] \\
&= \lambda_u\left[1-\left(1-\frac{\Theta^2}{2}-\frac{1+\frac{K^2}{2}}{2\gamma^2}+...\right)\right] \\
&\approx \lambda_u\left(\frac{\Theta^2}{2}+\frac{1+\frac{K^2}{2}}{2\gamma^2}\right)
\end{aligned}
\tag{4.17}
$$

Here we introduced the dimensionless undulator strength parameter K

$$
K = \frac{eB\lambda_u}{2\pi m_e c}
\tag{4.18}
$$

where e and m_e are the electron's charge and mass, B the maximum magnetic field and λ_u the undulator period length. Equation 4.17 yields the coherence condition for undulator radiation (and Thomson backscattering, respectively):

$$
\lambda_{X-ray} = \frac{\lambda_u}{2\gamma^2}\left(1+\frac{K^2}{2}+\gamma^2\Theta^2\right) \quad \text{or} \quad \lambda_{X-ray} = \frac{\lambda_L}{4\gamma^2}\left(1+\frac{a_0^2}{2}+\gamma^2\Theta^2\right)
\tag{4.19}
$$

In the latter, $a_0 = e|\vec{E}|/(m_e\omega_L c)$ is the laser's dimensionless amplitude and ω_L its frequency. These expressions show that the X-ray spectrum is, on the one hand, governed by the electron energy and direction relative to the observer (and therefore the emittance in case of a bunch), and on the other hand, by the wiggling field strength, period and number of oscillations. For an ideal undulator with N_p periods, the spectrum emitted on axis ($\Theta = 0$) into a solid angle $d\Omega$ can be written as [8]

$$
\frac{d^2W}{d\Omega d\omega} = \frac{e^2\gamma^2N_p^2}{4\pi\epsilon_0 c}\cdot L(N_p,\omega)\cdot F(n,K) \quad \text{with}
$$

$$
L(N_p,\omega) = \frac{\sin^2(N_p\pi\Delta\omega/\omega_{X-ray})}{N_p^2(\pi\Delta\omega/\omega_{X-ray})^2} \qquad F_n(K) = \frac{n^2K^2}{(1+K^2/2)}\left[J_{\frac{n+1}{2}}(Z)-J_{\frac{n+1}{2}}(Z)\right]^2
\tag{4.20}
$$

$$
Z = \frac{nK^2}{4(1+K^2/2)};
$$

Here, $d^2W/(d\Omega d\omega)$ is the energy radiated per frequency interval and solid angle, and $\Delta\omega = \omega - \omega_{X-ray}$ is the difference between the observed frequency and the resonant frequency as defined in Equation 4.19.

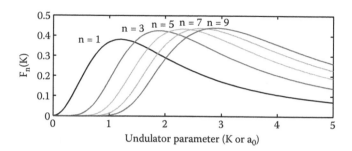

FIGURE 4.7 Increase of energy content of the odd harmonics with undulator strength K. For $K \ll 1$, the photon number grows quadratically, for $K \gg 1$ it decays.

Equation 4.20 confirms that the photon number scales as N_p^2 as expected from a coherent radiation process. $L(N_p, \omega)$ describes the spectral shape of a harmonic line as a sinc^2-function whose width scales as the inverse of the number of periods N_p. $F_n(K)$ determines the radiated energy, depending on the field strength K and the harmonic number n (see Figure 4.7). Only for weak transverse deflection can the electron forward momentum be treated as a constant. For $\gamma \gg 1$ and a maximum deflection angle α, the electron forward velocity oscillates between c and $c \cdot \cos(\alpha)$ and the electron falls behind or overtakes its mean velocity twice every wiggling period. This anharmonic motion leads to the emission of odd harmonics on-axis as multiples of the 2ω-longitudinal oscillation. Even harmonics only occur off-axis.

After these basic (and by no means complete) considerations for an undulator, we will now move on towards betatron and Thomson emission. Here, the wiggling field is not constant anymore, and a numerical treatment (see Appendix) is required. This changing field has profound impacts on the photon number, spectral shape and geometric distribution.

4.3 Thomson Scattering

In Thomson scattering, a light pulse scatters off a moving electron, and for simplicity, we will focus on a backscattering geometry. The wiggling force on the electron is now provided by a light pulse with spatially varying E and B-field instead of a constant-amplitude undulator. Its shorter wavelength creates harder X-rays for a given electron energy. Since even few fs LWFA electron bunches are much longer than the scattered radiation period, the process is incoherent* and can be treated as an incoherent sum of the radiation generated by all electrons in the bunch. Therefore, we will first discuss the radiation of a single electron. Most of this section follows Esarey et al. [1] and Ride et al. [27].

Spectrum A linearly polarized intense colliding field will cause a figure-eight motion of the electron in its average rest frame. This anharmonic motion emits harmonics at frequencies:

$$\hbar\omega_{X-ray} = \hbar\omega_L \frac{4\gamma^2 n}{1 + a_0^2/2 + \gamma^2 \Theta^2} \tag{4.21}$$

Here, ω_x and ω_L are the frequencies of the scattered and laser field, respectively; n is the harmonic number. The strengths of the individual harmonics are (Esarey et al. [1], Equation 40):

$$F(n, a_0) = \frac{n^2 a_0^2}{4\left(1 + a_0^2/2\right)} \left[J_{(n-1)/2}\left(\frac{na_0^2}{4(1 + a_0^2/2)}\right) - J_{(n+1)/2}\left(\frac{na_0^2}{4(1 + a_0^2/2)}\right) \right]^2 \tag{4.22}$$

---

* A strategy for generating FEL-type hard X-rays can be found in [12].

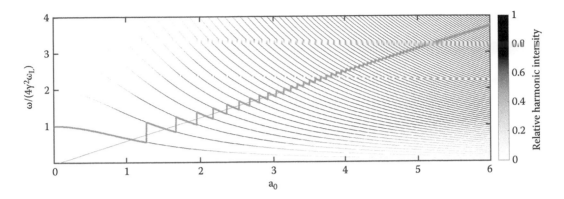

FIGURE 4.8 On-axis Thomson backscattering spectrum: individual harmonics depending on the wiggling field strength a_0. The frequency is normalized to the fundamental at $4\gamma^2\omega_L$. The dominant harmonic number n is marked blue, and a linear fit of the spectral peak as a function of a_0 is shown green. The colour code represents the spectral intensity, not the photon number.

where $J_\nu(x)$ is a Bessel function of the first kind. Figure 4.8 illustrates Equations 4.21 and 4.22. With increasing a_0 the fundamental frequency and the harmonic spacing are reduced, while the power radiated to higher harmonics increases. For $a_0 \gg 1$ the discrete spectrum merges into a synchrotron-like continuum, whose peak frequency scales as $\omega_{max} \approx 2.5a_0\gamma^2\omega_L$. For a temporally flat, top-head colliding pulse, the width of a harmonic line is inversely proportional to the number of oscillations N_p, as in the undulator case:

$$\Delta\omega \approx \frac{2.78}{\pi N_p} \frac{\omega_L 4\gamma^2}{1 + a_0^2/2} = \frac{2.78}{\pi N_p} \omega_{Xray} \qquad (4.23)$$

So far, we have treated single electrons in wiggling fields. In practice, we use electron bunches with a charge in the range of 10–100 pC, i.e. consisting of 10^7–10^8 electrons in a 6D phase space. To obtain the full bunch emission, we can numerically calculate the fields for a sufficiently large random subset of electrons and add the field vectors. This has been done using the code SPECTRA 10.0 [35] in Figures 4.9–4.11 for different cases assuming a zero-emittance electron bunch at 70 MeV.

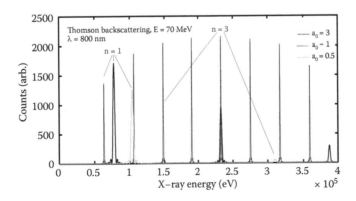

FIGURE 4.9 Thomson backscattering spectrum for a 70 MeV, zero-emittance and divergence electron bunch and a 25 period, $\lambda = 800$ mm top-hat optical undulator at three different strengths, given by a_0.

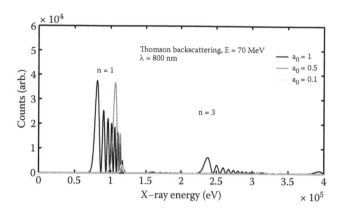

FIGURE 4.10   Gaussian 30 fs, 800 nm laser pulses with different peak a_0 backscattering off an electron bunch as in Figure 4.9. The spectrum is broadened by the varying field amplitude.

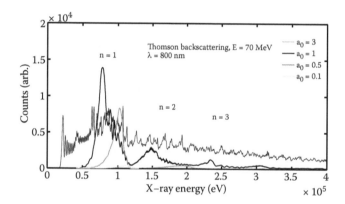

FIGURE 4.11   Same as Figure 4.10, but for a 5-mrad divergent bunch. Spectra smear out and even harmonics appear due to the observation of x-rays from off-axis electron trajectories.

For $a_0 \ll 1$ the emission occurs at the fundamental, but as soon as a_0 approaches unity, the down-shift in X-ray energy and the emission of harmonics become evident. For a constant-amplitude wiggling field, the emission spectrum of a single line resembles a sinc²-function as in an undulator (Figure 4.9). However, the typical laser pulse for Thomson scattering is a Gaussian with a different peak field at each oscillation period. Therefore, the X-ray emission downshifts during the first half on the pulse, and upshifts afterwards (Figure 4.10). If the electron beam is divergent, an on-axis observer will see spectrally-broadened off-axis emission and even harmonics from off-axis electrons (Figure 4.11). Bunches with non-zero energy spread will further smear out the emitted spectrum according to Equation 4.19. A good approximation for the emission linewidth, taking into account most of the relevant nonideal parameters, is given by Rykovanov et al [29], Equation 18:

$$\sqrt{\frac{\gamma^4 \Delta\theta^4}{16} + \frac{4\Delta\gamma^2}{\gamma^2} + \frac{a_0^4}{4} + \left[N_{sc} \cdot 2\gamma \frac{\hbar\omega_L}{m_e c^2} \right]} < \kappa \qquad (4.24)$$

Here $\kappa = \Delta\omega_X / \omega_{X,max}$ is the normalized bandwidth of the x-ray pulse, $\Delta\gamma$ is the FWHM-width of the electron energy distribution with mean energy γ and $\Delta\theta$ is the FWHM-divergence of the electron beam.

N_{sc} is the mean number of scattering events per electron, which can be estimated as [29]

$$N_{sc} = \frac{N_\gamma}{N_e} \approx 4.7 \cdot a_{0,\max} \cdot \sqrt{\frac{E_{L,J}}{\lambda_{L,\mu m}}} \tag{4.25}$$

provided that the duration of the laser is matched to the laser's confocal parameter $b = 2z_R$, where z_R is the Rayleigh length. With the wavelength $\lambda_{L,\mu m}$ in units of μm, the focusing geometry (FWHM diameter $d_{L,\mu m}$ and FWHM pulse duration $\Delta t_{L,fs}$) depends on the scattering pulse energy $E_{L,J}$ in Joules and the desired peak a_0 [29],

$$d_{L,\mu m} \rightarrow \frac{6.48 \lambda_{L,\mu m}^{3/4}}{\sqrt{a_{0,\max}}} E_{L,J}^{1/4} \quad \text{and} \quad \Delta t_{L,fs} \rightarrow 135.6 \sqrt{\frac{\lambda_{L,\mu m} E_{L,J}}{a_{0,\max}}} \tag{4.26}$$

This allows selecting the optimum pulse parameters for a given laser energy and harmonic content.

Pulse Duration The emission duration in the laboratory frame is simply marked by the time between the first and last electron to exit the wiggling field. Additionally we have to account for the path difference between the straight light field and the wiggling electron trajectory. Due to the coherence condition of undulator radiation, the electron slips one wavelength with respect to the light field for each oscillation in the proper frame. Assuming Gaussian electron and laser pulses with durations of Δt_{el} and Δt_L, respectively, in the proper frame, the pulse duration is therefore

$$\Delta t'_{rad} = \sqrt{\left(\Delta t'_{el}\right)^2 + \left(\Delta t'_L\right)^2} = \sqrt{(1+\beta)^2 \gamma^2 \Delta t_{el}^2 + (1+\beta \cos(\theta))^2 \gamma^2 \frac{N^2 \lambda_L^2}{c^2}} \tag{4.27}$$

Transformed back to the laboratory frame, this gives

$$\Delta t_{rad} = \frac{\sqrt{(1+\beta)^2 \gamma^2 \Delta t_{el}^2 + (1+\beta \cos(\theta))^2 \gamma^2 \frac{N^2 \lambda_L^2}{c^2}}}{(1+\beta)\gamma} = \sqrt{\Delta t_{el}^2 + \frac{(1+\beta \cos(\theta))^2}{(1+\beta)^2} \frac{N^2 \lambda_L^2}{c^2}} \tag{4.28}$$

In the case of a near head-on collision ($\cos\theta \approx \pi$), this simplifies to

$$\Delta t_{rad} \approx \sqrt{\Delta t_{el}^2 + \frac{(1-\beta)^2}{(1+\beta)^2} \frac{N^2 \lambda_L^2}{c^2}} \approx \sqrt{\Delta t_{el}^2 + \left(\frac{1}{4\gamma^2} \frac{N \lambda_L}{c}\right)^2} \approx \sqrt{\Delta t_{el}^2 + \frac{\Delta t_L^2}{4\gamma^2}} \tag{4.29}$$

This scaling allows one to tailor the scattering pulse without having to worry about pulse duration. In order to reach high brilliance, one strategy is to keep a_0 close to unity by stretching the pulse [16], which due to the $1/4\gamma^2$ contraction in time has little influence on the final pulse duration.

Scattered Power The total energy scattered by a single electron from a rectangular laser pulse during its duration Δt_L is given by Ride et al. [27], Equation 43*

$$W \approx \frac{8\pi^2}{3\lambda_L^2} e^2 c a_0^2 \gamma^2 \Delta t_L \approx \left[2 \times 10^{-31} \gamma^2 a_0^2 \frac{\Delta t_L[fs]}{\lambda_L[\mu m]}\right] [J] \tag{4.30}$$

---

* Ride et al. evaluated scattered power, which we multiplied with the interaction time $\Delta t_L/(1 + \beta)$, with $1 + \beta \approx 2$.

A more accurate result for a non-rectangular laser pulse is found by evaluating the integral $W \approx \frac{8\pi^2}{3\lambda_L}e^2c\gamma^2\int a_0^2(t)dt$ where $a_0(t)$ is the envelope of the colliding field. For a Gaussian envelope, this results in a replacement of the duration Δt_L in Equation 4.30 with $\sqrt{\pi/2}\sigma_t$ or $\sqrt{\pi/(16\ln(2))}\Delta t_{L,FWHM,int} \approx 0.53\Delta t_{L,FWHM,int}$ of the FWHM intensity duration. For $a_0 \leq 1$ and the predominant emission of the fundamental, we obtain the number of scattered photons per electron by dividing the scattered energy with the mean photon energy of the fundamental $\hbar\omega_X = \hbar 4\gamma^2\left(1+a_0^2/2\right)^{-1}\omega_L$:

$$N_X \approx \frac{\pi}{3\hbar\lambda_L}e^2a_0^2\left(1+a_0^2/2\right)\Delta t_L \approx 2.54\times 10^{-4}\frac{\Delta t_L[fs]}{\lambda_L[\mu m]}a_0^2\left(1+a_0^2/2\right) \tag{4.31}$$

Since the number of scattered photons does *not* depend on their energy, Thomson scattering is very attractive for high-energy photon generation. Finally, by employing the Thomson cross-section formula, one can deduce $N_{x\text{-}ray}$ for arbitrary laser and electron spot sizes. Assuming Gaussian beams, by integrating over the local electron and photon densities one arrives at:

$$N_{X-ray} = \sigma_t\frac{2}{\pi}N_eN_Lr_0^{-2} \quad \overset{\text{integrate over}}{\underset{\text{beam profiles}}{\Rightarrow}} \quad N_{X-ray,\Sigma} = \frac{\sigma_T N_L N_e}{\pi^2\left(\sigma_e^2+\sigma_L^2\right)} \tag{4.32}$$

On the left, N_L photons scatter off an electron areal density N_e/r_0^2, where r_0 is the interaction radius and $\sigma_T = 8/3\pi r_e^2$ is the Thomson cross section (with the classical electron radius $r_e = e^2/(4\pi\epsilon_0 m_e c^2)$). On the right, $\sigma_e \approx FWHM_e/2.35$ and $\sigma_L \approx FWHM_L/2.35$ are the r.m.s. sizes of electron bunch and laser focus at the interaction position, respectively.

Current Experimental Results Colliding two ultra-short, high energy laser pulses in a $10 \times 10 \times 10\mu m^3$ volume in space and time is a nontrivial task. An efficient solution to this problem has been found by Ta Phuoc et al. [24], by placing a foil as a plasma mirror into the plasma accelerator. The driving laser pulse back-reflects off the foil, hitting the electron pulse on its way back. This setup is automatically synchronized, but there is no possibility to shape or stretch the colliding pulse, and its intensity might be enhanced by self-focusing considerably beyond the $a_0 \approx 1$ limit. For applications not requiring a high degree of monochromaticity, this approach is a simple and efficient way for creating a brilliant source of hard X-rays with the smallest possible source size. True head-on collision experiments have been carried out with 800 nm Ti:Sa collision pulses by several groups, first by Schwoerer et al. [34] in 2006, and later by Powers et al. [26], Sarri et al. [30] and Khrennikov et al. [18]. One of the main challenges is ensuring the spatial and temporal overlap of the electron bunch and colliding laser pulse. A simple way to solve this problem is to image the Thomson side-scattering emission of the ion channels caused by both beams. This allows for the overlap of both channels in a camera image. The timing information is gathered by a separate transverse probe pulse using a shadowgraphy setup, imaging the ionization fronts of both pulses in the plasma. Schwoerer and Sarri used broadband electron bunches and relativistic-intensity collision pulses ($a_0 = 0.8$ and 2.5, respectively), leading to broadband X-ray spectra. In the pioneering case of Schwoerer, the electron spectra were quasi-Maxwellian with a temperature of 6–7 MeV, yielding X-rays in the few keV range, which are readily detected and spectroscopically characterized directly by an X-ray CCD in a single-photon counting mode. Sarri used electron energies of up to 800 MeV, necessitating the use of an indirect Li-based Compton electron spectrometer to detect and retrieve spectral information on the generated 6-20 MeV γ-pulses. They established the onset of the non-linear Thomson scattering regime by the number of photons in the detector's sensitivity window, and reported a world record peak brilliance in that energy range.

Powers used tunable, mostly quasi-monochromatic electron bunches with energies between 50 and 300 MeV and subrelativistic $a_0 = 0.3$ collision pulses to create X-rays tunable between 50 keV and 1 MeV, which were characterized by a filter array in front of a scintillator-based X-ray CCD camera. They verified the peaked nature of the X-ray spectra, and reported a record brilliance in the high-energy part of their observation range. Khrennikov collided weakly relativistic $a_0 = 0.85$ laser pulses with tunable quasi-monochromatic electron bunches in the energy range between 15 MeV and 70 MeV from a shock-front injection LWFA scheme [4]. For up to 45 MeV electron energy, the spectra were recorded by a single photon counting CCD allowing to record single-shot spectra (see Figure 4.12). They indicated the onset of the nonlinear regime both by the characteristic energy downshift (see Equation 4.19) attributed to a_0, approaching unity as well as the emission of the second and third harmonics. The former is detected on-axis due to the finite divergence of the electron beam. In summary, Thomson backscattering provides the highest energy X- (or γ-)rays to date for a given electron bunch energy when one neglects low-brilliance Bremsstrahlung and line emission sources. All-optical Compton-backscattering sources achieve peak brilliances many orders above other hard (100 keV-MeV) X-ray sources on account of their few-fs duration, small source size and directed emission. Moreover, their pulses are naturally synchronized to intense laser pulses, making them a potentially unique tool for new, ultrafast pump-probe experiments. Their high energy, potentially narrow bandwidth and tunability also makes them suited for phase-contrast imaging of the human body, X-ray fluorescence and absorption spectroscopy and two-colour X-ray imaging, if their average brightness can be substantially increased.

Perspectives for Further Development So far, the experiments realizing Thomson scattering have not exploited the optimum condition for highest brilliance, mainly due to circumstantial conditions of the LWFA's and scattering pulses they used. According to Figure 4.8, the maximum photon number in the fundamental line is achieved at $a_0 \approx 1.25$. Higher colliding pulse intensities should be avoided, since they would distribute the peak emission among several harmonics at a time and broaden the emission cone in polarization direction. However, as evident from Figure 4.10, the variation of $a(t)$ during the pulse causes significant spectral broadening even for $a_0 \approx 1$. Therefore, just increasing the colliding pulse energy without scaling the focus size or duration does raise the total photon number, but not the X-ray brilliance; $a_0 \approx 1$ is a natural limit beyond which spectral broadening and harmonic emission curb further brilliance improvement. Ghebregziabher et al. [16] proposed temporal stretching of the colliding pulse to limit a_0. We can take their concept a step further in order to optimize the performance for large laser systems.

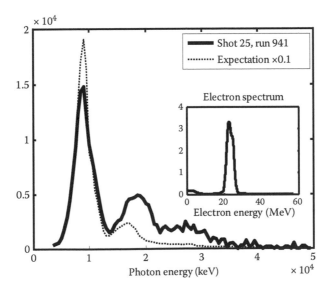

FIGURE 4.12 Measured spectrum of weakly nonlinear Thomson backscattering X-rays including harmonics.

The frequency offset caused by the temporal shape of $a(t)$ can be compensated by an appropriately shaped chirp of the colliding pulse. According to Equation 4.19, we can always find a pair of $\lambda_L(t)$- and $a(t)$-values to yield a fixed X-ray wavelength if

$$a(t) = \sqrt{\frac{8\gamma^2\lambda_{x-ray}}{\lambda_L(t)} - 2}$$

(4.33)

is fulfilled. If the stretching exceeds the Rayleigh length $z_r = \pi\sigma_0^2/\lambda_0$, where σ_0 is the $1/e^2$ spot radius, the divergence of the laser cannot be neglected anymore. This places the additional constraint on the $a(t)$ shape:

$$a(t) = a_0(2t) \frac{1}{\sqrt{1 + \frac{c^2 t^2}{z_R^2}}}$$

(4.34)

This relates the field $a(t)$ that the electron sees upon propagating through the focal region along the laser axis to the laser-strength at focus $a_0(2t)$. The $2t$ in the argument of a_0 is needed because of the head-on geometry where both the electron and the laser propagate against each other with c. For the sake of simplicity, we shall omit this factor 2 in the following, keeping in mind that the actual laser pulse is two times longer than the one that will be calculated. The on-axis a_0 in the focal plane itself depends on the instantaneous intensity on axis $I_m(t)$ and instantaneous frequency $\omega_L(t)$:

$$a_0(t) = \frac{e\sqrt{2I_m(t)\mu_0 c}}{m_e\omega_L(t)c}$$

(4.35)

Using the results from Equations 4.34 and 4.35 in Equation 4.33 yields the relationship between instantaneous laser frequency $\omega_L(t)$ and instantaneous laser intensity $I_m(t)$ that fixes the scattered X-ray frequency to its value for given constants ω_L and a_0 during the whole interaction:

$$\omega_L(t) = \frac{\omega_L}{2 + a_0^2}\left[2 + \frac{2e^2 I_m(t)\mu_0}{m_e^2\omega_L^2(t)c\left(1 + \frac{c^2 t^2}{z_R^2}\right)}\right]$$

(4.36)

What remains is simply to find $\omega_L(t)$ and corresponding $I_m(t)$ values while maximising z_R. Increase in z_R demands an increase in the total laser energy W_Σ. This yields the constraint on $I_m(t)$:

$$W_\Sigma = 2\int_{-z_R/c}^{z_R/c} I_m(t)d_L^2 2\pi\, dt$$

(4.37)

where we imply that the length of the stretched laser pulse equals twice its full Rayleigh-range of $[-z_R; z_R]$. d_L is again the FWHM diameter of the laser pulse. It does not seem practical to look for an analytical solution to Equation 4.36, so one has to rely on numerical maximization of z_R under the constraints of Equation 4.37 while keeping in mind the available bandwidth:

$$\omega_L(z_R/c) - \omega_L(-z_R/c) \leq BW$$

(4.38)

for obtaining the best $\omega_L(t)$ - function. Modern spectral shaping techniques [23] allow fine-tuning the spectrum of a CPA laser pulse. The spectral amplitude $a(\omega)$ of a pulse will be transformed into a temporal shape $a(t)$ by chirping. A spectral filter to create the correct $a(t)$ for a given chirp $\lambda_L(t)$ is all that

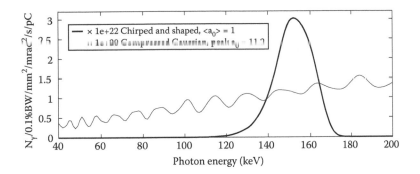

FIGURE 4.13 Thomson spectra for 100 MeV bunches with 1 μm size, 1 mrad divergence and 1 % energy spread (all values r.m.s.) interacting with 30 J, 800 nm laser pulses in a 17.7 μm FWHM spot. The chirped/compressed pulses are 3.8 ps/30 fs FWHM (a_0 = 11.2/1.0) long, respectively. The multiplication factor in the legend indicates about a 250-fold increase in peak brilliance for chirped pulses and 5.6×10^7 photons/mrad^2 for 100 pC bunch charge.

is needed to fulfil Equation 4.33. The effect on the achievable X-ray spectral brightness is dramatic, as depicted in Figure 4.13, again produced by SPECTRA 10.0 [35].

Other possibilities to circumvent the intensity-induced spectral broadening include travelling-wave Thomson scattering with a pulse-front tilted collision pulse [12], which spatially reduces the scattering intensity while keeping the field temporally synchronized with the electron bunch. Finally, new laser technology might raise interesting possibilities. Besides the general quest for higher repetition rates in the multi-TW regime, spearheaded by thin-disk and fibre laser technology, one architecture in particular lends itself to high average power Thomson scattering: multi-TW optical parametric chirped pulse amplifiers (OPCPA) pumped by J-class, kHz, frequency-doubled few-ps CPA disk lasers. Here, the ultra-short OPCPA pulse can drive the electrons and the unconverted, energetic pump light can be used as a ps-long collision pulse, the latter ideally matching the optimum collision pulse parameters.

In summary, Thomson sources currently have demonstrated limited-bandwidth (10–20%), tunable hard X-rays from several 10 keV to the multi-MeV range, along with extremely short pulse duration (few fs). However, they do suffer a total flux that is one to two orders of magnitude lower than simple betatron sources (see Chapter 5), which in turn struggle to reach photon energies beyond a few 10 keV. For an application in medical phase contrast imaging, for example, one needs Thomson-type photon energy (>50 keV) and betatron-like flux, which can only be achieved by further optimized sources.

4.4 Betatron Emission

Instead of wiggling electron bunches in an external field by undulators or lasers, laser-wakefield accelerators offer a third way for generating X-rays by virtue of the ultrastrong radial field gradients in the plasma accelerator itself. In an LWFA, electrons are injected in the plasma cavity by induced or spontaneous wavebreaking, ionization or external laser pulses, but for all injection methods, the individual electrons have a distribution of injection radii, i.e. the injection does not take place on-axis. Any off-axis injected electron will undergo a wiggling motion around the central axis, focused by the radial fields of the wake during the acceleration process. Therefore, betatron emission in an LWFA is a highly non-stationary and chaotic process.

Formalism In order to gain a basic understanding, we will first assume that during the wiggling motion no acceleration or deceleration takes place. The wiggling motion is caused by the focusing fields E_r of the plasma ion cavity where the plasma wavelength λ_p dictates the field gradient: [13]:

$$E_r = \frac{m_e c^2 2\pi^2}{e\lambda_p^2}r; \quad \lambda_p \approx \frac{2\pi c}{\omega_p} = 2\pi c\sqrt{\frac{m_e \epsilon_0}{n_e e^2}} \approx \frac{3.34\times10^4}{\sqrt{n_e[cm^3]}}[m] \tag{4.39}$$

A relativistic electron trapped with a non-vanishing, off-axis distance r then oscillates with the betatron frequency:

$$\omega_\beta = \omega_p/\sqrt{2\gamma} = 2\pi c/(\lambda_p\sqrt{2\gamma}) \tag{4.40}$$

It emits radiation at a wavelength $\lambda_{x\text{-}ray}$ governed by the betatron strength parameter K_β

$$\lambda_{X-ray}(n) = \frac{\lambda_\beta}{n2\gamma^2}\left(1 + \frac{K_\beta^2}{2} + \gamma^2\theta^2\right); \quad K_\beta = \gamma r_\beta \frac{2\pi}{\lambda_\beta} \tag{4.41}$$

For typical laser-wakefield scenarios, $K \approx 10$ and the emission takes place predominantly as high harmonics. The critical harmonic number $n_c \approx 3/4K_\beta^3$ is a measure for the high anharmonicity of the betatron process. An expression of the asymptotic, on-axis spectral shape for high harmonic numbers is given in Esarey et al. [13], Equation 64.

$$\frac{d^2I(\theta=0)}{d\omega d\Omega} \propto \left(\frac{\omega}{\omega_c}\right)^2 K_{2/3}^2\left(\frac{\omega}{\omega_c}\right) \tag{4.42}$$

where the critical frequency is defined as

$$\omega_c = \frac{2\pi c}{\lambda_X(n_c)} \overset{K_\beta^2/2 \gg 1}{\approx} n_c\frac{4\gamma^2}{K_\beta^2}\omega_\beta \overset{K_\beta = \gamma r_\beta \omega_\beta/c}{\approx} 3r_\beta\frac{1}{c}\gamma^3\omega_\beta^2 \approx r_\beta\gamma^2\frac{6\pi^2 c}{\lambda_p^2} \tag{4.43}$$

The oscillation amplitude $r_\beta \approx r_{\beta0}(\gamma_0/\gamma)^{1/4}$ weakly decreases with increasing electron energy [11]. Here $r_{\beta0}$ and γ_0 are the electron's transverse position and energy at injection. A value $r_\beta \approx 0.5 - \mu m$ is suggested by experimental and numerical source size estimates. Corde et al. [11] gives the cut-off energy in practical units:

$$\hbar\omega_c[eV] \approx 5.24\times10^{-21}\gamma^2 n_e[cm^{-3}]r_\beta[\mu m] \tag{4.44}$$

However, these numbers can only serve as very coarse estimates of the true spectrum. Since the emission takes place *during* the LWFA process, the electron energy and γ vary within one betatron period. Hence the emitted spectrum has to be calculated from the Lienard–Wiechert potentials using the actual electron trajectories, as detailed in the Appendix. The number of photons scales linearly with K and the number of oscillations. The total number at the averaged frequency $0.3\omega_c$ per oscillation period and electron can be given as [11, Equation 53]

$$N_X \approx 3.31\times10^{-2}K_\beta \tag{4.45}$$

At $K_\beta \approx 30$, each electron will produce one photon per oscillation period. In typical experimental conditions, the electron performs only one to two betatron oscillations with the maximum energy before they are rapidly decelerated again by dephasing. The radiation is confined in a cone with an half-angle of

$\theta = K_\beta/\gamma$ [11]. It is worth mentioning that with increasing K_β, the radiated energy per unit bandwidth stays constant. The spectral peak merely shifts towards higher frequencies and the spectrum gets broader, both being proportional to K_β. So, the total radiated energy in the forward direction scales with K_β. Since the radiation cone also scales with K_β, the total photon number scales with K_β^2. This doesn't contradict the linear dependence of the photon number Nx, as given in Equation 4.45, since the mean photon energy is also linear with K_β. As an illustration, we show a numerical calculation of the single-electron betatron spectrum along with the electron trajectory evolution in an idealized bubble accelerator scenario for three different values of initial transverse position in Figure 4.14. This illustrates the strong dependence of the betatron spectrum on initial conditions, despite not even taking into account energy or angle distributions in a real bunch. Therefore, meaningful predictions of betatron performance have to rely on the accuracy of large-scale particle-in-cell simulations for calculating individual electron trajectories. Figure 4.15 shows a selection of betatron spectra from electron bunches whose spectral and angular distribution was assumed flat up to a maximum cut-off energy and angle, respectively.

Figure 4.15 neatly illustrates the γ^2-energy scaling despite the large energy spread, indicating that the strongest radiation is emitted around the dephasing point, when the electrons reach their highest energy. The right-hand side plots the 1% cut-off energy as a function of electron energy and maximum betatron radius (reflecting an effective K). The downshift in fundamental photon energy for large K is obviously overcompensated for by the emission of higher harmonics. This illustrates the benefits of asymmetric injection to boost the output.

Recent Experiments Several groups have used Betatron radiation for X-ray imaging purposes, demonstrating, for example, that a single laser shot is sufficient for exposing a phase-contrast image of an insect [14,19] or an absorption contrast image of a human bone sample [10]. In the latter case, a 500-TW laser system produced 20–30 keV betatron photons that were needed to penetrate bone. Multiple absorption images from different angles were used to reconstruct a 3D tomogram of the sample. Similarly, Wenz et al. [38] reconstructed a phase-contrast tomogram of an insect using a 60-TW laser; 5 keV photons penetrated the insect with little absorption and yielded almost pure phase-contrast images (see Figure 4.16).

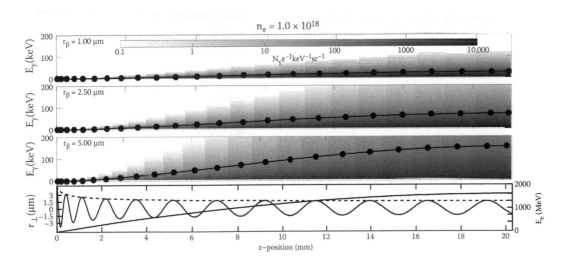

FIGURE 4.14 On-axis betatron spectra of a single e⁻ undergoing acceleration until de-phasing in an idealized plasma bubble at a plasma density of $n_e = 10^{18}\text{cm}^{-1}$. The upper three plots show the evolution of the radiated spectrum during the acceleration for the initial transverse positions $r_{\beta0} = 1, 2.5, 5$ µm. The logarithmic spectral intensity is color-coded and normalized to the photon energy $N_\gamma/keV/sr$. The black curves show the cut-off energy from Equation 4.43 for the final value of γ. These curves overestimate the simulated spectral cut-off because of the rapid evolution of the electron γ during the acceleration. The lowest plot shows the transverse trajectory of the e⁻ for $r_{\beta0} = 5$ µm in blue and its energy evolution in green.

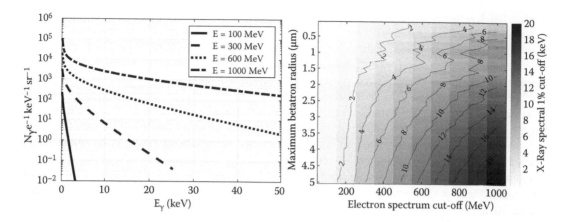

FIGURE 4.15   Left: Betatron spectra from an e^- bunch during acceleration with a flattop final spectrum extending to the stated cut-off energy. The distribution of transverse injection position is flat between 0 μm and 5 μm. Right: 1% intensity X-ray cut-off energy as a function of maximum e^- energy E and maximum transverse injection radius $r_{\beta 0}$.

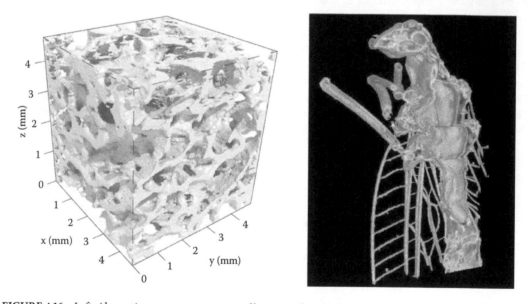

FIGURE 4.16   Left: Absorption contrast tomogram of human trabecular bone recorded at E_{crit} = 33 keV. (Originally published in *Nature Scientific Reports*, "Laser-wakefield accelerators as hard x-ray sources for 3D medical imaging of human bone" by J.M. Cole et al. Reused under the Creative Commons 4.0 licence https://creativecommons.org/licenses /by/4.0/) Right: Phase-contrast tomogram of a green lacewing at 5 keV. (Originally published in *Nature Communications*, "Quantitative X-ray phase-contrast microtomography from a compact laser-driven betatron source" by J. Wenz et al. Reused under the Creative Commons CC-BY licence https://creativecommons.org/licenses/by/4.0/)

In most current experiments, the betatron output is optimized phenomenologically, i.e. by adjusting the LWFA parameters until an optimum exposure is reached on the X-ray detector.

A more systematic approach draws on the fact that in an LWFA the effective K_β scales linearly with the electron injection radius. The latter can be controlled by imprinting a transverse asymmetry onto the wakefield by the drive pulse, either by introducing a coma to the focal spot [22] or by tilting the pulse front [25,33]. Mangles et al. [22] experimentally demonstrated the expected increase of photon energy caused by the larger effective K_β.

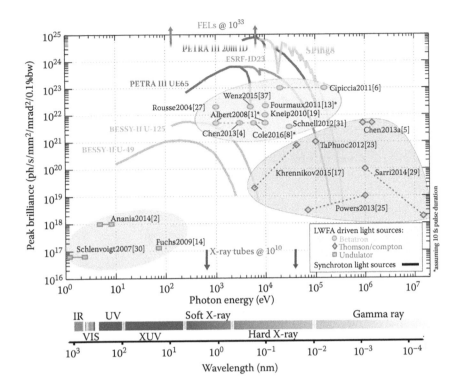

FIGURE 4.17 Landscape of laser-driven, free-electron X-ray sources. (Courtesy of Johannes Wenz. Reproduced with permission.)

The reported peak brilliance of Thomson, betatron and all-optical undulator sources is plotted in Figure 4.17. Obviously some Betatron and most Thomson sources exceed large 3rd generation RF-accelerator-driven light sources in terms of peak brilliance, due to their source size, duration and single-pulse photon number. While this impressively shows the potential of all-optical sources, it should not be forgotten that for average brilliance the prospect is a lot bleaker due to the low repetition rate of current laser systems. Therefore, before new laser technology establishes itself, all-optical sources will mainly find applications that use their unique shortness and exact (few-fs-scale) synchronization with laser pulses. To the best of our knowledge, no such experiments have been reported yet. The growing maturity of laser-driven technology, which we demonstrated in this chapter, may foster creativity of potential applicants.

4.5 Appendix: Numerical Treatment, a Do-It-Yourself Approach

The purpose of this appendix is to give a step-by-step tutorial of how to numerically compute the radiation fields of single particles in external fields. This will result in MATLAB routines that can form the basis for the reader's own simulations. First we numerically solve the relativistic equation of motion of a particle in a given external field, such as an undulator or colliding laser pulse. The resulting trajectory is then used to compute the radiation field via the Lienard–Wiechert potentials.

Particle trajectory for given fields: The particle trajectory in external electric $\vec{E}(\vec{r},t)$ and magnetic $\vec{B}(\vec{r},t)$ fields is governed by the relativistic equation of motion:

$$\vec{F} = \frac{d\vec{p}}{dt} = e(\vec{E} + \vec{v} \times \vec{B}) \tag{4.46}$$

In order to solve it in a relativistically correct way, we rewrite this equation. In order to save space, we use primes to denote time derivatives:

$$\frac{d\vec{p}}{dt} = mc\frac{d}{dt}(\gamma\vec{\beta}) \tag{4.47}$$

$$= mc\left[\gamma\vec{\beta}' + \vec{\beta}\gamma'\right] \quad \left|\gamma' = \left[\frac{1}{\sqrt{1-\beta^2}}\right]' = \frac{(\vec{\beta}\cdot\vec{\beta}')}{\sqrt{1-\beta^2(1-\beta^2)}}\right| \tag{4.48}$$

$$= \frac{mc}{\sqrt{1-\beta^2(1-\beta^2)}}\left(\vec{\beta}'(1-\beta^2) + (\vec{\beta}\cdot\vec{\beta}')\vec{\beta}\right) \tag{4.49}$$

Since radiation is produced by accelerated charges, the equation for $\vec{\beta}'$ is of highest interest. Rewriting Equation 4.49 component-wise and executing the vector multiplications leads to the following matrix equation:

$$\vec{p}' = mc(\gamma\vec{\beta})' = \frac{mc}{(1-\beta^2)^{3/2}} \cdot \begin{pmatrix} 1-\beta_y^2-\beta_z^2 & \beta_x\beta_y & \beta_x\beta_z \\ \beta_x\beta_y & 1-\beta_x^2-\beta_z^2 & \beta_y\beta_z \\ \beta_x\beta_z & \beta_y\beta_z & 1-\beta_x^2-\beta_y^2 \end{pmatrix}\vec{\beta}' = e(\vec{E} + c\vec{\beta}\times\vec{B}) \tag{4.50}$$

The only step remaining to obtain an expression for $\vec{\beta}'$ is to multiply both sides of the equation above with the inverse matrix from the left. The inverse matrix is:

$$\begin{pmatrix} 1-\beta_y^2-\beta_z^2 & \beta_x\beta_y & \beta_x\beta_z \\ \beta_x\beta_y & 1-\beta_x^2-\beta_z^2 & \beta_y\beta_z \\ \beta_x\beta_z & \beta_y\beta_z & 1-\beta_x^2-\beta_y^2 \end{pmatrix}^{-1} = \frac{1}{\beta^2-1}\begin{pmatrix} \beta_x^2-1 & \beta_x\beta_y & \beta_x\beta_z \\ \beta_x\beta_y & \beta_y^2-1 & \beta_y\beta_z \\ \beta_x\beta_z & \beta_y\beta_z & \beta_z^2-1 \end{pmatrix} \tag{4.51}$$

This finally leads to a lean version of the Lorentz equation ready to be solved numerically:

$$\frac{d\vec{\beta}}{dt} = -\frac{e\sqrt{1-\beta^2}}{mc}\begin{pmatrix} \beta_x^2-1 & \beta_x\beta_y & \beta_x\beta_z \\ \beta_x\beta_y & \beta_y^2-1 & \beta_y\beta_z \\ \beta_x\beta_z & \beta_y\beta_z & \beta_z^2-1 \end{pmatrix}(\vec{E} + c\vec{\beta}\times\vec{B}) \tag{4.52}$$

Remark on numerical integration: A numerical problem that needs to be avoided is the apparent gain of energy when naively trying to calculate a circular trajectory by applying a normal acceleration in finite time steps. Instead of rotating the particle, this just adds a normal velocity component, which leads to an increase in the overall energy. In order to overcome this problem, the following approach may be used. At each iteration of the trajectory integration we use the velocity $\vec{\beta}_n$ of the particle at time instant n, and the acceleration $\vec{\beta}'$ in order to calculate the velocity $\vec{\beta}_{n+1}$ at the next time step. The direction of $\vec{\beta}_{n+1}$ obviously is $\dfrac{\vec{\beta}_n + \vec{\beta}'\Delta t}{\|\vec{\beta}_n + \vec{\beta}'\Delta t\|}$ and its norm $\|\vec{\beta}_n\| + \dfrac{(\vec{\beta}'\Delta t \cdot \vec{\beta}n)}{\|\vec{\beta}_n\|}$ Combining both, we obtain

$$\vec{\beta}_{n+1} = \left(\frac{\vec{\beta}_n + \vec{\beta}'\Delta t}{\|\vec{\beta}_n + \vec{\beta}'\Delta t\|}\right) \cdot \left(\|\vec{\beta}_n\| + \frac{(\vec{\beta}'\Delta t \cdot \vec{\beta}_n)}{\|\vec{\beta}_n\|}\right) \tag{4.53}$$

This is a bit more complicated than just using $\vec{\beta}_{n+1} = \vec{\beta}_n + \vec{\beta}'\Delta t$, but Figure 4.18 demonstrates the merit: In a magnetic undulator, the total energy of the particle does not change if, as above, the radiation loss is neglected. However, in the top row, an increase of β by $\approx 2 \times 10^{-5}$ is evident, and is equivalent to a 13 MeV energy boost, along with a reduction of the oscillation amplitude due to the relativistic mass increase. In the bottom row, using the correct Equation 4.53, the deviation is in the 10^{-14}-range, and limited by the finite precision of *double-type* variables.

MATLAB©Code Sample: The following sample evaluates the trajectory of an electron in a helical undulator with wave-number ku, magnetic field amplitude BO, period length Lu and number of periods NN. ee, me and cc are electron charge, mass and speed of light, respectively. All units are SI. This code sample calculates the position pos and velocity beta of the particle depending on the time tt; tt is an array of fixed time points at which the coordinates of the particle are to be evaluated. This approach requires adequate sampling which is best found heuristically: double the sampling until the results remain the same.

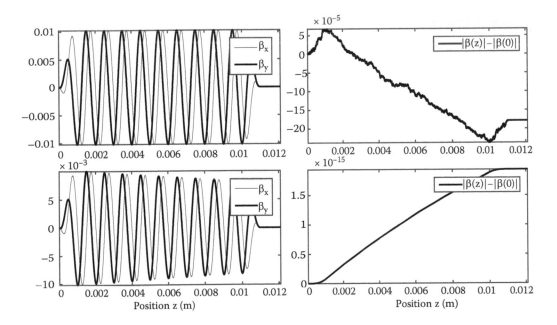

FIGURE 4.18 Numerical integration example. Left: Trajectory of a 50 MeV electron in a helical undulator with $K = 1$. Right: deviation of particle velocity from its initial value. Top row: brute-force integration ($\vec{\beta}_{n+1} = \vec{\beta}_n + \vec{\beta}'\Delta t$) of equation of motion. Bottom row: 'correct' integration following Equation 4.53.

```
Listing 4.1: Evaluation of particle trajectory in external fields
tt=0: dt :1.5*NN*Lu/cc;
for it=2: length (tt)

% evaluate current position of the particle
pos(:, it)= pos (:,it -1) + beta(:,it -1)*cc*dt;

% evaluate the fields at this position and time
EE=[0;0;0];
if(pos (3, it) >(NN*Lu))
BB=[0;0;0];
else
BB=B0*[ sin (ku*pos (3, it));
cos (ku*pos (3,it));
0];
if(pos(3, it) < Lu), BB=BB*(pos(3, it))/Lu; end
if(pos(3, it) > (NN -1)*Lu), BB=BB*(NN*Lu -(pos(3, it)))/Lu; end
end

% evaluate the increment of beta=dbeta *dt
dbeta=ee/me/cc*(sqrt(1 - bx*bx - by*by - bz*bz))* ...
([bx ^2 - 1, bx*by , bx*bz;
bx*by, by ^2-1, by*bz;
bx*bz, by*bz, bz ^2-1]* ...
(EE+cc* cross ([bx;by;bz],BB))*dt);

% evaluate new beta according to its increment
beta (:, it)= ([bx;by;bz] + dbeta)/norm([bx;by;bz] + dbeta)* ...
(bNorm + dot([bx;by;bz], dbeta)/bNorm);

% save its individual components to separate variables simply to
% enhance readability
bx=beta (1, it);
by=beta (2, it);
bz=beta (3, it);
bNorm=sqrt (bx ^2 + by ^2 + bz ^2) ;

end
```

Radiated spectrum for a given trajectory: The Lienard–Wiechert potentials are the solution of Maxwell's equations of a moving single charge q, where $\vec{n}$ is the observation direction:

$$\vec{E}(\vec{x},t)=\frac{q}{4\pi\epsilon_0 c}\left[\frac{c(\vec{n}-\vec{\beta})}{\gamma^2(1-\vec{\beta}\cdot\vec{n})^3 R^2}+\frac{\vec{n}\times\{(\vec{n}-\vec{\beta})\times\vec{\dot{\beta}}\}}{(1-\vec{\beta}\cdot\vec{n})^3 R}\right]_\tau \tag{4.54}$$

$$\vec{B}(\vec{x},t)=\frac{\vec{n}_{tret}}{c}\times\vec{E}(\vec{x},t) \tag{4.55}$$

The first component of the field $\vec{E}$ at the *retarded time* $\tau = t - R(\tau)/c$ is proportional to the particle velocity $\vec{\beta}$ alone which describes the particle's *self-* or *Coulomb-field*, whereas the second also depends on its acceleration $\dot{\vec{\beta}}$ and describes the *radiated field*. The radiated field decays much slower with distance $R=|\vec{R}(t)|$ between the charge and the observer, and $\vec{n}=\vec{R}/|R|$ gives the observation direction. Since we are interested in the radiated field, in the following we will neglect the self-field.

Getting rid of the retarded time: Given the fields in retarded time calculating the spectrum requires a Fourier transform in the retarded frame ($t_{ret} = \tau - R(t_{ret})/c$):

$$\frac{d^2W(\omega,(\theta,\phi))}{d\omega d\Omega} = \left[\frac{q}{4\pi\epsilon_0 c}\right]^2 \left|\int \frac{\vec{n}\times\{(\vec{n}-\vec{\beta})\times\vec{\dot{\beta}}\}}{(1-\vec{\beta}\cdot\vec{n})^2}\cdot e^{i\omega(\tau+R/c)}\,d\tau\right|^2 \tag{4.56}$$

This integral is computationally expensive: one has to do one full integration to obtain the spectral intensity at one frequency ω. A more numerically appealing approach is to evaluate both, the observer field $\vec{E}_k$ and time tk associated with every charge trajectory point $(\vec{\beta}(\tau),\vec{\dot{\beta}}(\tau),\tau)_k$, $k \in 1..N$:

$$\vec{E}_k = \frac{q}{4\pi\epsilon_0 c}\left[\frac{\vec{n}\times\{(\vec{n}-\vec{\beta}_k)\times\vec{\dot{\beta}}_k\}}{(1-\vec{\beta}_k\cdot\vec{n})^3}\right] \tag{4.57}$$

$$t_k = (\tau)_k + R_k/c \tag{4.58}$$

This will give us the field that the observer sees in his time, which we can then Fourier transform to get the spectrum. We omitted the $1/R$ dependence of the field amplitude. This is justified if the change in distance ΔR during the radiation process is negligible compared to the absolute distance R_o. For the same reason, we set the direction from the charge to observer $\vec{n}$ to be constant during the radiation process. We also remove the independence of time since we're merely interested in the direction-resolved spectrum. This can be done by integrating the time derivative itself:

$$t(\tau) = t(0) + \int_0^\tau \frac{dt}{d\tau'}\,d\tau' = \int_0^\tau [1-\vec{\beta}(\tau')\cdot\vec{n}]d\tau' \tag{4.59}$$

where we set $t(0) = 0$ without loss of generality. The integral in Equation 4.59 may at first glance seem computationally expensive to evaluate (necessary for every single trajectory point $(\vec{\beta}_k,t_k)$); but in reality this not the case: one computes subsequent values of t_{k+1} just by integrating the current steps' time. In the trapezoidal approximation, we get:

$$t_{k+1} = t_k + (1-\vec{\beta}_k\cdot\vec{n})\Delta\tau \tag{4.60}$$

For computational efficiency we use the fast Fourier transform, which requires equidistant sampling. This introduces one more calculation step: the interpolation of the pairs $(\vec{E}_k,t_k)$ onto the evenly spaced time grid $(t_{eq})_j = \Delta t\cdot j, j \in 0..M$. The required number of equidistant time points M would obviously be $M = t_N/\min[t_{k+1}-t_k, k \in 0..(N-1)]$, but for most situations that is too fine. A good strategy starts with $M = N$ points and increases the sampling until the difference between the resulting spectra becomes insignificant.

Absolute field scaling for Fast Fourier Transform: Knowledge of the equally spaced E_j and t_j leaves just one remaining step to obtain the spectrum:

$$W_n = c\epsilon_0 \left|FFT(E_j)_n\right|^2; n \in [1,M/2] \tag{4.61}$$

Since Equation 4.61 employs the discrete fast Fourier transform (FFT) it yields the quantity

$$W_n = \frac{dW(\omega_n,(\theta,\phi))}{d\Omega} \tag{4.62}$$

Therefore, Equation 4.62 describes the radiated *power in discrete modes at frequencies* ω_n having the units of $Jm^{-2}s^{-1}$. This is important for evaluating the energy (or number of photons) within some frequency interval:

$$\frac{dW_X}{d\Omega}(\omega_0,\Delta\omega) = \Delta T \sum_{i:\,\omega_i \in (\omega_0 \pm \Delta\omega)} W_i \tag{4.63}$$

Here, $\Delta T = t_M - t_0$ is the entire time window over which the FFT in Equation 4.61 was performed. The above equation is easily transformed to a handy 'engineering' formula to calculate the radiation measured on a fixed detector:

$$W_{Det}((\omega,\Delta\omega),(r,\vec{dS})) = \frac{dW_X}{d\Omega}(\omega,\Delta\omega) \cdot \frac{\alpha(\vec{S},\vec{n})}{r^2} \vec{dS} \tag{4.64}$$

where $(\omega, \Delta\omega)$ define the spectral range and r is the distance from the charge to detector with a surface element $\vec{dS}$. α is the angle between the particle-observer direction $\vec{n}$ and the normal to the observer screen pixel - $\vec{dS}$.

MATLAB©Code Sample: The following MATLAB© code implements the previous considerations:

```
Listing 4.2: Evaluation of the radiated spectrum
%Inputs:
%tt-array of trajectory-point time stemps
%dt=tt(2)-tt(1)-trajectory time step
%beta : array [3, length (tt)]-norm velocity
%nx, ny, nz-direction to which to calculate spect

%Direction matrix
%(pointing to the same direction at each trajectory point):
nArr=[nx* ones (1, length (tt) -1);
ny* ones (1, length (tt) -1);
nz* ones (1, length (tt) -1)];

%dt/dt_ret :
rTime=1- dot (beta (:, 1: end -1), nArr, 1);

%Electric field:
Efield=ee/4/pi/eps0/cc*...
cross (nArr ,...
cross (nArr-beta (:, 1: end -1), diff (beta, 1, 2)/dt, 1)...
,1) ./ repmat (rTime, 3, 1).^3;

%observer time:
tt_obs=cumsum (rTime)*dt;

%resample the time to make it equidistant:
tt_obs_eq=linspace (tt_obs (2), tt_obs (end -1), length (tt_obs));
dt_obs=tt_obs_eq (2)-tt_obs_eq (1);

%interpolate the E-Field on the equidistant time mesh:
Efield_eq=interp1 (tt_obs, Efield (1, :), tt_obs_eq);
```

```
%Zero-pad the array to increase the spectral resolution:
tt_obs_eq=tt_obs_eq (1) : dt_obs: 20* tt obs_eq (end);
Efield_eq=[Efield_eq, zeros (1, size (tt_obs_eq, 2)-size (Efield_eq, 2))];

%FFT to get spectrum:
ESpect=fft (Efield_eq, [], 2)/length (Efield_eq);

ESpect=cc* eps0 *(ESpect. * conj (ESpect))...
*(tt_obs_eq (end)-tt_obs_eq (1));

%Ignore negative frequencies, add their power to positive
ESpect=2* ESpect (1: ceil (length (ESpect/2)));

%And the frequency axis in [eV]:
ESpect_x=(1: length (ESpect))/...
(tt_obs_eq (end)-tt_obs_eq (1))*h_eVs;

%Return spectrum matrix!
mSpect=horzcat ( ESpect_x', ESpect');

%Done ;-)
```

References for Chapter 4

1. E. Esarey, S. K. Ride and P. Sprangle, "Nonlinear Thomson scattering of intense laser pulses from beams and plasmas", *Physical Review E*, 48, 3003, (1993).
2. F. Albert, R. Shah, K. Ta Phuoc, R. Fitour, F. Burgy, J. P. Rousseau, A. Tafzi, D. Douillet, T. Lefrou, A. Rousse, "Betatron oscillations of electrons accelerated in laser wakefields characterized by spectral x-ray analysis", *Physical Review E—Statistical, Nonlinear, and Soft Matter Physics*, 77, 5, 1–6, (2008).
3. M. P. Anania, E. Brunetti, S. M. Wiggins, D. W. Grant, G. H. Welsh, R. C. Issac, S. Cipiccia, R. P. Shanks, G. G. Manahan, C. Aniculaesei, S. B. Van Der Geer, M. J. De Loos, M, W. Poole, B. J A Shepherd, J. A. Clarke, W. A. Gillespie, A. M. Macleod, D. A. Jaroszynski, "An ultrashort pulse ultra-violet radiation undulator source driven by a laser plasma wakefield accelerator", *Applied Physics Letters*, 104, 26, (2014).
4. A. Buck, J. Wenz, J. Xu, K. Khrennikov, K. Schmid, M. Heigoldt, J. M. Mikhailova, M. Geissler, B. Shen, F. Krausz, S. Karsch, L. Veisz, "Shock-front injector for high-quality laser-plasma acceleration", *Physical Review Letters*, 110, 18, 185006, (2013).
5. L. M. Chen, W. C. Yan, D. Z. Li, Z. D. Hu, L. Zhang, W. M. Wang, N. Hafz, J. Y. Mao, K. Huang, Y. Ma, J. R. Zhao, J. L. Ma, Y. T. Li, X. Lu, Z. M. Sheng, Z. Y. Wei, J. Gao, J. Zhang, "Bright betatron X-ray radiation from a laser-driven-clustering gas target.", *Scientific Reports*, 3, 1912, (2013).
6. S. Chen, N. D. Powers, I. Ghebregziabher, C. M. Maharjan, C. Liu, G. Golovin, S. Banerjee, J. Zhang, N. Cunningham, A. Moorti, S. Clarke, S. Pozzi, D. P. Um-stadter, "MeV-energy X rays from inverse Compton scattering with laser-wakefield accelerated electrons", *Physical Review Letters*, 110, 15, 155003, (2013).
7. S. Cipiccia, M. R. Islam, B. Ersfeld, R. P. Shanks, E. Brunetti, G. Vieux, X. Yang, R. C. Issac, S. M. Wiggins, G. H. Welsh, M.-P. Anania, D. Maneuski, R. Montgomery, G. Smith, M. Hoek, D. J. Hamilton, N. R. C. Lemos, D. Symes, P. P. Rajeev, V. O. Shea, J. M. Dias, D. A. Jaroszynski, "Gamma-rays from harmonically resonant betatron oscillations in a plasma wake", *Nature Physics*, 7, 11, 867–871, (2011).

8. J. A. Clarke, "The science and technology of undulators and wigglers", *Oxford Series on Synchrotron Radiation*, 248, (2004).

9. J. M. Cole, J. C. Wood, N. C. Lopes, K. Poder, R. L. Abel, S. Alatabi, J. S. J. Bryant, A. Jin, S. Kneip, K. Mecseki, S. Parker, D. R. Symes, M. A. Sandholzer, S. P. D. Mangles, Z. Najmudin, "Tomography of human trabecular bone with a laser-wakefield driven x-ray source", *Plasma Physics and Controlled Fusion*, 14008, (2016).

10. J. M. Cole, J. C. Wood, N. C. Lopes, K. Poder, R. L. Abel, S. Alatabi, J. S. J. Bryant, A. Jin, S. Kneip, K. Mecseki, D. R. Symes, S. P. D. Mangles, Z. Najmudin, "Laser-wakefield accelerators as hard x-ray sources for 3D medical imaging of human bone", *SCIENTIFIC reports*, 5, 13244, (2015).

11. S. Corde, K. Ta Phuoc, G. Lambert, R. Fitour, V. Malka, A. Rousse, A. Beck, E. Lefebvre, "Femtosecond x rays from laser-plasma accelerators", *Reviews of Modern Physics*, 85, 1, 1–48, (2013).

12. A. D. Debus, M. Bussmann, M. Siebold, A. Jochmann, U. Schramm, T. E. Cowan, R. Sauerbrey, "Traveling-wave Thomson scattering and optical undulators for high-yield EUV and X-ray sources", *Applied Physics B*, 100, 1, 61–76, (2010).

13. E. Esarey, B. A. Shadwick, P. Catravas, W. P. Leemans, "Synchrotron radiation from electron beams in plasma-focusing channels", *Physical Review E—Statistical, Nonlinear, and Soft Matter Physics*, 65, 5, 1–15, (2002).

14. S. Fourmaux, S. Corde, K. Ta Phuoc, P. Lassonde, G. Lebrun, S. Payeur, F. Martin, S. Sebban, V. Malka, A. Rousse, J. C. Kieffer, "Single shot phase contrast imaging using laser-produced Betatron x-ray beams", *Optics Letters*, 36, 13, 2426–2428, (2011).

15. M. Fuchs, R. Weingartner, A. Popp, Z. Major, S. Becker, J. Osterhoff, I. Cortrie, B. Zeitler, R. Hrlein, G. D. Tsakiris, U. Schramm, T. P. Rowlands-Rees, S. M. Hooker, D. Habs, F. Krausz, S. Karsch, F. Grner, "Laser-driven soft-X-ray undulator source", *Nature Physics*, 5, 11, 826–829, (2009).

16. I. Ghebregziabher, B. A. Shadwick, D. Umstadter, "Spectral bandwidth reduction of Thomson scattered light by pulse chirping", *Physical Review Special Topics—Accelerators and Beams*, 16, 3, 030705, (2013).

17. J. D. Jackson, *Classical Electrodynamics*, Wiley & Sons, 3rd Edition, ISBN 0471309321, (1998).

18. K. Khrennikov, J. Wenz, A. Buck, J. Xu, M. Heigoldt, L. Veisz, S. Karsch, "Tunable all-optical quasimonochromatic Thomson X-ray source in the nonlinear regime", *Physical Review Letters*, 114, 19, 1–5, (2015).

19. S. Kneip, C. McGuffey, F. Dollar, M. S. Bloom, V. Chvykov, G. Kalintchenko, K. Krushelnick, A. Maksimchuk, S. P. D. Mangles, T. Matsuoka, Z. Najmudin, C. A. J. Palmer, J. Schreiber, W. Schumaker, A. G. R. Thomas, V. Yanovsky, "X-ray phase contrast imaging of biological specimens with femtosecond pulses of betatron radiation from a compact laser plasma wakefield accelerator", *Applied Physics Letters*, 99, 9, 1–4, (2011).

20. S. Kneip, C. McGuffey, J. L. Martins, S. F. Martins, C. Bellei, V. Chvykov, F. Dollar, R. Fonseca, C. Huntington, G. Kalintchenko, A. Maksimchuk, S. P. D. Mangles, T. Matsuoka, S. R. Nagel, C. A. J. Palmer, J. Schreiber, K. Ta Phuoc, a. G. R. Thomas, V. Yanovsky, L. O. Silva, K. Krushelnick, Z. Najmudin, "Bright spatially coherent synchrotron X-rays from a table-top source", *Nature Physics*, 6, 12, 980–983, (2010).

21. W. Lu, M. Tzoufras, C. Joshi, F. Tsung, W. Mori, J. Vieira, R. Fonseca, L. Silva, "Generating multi-GeV electron bunches using single stage laser wakefield acceleration in a 3D nonlinear regime", *Physical Review Special Topics—Accelerators and Beams*, 10, 6, 061301, (2007).

22. S. P. D. Mangles, G. Genoud, S. Kneip, M. Burza, K. Cassou, B. Cros, N. P. Dover, C. Kamperidis, Z. Najmudin, A. Persson, J. Schreiber, F. Wojda, C. G. Wahlstrm, "Controlling the spectrum of x-rays generated in a laser-plasma accelerator by tailoring the laser wavefront", *Applied Physics Letters*, 95, 18, (2009).

23. T. Oksenhendler, D. Kaplan, P. Tournois, G. M. Greethan, F. Estable, "Intracavity acousto-optic programmable gain control for ultra-wide-band regenerative amplifiers", *Applied Physics B*, 495, 491–494, (2006).

24. K. Ta Phuoc, S. Corde, C. Thaury, V. Malka, A. Tafzi, J. P. Goddet, R. C. Shah, S. Sebban, A. Rousse, "All-optical Compton gamma-ray source", *Nature Photonics*, 6, April, 1–4, (2012).

25. A. Popp, J. Vieira, J. Osterhoff, Zs. Major, R. Hrlein, M. Fuchs, R. Weingartner, T. P. Rowlands-Rees, M. Marti, R. A. Fonseca, S. F. Martins, L. O. Silva, S. M. Hooker, F. Krausz, F. Grner, S. Karsch, "All-optical steering of laser-wakefield-accelerated electron beams", *Physical Review Letters*, 105, 21, 215001, (2010).

26. N. D. Powers, I. Ghebregziabher, G. Golovin, C. Liu, S. Chen, S. Banerjee, J. Zhang, D. P. Umstadter, "Quasi-monoenergetic and tunable X-rays from a laser-driven Compton light source", *Nature Photonics*, 8, 1, 28–31, (2013).

27. K. Ride, E. Esarey, M. Baine, "Thomson scattering of intense lasers from electron beams at arbitrary interaction angles", *PR E*, 52, 5, (1995).

28. A. Rousse, K. Phuoc, R. Shah, A. Pukhov, E. Lefebvre, V. Malka, S. Kiselev, F. Burgy, J.-P. Rousseau, D. Umstadter, D. Hulin, "Production of a keV X-ray beam from synchrotron radiation in relativistic laser–plasma interaction", *Physical Review Letters*, 93, 13, 135005, (2004).

29. S. G. Rykovanov, C. G. R. Geddes, J.-L. Vay, C. B. Schroeder, E. Esarey, W. P. Leemans, "Quasi-monoenergetic femtosecond photon sources from Thomson scattering using laser plasma accelerators and plasma channels", *Journal of Physics B: Atomic, Molecular and Optical Physics*, 47, 23, 234013, (2014).

30. G. Sarri, D. J. Corvan, W. Schumaker, J. M. Cole, A. Di Piazza, H. Ahmed, C. Harvey, C. H. Keitel, K. Krushelnick, S. P. D. Mangles, Z. Najmudin, D. Symes, A. G. R. Thomas, M. Yeung, Z. Zhao, M. Zepf, "Ultrahigh brilliance multi-MeV ??-ray beams from nonlinear relativistic Thomson scattering", *Physical Review Letters*, 113, 22, 1–5, (2014).

31. H.-P. Schlenvoigt, K. Haupt, A. Debus, F. Budde, O. Jckel, S. Pfotenhauer, H. Schwoerer, E. Rohwer, J. G. Gallacher, E. Brunetti, R. P. Shanks, S. M. Wiggins, D. A. Jaroszynski, "A compact synchrotron radiation source driven by a laser-plasma wakefield accelerator", *Nature Physics*, 4, 2, 130–133, (2007).

32. M. Schnell, A. Svert, B. Landgraf, M. Reuter, M. Nicolai, O. Jckel, C. Peth, T. Thiele, O. Jansen, A. Pukhov, O. Willi, M. C. Kaluza, C. Spielmann, "Deducing the electron-beam diameter in a laser-plasma accelerator using x-ray betatron radiation", *Physical Review Letters*, 108, 7, 1–5, (2012).

33. M. Schnell, A. Svert, I. Uschmann, M. Reuter, M. Nicolai, T. Kmpfer, B. Landgraf, O. Jckel, O. Jansen, A. Pukhov, M. C. Kaluza, C. Spielmann, "Optical control of hard X-ray polarization by electron injection in a laser wakefield accelerator", *Nature Communications*, 4, May, 2421, (2013).

34. H. Schwoerer, B. Liesfeld, H.-P. Schlenvoigt, K.-U. Amthor, R. Sauerbrey, "Thomson-backscattered X rays from laser-accelerated electrons", *Physical Review Letters*, 96, 1, 014802, (2006).

35. T. Tanaka, H. Kitamura, "SPECTRA: A synchrotron radiation calculation code", *Journal of Synchrotron Radiation*, 8, 6, 1221–1228, (2001).

36. C. Thaury, E. Guillaume, S. Corde, R. Lehe, M. Le Bouteiller, K. Ta Phuoc, X. Davoine, J. M. Rax, A. Rousse, V. Malka, "Angular-momentum evolution in laser-plasma accelerators", *Physical Review Letters*, 111, 13, 1–5, (2013).

37. J. Ullrich, A. Rudenko, R. Moshammer, "Free-electron lasers: New avenues in molecular physics and photochemistry", *Annual Review of Physical Chemistry*, 63, 635–660, (2012).

38. J. Wenz, S. Schleede, K. Khrennikov, M. Bech, P. Thibault, M. Heigoldt, F. Pfeiffer, S. Karsch, "Quantitative X-ray phase-contrast microtomography from a compact laser-driven betatron source", *Nature Communications*, 6, May, 7568, (2015).

39. K. Wille, *Physik der Teilchen-beschleuniger und Synchrotron-strahlungsquellen*, Teubner, ISBN 3519130874, (1996).

5

Laser-Driven Ion
Acceleration

Andrea Macchi

5.1 Introduction

The observation of intense multi–MeV proton emission from solid targets irradiated at ultra-high intensities in three independent experiments [Clark et al., 2000; Maksimchuk et al., 2000; Snavely et al., 2000] performed in the year 2000 promptly boosted a great research effort on laser-driven (or laser–plasma) ion accelerators. Such research has been oriented to several foreseen applications in nuclear fusion, medicine and high energy density science. Moreover, it has represented a major motivation for the development of laser systems with increasing peak power, towards the multi-petawatt (PW) frontier [Danson et al., 2015].

Figure 5.1 shows a very schematic representation of the experimental scenario: a high intensity laser pulse is focused on one side of the target (commonly referred to as the 'front' side) and a bunch of energetic protons* (or heavier ions) is detected on the opposite ('rear') side. The 'black box' in Figure 5.1 conceals a complex acceleration physics, involving several possible mechanisms.

This chapter aims to give a concise overview on the motivations, principles, state of the art, physics and perspectives of laser-driven ion acceleration. The goal is to provide an introduction to this field accessible to the non-specialist reader and to summarize major achievements and most recent developments. More complete and detailed presentations may be found in several review papers both by the present author and co-workers [Macchi et al., 2013a,b; Borghesi and Macchi, 2016] and by others [Daido et al., 2012; Fernández et al., 2014; Schreiber et al., 2016].

---

* Proton emission was also observed using targets with no hydrogen in their chemical compositions (e.g. metallic targets). This is because hydrogen-containing impurities (water, hydrocarbons) are typically present on the target surface.

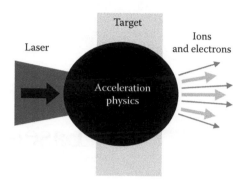

FIGURE 5.1   Schematic representation of laser-driven (or laser–plasma) acceleration of ions, showing the essential elements: the laser pulse, the target and the ion beam (charge-neutralized by co-moving electrons). All the physics which leads to conversion of the energy and momentum of the laser pulse into energy and momentum of the ions is concealed in the central 'black box'.

5.1.1 Properties of Laser-Accelerated Protons

The reason for a sudden, enormous interest in laser-accelerated protons has been in their foreseen use in several applications. The latter exploit the strongly localized energy deposition (the Bragg peak) of high energy ions [Knoll, 2010; Ziegler et al., 2008] combined with the unique properties which were apparent in the early experiments with intense lasers. Such properties include high laser-to-proton energy conversion efficiency (>10% estimated for a petawatt driver [Snavely et al., 2000]) resulting in up to a few tens of joules total energy of the protons; the very large number of protons per shot (up to ~10^{13}); the proton collimation within a typical cone angle of few tens of degrees; the focusability of the protons by a simple shaping of the target rear surface; the very low emittance* (down to ~10^{-3} mm mrad [Cowan et al., 2004; Nuernberg et al., 2009]); and the inferred ultrashort duration in the ps (10^{-12} s) regime. In addition, the protons are accompanied by a co-moving electron cloud, which ensures charge neutralization and prevents the proton bunch from exploding due to electrostatic repulsion ('Coulomb explosion'), so that one could even consider this to be a quasi-neutral 'plasma bunch'.

Some of the differences with the protons delivered by standard accelerators (such as the compact cyclotrons, synchrotrons and linacs used in medical physics or other applications) appear to be very remarkable. The accelerating gradient, i.e. the typical value of the electric field on ions, is of the order of MV/µm which is some four orders of magnitude greater than the value in a linac (up to 100 MV/m). A pulse duration in the picosecond range is at least some three orders of magnitude shorter than what is achievable with standard accelerators, which are in the range between ns and µs. Combined with the high number of particles per bunch, this yields a peak current of several kA (or even higher), which may be compared with the mA steady current in a cyclotron. Finally, the transverse emittance is typically three orders of magnitude lower than the typical value in a linac (~1 mm mrad), and the longitudinal emittance is very low as well.

The above mentioned properties make laser-accelerated protons highly promising for any application requiring an extremely localized (both in space and time) energy deposition in dense matter. The first observations were promptly followed by the proposal of using laser-accelerated protons to create a hot spot in an inertial confinement fusion (ICF) target [Roth et al., 2001; Ruhl et al., 2001; Atzeni et al., 2002], thus providing an alternative ignitor for the fast ignition concept [Tabak et al., 1994] in ICF. Soon after, laser-accelerated protons were proposed for oncological ion beam therapy (IBT) as a possibly more compact and cheaper option than that using conventional accelerators [Bulanov et al., 2002; Bulanov and Khoroshkov, 2002; Fourkal et al., 2003; Malka et al., 2004]. As another medical

---

* The emittance needs to be properly defined and measured for non-monoenergetic beams [Nuernberg et al., 2009].

application, the production of short-lived radioisotopes was also proposed and preliminary investigations were performed [Nemoto et al., 2001; Santala et al., 2001; Spencer et al., 2001; Fritzler et al., 2003; Ledingham et al., 2003].

The low emittance favoured the application of laser-accelerated protons for radiography and imaging [Roth et al., 2002; Cobble et al., 2002; Borghesi et al., 2004]. In particular, the broad energy spectrum combined with the short duration of the proton bunch has enabled the development of single-shot detection of electromagnetic fields in laser–plasma interaction phenomena with picosecond temporal resolution [Borghesi et al., 2002; Mackinnon et al., 2004]. This innovative application has yielded much valuable information on the nonlinear dynamics of plasmas, including the proton acceleration mechanism itself [Romagnani et al., 2005].

5.1.2 Principles of Ion Acceleration

The special properties of laser-accelerated proton bunches have their roots in the *coherent* (in the sense of *collective* or *cooperative*) nature of the acceleration process, which is basically different from conventional technologies. The coherent acceleration paradigm was outlined by Veksler [1957], before the invention of the laser. Key features of this paradigm include: 1) the accelerating field on each particle is proportional to the number of accelerated particles (the larger their number, the higher the kinetic energy of the single particle); 2) the field is localized in space and synchronized in time with the accelerated particles; 3) eventually the acceleration process produces globally charge-neutralized bunches. These latter three features are realized in the acceleration of ions occurring via the interaction with sufficiently dense targets, i.e. in *laser–plasma* acceleration.

While collective plasma dynamics is the basis of the unique properties and potential applications of laser-accelerated ion bunches, its complex nonlinearity poses great challenges of control, stability and modelling with respect to traditional approaches. The basic acceleration mechanism underlying most of the experiments reported so far, commonly named target normal sheath acceleration (TNSA), has reached a good level of reliability and robustness, and provides a framework for further developments, such as 'all-optical' (i.e. laser-controlled) bunch control and post-acceleration. However, foreseen applications have stringent requirements on properties such as the energy per particle, the spectral distribution and the suitability for high-repetition rate operation, as well as others. The parameters and characteristics of laser-accelerated protons have still to reach such requirements, and it is still uncertain whether the availability of multi-PW lasers in the next few years combined with developments of innovative targets will be sufficient for such aims, at least for what concerns TNSA-based approaches. This issue has stimulated the proposal of alternative schemes, such as radiation pressure acceleration (RPA) or collisionless shock acceleration (CSA) whose investigation is still relatively preliminary compared to that for TNSA. The physics of such mechanisms will be discussed in Section 5.3.

5.2 State of the Art

Reporting on the state of the art in laser-driven ion acceleration is not straightforward for several reasons. For instance, progress has been achieved on several properties (e.g. the maximum proton energy, the conversion efficiency, the spectral width, etc.) but in different experiments and, in most cases, for different acceleration mechanisms (Section 5.3). Moreover, the experimental characterization of laser-driven ion bunches is not easy. Established methods and instruments have required modifications and adaption, and new diagnostic techniques have been developed (for reviews see, e.g. Bolton et al. [2014] and section II.E of Macchi et al. [2013a]). While such effort has allowed a study of laser-driven ion acceleration over a wide range of laser parameters (intensity, energy, duration, polarization, etc.), not all the relevant properties of the accelerated ions are usually measured in a single experiment. In addition, a precise control of the experimental conditions is also challenging with high-power lasers, and this may account for variations observed between experiments performed in conditions which would seem similar at a first glance. A likely consequence of these issues is that the scaling of the most important characteristics (such as the energy per particle) with laser and target parameters is still unclear to a large extent, despite the large number of investigations performed. The reader should keep in mind these issues and remarks in the following description of experimental achievements.

Most of the experiments performed so far have dealt with the acceleration of protons from solid targets. In the literature, progress in the field has been mostly monitored and claimed on the basis of the maximum proton energy observed, since for instance reaching the energy window for IBT applications (60–250 MeV, where at least 150 MeV is required for deeply seated tumours) has been considered to be a major goal. The energy spectra of protons are typically broad and in most cases exponential-like, and the maximum energy corresponds to the upper cut-off in such spectra. However, the observed cut-off in the spectrum may depend on the sensitivity of the detector used and on the level of background noise, so that it is not always clear how precisely a maximum energy can be measured, and to what extent the value is affected by diagnostic factors. A comparison of absolute differential spectra, obtained with calibrated detectors yielding the number of particles per energy interval, can be (when such spectra are available) less prone to diagnostic factors and thus give a safer indicator of the acceleration performance. Moreover, the number of protons per energy slice (and possibly per opening angle) is an important parameter for foreseen applications which require a sufficient particle flux in addition to a given energy range.

5.2.1 Progress in Proton Energy Enhancement

Figure 5.2 shows a comparison of calibrated spectra from three experiments [Snavely et al., 2000; Gaillard et al., 2011; Wagner et al., 2016] which established new world records for the cut-off energy at the time of the publication of the results in the period between 2000 and 2016. All three experiments were performed with 'large' laser systems, delivering several tens of joules onto a solid target, with a typical pulse duration of several hundreds of femtoseconds. The facilities where the experiments were performed and the corresponding reported values of the energy 'on target' (i.e. contained into the focal spot) and the pulse duration were, respectively, the petawatt system at Lawrence Livermore National

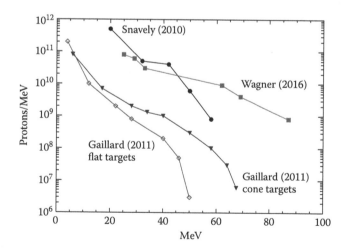

FIGURE 5.2 Proton energy spectra from the three experiments reporting new world records in the cut-off energy from 2000 to 2016. For the experiment of Gaillard et al. [2011] the record energy was achieved using specially shaped 'cone' targets; the spectrum obtained with standard 'flat' targets is reported for comparison. See text for parameters and details. (Data were taken from R. A. Snavely et al., *Phys. Rev. Lett.*, 85:2945–2948, 2000; S. A. Gaillard et al., *Phys. Plasmas*, 18:056710, 2011; F. Wagner et al., *Phys. Rev. Lett.*, 116:205002, 2016.)

Laboratory (LLNL) in the year 2000 (150 J, 500 fs) [Snavely et al., 2000], the TRIDENT laser at Los Alamos National Laboratory (LANL) in 2011 (39 ± 7 J, 670 ± 130 fs) [Gaillard et al., 2011], and the PHELIX laser at GSI Helmholtzzentrum fuer Schwerionenforschung Darmstadt in 2010 (48–60 J, 500–600 fs) [Wagner et al., 2016]. The corresponding values of the conversion efficiency of laser energy into proton energy were 12% (LLNL) obtained integrating the proton spectrum for energies >10 MeV, 1.75% (LANL) and 7% ± 3% (GSI) both for proton energies >4 MeV. The LLNL and GSI data were both obtained with plastic targets (CH and CH_2, respectively) but with very different target thickness values (100 μm and 0.9 μm, respectively). The LANL data were obtained with copper (Cu) targets with 10 μm thickness, and the highest energies were obtained with special targets where the laser was focused into a microcone placed at the front side (i.e. the target side exposed to the laser irradiation), obtaining an increase to 67.5 MeV as compared to 50 MeV from flat targets (see Figure 5.2).

Comparing the spectra from the three experiments in Figure 5.2 suggests that the cut-off increase obtained at LANL with respect to LLNL data may be effectively due to an increased sensitivity of the detector, since the number of protons at the cut-off is smaller by about two orders of magnitude. However, it is remarkable that the cone targets yielded similar proton energies with much less laser energy. The reduction of the target thickness in the GSI experiment appears to produce a substantial progress with respect to the LLNL data (also at considerably less energy), since the energy cut-off is increased by nearly 30 MeV at almost the same number of particles.

The above analysis also indicates that the cut-off energy, apart from the difficulties related to its definition and measurement, is not the only parameter by which the performance of ion acceleration should be measured; an increase of only 30 MeV over 16 years might appear as slow progress, but one should consider that, while the full energy available with large systems delivering picosecond pulses has not increased during this time period, significantly less energy has been used to obtain similar or even higher proton energies.

One may also argue that the large systems, producing hundreds of Joule pulses, used for the experiments reported in Figure 5.2 are far from being compact and are generally unsuitable for high repetition rate operation, which is a key requirement for most applications. For this reason, it is of interest to evaluate the progress obtained with 'smaller' laser systems which may operate at $10–10^3$ Hz rate and typically have a pulse duration of a few tens of fs and an available energy $\lesssim 10$ J.

5.2.2 Proton Energy Scaling with Short-Pulse Drivers

In reviewing data obtained with smaller short-pulse systems, we also select experiments where a calibrated energy spectrum is available. In addition, we only consider experiments using 'high-contrast' pulses as an additional criterion. High power laser systems typically do not produce 'clean', isolated short pulses; indeed, the 'main' pulse of sub-picosecond duration is preceded by other pulses of the same duration and lower power, a few nanoseconds 'pedestal' pulse, and another pedestal of picosecond duration produced by imperfect recompression of the main pulse. When aiming at the highest intensity of the main short pulse, the prepulses may be already intense enough to cause ionization and heating in the target, producing a 'preplasma' at the interaction surface. A controlled short-pulse interaction requires pulses with a sufficiently high 'contrast' ratio between the intensities of the main pulse and the prepulse(s) must be high enough to prevent target damage and preplasma formation, to which the laser–plasma coupling is highly sensitive. In recent years the development of optical devices such as the plasma mirror [Dromey et al., 2004; Thaury et al., 2007] made possible achieving pulse contrast values of 10^{10} (typically measured a few ps before the short fs pulse) and beyond. This means that even at the highest short pulse intensities $\sim 10^{21}$ W cm^{-2} available today, the prepulse intensity is $\sim 10^{+11}$ W cm^{-2}, which is under the ionization threshold of most target materials [von der Linde and Schüler, 1996]. It is worth noting, however, that our choice to consider high-contrast experiments is only to have similar interaction conditions, and it does not imply that high contrast always favours the enhancement of the proton energy or other properties.

The selected experiments were performed in different laboratories using laser systems of different nominal power, i.e. the 3 TW laser at I3M Valencia [Seimetz et al., 2016], the LLC 30 TW laser in Lund [Neely et al., 2006], the DRACO 150 TW laser at HZDR Dresden [Zeil et al., 2010, 2012, 2014], the LiFSA 100 TW [Choi et al., 2011; Margarone et al., 2012] and PULSER 1 PW [Kim et al., 2013; Margarone et al., 2015; Passoni et al., 2016] lasers at GIST Gwangju, the JKAREN 200 TW laser at JAEA/KPSI Kyoto [Ogura et al., 2012] and the GEMINI 200 TW laser at RAL [Green et al., 2014]. The corresponding ranges of pulse duration (τ_L) and intensity on target (I_L) are 25–40 fs and $4 \times 10^{18} - 2 \times 10^{21}$ W cm^{-2}. In order to reduce possible uncertainties due to different optical transport and focusing systems, for each experiment the energy *on target*, contained within the full width at half-maximum (FWHM) of the intensity distribution in the focal spot has been considered. The targets have different thicknesses in the range 0.01 – 4.0 μm. To simplify the analysis, we only consider the higher energy tail of each experimental spectrum, which we found to be satisfactorily approximated by a simple exponential function $N_p(\mathcal{E}) = N_{p0} \exp(-\mathcal{E}/T_p)$ with $\mathcal{E} < \mathcal{E}_{co}$, the cut-off energy. The 'temperature' T_p is a parameter giving information on the mean proton energy and the spectral roll-off with increasing $\mathcal{E}$.

Figure 5.3 shows that the results of Kim et al. [2013] and Ogura et al. [2012] are at variance with other experiments for what concerns the proton spectral density N_p, the cut-off energy $\mathcal{E}_{co}$ and the 'temperature' T_p (out of scale in the plot). The rest of the data show an increasing trend for these parameters as a function of the energy on target U_L, with an apparent almost linear scaling of $\mathcal{E}_{co} = kU_L$ with $k \simeq$ 8.6 MeV/J. Such linear scaling is similar to that proposed by Zeil et al. [2010] on the basis of a parametric investigation with the DRACO laser and extends the latter to higher values of the pulse energy. The linear scaling is different from, and more favourable than the intensity dependence $\mathcal{E}_{co} \sim I_L^{1/2}$ which has been inferred in previous reviews from data obtained using short-pulse lasers [Fuchs et al., 2006; Borghesi et al., 2008; Daido et al., 2012; Kiefer et al., 2013]. Also T_p shows a scaling which is almost

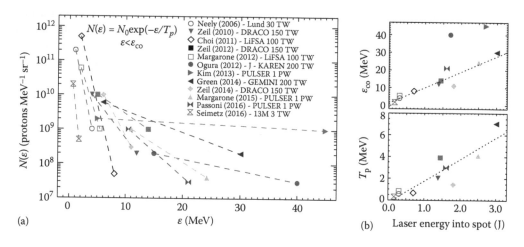

FIGURE 5.3 Frame (a): proton energy spectra from experiments using high contrast, sub-100 fs, sub-10 J laser pulses and thin solid targets, shown as simple exponential interpolations (dashed lines) $N_p(\mathcal{E}) = N_{p0} \exp(-\mathcal{E}/T_p)$ (with $\mathcal{E} \leq \mathcal{E}_{co}$, the cut-off energy) of the high-energy tail of experimentally measured spectra. Frame b): the parameters $\mathcal{E}_{co}$ (top) and T_p (bottom) as a function of the laser pulse energy in the focal spot. (Data are from I. J. Kim et al., *Phys. Rev. Lett.*, 111:165003, 2013; M. Passoni et al., *Phys. Rev. Accel. Beams*, 19: 061301, 2016; M. Seimetz et al., *J. Inst.*, 11: C11012, 2016; D. Neely et al., *Appl. Phys. Lett.*, 89:021502, 2006; K. Zeil et al., *N J. Phys.*, 12:045015, 2010; K. Zeil et al., *Nat. Comm.*, 3:874, 2012; K. Zeil et al., *Plasma Phys. Control. Fusion*, 56:084004, 2014; I. W. Choi et al., *Appl. Phys. Lett.*, 99:181501, 2011; D. Margarone et al., *Phys. Rev. Lett.*, 109:234801, 2012; K. Ogura et al., *Opt. Lett.*, 37:2868–2870, 2012; J. S. Green et al., *Appl. Phys. Lett.*, 104:214101, 2014; D. Margarone et al., *Phys. Rev. ST Accel. Beams*, 18:071304, 2015 as indicated in the text.) Empty and filled symbols are for intensities in the $I_L = (0.4 - 5) \times 10^{19}$ W cm^{-2} and $I_L = (0.8 - 2) \times 10^{21}$ W cm^{-2} ranges, respectively. The pulse durations are in the $\tau_L = (25 - 40)$ fs range. All the targets are planar foils of various thickness (in the 0.05 – 4.0 μm range).

linear with the energy in the focal spot. It is worth stressing that, although previous reviews of energy scaling vs. laser intensity or energy use a larger set of data than in Figure 5.3, the scatter with respect to the proposed scalings appear to be larger than for the data selected in Figure 5.3 (also note that, unlike Figure 5.3b, data are most often represented on a logarithmic scale). The residual data scatter in Figure 5.3 may be ascribed to the variation of other parameters (laser pulse duration, target thickness and material, incidence angle, etc.). The scaling of proton energy with such parameters is less clear and is not discussed here. Some comments on the 'anomalous' data in Figure 5.3 [Kim et al., 2013; Ogura et al., 2012] will be given at the end of Section 5.5, on the basis of the discussion of the acceleration mechanism in Section 5.3.

5.2.3 Progress in Ion Bunch Properties

The experimental effort for characterization and optimization of ion bunch properties different from the maximum energy per particle has been less extended and systematic. It is worth noting that optimization of the bunch properties needs to be guided by an understanding of the acceleration process. This is particularly needed for aims such as obtaining a narrow energy spectrum, which is *qualitatively* different from the data shown in the preceding sections. Progress on this side has been mostly obtained via the engineering of target structure and chemical composition (see Hegelich et al. [2006]; Schwoerer et al. [2006]; Pfotenhauer et al. [2008]) or via the exploration of advanced acceleration mechanisms (see e.g. Palmer et al. [2011]; Haberberger et al. [2012]; Kar et al. [2012]; Palaniyappan et al. [2015]). In most of these cases the results are either preliminary or need further confirmation. For instance a spectral width as low as 1% has been measured using CO_2 lasers and gas jet targets [Haberberger et al., 2012] but the results are at variance with others obtained in apparently similar conditions (see Section 5.3.4). Perspectives will be discussed in Section 5.5 after the acceleration physics has been discussed.

For the case of broad energy spectra which are observed in most experiments, the bunch properties may be only defined with reference to a given energy range. Here we only mention recent progress related to characterization and optimization of selected properties, such as ion bunch duration and conversion efficiency. More complete lists of achieved values can be found in the review papers cited in Section 5.1.

One of the most peculiar and possibly unique properties is the ultra-short duration of the ion bunch. In early experiments, a picosecond duration at the source has been inferred and roughly estimated from the analysis of proton probing experiments [Borghesi et al., 2002; Mackinnon et al., 2004]. However, until recently, no direct measurements were reported. Of course the measurement of bunch duration must be associated not only to a well-defined energy band but also to a value of the distance from the source, since velocity dispersion will cause the bunch duration to increase along the propagation path. In recent experiments with the TARANIS laser (~600 fs pulse duration) at the Queen's University of Belfast, the bunch duration has been measured via observations of proton-generated ionization in SiO_2 [Dromey et al., 2016]. For a bunch of protons with energy of 10 ± 0.5 MeV, the effective duration measured in a SiO2 sample at a distance of ~5 mm was 3.5 ± 0.7 ps. This value opens up perspectives for ultrafast studies of ion-induced damage and energy deposition in dense matter. It might be possible to generate even shorter ion bunches by employing a laser driver with few tens of fs duration and specifically tailored bunch modulators.

The ~12% conversion efficiency of laser energy into protons in the 10–58 MeV range observed at LLNL in 2000 [Snavely et al., 2000] had an even more long-lasting record than the cut-off energy. An experiment performed on VULCAN in 2014 [Brenner et al., 2014] obtained a ~15% efficiency for protons in the 3–30 MeV energy range. In order to optimize the acceleration process the experiment used a controlled short prepulse, which was shown in previous experiments to enhance the cut-off energy and conversion efficiency [Markey et al., 2010] and to produce modulations in the energy spectrum [Dollar et al., 2011]. These experiments provide examples of all-optical techniques for manipulation and optimization of proton acceleration, which will be discussed in Section 5.4.2.

5.3 Ion Acceleration Physics

The present section is devoted to a basic description of laser–plasma interactions at high intensities and of the main ion acceleration mechanisms,* whose understanding is essential for developments of laser-driven ion sources. For each mechanism, most significant experimental confirmations are also mentioned.

5.3.1 Laser–Plasma Interaction Scenario

In the interaction regimes relevant to the present context, the laser pulse is intense enough to ionize matter almost instantaneously, and couples with the freed electrons which absorb energy and momentum from the electromagnetic field. The interaction typically leads to the generation of suprathermal electrons of high energy; such 'fast' electrons tend to escape from the target generating regions of charge separation and related electrostatic fields in the presence of density gradients, particularly at target boundaries ('sheath' regions), and in turn, the electrostatic fields accelerate ions and drive the expansion of the plasma. Momentum absorption occurs due to a secular ponderomotive force (corresponding to the local flow of electromagnetic momentum) which modifies the electron density and consequently the electrostatic fields, leading to radiation pressure action on the plasma. Under suitable conditions, the combination of heating and radiation pressure can drive nonlinear shock waves which also lead to ion acceleration. The basic mechanisms described below originate from the dominance of each of these effects, which may generally coexist in experiments leading to a complex acceleration scenario.

In the following, the laser and target requirements and developments needed to advance each mechanism are emphasized. In order to characterize the interaction regime, two dimensionless parameters are particularly useful and important. The first one is the ratio between the free electron density n_e in the target and the cut-off or 'critical' density n_c, i.e. the maximum value of the electron density above which the laser pulse does not propagate:

$$n_c = n_c(\omega_L) = \frac{m_e \omega_L^2}{4\pi e^2} \tag{5.1}$$

where ω_L is the laser frequency. Equation 5.1 originates from the expression of the refractive index of a collisionless, unmagnetized plasma $n(\omega) = (1 - \omega_p^2/\omega^2)^{1/2} = (1 - n_e/n_c)^{1/2}$ where $\omega_p = (4\pi e^2 n_e/m_e)^{1/2}$ is the plasma frequency. Plasmas with density $n_e > n_c$ ($n_e < n_c$) are called overdense (underdense) and are opaque (transparent) to the laser light. For practical reasons it is useful to write n_c as a function of the laser wavelength $\lambda_L = 2\pi c/\omega_L$,

$$n_c = \frac{\pi m_e c^2}{e^2 \lambda_L^2} = \frac{1.1 \times 10^{21} \text{cm}^{-3}}{(\lambda_L/1\,\mu\text{m})^2}. \tag{5.2}$$

This expression makes clear that for optical or near-infrared lasers with $\lambda_L \simeq 1\,\mu$m the cut-off density is about one hundredth (or less) of the electron density of solid targets.

The second parameter is the dimensionless amplitude of the laser a_0, which corresponds to the oscillation momentum in the electric field of the laser in units of $m_e c$, i.e.

$$a_0 = \frac{eE_L}{m_e c \omega_L} = \left(\frac{e^2 I_L \lambda_L^2}{\pi m_e^2 c^5} \right)^{1/2} = \left(\frac{I_L}{m_e c^3 n_c} \right)^{1/2} \tag{5.3}$$

---

* A more detailed tutorial introduction to the acceleration physics at a (mild) postgraduate level can be found in Chapter 5 of the author's textbook [Macchi, 2013].

where $\omega_L = 2\pi c/\lambda_L$, E_L and $I_L = cE_L^2/4\pi$ are the laser frequency, electric field amplitude, and intensity, respectively. A practical formula for a_0 as a function of I_L and λ_L is given by

$$a_0 = 0.85 \left(\frac{I_L \lambda_L^2}{10^{18} \, \mathrm{W\,cm^{-2}\,\mu m^2}} \right)^{1/2} \tag{5.4}$$

When $a_0 > 1$, the electron dynamics in the laser field is relativistic. Most of the experiments on ion acceleration have been performed with optical or near-infrared lasers, and in the intensity range $I_L = 10^{18} - 10^{21}$ W cm^{-2}, which corresponds to $a_0 \simeq 0.85 - 28$ for $\lambda_L = 1$ μm. CO$_2$ lasers with $\lambda_L \simeq 10$ μm have been also used with typical intensities of $I_L = 10^{16}$ W cm^{-2}, yielding a mildly relativistic interaction regime.

The transmission of a laser pulse through a plasma is modified by relativistic effects on electron motion, which favour pulse penetration at densities higher than n_c, a phenomenon known as 'relativistic transparency'. Details depend on the laser and target parameters. A simple, although far from rigorous criterion applicable to targets much thicker than the laser wavelength λ_L consists of assuming an increase of the cut-off density from n_c to $n_c\gamma$ with $\gamma = \left(1 + a_0^2/2\right)^{1/2}$, which is equivalent to assuming an effective electron mass equal to $m_e\gamma$ due to the oscillation energy of electrons in the laser field. Notice that, in general, the effective mass will be a function of time and position since it will depend on the local amplitude of the electromagnetic field. This dependence produces a class of nonlinear optical effects, such as self-focusing and channelling of the laser pulse (see Macchi [2013], Chapter 3). In addition, boundary effects are crucial for relativistic transparency: at the laser–plasma interface, the radiation pressure of the laser pulse pushes the plasma electrons and produces a local increase of the electron density, which counteracts the relativistic effect. As a result, the transition to transparency has a higher threshold than would be obtained by simply assuming $n_e = \gamma n_c$ (see Cattani et al. [2000] and Macchi [2013], Section 3.4.1).

For a target thinner than λ_L, the transparency threshold also depends on the target thickness l. As a case of particular importance for the following, the nonlinear reflectivity of a thin foil taking relativistic effects into account can be calculated in the limit of a Dirac-delta density profile (Vshivkov et al. [1998]; see also Macchi [2013], Section 3.4.2). The onset of relativistic transparency is found to occur for

$$a_0 > \zeta \equiv \pi \frac{n_e l}{n_c \lambda_L}. \tag{5.5}$$

The generation of populations of 'suprathermal' or 'fast' electrons, i.e. populations of electrons having energy higher than the average energy of the bulk electrons, is of key importance. A theoretical analysis of fast electron generation mechanisms is outside the scope of the present paper (the reader may consult Gibbon [2005]; Macchi [2013] for introductory tutorials), thus we only resume the main features. Typically, fast electrons have broad energy spectra reasonably described with an exponential function $f(\mathcal{E}_f) \propto \exp(-\mathcal{E}_f/T_f)$, extended up to a cut-off of a few times the 'temperature' T_f (energy units are used). The latter is often assumed to be of the order of the so-called "ponderomotive energy" $\mathcal{E}_p$, i.e. the oscillation energy in the electric field of the laser:

$$\mathcal{E}_p = m_e c^2 \left(\left(1 + a_0^2/2\right)^{1/2} - 1 \right) \tag{5.6}$$

In current experiments on ion acceleration, the range of laser intensity roughly corresponds to $a_0 = 1 - 30$, so that $T_f \simeq 0.1 - 10$ MeV may be expected (note, however, that measurements of T_f are difficult, so that there is not a strong experimental background for the $T_f = \mathcal{E}_p$ assumption).

For short laser pulses and sharp boundary targets (a situation typical of the interaction of femtosecond, high-contrast pulses with solid targets) a simple picture of fast electron generation which supports Equation 5.6 is based on the so-called 'vacuum heating' (VH) mechanism: the component of the Lorentz force perpendicular to the target surface periodically pulls electrons from the plasma into the vacuum side and then pushes them back into the plasma with an energy of the order of $\mathcal{E}_p$. In the non-relativistic regime ($a_0 < 1$) the VH mechanism is efficient at oblique laser incidence and for P-polarization: a simple model of the VH absorption coefficient $\mathcal{A}_f$ (i.e. the fraction of laser energy which is converted into fast electron energy via VH) in the $\mathcal{A}_f \ll 1$ limits yields $\mathcal{A}_f \simeq a_0 \sin^3 \theta / \cos \theta$, where θ is the angle of incidence. In the $a_0 > 1$ regime, the expression for $\mathcal{A}_f$ is modified accounting for relativistic electron energies and laser pulse depletion (Gibbon [2005], Section 5.5.2), showing that the dependence of $\mathcal{A}_f$ on θ and a_0 becomes weaker than for $a_0 < 1$. In addition, in the relativistic regime, the magnetic part ($-e\mathbf{v} \times \mathbf{B}$) of the Lorentz force plays an important role and leads to efficient absorption and fast electron generation also for normal incidence. However, fast electron generation can be strongly quenched for normal incidence and circular polarization [Macchi et al., 2005] since in such conditions there is no component of the Lorentz force perpendicular to the target surface, suppressing the VH effect.

Assuming that the absorption is mostly due to fast electrons, a balance condition between the absorbed laser intensity and the flux of fast electron energy through the target may be written in order to estimate the density of fast electrons (n_f):

$$\mathcal{A}_f I_L \simeq n_f \upsilon_f T_f, \qquad (5.7)$$

where υ_f is the fast electron velocity. In the non-relativistic regime, $\upsilon_f \simeq (2T_f/m_e)^{1/2}$ while $\upsilon_f \simeq c$ in the strongly relativistic case. Commonly, empirical values of $\mathcal{A}_f$ (typically of the order of 10%) and the 'ponderomotive' estimate $T_f \simeq \mathcal{E}_p$ are used in Equation 5.7 to evaluate n_f. The values obtained are typically equal to a fraction of the cut-off density n_c, e.g. $n_f \sim 10^{20}$ cm^{-3} for $\lambda_L \simeq 1$ μm. We remark again that these numbers should be considered only as gross estimates.

Note that both Equation 5.6 and the VH absorption coefficient $\mathcal{A}_f$ only depend on the laser intensity and not on target parameters. Actually, absorption and fast electron generation can be enhanced in targets which are weakly overdense, i.e. the electron density n_e is at most a few times n_c. A rough explanation is that for such densities the laser frequency ω_L gets closer to the plasma frequency ω_p and thus the plasma response becomes more 'resonant' (see Mulser et al. [2008] and Mulser and Bauer [2010], Section 8.3.3 for a theoretical picture of absorption based on nonlinear, anharmonic plasma resonance). Moreover, the laser pulse may penetrate deeply into the plasma, leading to a stronger coupling. Eventually the low density favours the onset of relativistic transparency which is usually correlated with strong, turbulent heating of electrons; attempts to exploit this regime for ion acceleration are described in Section 5.3.5.

Absorption and fast electron generation in solid targets also turn out to be sensitive to sub-wavelength density gradients and structuring of the interaction surface. Therefore, the pulse contrast (see Section 5.2.2) plays an important part. In particular, high contrast systems allow the use of suitable micro- and nano-structured targets to enhance and optimize ion acceleration, as will be discussed in Section 5.4.1.

5.3.2 Target Normal Sheath Acceleration

The interpretation of the acceleration mechanism underlying the early observation of protons from solid targets [Clark et al., 2000; Maksimchuk et al., 2000; Snavely et al., 2000] was the subject of some debate. Ultimately, the most successful description was that proposed by the LLNL group [Snavely et al., 2000; Wilks et al., 2001], which was the basis of the so-called TNSA model. It is widely recognized that TNSA was the dominant mechanism for proton acceleration in most experiments reported so far (see Macchi et al. 2013a, Section III, for a detailed overview and list of references), including both the 'long' and 'short' pulse experiments included in Figures 5.2 and 5.3.

TNSA is based on the efficient generation of fast electrons (Section 5.3.1) in the relativistic regime $(a_0 > 1$ or $I_L \lambda_L^2 > 10^{18}$ W $\mu m^2 cm^{-2})$. Fast electron generation produces very intense electrical currents into the target: the current density may be of the order of $J_f \simeq e n_f c \sim 10^{11}$ A cm^{-2} which may correspond to more than 10^5 A through the focal area. If the target is relatively thin (from a few tens of microns down to sub-micrometric values) the fast electrons reach the rear side of the target (opposite the laser plasma interaction side, see Figure 5.4) producing a sheath region. In the sheath, a space-charge electric field is generated with a back-holding effect for the fast electrons, which implies that the electric potential drop through the sheath is $\Delta\Phi \simeq T_f/e$. The field accelerates ions in the direction normal to the target surface. In metallic targets, protons are ordinarily present as surface impurities and are thus located near the peak of the sheath field; such localization, combined with the high charge-to-mass ratio, favours their acceleration with respect to heavier ions, unless the hydrogen containing layers are carefully removed [Hegelich et al, 2002, 2006]. A test proton crossing the sheath region will acquire an energy $\mathcal{E} = e\Delta\Phi \simeq T_f$ which provides a first rough estimate of the energy gain and, assuming $T_f \simeq \mathcal{E}_p$ (Equation 5.6), of the scaling of the proton energy with the pulse intensity.

Larger energy values may be achieved in the course of the expansion of the sheath plasma, where ultimately electrons and ions will reach the same drift velocity. In order to describe the acceleration mechanism, several theory papers have revisited the classic problem of plasma expansion into a vacuum in order to provide estimates for the proton energy as a function of the laser and target parameters (see Kiefer et al., 2013; Mora, 2003; Betti et al., 2005; Mora, 2005; Huang et al., 2013, and references therein). Experimentally, the expanding sheath has been visualized by using the proton imaging technique, i.e. using a probe proton beam (also generated via TNSA) directed transversely to the plasma expansion [Romagnani et al., 2005].

Establishing the scaling of the energy per nucleon $\mathcal{E}_n$ with laser and target parameters, and particularly with the laser pulse energy (U_L) and intensity (I_L), is of fundamental importance to evaluate the potential of TNSA-based schemes for applications and give directions for further developments. Scaling laws have been inferred both by reviewing data from different laboratories (see Borghesi et al., 2006, 2008; Daido et al., 2012; Kiefer et al., 2013) and by performing parametric studies on a single laser system (see Fuchs et al., 2006; Robson et al., 2007; Zeil et al., 2010). In both cases, theoretical and semi-empirical models have been compared to the data. Inferred power-law scalings $\left(\mathcal{E}_n \propto U_L^\alpha$ or $\mathcal{E}_n \propto I_L^\alpha\right)$, with a ranging from 1/3 to 1 and possibly depending on the pulse duration, have been proposed. As already discussed in Section 5.2, the uncertainty in establishing scaling laws might be ascribed both to the difficulties in the control and characterization of experimental conditions and to the several unknown quantities in models (such quantities are often used as parameters for fitting of experimental data).

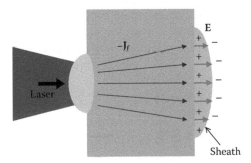

FIGURE 5.4 The basic scheme of TNSA. The laser pulse incident on the front side of the target generates an intense current J_f of 'fast' electrons which propagate through the target and produce a sheath at the rear side. The induced space-charge electric field E accelerates ions. Protons, which may be present either as a component of the target material or in hydrogen impurities present in a thin surface layer, are favoured by the high charge-to-mass ratio and are thus preferentially accelerated.

5.3.3 Radiation Pressure Acceleration

The incidence of an EM wave of intensity I_L on a plane target leads to absorption of EM momentum, producing a pressure (for normal incidence)

$$P_{\mathrm{rad}} = (1 + \mathcal{R} - \mathcal{T})\frac{I_L}{c} = (2\mathcal{R} + \mathcal{A})\frac{I_L}{c}, \tag{5.8}$$

where $\mathcal{R}$, $\mathcal{T}$ and $\mathcal{A}$ are the reflection, transmission and absorption coefficients, respectively (energy conservation imposes the constraint $\mathcal{R} + \mathcal{T} = 1 - \mathcal{A}$). A maximum pressure $P_{\mathrm{rad}} = 2I_L/c$ is obtained in the case of an ideal 'perfect' mirror with $\mathcal{R} = 1$ and $\mathcal{T} = \mathcal{A} = 0$.

In ultra-intense laser interactions with overdense plasmas, the radiation pressure may overcome the thermal pressure and push as a piston the plasma, steepening the density profile and driving the recession on the interaction surface (Figure 5.5a). In multi-dimensional geometry, such radiation pressure action bores a hole in the plasma, so that the velocity of the surface is commonly known as the 'hole boring' (HB) velocity u_{HB}. Assuming steady conditions, the balance between the flows of electromagnetic and kinetic momentum at the surface yields $u_{\mathrm{HB}} = (I_L/\rho c)^{1/2}$ (valid for $u_{\mathrm{HB}} \ll c$; see Robinson et al. [2009] for a relativistic expression), where ρ is the mass density. Moreover, the balance of mass and momentum flows at the moving piston surface implies that there must exist a flow of ions 'reflected' from the recession front at twice u_{HB}, resulting in a ion population with energy per nucleon

$$\mathcal{E}_{\mathrm{HB}} = \frac{m_p}{2}(2u_{\mathrm{HB}})^2 = \frac{2m_p I_L}{\rho c} = 2m_e c^2 \frac{Z n_c}{A n_e} a_0^2, \tag{5.9}$$

where we used $\rho \simeq A m_p n_i = (A/Z) m_p n_e$, being Z and A the ion charge and mass numbers, respectively. Equation 5.9 holds for a totally reflecting, cold plasma. Fast electron generation will reduce the HB efficiency by both decreasing $\mathcal{R}$ and producing a strong kinetic pressure which counteracts the radiation pressure. The quenching of fast electron generation by using circularly polarized pulses at normal incidence (Section 5.3.1) favours the radiation pressure action. The suppression of fast electrons also leads to reducing the intensity and energy of hard X-ray emission from the target, as experimentally observed [Aurand et al., 2015], so that a 'cleaner' source of energetic ions without secondary emissions may be obtained.

The scaling of $\mathcal{E}_{\mathrm{HB}}$ with the density implies that for values typical of solid targets ($n_e > 100n_c$) modest energies may be obtained. Higher energies may be obtained via HB acceleration if the target density is reduced down to values slightly exceeding the cut-off density n_c (lower density values are not suitable since the laser pulse would be transmitted through the target without any 'piston' action). Combining such low-density targets with laser pulses at foreseeable intensities $I_L > 10^{22}$ W cm^{-2} may allow reaching

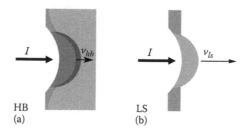

HB
(a)

LS
(b)

FIGURE 5.5 Radiation pressure acceleration (RPA): simple illustration of the hole boring (HB) regime for thick targets (frame a) and of the light sail (LS) regime for thin targets (frame b).

>100 MeV energies as investigated theoretically [Macchi and Benedetti, 2010; Robinson, 2011; Robinson et al., 2012].

Experimental evidence of HB acceleration has been provided in an experiment at the Accelerator Test Facility of Brookhaven National Laboratory (BNL) using a CO_2 infrared laser ($\lambda \simeq 10$ μm), for which $n_c \simeq 10^{19}$ cm^{-3}, and a hydrogen jet target having $n_e \gtrsim n_c$. Using pulses with circular polarization and $I_L \simeq 10^{16}$ W cm^{-2}, proton spectral peaks at energy $\mathcal{E} \simeq 1$ MeV have been observed [Palmer et al., 2011]. Recently, in a similar experiment, the HB velocity has been measured using interferometric techniques to map the plasma profile in time [Gong et al., 2016]; it is found that the HB dynamics is affected by the peculiar temporal structure of the CO_2 pulse, which is actually a train of 3 ps 'micropulses' separated by some tens of ps.

Using the long wavelength CO_2 laser as a driver enables the use of a flowing gas target, which may simplify high repetition rate operation compared to a solid target, since the latter needs to be mechanically replaced or displaced in a very short time. However, in order to reach the ion energy range suitable for medical applications, the CO_2 laser intensity should be increased by two orders of magnitude; development projects and required technical advances are discussed by Haberberger et al. [2010]; Bravy et al. [2012]; Pogorelsky et al. [2016].

Optical and near-IR lasers ($\lambda \simeq 0.8$–1 μm) presently produce much higher values of a_0 than CO_2 lasers. In order to maximize the energy obtainable via HB acceleration, a crucial step is the development of special targets with density values slightly above $n_c \simeq 10^{21}$ cm^{-3}, i.e. intermediate between those typical of gas jets and solid targets. Possible approaches include very high density gas jets [Sylla et al., 2012] and foam materials with low average density [Zani et al., 2013]. The production of a low-density 'preplasma' by a laser prepulse impinging on solid targets before the main short pulse might also be a suitable strategy. Signatures of HB acceleration in solid targets have been associated with the interferometric observation of collimated plasma jets in petawatt interactions with foil targets of few micron thickness [Kar et al., 2008a, 2013].

If the target thickness $l \ll u_{HB}\tau_p$, with τ_p the laser pulse duration, the HB front reaches the rear side of the target before the end of the pulse duration; the whole mass of the target is then accelerated (strictly speaking the central region, as in Figure 5.5b), and the acceleration process can be iterated obtaining high velocities. This regime can be modelled, in its simplest form, as a thin mirror boosted by radiation pressure, i.e. a 'light sail' (LS). Laser-driven sails were first proposed as a mechanism for interstellar travel propulsion [Marx, 1966; Forward, 1984], a concept recently brought back in press headlines because of the 'breakthrough starshot' project [Merali, 2016].

The 1D model of a mirror composed by a single species (with charge and mass numbers Z and A, respectively) having areal mass density ρl and boosted by a plane wave pulse of intensity I_L and duration τ_p gives for the energy per nucleon (see Macchi [2013], Section 5.7.3)

$$\mathcal{E}_{LS} = m_p c^2 \frac{\mathcal{F}^2}{2(1+\mathcal{F})}, \quad \mathcal{F} = \frac{2I_L\tau_p}{\rho l c^2} = 2\frac{Zn_c}{An_e}\frac{m_e}{m_p}\frac{\tau_p c}{l}a_0^2 \tag{5.10}$$

Since $\mathcal{F}$ is inversely proportional to the areal density $n_e l$, using the thinnest targets as possible is advantageous in order to increase the particle energy. However, this approach is limited by the onset of pulse transmission through the target, which reduces the radiation pressure boost. Assuming the foil to be much thinner than the laser wavelength, the threshold for pulse transmission due to relativistic transparency effects is given by Equation 5.5. Interestingly, a relation equivalent to Equation 5.5 is also obtained when the total radiation pressure is equal to the maximum electrostatic tension that can back-hold electrons in the foil; for higher intensities, the electrons are pushed away from the foil, and the ions undergo a Coulomb explosion, i.e. they are accelerated from their own space charge field which is unscreened by electrons. From both points of view, the condition 5.5 represents a compromise between maximizing the boosting radiation pressure on the foil and minimizing the foil mass, so it may be considered as an optimal working point for LS acceleration. It should be kept in mind, however, that Equation 5.5 is based on a simple model and plane geometry; in a realistic multi-dimensional geometry the expansion of the target in the transverse direction may lead to an earlier transition to transparency, limiting the energy gain. However, for high velocities of the target the reflectivity increases dynamically due to the increase of the laser wavelength in the frame co-moving with the foil, and (at least for relativistic ion velocities) the transverse expansion can reduce the areal mass on axis allowing to increase the maximum energy [Bulanov et al., 2010; Sgattoni et al., 2014]. After these remarks, we still use Equation 5.5 for a simple estimate of the energy gain in LS acceleration.

Inserting the optimal condition $a_0 = \zeta$ from Equation 5.5 in Equation 5.10 leads to an effective scaling (for non-relativistic ions, i.e. $\mathcal{E}_{LS} = m_p V^2 / 2 = m_p c^2 \mathcal{F}^2 / 2$, $\mathcal{F} \ll 1$)

$$\mathcal{E}_{LS}^{(opt)} = 2\pi^2 m_p c^2 \left(\frac{Z}{A} \frac{m_e}{m_p} \frac{c\tau_p}{\lambda} a_0 \right)^2 = 2\pi^2 m_e c^2 \left(\frac{m_e}{m_p} \right) \left(\frac{Z}{A} \frac{c\tau_p}{\lambda} a_0 \right)^2. \tag{5.11}$$

For currently reachable laser intensities, the optimal thickness condition $a_0 = \zeta$ requires $l \sim 10^{-2}\lambda$ or smaller for solid densities, which corresponds to nm-thick targets. Such ultrathin foils can be nowadays produced using, e.g. diamond-like carbon foil technology [Ma et al., 2011]. At the same time, systems producing laser pulses with ultrahigh pulse contrast, such that no significant target ionization and damage is produced before the short pulse interaction, have been developed. Such combination of target and laser technology enables the experimental investigation of RPA-LS, which is appealing because of the favourable scaling with laser parameters (especially for sub-relativistic ion energies), the expectation of monoenergetic spectra (as all ions in the sail should move coherently with the same velocity), and the remarkable efficiency which comes with high sail velocity: in fact the mirror model predicts a degree of conversion of laser energy into pulse energy equal to $\eta = 2\beta/(1 + \beta)$ where β is the sail velocity normalized to c, so that $\eta \simeq 40\%$ ($\beta \simeq 0.2$) for 100 MeV/nucleon ions. Such energies appear to be within reach with current laser and target technology: using a Ti:Sa laser ($\lambda = 0.8$ μm) delivering 40 fs pulses ($c\tau_p/\lambda = 15$) at an intensity of 10^{21} W cm^{-2} ($a_0 = 22$), Equation 5.11 gives $\mathcal{E}_{LS}^{(opt)} \simeq 150$ MeV.

At this point it is worth recalling that RPA-LS acceleration with ultra-intense lasers was first proposed in 2004 on the basis of 3D simulations [Esirkepov et al., 2004] showing that the acceleration of thin targets at intensities exceeding 10^{23} W cm^{-2} was well described by simple LS formulas. The extremely high intensity, still beyond current experimental capabilities, was considered to be necessary in order to enforce the dominance of RPA over other mechanisms (basically, the ions need to become relativistic within a laser cycle). However, later work showed that RPA is dominant at 'any' intensity when circularly polarized pulses at normal incidence are used [Macchi et al., 2005], which allows us to investigate LS acceleration using available laser systems as was proposed in theory papers [Zhang et al., 2007; Klimo et al., 2008; Robinson et al., 2008]. Later work also suggested a dominance of RPA for linearly polarized pulses at intensities of ~10^{21} W cm^{-2} [Qiao et al., 2012; Macchi, 2014]. Despite these findings, it is still often quoted that RPA requires extreme intensities.

At first, LS seems not particularly suitable for proton acceleration: while a thin foil of solid hydrogen appears hardly feasible, in a multispecies target one expects that all components should ultimately move with the same velocity and thus the same energy per nucleon. This is based on the following argument: if light ions overcame heavier ones, the trailing edge of the sail would screen the laser pulse at their location, stopping the acceleration by radiation pressure. While LS would remain, in any case, a preferred option for the acceleration of multiply charged ions, an analysis of the LS dynamics beyond the rigid mirror model shows that, for pulses of finite duration, only part of the target ions are accelerated as a monoenergetic bunch, and in proper conditions this part may contain only the target protons [Macchi et al., 2009]; in addition, the formation of a region of accelerating field ahead of the target related to 'leaking' transmission of the laser pulse may also accelerate protons [Qiao et al., 2010].

First experimental investigations of the RPA-LS regime [Kar et al., 2012; Henig et al., 2009; Palmer et al., 2012; Dollar et al., 2012; Aurand et al., 2013; Steinke et al., 2013] showed some promising results but also several issues, such as non-monoenergetic spectra, weak dependence on polarization and non-uniformity of the accelerated beam. Indications of the transition to the LS regime, but also of non-optimized conditions, have been found in using the VULCAN petawatt laser ($I_L = 0.5 - 3 \times 10^{20}$ W cm^{-2}, $\tau_p = 0.7 - 0.9$ ps and thin metallic targets ($l = 0.1 - 0.8$ µm) containing carbon and hydrogen impurities [Kar et al., 2012]. Narrow-band spectra (with energy spread ~20%) were observed centred at energies per nucleon approaching 10 MeV, displaying a scaling with $I_L \tau_p / \rho$ broadly consistent with the LS model prediction (Equation 5.10). The weak dependence on polarization is probably due to the relatively long laser pulse which lead to strong deformation of the target so that the incidence is not strictly normal anymore.

More recently, an experiment performed on the GEMINI laser with a shorter pulse ($\tau_p = 45$ fs, $I_L = 6 \times 10^{20}$ W cm^{-2}) and ultrathin ($l = 0.01$–0.1 µm) amorphous Carbon targets has given evidence of much higher Carbon ion energies for circular polarization (a 25 MeV cut-off energy to be compared with 10 MeV for linear polarization) [Scullion et al., 2017]. The analysis of the experiment, supported by 3D simulations, suggested that the energy gain was mostly limited by the onset of target transparency.

5.3.4 Collisionless Shock Acceleration

Under certain conditions, high intensity laser–plasma interactions lead to the generation of collisionless shock waves, i.e. sharp fronts of density and electric field which propagate in the plasma with supersonic velocity $V_s = Mc_s$, where the Mach number $M > 1$ and $c_s = (ZT_e/Am_p)^{1/2}$ is the 'speed of sound' (velocity of ion-acoustic waves) in a plasma. The term 'collisionless' originates from the fact that, contrary to standard hydrodynamics, collisional and viscosity effects are not needed for the formation of the shock front, which is sustained by charge separation effects.

The shock waves may be generated due to the piston action of the laser, i.e. the HB process (Section 5.3.3) in a hot plasma, so that the density perturbation produced by radiation pressure detaches from the interaction surface and propagates in the plasma (Figure 5.6). This requires the shock velocity to be higher than the HB velocity u_{hb}, which may occur if the shock wave is sustained by the fast electron population. In such case we may replace the bulk electron temperature T_e with the fast electron temperature T_f in the expression for V_s. As an alternative mechanism, the fast electron stream might be subject to instabilities which ultimately develop shock waves.

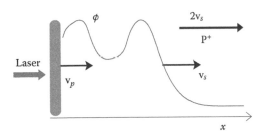

FIGURE 5.6 Collisionless shock acceleration (CSA): simple illustration of the driving of a shock wave by the piston action of the radiation pressure of the laser, and of the reflection of protons from the shock front at twice the shock velocity v_s. The ϕ curve is a sketch of the electric potential associated to the electrostatic, collisionless shock. (From D. A. Tidman and N. A. Krall. *Shock Waves in Collisionless Plasmas*, chapter 6. Wiley/Interscience, New York, 1971.)

The electric field at the shock front may act as a potential barrier for ions in the plasma, accelerating some ions 'by reflection' up to velocities $2V_S$: this is the basis for collisionless shock acceleration (CSA). The resulting energy per nucleon ε_{SA} is (assuming a non-relativistic shock velocity)

$$\varepsilon_{SA} = \frac{m_p}{2}(2V_s)^2 = 2\frac{Z}{A}M^2 T_f. \tag{5.12}$$

As far as the shock front propagates at constant velocity, the reflected ions are monoenergetic.

CSA has been invoked as the mechanism leading to the generation of highly monoenergetic proton spectra (with up to ~20 MeV peak energy) in an experiment performed using the Neptune CO_2 laser and gaseous hydrogen jet targets at the University of California at Los Angeles (UCLA) [Haberberger et al., 2012]. The energy spread of less than 1% is the narrowest one observed in laser–plasma accelera- tion experiments. In this experiment, the pulse train structure (a sequence of 3 ps micropulses with peak intensity $I \simeq 6 \times 10^{16}$ Wcm²) was found to be crucial to obtain the observed spectra: the density profile modification and plasma heating produced by the first micropulses in the train creates condi- tions favourable for the shock to be driven by the following micropulses. This suggests that a controlled, reproducible pulse sequence may be used to optimize CSA. Simulations of this scheme at higher intensi- ties [Haberberger et al., 2012] suggested that CSA driven by CO_2 pulses in hydrogen gas jets could pro- duce >200 MeV protons at intensities of the order of 10^{18} Wcm⁻², which would require a 20 times higher intensity compared to the currently most advanced CO_2 laser system.

So far, tailoring of the density profile for CSA optimization has been performed using a low inten- sity prepulse in experiments at BNL [Chen et al., 2015; Tresca et al., 2015]. With this approach, quasi- monoenergetic proton spectra with peak energy –1.2 MeV have been obtained with CO_2 pulses at an intensity of 2.5×10^{16} Wcm².

The observed number of accelerated protons in the UCLA experiment [Haberberger et al., 2012] is very low, about three orders of magnitude lower than produced via HB acceleration at BNL (see Section 5.3.3) in similar laser and target conditions [Palmer et al., 2011]. The comparison between the two experiments suggests that, for the same laser and plasma parameters, HB leads to lower energies than CSA but also to higher numbers of accelerated protons, which can be advantageous for some applications, such as iso- choric heating and the creation of warm dense matter. In CSA, the number of accelerated protons must be low in order to prevent excessive 'loading' of the shock wave: if too many protons are reflected from the shock, the latter loses energy and progressively reduces its velocity, which in turn causes the reflected proton energy to shift down to lower values, broadening the proton spectrum [Macchi et al., 2012].

The control of the density profile is also crucial for the demonstration of CSA with optical lasers, which allow much higher intensities but also requires higher densities. A recent parametric study based on 3D simulations predicts that a proper combination of two laser pulses may allow CSA to produce >100 MeV protons using petawatt power systems [Stockem Novo et al., 2016].

5.3.5 Relativistic Transparency and Other Mechanisms

Not all the observations of ion acceleration may be fully explained in terms of TNSA, RPA or CSA. Depending on the laser and target parameters, the mechanism may be of a hybrid nature, combining aspects of all the three 'basic' acceleration concepts. In addition, in the literature there are many pro- posals of particular schemes, employing, e.g. complex target configurations, and which are typically supported by numerical simulations. Only a minority of such proposals have been investigated in the laboratory so far. For the sake of brevity in this section, we restrict ourselves to experimental observa- tions and to the main trends in hybrid or alternative schemes.

As discussed in Section 5.3.3, the onset of relativistic transparency in thin targets is apparently the main factor limiting the RPA-LS scheme. However, a different approach to ion acceleration actually exploits transparency to reach high energies. The basic idea is that the transition from opacity to transparency is accompanied by an efficient, although somewhat turbulent, energy absorption by the plasma electrons (see Section 5.3.1), which may couple to ions also via collective plasma modes or instabilities. This regime has been investigated in particular at the TRIDENT laser facility at Los Alamos National Laboratory (LANL) [Henig et al., 2009; Hegelich et al., 2011; Jung et al., 2013a,b; Hegelich et al., 2013]. In the latest reported experiments at LANL, C and Al ions were accelerated up to 18 MeV/nucleon with up to 5% conversion efficiency and an energy spread as narrow as 7% for laser pulses of up to 80 J energy, 650 fs duration, and intensity up to 8×10^{20} W cm^{-2} [Palaniyappan et al., 2015].

Ion acceleration using foil targets undergoing transparency has been studied recently also using the PHELIX laser at GSI Darmstadt [Wagner et al., 2015] and at VULCAN [Powell et al., 2015], both laser systems having pulse durations of several hundreds of fs like TRIDENT. These experiments suggest a complex acceleration scenario, where different mechanisms (either TNSA-like or RPA-like) contribute at different stages and produce typical signatures in the energy spectrum and angular distribution of the ions. In some conditions, the cut-off energies achieved in this regime may be higher than those with respect to TNSA from thicker targets at similar laser parameters, but typically such energies are observed at some angle with respect to the laser axis.

The transparency regime is also accessible with ultra-short (tens of fs) pulses if ultra-thin (tens of nm) targets are used. Experiments in this regime performed with the PULSER I laser at the Advanced Photonics Research Institute (APRI) of GIST, Korea [Kim et al., 2013] and at MBI Berlin [Braenzel et al., 2015] have identified signatures of Coulomb explosion, which occurs above the transparency threshold (see Section 5.3.3). In particular, for the MBI experiments this regime seems favourable for the efficient acceleration of heavy (Gold) ions [Braenzel et al., 2015]. Ultrathin targets also enabled a different exploitation of relativistic transparency in controlling proton acceleration, which has been recently explored in an experiment on the GEMINI laser [Gonzalez-Izquierdo et al., 2016]: in the transparent regime, the proton spatial distribution is sensitive to the laser pulse polarization, which may be then used to control the proton beam.

An alternative approach to generate a near-transparent, or near-critical ($n_e \simeq n_c$) plasma for high absorption is to use special target material such as a foam. An experiment on VULCAN investigated relativistic transparency and related proton acceleration using foam targets [Willingale et al., 2009]. Experiments on foam-covered foils for enhancement in TNSA have been performed on PULSER [Prencipe et al., 2016; Passoni et al., 2016] and will be further discussed in Section 5.4.1.

A few experiments employed under-dense ($n_e < n_c$) plasmas produced from gas jets, which would be advantageous for high repetition rate operations (such as gas jets used in combination with CO_2 lasers for CSA, see Section 5.3.4). In this regime, an intense laser pulse may generate a charge-displacement channel with an electrostatic field accelerating ions mostly in the direction radial with respect to the propagation axis. However, He ions collimated along the propagation direction were also observed using VULCAN [Willingale et al., 2006] with a 40 MeV cut-off at 6×10^{20} W cm^{-2} intensity and 1 ps duration, and also using the JLITE-X laser at JAEA-KPSI at much lower intensity (7×10^{17} W cm^{-2} intensity, with 40 fs duration) with a surprising 20 MeV cut-off [Fukuda et al., 2009]. The interpretation of these experiments has stimulated the proposal of a particular mechanism, named magnetic vortex acceleration (MVA) where the accelerating electric field is generated by magnetic induction via the formation of electron vortices at the plasma-vacuum interface [Bulanov and Esirkepov, 2007].

5.4 Advanced Optimization Strategies

5.4.1 Target Engineering

Since its formulation, the TNSA model has provided a framework for optimization of ion acceleration by manipulating the target properties. Engineering of the chemical composition on the rear surface has been used for spectral manipulation. The effect is related to the expansion of multi-species plasmas, where depending on the relative concentration between heavy and light ions (typically protons) peaks are formed in the spectrum of the light species (see Tikhonchuk et al., 2005 and Section III.C.r of Macchi et al., 2013a). In addition, the concentration of hydrogen-containing molecules in thin dots coated on the rear side was exploited in order to reduce spectral broadening due to the transverse inhomogeneity of the sheath field, obtaining spectral peaks with ~10% spread [Schwoerer et al., 2006; Pfotenhauer et al., 2008]. As a specular approach, the removal of hydrogen impurities was used for TNSA of heavier species, e.g. carbon [Hegelich et al., 2002], also obtaining spectral peaks in particular conditions [Hegelich et al., 2006]. So far, however, these approaches have been limited to modest ion energies (a few MeV per nucleon) and progress appears to have been slow.

Engineering of the front side of the target has been oriented to increase the energy and conversion efficiency of fast electrons, in order to obtain enhanced TNSA. As already mentioned in Section 5.2, focusing the laser pulse into gold microcones placed on the front surface of thin foils yielded a ~30% increase in the cut-off energy and about one order of magnitude increase in proton number with respect to the flat case in an experiment performed at LANL (Figure 5.2).

It has also been observed that a non-planar surface, with structures having a size of the order of the laser wavelength, may allow for more efficient laser absorption and, in turn, higher proton cut-off energies [Margarone et al., 2012; Floquet et al., 2013; Margarone et al., 2015; Prencipe et al., 2016; Passoni et al., 2016]. Clearly, the exploitation of micro- or nano-structures requires the use of very high contrast, femtosecond pulses, otherwise the structuring would be washed out by the prepulse or, in any case, well before the peak pulse intensity.

Another strategy is to use 'mass-limited' targets, i.e. using targets with a reduced volume to confine the absorbed energy and obtaining higher temperatures: for example, the enhancement of conversion efficiency and proton cut-off energy has been observed in foils with limited (tens of microns) transverse extension [Buffechoux et al., 2010; Zeil et al., 2014]. In some cases, isolated targets with no mechanical support, such as droplets, have been used [Ter-Avetisyan et al., 2006; Sokollik et al., 2009; Ter-Avetisyan et al., 2012; Ostermayr et al., 2016].

For some of the experiments mentioned in Section 5.2.2 with data shown in Figure 5.3a, the flat target results were compared with those obtained with engineered targets. The latter include foil targets with the irradiated surface either covered by micro- or nano-sphere layers (Figure 5.3b) [Margarone et al., 2012, 2015] or by foam layers (Figure 5.3c) [Passoni et al., 2016], and transversely limited foils [Zeil et al., 2014]. In all these cases, the comparison with the flat target was made at the same value of foil thickness. Figure 5.7 shows that structured targets may yield up to a ~50% enhancement of the energy cut-off and to more than one order of magnitude increase in the spectral density of protons at a given energy. The enhancement effect produced by the microsphere layer is further discussed by Floquet et al. [2013]. Additional observations in the foam-covered target experiments [Passoni et al., 2016], such as independence on pulse polarization, suggest that the enhancement is due to a geometrical effect related to the microscopic structure of the foam which allows efficient volumetric heating.

Special targets may be also designed in order to allow high absorption via the excitation of resonances in the laser-produced plasma (Section 5.3.1). For example, foam targets can be produced with average density $n_e \simeq n_c$, so that the laser field would couple with bulk plasma oscillations or plasmons. However, this interaction scenario probably requires some preheating of the foam target in order to smooth its inhomogeneous structure, which is preserved in high contrast interactions (see the discussion in the preceding paragraph). In 'grating' targets with a periodic modulation at the laser-irradiated surface, the laser pulse can couple instead to surface plasmons at angles of incidence θ_i, such that $\sin \theta_i \sim 1 - j\lambda/d$, where d is the spatial period and j is an integer number. In an experiment with the UHI laser (25 fs, 80TW) at the SLIC facility of CEA Saclay, high contrast pulse interaction with grating targets at angles around the $j = 1$ resonance has shown to produce a factor of ~2.5 increase in the proton cut-off energy with respect to flat targets at the same angle of incidence [Ceccotti et al., 2013].

Nanostructured targets may also be used with acceleration mechanisms different from TNSA. Recently, in order to investigate RPA-LS with extremely intense and sharp rising fs pulses, thin foil

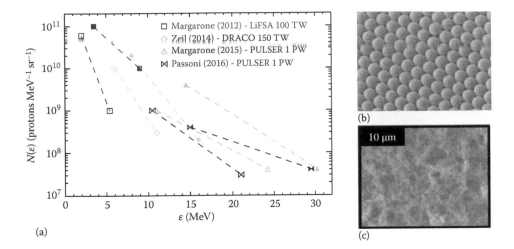

(a)

(b)

(c)

FIGURE 5.7 (a) Proton energy spectra from experiments comparing flat and structured targets. The flat target data are already included in Figure 5.3 and shown here with empty symbols: filled symbols correspond to the same experimental parameters, but using structured targets. The latter included thin foils covered by sub-micron size spheres [Margarone et al., 2012], mass-limited targets [Zeil et al., 2014] and foam-covered foils [Passoni et al., 2016]. (b) Scanning electron microscope (SEM) image of a layer of 0.94 μm spheres. (Adapted from V. Floquet et al., *J. Appl. Phys.* 114, 2013. With permission.) (c) SEM image of a carbon foam. (Adapted from M. Passoni et al., *Plasma Phys. Control. Fusion* 56:045001, 2014. With permission.)

targets covered by a few-micron Carbon nanotube foam (CNF) on the interaction side have been used in order to generate self-focusing and self-steepening of the laser pulse in a plasma of density close to n_c. Using such a technique on the GEMINI laser, delivering 50 fs pulses at $I = 2 \times 10^{20}$ W cm^{-2}, enhanced acceleration of carbon ions (up to ~20 MeV energy per nucleon) with RPA features has been observed [Bin et al., 2015].

Very recently, first experiments have been performed employing cryogenic targets. The advantages of such technology include the low electron density (down to a few tens of n_c) and the suitability for a high repetition rate, since the target is flowing. In addition, the use of cryogenic hydrogen affords obtaining pure proton spectra. Preliminary results have been obtained with the TITAN laser at LLNL [Gauthier et al., 2016] and with the PALS laser (600 ps duration) in Prague [Margarone et al., 2016].

5.4.2 Optical Control and Post-Acceleration

In principle, laser-accelerated ions can be handled using conventional accelerator techniques for energy selection, transport and focusing. However, it is of interest to investigate handling techniques which can be integrated in the laser and target configuration, i.e. on a very short and compact scale similar to that over which the acceleration occurs. Most of these techniques has been developed in the TNSA framework, although they might be adapted to other laser acceleration schemes.

Already in the first experiments on proton acceleration it was observed that the direction of protons could be simply controlled by shaping and orientation of the rear surface of the target, since protons are emitted in the direction normal to such a surface. It was thus apparent that a focusing of the proton bunch could be obtained by using a target with a spherically shaped rear surface (see Figure 5.8a). This static approach has been used for the first time by Patel et al. [2003], where proton focusing was exploited to isochorically heat matter.

Further developments of the target shaping concept have been investigated by Kar et al. [2008b], Burza et al. [2011], Kar et al. [2011], Bartal et al. [2012] and Chen et al. [2012]. Some of these target configurations were designed in order to exploit the transient nature of the TNSA field for dynamic focusing:

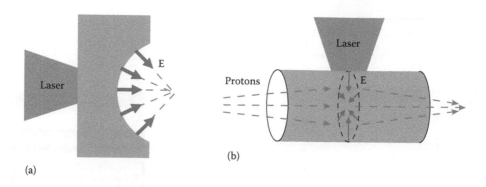

(a)

(b)

FIGURE 5.8 (a) 'Static' focusing of TNSA-produced protons by a spherical target. (b) 'Dynamic' focusing of protons by the electric field generated inside a laser-irradiated cylinder. (From T. Toncian et al., *Science*, 312:410, 2006.)

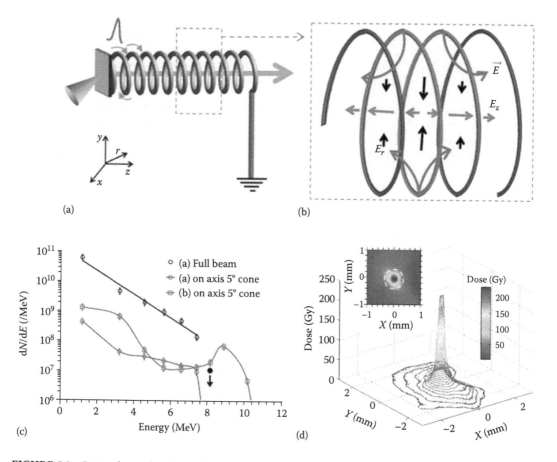

FIGURE 5.9 Proton beam focusing and post-acceleration using an helical coil attached to the rear side of the target. (a) Scheme of the device. (b) Sketch of the electric field distribution inside the coil, in correspondence of the travelling signal. (c) Comparison of spectra with and without the coil, showing the energy increase and spectral bunching near cut-off for protons travelling near the axis. (d) Narrow spatial distribution of 8.9 MeV protons after focusing by the coil. (Adapted from S. Kar et al., *Nat. Comm.* 7:10792, 2016. With permission.)

since the field travels along the target as a unipolar surface wave with picosecond duration [Quinn et al., 2009; Tokita et al., 2015], the target can be designed in a way that the fields interacts at its peak with protons of a given energy, yielding energy selection capability. Alternatively, a double-pulse scheme may be used as in the experiment of Toncian et al. [2006], where a TNSA bunch was focused by the field generated inside a hollow cylinder by a secondary pulse (Figure 5.8b): by changing the synchronization between the two pulses, a particular energy slice can be focused and selected.

Recently, the above ideas have been further developed in a novel device for post-acceleration with chromatic focusing and energy enhancement capability [Kar et al., 2016; Kar, 2016]. The design of the device is shown in Figure 5.9a. A metallic wire attached to the rear side of the target is bent in order to form a helical coil coaxial with the proton beam. After the interaction, while the protons are accelerated in the sheath, the unipolar pulse propagates along the coil, its velocity along the longitudinal direction being determined by the coil radius and pitch. This allows the unipolar pulse to be synchronized with protons of a given velocity. The electric field generated near the axis of the coil has both radial components directed towards the axis, and longitudinal components that are either parallel or anti-parallel to the axis as shown in Figure 5.9b. Thus, a fraction of the protons may be simultaneously focused and further accelerated by the electric field of the traveling SW, as it has been demonstrated in proof-of-principle experiments [Kar et al., 2016] using the ARCTURUS laser system at Heinrich Heine University in Düsseldorf. The main results are summarized in Figure 5.9c and d: an increase of the cut-off energy by ~2 MeV is obtained in combination with a tight focusing of the protons along the axis of the system.

5.5 Discussion and Outlook

As already mentioned in this chapter, evaluating the progress made in laser-driven ion acceleration requires critical insight. For instance, let us focus for the moment on the enhancement of the cut-off energy. While the latter has been considered the most relevant parameter to assess the progress in the field, its experimental determination may be ambiguous, particularly for broad exponential-like spectra, as it depends on the detection method and its sensitivity as well as on the practices for signal discrimination from background noise. While establishing reliable and agreed experimental procedures remains a priority for the community, the analysis in Section 5.2 has made an attempt to perform a meaningful comparison by taking full, calibrated spectra into account.

The most relevant result emerging from such analysis is, in our opinion, the enhancement of the cut-off energy achieved with 'small', short-pulse laser systems with a scaling law which appears more promising than inferred in previous work, as shown in Figure 5.3; roughly, the scaling is linear in the pulse energy *in the focal spot* with a ~9 MeV/J slope. Establishing whether this scaling will be maintained for larger pulse energies, delivering such energies in the focal spot and providing a theoretical support for this scaling are possible near-term research goals.

As already mentioned, two experiments in our selection [Kim et al., 2013; Ogura et al., 2012] have made claims of cut-off energies significantly higher than the values predicted from the scaling law at the same laser pulse energy. Possible explanations for these 'anomalous' data might be searched in differences in the experimental set-up. For the experiment of Ogura et al. [2012], where 0.8 μm thick targets were used, the only apparent difference with respect to other references is a lower pulse contrast, since plasma mirrors were not used; in principle this might favour the formation of a small preplasma, which may already affect the interaction.

The experiment of Kim et al. [2013] was performed on the PULSER laser using ultrathin targets (in the 10–100 nm range) and intensities up to 3.3×10^{20} W cm^{-2} which might lead to an acceleration regime different from TNSA, with strong effects of both RPA (Section 5.3.3) and transparency (Section 5.3.5). A hybrid acceleration mechanism with overlapping contributions by both TNSA and RPA was also proposed on the basis of a more recent experiment with the ATLAS laser system at MPI Garching ($I_L = 8 \times 10^{19}$ W cm^{-2}, $\tau_L = 30$ fs), also using ultrathin targets (5 – 20 nm). Measured the proton cut-off energy ε_{co} as a function of the absorption coefficient $\mathcal{A}$, obtaining a scaling $\varepsilon_{co} \propto \mathcal{A}$ in agreement with models predicting a linear scaling of $\mathcal{E}_{co}$ with the absorbed laser energy [Schreiber et al., 2006; Zeil et al., 2010]. The extremely thin targets used (5–20 nm) favoured a contribution of RPA (Section 5.3.3) to the proton energy, so that the acceleration mechanism may be considered of hybrid nature.

Hybrid regimes of accelerations (Section 5.3.5) are possibly promising for proton acceleration but still require thorough investigations before applications can be tackled on their basis. Similarly, mechanisms such as CSA (Section 5.3.4) or magnetic vortex acceleration (Section 5.3.5) are still in a very early stage of investigation and more experiments are needed to confirm both preliminary findings and theoretical predictions.

Recently, a proton energy cut-off of 93 MeV has been reported from another campaign on PULSER [Kim et al., 2016] using the thinnest targets (10 nm) employed by Kim et al. [2013]. At the highest intensity (7×10^{20} W cm^{-2}) only, the cut-off energy was larger for circular polarization than for linear polarization, so that efficient RPA-LS acceleration (Section 5.3.3) was claimed although the proton spectrum was broad (with no clear spectral peak) and anomalously modulated. The accuracy of determining the maximum energy (strongly at variance with experiments performed for similar parameters in different laboratories) was at the limit of the detector (Thomson parabola) range with possible issues of low resolution, noise floor and trace discrimination. For these reasons, further support by additional experiments will be required. An expansion of the data on RPA-LS is expected with growing number of petawatt-laser systems starting their operation, since the RPA-LS mechanism remains very attractive because of its scaling properties; the control and understanding of the onset of transparency and the target stability are expected challenges on the route to RPA optimization.

Enhancement of the energy per nucleon is by no means the only required development to make laser-driven ion acceleration suitable for applications. For instance, while at least for 'large' laser drivers (Figure 5.2) the cut-off energy now falls in the therapeutic window for ion beam therapy in oncology,* laser-driven acceleration remains quite far from the other stringent requirements [Linz and Alonso, 2007, 2016]. At present it appears not possible to predict either when such requirements will be reached or if such developments in laser-driven schemes will be faster than those in approaches based on conventional accelerators, which are becoming more compact and cheaper. Probably, the key issue will be whether the unique properties of laser-accelerated ions may be advantageously exploited for therapeutic benefits. For example, the availability of an extremely high dose rate could enable the first investigation of 'collective' regimes in the biological response to irradiation [Fourkal et al., 2011], and the fast optical control of the short-duration ion bunches might be useful for the irradiation of moving organs [Hofmann et al., 2012].

While medical applications remain a long-term challenge, laser-accelerated protons have already had a major impact as a diagnostic of laser–plasma interactions, providing data which are also of broad interest for nonlinear science and astrophysically relevant phenomena. An increasing use of laser-accelerated protons and ions in production of warm dense matter and time-resolved studies of material damage is also likely (see Chapter 9 by Dromey and Chapter 16 by Nordlund and Djurabekova). These applications exploit the short duration of the ion bunch and its natural synchronization with laser pulses, which open the way to pump-probe experiments.

We expect that such perspectives will stimulate further developments in laser-driven ion acceleration. Such developments may be supported both by the expected increase in available laser intensities and by progress in target engineering, including micro- and nano-structuring and the use of special materials. With the support of theory and massively parallel numerical simulations, hopefully such advances will allow us to reach significant milestones in the next decade.

While the present review paper was in the editorial production stage, a publication reported on the observation of proton cut-off energies exceeding 94 MeV, which sets a new world record with respect to

---

* Ocular tumours may be treated with ~60 MeV protons; however, most hadron therapy treatments require 200–225 MeV protons at the surface of the patient.

the data shown in Figure 5.2. The experiment was performed in the VULCAN facility using ultrathin targets and ~1 ps pulses with ~60 J energy in the focal spot. The proposed acceleration mechanism is of hybrid nature, combining TNSA and RPA in a regime of target transparency [Higginson et al. 2010].

While the present review paper was in the editorial production stage, a publication reported on the observation of proton cut-off energies exceeding 94 MeV, that sets a new world record with respect to the data shown in Fig.5.2. The experiment was performed on the VULCAN facility using ultrathin targets and ~1 ps pulses with ~60 J energy in the focal spot. The proposed acceleration mechanism is of hybrid nature, combining TNSA and RPA in a regime of target transparency. A. Higginson, R. J. Gray, M. King, R. J. Dance, S. D. R. Williamson, N. M. H. Butler, R. Wilson, R. Capdessus, C. Armstrong, J. S. Green, S. J. Hawkes, P. Martin, W. Q. Wei, S. R. Mirfayzi, X. H. Yuan, S. Kar, M. Borghesi, R. J. Clarke, D. Neely, and P. McKenna. Near-100 MeV protons via a laser-driven transparency-enhanced hybrid acceleration scheme. Nature Communications (2018), at press.

References for Chapter 5

S. Atzeni, M. Temporal, and J. Honrubia. A first analysis of fast ignition of precompressed ICF fuel by laser-accelerated protons. *Nucl. Fusion*, 42:L1, 2002. doi: 10.1088/0029-5515/42/3/101.

B. Aurand, S. Kuschel, O. Jaeckel, C. Roedel, H. Y. Zhao, S. Herzer, A. E. Paz, J. Bierbach, J. Polz, B. Elkin, G. G. Paulus, A. Karmakar, P. Gibbon, T. Kuehl, and M. C. Kaluza. Radiation pressure-assisted acceleration of ions using multi-component foils in high-intensity laser-matter interactions. *N. J. Phys.*, 15:033031, 2013. doi: 10.1088/1367-2630/15/3/033031.

B. Aurand, S. Kuschel, C. Rdel, O. Jckel, J. Polz, B. Elkin, H. Zhao, A. Karmakar, P. Gibbon, M. Kaluza, and T. Kuehl. Reduction of X-ray generation in high-intensity laser ion acceleration. *Appl. Phys. B: Lasers Opt.*, 118:247–251, 2015. doi: 10.1007/s00340-014-5979-7.

T. Bartal, M. E. Foord, C. Bellei, M. H. Key, K. A. Flippo, S. A. Gaillard, D. T. Offermann, P. K. Patel, L. C. Jarrott, D. P. Higginson, M. Roth, A. Otten, D. Kraus, R. B. Stephens, H. S. McLean, E. M. Giraldez, M. S. Wei, D. C. Gautier, and F. N. Beg. Focusing of short-pulse high-intensity laser-accelerated proton beams. *Nat. Phys.*, 8:139–142, 2012. doi: 10.1038/nphys2153.

S. Betti, F. Ceccherini, F. Cornolti, and F. Pegoraro. Expansion of a finite-size plasma in vacuum. *Plasma Phys. Contr. Fusion*, 47:521, 2005. doi: 10.1088/0741-3335/47/3/008.

J. H. Bin, W. J. Ma, H. Y. Wang, M. J. V. Streeter, C. Kreuzer, D. Kiefer, M. Yeung, S. Cousens, P. S. Foster, B. Dromey, X. Q. Yan, R. Ramis, J. Meyer-ter Vehn, M. Zepf, and J. Schreiber. Ion acceleration using relativistic pulse shaping in near-critical-density plasmas. *Phys. Rev. Lett.*, 115:064801, 2015. doi: 10.1103/PhysRevLett.115.064801.

P. Bolton, M. Borghesi, C. Brenner, D. Carroll, C. D. Martinis, F. Fiorini, A. Flacco, V. Floquet, J. Fuchs, P. Gallegos, D. Giove, J. Green, S. Green, B. Jones, D. Kirby, P. McKenna, D. Neely, F. Nuesslin, R. Prasad, S. Reinhardt, M. Roth, U. Schramm, G. Scott, S. Ter-Avetisyan, M. Tolley, G. Turchetti, and J. Wilkens. Instrumentation for diagnostics and control of laser-accelerated proton (ion) beams. *Phys. Med.*, 30:255–270, 2014. doi: 10.1016/j.ejmp.2013.09.002.

M. Borghesi, A. Bigongiari, S. Kar, A. Macchi, L. Romagnani, P. Audebert, J. Fuchs, T. Toncian, O. Willi, S. V. Bulanov, A. J. Mackinnon, and J. C. Gauthier. Laser-driven proton acceleration: Source optimization and radiographic applications. *Plasma Phys. Contr. Fusion*, 50:124040, 2008. doi: 10.1088/0741-3335/50/12/124040.

M. Borghesi, D. H. Campbell, A. Schiavi, M. G. Haines, O. Willi, A. J. MacKinnon, P. Patel, L. A. Gizzi, M. Galimberti, R. J. Clarke, F. Pegoraro, H. Ruhl, and S. Bulanov. Electric field detection in laser–plasma interaction experiments via the proton imaging technique. *Phys. Plasmas*, 9:2214–2220, 2002. doi: 10.1063/1.1459457.

M. Borghesi, J. Fuchs, S. V. Bulanov, A. J. MacKinnon, P. K. Patel, and M. Roth. Fast ion generation by high-intensity laser irradiation of solid targets and applications. *Fus. Sci. Techn.*, 49:412, 2006.

M. Borghesi and A. Macchi. Laser-driven ion accelerators: State of the art and applications. In A. Giulietti, editor, *Laser-Driven Particle Acceleration towards Radiobiology and Medicine*, pages 221–247. Springer International Publishing, Cham, 2016. doi: 10.1007/978-3-319-31563-810.

M. Borghesi, A. J. Mackinnon, D. H. Campbell, D. G. Hicks, S. Kar, P. K. Patel, D. Price, L. Romagnani, A. Schiavi, and O. Willi. Multi-MeV proton source investigations in ultraintense laser-foil interactions. *Phys. Rev. Lett.*, 92:055003, 2004. doi: 10.1103/Phys-RevLett.92.055003.

J. Braenzel, A. A. Andreev, K. Platonov, M. Klingsporn, L. Ehrentraut, W. Sandner, and M. Schnürer. Coulomb-driven energy boost of heavy ions for laser–plasma acceleration. *Phys. Rev. Lett.*, 114:124801, 2015. doi: 10.1103/PhysRevLett.114.124801.

B. G. Bravy, Y. A. Chernyshev, V. M. Gordienko, E. F. Makarov, V. Y. Panchenko, V. T. Platonenko, and G. K. Vasil'ev. Multi-terawatt picoseconds $10\mu m$, O_2 laser system: Design and parameters' control. *Opt. Express*, 20:25536–25544, 2012. doi: 10.1364/OE.20.025536.

C. M. Brenner, A. P. L. Robinson, K. Markey, R. H. H. Scott, R. J. Gray, M. Rosinski, O. Deppert, J. Badziak, D. Batani, J. R. Davies, S. M. Hassan, K. L. Lancaster, K. Li, I. O. Musgrave, P. A. Norreys, J. Pasley, M. Roth, H.-P. Schlenvoigt, C. Spindloe, M. Tatarakis, T. Winstone, J. Wolowski, D. Wyatt, P. McKenna, and D. Neely. High energy conversion efficiency in laser-proton acceleration by controlling laser-energy deposition onto thin foil targets. *Appl. Phys. Lett.*, 104:081123, 2014. doi: dx.doi.org/10.1063/1.4865812.

S. Buffechoux, J. Psikal, M. Nakatsutsumi, L. Romagnani, A. Andreev, K. Zeil, M. Amin, P. Antici, T. Burris-Mog, A. Compant-La-Fontaine, E. d'Humières, S. Fourmaux, S. Gaillard, F. Gobet, F. Hannachi, S. Kraft, A. Mancic, C. Plaisir, G. Sarri, M. Tarisien, T. Toncian, U. Schramm, M. Tampo, P. Audebert, O. Willi, T. E. Cowan, H. Pépin, V. Tikhonchuk, M. Borghesi, and J. Fuchs. Hot electrons transverse refluxing in ultraintense laser-solid interactions. *Phys. Rev. Lett.*, 105:015005, 2010. doi: 10.1103/Phys-RevLett.105.015005.

S. V. Bulanov, E. Y. Echkina, T. Z. Esirkepov, I. N. Inovenkov, M. Kando, F. Pegoraro, and G. Korn. Unlimited ion acceleration by radiation pressure. *Phys. Rev. Lett.*, 104:135003, 2010. doi: 10.1103/PhysRevLett.104.135003.

S. V. Bulanov and T. Z. Esirkepov. Comment on "collimated multi-MeV ion beams from high-intensity laser interactions with underdense plasma". *Phys. Rev. Lett.*, 98:049503, 2007. doi: 10.1103/PhysRevLett.98.049503.

S. V. Bulanov, T. Z. Esirkepov, V. S. Khoroshkov, A. V. Kuznetsov, and F. Pegoraro. Oncological hadron-therapy with laser ion accelerators. *Phys. Lett. A*, 299:240–247, 2002. doi: 10.1016/S0375-9601(02)00521-2.

S. V. Bulanov and V. Khoroshkov. Feasibility of using laser ion accelerators in proton therapy. *Plasma Phys. Reports*, 28:453–456, 2002. doi: 10.1134/1.1478534.

M. Burza, A. Gonoskov, G. Genoud, A. Persson, K. Svensson, M. Quinn, P. McKenna, M. Marklund, and C.-G. Wahlström. Hollow microspheres as targets for staged laser-driven proton acceleration. *N. J. Phys.*, 13:013030, 2011. doi: 10.1088/1367-2630/13/1/013030.

F. Cattani, A. Kim, D. Anderson, and M. Lisak. Threshold of induced transparency in the relativistic interaction of an electromagnetic wave with overdense plasmas. *Phys. Rev. E*, 62:1234–1237, 2000. doi: 10.1103/PhysRevE.62.1234.

T. Ceccotti, V. Floquet, A. Sgattoni, A. Bigongiari, O. Klimo, M. Raynaud, C. Riconda, A. Heron, F. Baffigi, L. Labate, L. A. Gizzi, L. Vassura, J. Fuchs, M. Passoni, M. Květon, F. Novotny, M. Possolt, J. Prokůpek, J. Proška, J. Pšikal, L. Stolcová, A. Velyhan, M. Bougeard, P. D'Oliveira, O. Tcherbakoff, F. Réau, P. Martin, and A. Macchi. Evidence of resonant surface-wave excitation in the relativistic regime through measurements of proton acceleration from grating targets. *Phys. Rev. Lett.*, 111:185001, 2013. doi: 10.1103/PhysRevLett.111.185001.

S. N. Chen, E. d'Humières, E. Lefebvre, L. Romagnani, T. Toncian, P. Antici, P. Audebert, E. Brambrink, C. A. Cecchetti, T. Kudyakov, A. Pipahl, Y. Sentoku, M. Borghesi, O. Willi, and J. Fuchs. Focusing dynamics of high-energy density, laser-driven ion beams. *Phys. Rev. Lett.*, 108:055001, 2012. doi: 10.1103/PhysRevLett.108.055001.

Y.-H. Chen, M. H. Helle, A. Ting, D. F. Gordon, M. N. Polyanskiy, I. Pogorelsky, M. Babzien, and Z. Najmudin. Observation of monoenergetic protons from a near-critical gas target tailored by a hydrodynamic shock. *Proc. SPIE*, 9314.93140C=93140C=6, 2015. doi: 10.1117/12.2102091.

I. W. Choi, I. J. Kim, K. H. Pae, K. H. Nam, C.-L. Lee, H. Yun, H. T. Kim, S. K. Lee, T. J. Yu, J. H. Sung, A. S. Pirozhkov, K. Ogura, S. Orimo, H. Daido, and J. Lee. Simultaneous generation of ions and high-order harmonics from thin conjugated polymer foil irradiated with ultrahigh contrast laser. *Appl. Phys. Lett.*, 99:181501, 2011. doi: 10.1063/1.3656338.

E. L. Clark, K. Krushelnick, J. R. Davies, M. Zepf, M. Tatarakis, F. N. Beg, A. Machacek, P. A. Norreys, M. I. K. Santala, I. Watts, and A. E. Dangor. Measurements of energetic proton transport through magnetized plasma from intense laser interactions with solids. *Phys. Rev. Lett.*, 84:670–673, 2000. doi: 10.1103/PhysRevLett.84.670.

J. A. Cobble, R. P. Johnson, T. E. Cowan, N. R.-L. Galloudec, and M. Allen. High resolution laser-driven proton radiography. *J. Appl. Phys.*, 92:1775–1779, 2002. doi: 10.1063/1.1494128.

T. E. Cowan, J. Fuchs, H. Ruhl, A. Kemp, P. Audebert, M. Roth, R. Stephens, I. Barton, A. Blazevic, E. Brambrink, J. Cobble, J. Fernández, J.-C. Gauthier, M. Geissel, M. Hegelich, J. Kaae, S. Karsch, G. P. Le Sage, S. Letzring, M. Manclossi, S. Meyroneinc, A. Newkirk, H. Pépin, and N. Renard-LeGalloudec. Ultralow emittance, multi-MeV proton beams from a laser virtual-cathode plasma accelerator. *Phys. Rev. Lett.*, 92:204801, 2004. doi: 10.1103/PhysRevLett.92.204801.

H. Daido, M. Nishiuchi, and A. S. Pirozhkov. Review of laser-driven ion sources and their applications. *Rep. Prog. Phys.*, 75:056401, 2012. doi: 10.1088/0034-4885/75/5/056401.

C. Danson, D. Hillier, N. Hopps, and D. Neely. Petawatt class lasers worldwide. *High Power Laser Sci. Eng.*, 3:e3, 2015. doi: 10.1017/hpl.2014.52.

F. Dollar, C. Zulick, A. G. R. Thomas, V. Chvykov, J. Davis, G. Kalinchenko, T. Matsuoka, C. McGuffey, G. M. Petrov, L. Willingale, V. Yanovsky, A. Maksimchuk, and K. Krushel-nick. Finite spot effects on radiation pressure acceleration from intense high-contrast laser interactions with thin targets. *Phys. Rev. Lett.*, 108:175005, 2012. doi: 10.1103/Phys-RevLett.108.175005.

F. Dollar, T. Matsuoka, G. M. Petrov, A. G. R. Thomas, S. S. Bulanov, V. Chvykov, J. Davis, G. Kalinchenko, C. McGuffey, L. Willingale, V. Yanovsky, A. Maksimchuk, and K. Krushelnick. Control of energy spread and dark current in proton and ion beams generated in high-contrast laser solid interactions. *Phys. Rev. Lett.*, 107.065003, 2011. doi: 10.1103/PhysRevLett.107.065003.

B. Dromey, M. Coughlan, L. Senje, M. Taylor, S. Kuschel, B. Villagomez-Bernabe, R. Stefanuik, G. Nersisyan, L. Stella, J. Kohanoff, M. Borghesi, F. Currell, D. Riley, D. Jung, C. Wahlström, C. Lewis, and M. Zepf. Picosecond metrology of laser-driven proton bursts. *Nat. Comm.*, 7:10642, 2016. doi: 10.1038/ncomms10642.

B. Dromey, S. Kar, M. Zepf, and P. Foster. The plasma mirror—a subpicosecond optical switch for ultra-high power lasers. *Rev. Scientif. Instrum.*, 75:645–649, 2004. doi: 10.1063/1.1646737.

T. Esirkepov, M. Borghesi, S. V. Bulanov, G. Mourou, and T. Tajima. Highly efficient relativistic-ion generation in the laser-piston regime. *Phys. Rev. Lett.*, 92:175003, 2004. doi: 10.1103/PhysRevLett.92.175003.

J. Fernández, B. Albright, F. Beg, M. Foord, B. Hegelich, J. Honrubia, M. Roth, R. Stephens, and L. Yin. Fast ignition with laser-driven proton and ion beams. *Nuclear Fusion*, 54: 054006, 2014.

V. Floquet, O. Klimo, J. Psikal, A. Velyhan, J. Limpouch, J. Proska, F. Novotny, L. Stolcova, A. Macchi, A. Sgattoni, L. Vassura, L. Labate, F. Baffigi, L. A. Gizzi, P. Martin, and T. Ceccotti. Micro-sphere layered targets efficiency in laser driven proton acceleration. *J. Appl. Phys.*, 114:083305, 2013. doi: 10.1063/1.4819239.

R. L. Forward. Roundtrip interstellar travel using laser-pushed lightsails. *J. Spacecraft*, 21: 187, 1984.

E. Fourkal, I. Velchev, C.-M. Ma, and J. Fan. Linear energy transfer of proton clusters. *Phys. Med. Biol.*, 56:3123, 2011. doi: 10.1088/0031-9155/56/10/015.

E. Fourkal, J. S. Li, W. Xiong, A. Nahum, and C. Ma. Intensity modulated radiation therapy using laser-accelerated protons: A Monte Carlo dosimetric study. *Phys. Med. Biol.*, 48:3977, 2003. doi: 10.1088/0031-9155/48/24/001.

S. Fritzler, V. Malka, G. Grillon, J. P. Rousseau, F. Burgy, E. Lefebvre, E. d'Humiéres, P. McKenna, and K. W. D. Ledingham. Proton beams generated with high-intensity lasers: Applications to medical isotope production. *Appl. Phys. Lett.*, 83:3039–3041, 2003. doi: 10.1063/1.1616661.

J. Fuchs, P. Antici, E. d'Humiéres, E. Lefebvre, M. Borghesi, E. Brambrink, C. A. Cecchetti, M. Kaluza, V. Malka, M. Manclossi, S. Meyroneinc, P. Mora, J. Schreiber, T. Toncian, H. Pepin, and P. Audebert. Laser-driven proton scaling laws and new paths towards energy increase. *Nat. Phys.*, 2:48, 2006. doi: 10.1038/nphys199.

Y. Fukuda, A. Y. Faenov, M. Tampo, T. A. Pikuz, T. Nakamura, M. Kando, Y. Hayashi, A. Yogo, H. Sakaki, T. Kameshima, A. S. Pirozhkov, K. Ogura, M. Mori, T. Z. Esirkepov, J. Koga, A. S. Boldarev, V. A. Gasilov, A. I. Magunov, T. Yamauchi, R. Kodama, P. R. Bolton, Y. Kato, T. Tajima, H. Daido, and S. V. Bulanov. Energy increase in multi-MeV ion acceleration in the interaction of a short pulse laser with a cluster-gas target. *Phys. Rev. Lett.*, 103:165002, 2009. doi: 10.1103/PhysRevLett.103.165002.

S. A. Gaillard, T. Kluge, K. A. Flippo, M. Bussmann, B. Gall, T. Lockard, M. Geissel, D. T. Offermann, M. Schollmeier, Y. Sentoku, and T. E. Cowan. Increased laser-accelerated proton energies via direct laser-light-pressure acceleration of electrons in microcone targets. *Phys. Plasmas*, 18:056710, 2011. doi: 10.1063/1.3575624.

M. Gauthier, J. B. Kim, C. B. Curry, B. Aurand, E. J. Gamboa, S. Gde, C. Goyon, A. Hazi, S. Kerr, A. Pak, A. Propp, B. Ramakrishna, J. Ruby, O. Willi, G. J. Williams, C. Rdel, and S. H. Glenzer. High-intensity laser-accelerated ion beam produced from cryogenic micro-jet target. *Rev. Sci. Instrum.*, 87, 2016. doi: 10.1063/1.4961270.

P. Gibbon. *Short Pulse Laser Interaction with Matter.* Imperial College Press, 2005.

C. Gong, S. Y. Tochitsky, F. Fiuza, J. J. Pigeon, and C. Joshi. Plasma dynamics near critical density inferred from direct measurements of laser hole boring. *Phys. Rev. E*, 93:061202, 2016. doi: 10.1103/PhysRevE.93.061202.

B. Gonzalez-Izquierdo, M. King, R. J. Gray, R. Wilson, R. J. Dance, H. Powell, D. A. Maclellan, J. McCreadie, N. M. H. Butler, S. Hawkes, J. S. Green, C. D. Murphy, L. C. Stockhausen, D. C. Carroll, N. Booth, G. G. Scott, M. Borghesi, D. Neely, and P. McKenna. Towards optical polarization control of laser-driven proton acceleration in foils undergoing relativistic transparency. *Nat. Comm.*, 7:12891, 2016. doi: 10.1038/ncomms12891.

J. S. Green, A. P. L. Robinson, N. Booth, D. C. Carroll, R. J. Dance, R. J. Gray, D. A. MacLellan, P. McKenna, C. D. Murphy, D. Rusby, and L. Wilson. High efficiency proton beam generation through target thickness control in femtosecond laser–plasma interactions. *Appl. Phys. Lett.*, 104:214101, 2014. doi: 10.1063/1.4879641.

D. Haberberger, S. Tochitsky, F. Fiuza, C. Gong, R. A. Fonseca, L. O. Silva, W. B. Mori, and C. Joshi. Collisionless shocks in laser-produced plasma generate monoenergetic high-energy proton beams. *Nat. Phys.*, 8:95–99, 2012. doi: 10.1038/nphys2130.

D. Haberberger, S. Tochitsky, and C. Joshi. Fifteen terawatt picosecond CO_2 laser system. *Opt. Express*, 18:17865–17875, 2010. doi: 10.1364/OE.18.017865.

B. M. Hegelich, B. J. Albright, J. Cobble, K. Flippo, S. Letring, M. Paffett, H. Ruhl, J. Schreiber, R. K. Schulze, and J. C. Fernández. Laser acceleration of quasi-monoenergetic MeV ion beams. *Nature*, 439:441, 2006. doi: 10.1038/nature04400.

B. Hegelich, D. Jung, B. Albright, J. Fernández, D. Gautier, C. Huang, T. Kwan, S. Letzring, S. Palaniyappan, R. Shah, H.-C. Wu, L. Yin, A. Henig, R. Hrlein, D. Kiefer, J. Schreiber, X. Yan, T. Tajima, D. Habs, B. Dromey, and J. Honrubia. Experimental demonstration of particle energy, conversion efficiency and spectral shape required for ion-based fast ignition. *Nucl. Fusion*, 51:083011, 2011. doi: 10.1088/0029-5515/51/8/083011.

B. M. Hegelich, I. Pomerantz, L. Yin, H. C. Wu, D. Jung, B. J. Albright, D. C. Gautier, S. Letzring, S. Palaniyappan, R. Shah, K. Allinger, R. Hrlein, J. Schreiber, D. Habs, J. Blakeney, G. Dyer, L. Fuller, E. Gaul, E. Mccary, A. R. Meadows, C. Wang, T. Ditmire, and J. C. Fernández. Laser-driven ion acceleration from relativistically transparent nanotargets. *N. J. Phys.*, 15:085015, 2013. doi: 10.1088/1367-2630/15/8/085015.

M. Hegelich, S. Karsch, G. Pretzler, D. Habs, K. Witte, W. Guenther, M. Allen, A. Blazevic, J. Fuchs, J. C. Gauthier, M. Geissel, P. Audebert, T. Cowan, and M. Roth. Mev ion jets from short-pulse-laser interaction with thin foils. *Phys. Rev. Lett.*, 89:085002, 2002. doi: 10.1103/PhysRevLett.09.005002.

A. Henig, D. Kiefer, K. Markey, D. C. Gautier, K. A. Flippo, S. Letzring, R. P. Johnson, T. Shimada, L. Yin, B. J. Albright, K. J. Bowers, J. C. Fernández, S. G. Rykovanov, H.-C. Wu, M. Zepf, D. Jung, V. K. Liechtenstein, J. Schreiber, D. Habs, and B. M. Hegelich. Enhanced laser-driven ion acceleration in the relativistic transparency regime. *Phys. Rev. Lett.*, 103:045002, 2009. doi: 10.1103/PhysRevLett.103.045002.

A. Higginson, R. J. Gray, M. King, R. J. Dance, S. D. R. Williamson, N. M. H. Butler, R. Wilson, R. Capdessus, C. Armstrong, J. S. Green, S. J. Hawkes, P. Martin, W. Q. Wei, S. R. Mirfayzi, X. H. Yuan, S. Kar, M. Borghesi, R. J. Clarke, D. Neely, and P. McKenna. Near-100 MeV protons via a laser-driven transparency-enhanced hybrid acceleration scheme. *Nature Communications*, 2018, in press.

K. M. Hofmann, S. Schell, and J. J. Wilkens. Laser-driven beam lines for delivering intensity modulated radiation therapy with particle beams. *J. Biophoton.*, 5(11–12):903–911, 2012. doi: 10.1002/jbio.201200078.

Y. Huang, N. Wang, X. Tang, and Y. Shi. Relativistic plasma expansion with Maxwell–Jüttner distribution. *Phys. Plasmas*, 20:113108, 2013. doi: 10.1063/1.4834496.

D. Jung, B. J. Albright, L. Yin, D. C. Gautier, R. Shah, S. Palaniyappan, S. Letzring, B. Dromey, H.-C. Wu, T. Shimada, R. P. Johnson, M. Roth, J. C. Fernández, D. Habs, and B. M. Hegelich. Beam profiles of proton and carbon ions in the relativistic transparency regime. *N. J. Phys.*, 15:123035, 2013a. doi: 10.1088/1367-2630/15/12/123035.

D. Jung, L. Yin, B. J. Albright, D. C. Gautier, S. Letzring, B. Dromey, M. Yeung, R. Hrlein, R. Shah, S. Palaniyappan, K. Allinger, J. Schreiber, K. J. Bowers, H.-C. Wu, J. C. Fernández, D. Habs, and B. M. Hegelich. Efficient carbon ion beam generation from laser-driven volume acceleration. *N. J. Phys.*, 15:023007, 2013b. doi: 10.1088/1367-2630/15/2/023007.

S. Kar. Beam focusing and accelerating system (patent 20160379793), December 2016. URL http://www.freepatentsonline.com/y2016/0379793.html.

S. Kar, H. Ahmed, R. Prasad, M. Cerchez, S. Brauckmann, B. Aurand, G. Cantono, P. Hadjisolomou, C. L. S. Lewis, A. Macchi, G. Nersisyan, A. P. L. Robinson, A. Schroer, M. Swantusch, M. Zepf, O. Willi, and M. Borghesi. Guided post-acceleration of laser driven ions by a miniature modular structure. *Nat. Comm.*, 7:10792, 2016. doi: 10.1038/ncomms10792.

S. Kar, M. Borghesi, S. V. Bulanov, M. H. Key, T. V. Liseykina, A. Macchi, A. J. Mackinnon, P. K. Patel, L. Romagnani, A. Schiavi, and O. Willi. Plasma jets driven by ultraintense-laser interaction with thin foils. *Phys. Rev. Lett.*, 100:225004, 2008a. doi: 10.1103/Phys-RevLett.100.225004.

S. Kar, K. F. Kakolee, M. Cerchez, D. Doria, A. Macchi, P. McKenna, D. Neely, J. Osterholz, K. Quinn, B. Ramakrishna, G. Sarri, O. Willi, X. H. Yuan, M. Zepf, and M. Borghesi. Experimental investigation of hole boring and light sail regimes of rpa by varying laser and target parameters. *Plasma Phys. Controll. Fusion*, 55:124030, 2013. doi: 10.1088/0741-3335/55/12/124030.

S. Kar, K. F. Kakolee, B. Qiao, A. Macchi, M. Cerchez, D. Doria, M. Geissler, P. McKenna, D. Neely, J. Osterholz, R. Prasad, K. Quinn, B. Ramakrishna, G. Sarri, O. Willi, X. Y. Yuan, M. Zepf, and M. Borghesi. Ion acceleration in multispecies targets driven by intense laser radiation pressure. *Phys. Rev. Lett.*, 109:185006, 2012. doi: 10.1103/Phys-RevLett.109.185006.

S. Kar, K. Markey, M. Borghesi, D. C. Carroll, P. McKenna, D. Neely, M. N. Quinn, and M. Zepf. Ballistic focusing of polyenergetic protons driven by petawatt laser pulses. *Phys. Rev. Lett.*, 106:225003, 2011. doi: 10.1103/PhysRevLett.106.225003.

S. Kar, K. Markey, P. T. Simpson, C. Bellei, J. S. Green, S. R. Nagel, S. Kneip, D. C. Carroll, B. Dromey, L. Willingale, E. L. Clark, P. McKenna, Z. Najmudin, K. Krushelnick, P. Norreys, R. J. Clarke, D. Neely, M. Borghesi, and M. Zepf. Dynamic control of laser-produced proton beams. *Phys. Rev. Lett.*, 100:105004, 2008b. doi: 10.1103/PhysRevLett.100.105004.

T. Kiefer, T. Schlegel, and M. C. Kaluza. Plasma expansion into vacuum assuming a step-like electron energy distribution. *Phys. Rev. E*, 87:043110, 2013. doi: 10.1103/Phys-RevE.87.043110.

I. J. Kim, K. H. Pae, I. W. Choi, C.-L. Lee, H. T. Kim, H. Singhal, J. H. Sung, S. K. Lee, H. W. Lee, P. V. Nickles, T. M. Jeong, C. M. Kim, and C. H. Nam. Radiation pressure acceleration of protons to 93MeV with circularly polarized petawatt laser pulses. *Phys. Plasmas*, 23:070701, 2016. doi: 10.1063/1.4958654.

I. J. Kim, K. H. Pae, C. M. Kim, H. T. Kim, J. H. Sung, S. K. Lee, T. J. Yu, I. W. Choi, C.-L. Lee, K. H. Nam, P. V. Nickles, T. M. Jeong, and J. Lee. Transition of proton energy scaling using an ultrathin target irradiated by linearly polarized femtosecond laser pulses. *Phys. Rev. Lett.*, 111:165003, 2013. doi: 10.1103/PhysRevLett.111.165003.

O. Klimo, J. Psikal, J. Limpouch, and V. T. Tikhonchuk. Monoenergetic ion beams from ultrathin foils irradiated by ultrahigh-contrast circularly polarized laser pulses. *Phys. Rev. ST Accel. Beams*, 11:031301, 2008. doi: 10.1103/PhysRevSTAB.11.031301.

G. F. Knoll. *Radiation Detection and Measurement.* John Wiley & Sons, New York, 4th ed. edition, 2010. ISBN 0470131489.

K. W. D. Ledingham, P. McKenna, and R. P. Singhal. Applications for nuclear phenomena generated by ultra-intense lasers. *Science*, 300:1107–1111, 2003. doi: 10.1126/sci-ence.1080552.

U. Linz and J. Alonso. What will it take for laser driven proton accelerators to be applied to tumor therapy? *Phys. Rev. ST Accel. Beams*, 10:094801, 2007. doi: 10.1103/Phys-RevSTAB.10.094801.

U. Linz and J. Alonso. Laser-driven ion accelerators for tumor therapy revisited. *Phys. Rev. Accel. Beams*, 19:124802, Dec 2016. doi: 10.1103/PhysRevAccelBeams.19.124802.

W. Ma, V. Liechtenstein, J. Szerypo, D. Jung, P. Hilz, B. Hegelich, H. Maier, J. Schreiber, and D. Habs. Preparation of self-supporting diamond-like carbon nanofoils with thickness less than 5 nm for laser-driven ion acceleration. *Nucl. Inst. Meth. Phys. Res. A*, 655:53–56, 2011. doi: 10.1016/j.nima.2011.06.019.

A. Macchi. *A Superintense Laser–Plasma Interaction Theory Primer.* SpringerBriefs in Physics. Springer, 2013. doi: 10.1007/978-94-007-6125-4.

A. Macchi. Theory of light sail acceleration by intense lasers: An overview. *High Power Laser Sci. Eng.*, 2:e10, 2014. doi: 10.1017/hpl.2014.13.

A. Macchi and C. Benedetti. Ion acceleration by radiation pressure in thin and thick targets. *Nucl. Inst. Meth. Phys. Res. A*, 620:41–45, 2010. doi: 10.1016/j.nima.2010.01.057.

A. Macchi, M. Borghesi, and M. Passoni. Ion acceleration by superintense laser–plasma interaction. *Rev. Mod. Phys.*, 85:751–793, 2013a. doi: 10.1103/RevModPhys.85.751.

A. Macchi, F. Cattani, T. V. Liseykina, and F. Cornolti. Laser acceleration of ion bunches at the front surface of overdense plasmas. *Phys. Rev. Lett.*, 94:165003, 2005. doi: 10.1103/PhysRevLett.94.165003.

A. Macchi, A. S. Nindrayog, and F. Pegoraro. Solitary versus shock wave acceleration in laser–plasma interactions. *Phys. Rev. E*, 85:046402, 2012. doi: 10.1103/PhysRevE.85.046402.

A. Macchi, A. Sgattoni, S. Sinigardi, M. Borghesi, and M. Passoni. Advanced strategies for ion acceleration using high-power lasers. *Plasma Phys. Contr. Fusion*, 55:124020, 2013b. doi: 10.1088/0741-3335/55/12/124020.

A. Macchi, S. Veghini, and F. Pegoraro. "Light Sail" acceleration reexamined. *Phys. Rev. Lett.*, 103:085003, 2009. doi: 10.1103/PhysRevLett.103.085003.

A. J. Mackinnon, P. K. Patel, R. P. Town, M. J. Edwards, T. Phillips, S. C. Lerner, D. W. Price, D. Hicks, M. H. Key, S. Hatchett, S. C. Wilks, M. Borghesi, L. Romagnani, S. Kar, T. Toncian, G. Pretzler, O. Willi, M. Koenig, E. Martinolli, S. Lepape, A. Benuzzi-Mounaix, P. Audebert, J. C. Gauthier, J. King, R. Snavely, R. R. Freeman, and T. Boehlly. Proton radiography as an electromagnetic field and density perturbation diagnostic (invited). *Rev. Sci. Instrum.*, 75:3531–3536, 2004. doi: 10.1063/1.1788893.

A. Maksimchuk, S. Gu, K. Flippo, D. Umstadter, and V. Y. Bychenkov. Forward ion acceleration in thin films driven by a high-intensity laser. *Phys. Rev. Lett.*, 84:4108–4111, 2000. doi: 10.1103/PhysRevLett.84.4108.

V. Malka, S. Fritzler, E. Lefebvre, E. d'Humières, R. Ferrand, G. Grillon, C. Albaret, S. Meyroneinc, J.-P. Chambaret, A. Antonetti, and D. Hulin. Practicability of proton therapy using compact laser systems. *Med. Phys.*, 31:1587–1592, 2004. doi: 10.1118/1.1747751.

D. Margarone, I. J. Kim, J. Psikal, J. Kaufman, T. Mocek, I. W. Choi, L. Stolcova, J. Proska, A. Choukourov, I. Melnichuk, O. Klimo, J. Limpouch, J. H. Sung, S. K. Lee, G. Korn, and T. M. Jeong. Laser-driven high-energy proton beam with homogeneous spatial profile from a nanosphere target. *Phys. Rev. ST Accel. Beams*, 18:071304, 2015. doi: 10.1103/Phys-RevSTAB.18.071304.

D. Margarone, O. Klimo, I. J. Kim, J. Prokůpek, J. Limpouch, T. M. Jeong, T. Mocek, J. Pšikal, H. T. Kim, J. Proška, K. H. Nam, L. Stolcová, I. W. Choi, S. K. Lee, J. H. Sung, T. J. Yu, and G. Korn. Laser-driven proton acceleration enhancement by nanostructured foils. *Phys. Rev. Lett.*, 109:234801, 2012. doi: 10.1103/PhysRevLett.109.234801.

D. Margarone, A. Velyhan, J. Dostal, J. Ullschmied, J. P. Perin, D. Chatain, S. Garcia, P. Bonnay, T. Pisarczyk, R. Dudzak, M. Rosinski, J. Krasa, L. Giuffrida, J. Prokupek, V. Scuderi, J. Psikal, M. Kucharik, M. De Marco, J. Cikhardt, E. Krousky, Z. Kalinowska, T. Chodukowski, G. A. P. Cirrone, and G. Korn. Proton acceleration driven by a nanosecond laser from a cryogenic thin solid-hydrogen ribbon. *Phys. Rev. X*, 6:041030, 2016. doi: 10.1103/PhysRevX.6.041030.

K. Markey, P. McKenna, C. M. Brenner, D. C. Carroll, M. M. Günther, K. Harres, S. Kar, K. Lancaster, F. Nürnberg, M. N. Quinn, A. P. L. Robinson, M. Roth, M. Zepf, and D. Neely. Spectral enhancement in the double pulse regime of laser proton acceleration. *Phys. Rev. Lett.*, 105:195008, 2010. doi: 10.1103/PhysRevLett.105.195008.

G. Marx. Interstellar vehicle propelled by terrestrial laser beam. *Nature*, 211:22–23, 1966. doi: 10.1038/211022a0.

Z. Merali. Shooting for a star. *Science*, 352(6289):1040–1041, 2016. doi: 10.1126/sci-ence.352.6289.1040.

P. Mora. Plasma expansion into a vacuum. *Phys. Rev. Lett.*, 90:185002, 2003. doi: 10.1103/PhysRevLett.90.185002.

P. Mora. Thin-foil expansion into a vacuum. *Phys. Rev. E*, 72:056401, 2005. doi: 10.1103/PhysRevE.72.056401.

P. Mulser and D. Bauer. *High Power Laser-Matter Interaction*, volume 238 of *Springer Tracts in Modern Physics*. Springer, 2010.

P. Mulser, D. Bauer, and H. Ruhl. Collisionless laser-energy conversion by anharmonic resonance. *Phys. Rev. Lett.*, 101:225002, 2008. doi: 10.1103/PhysRevLett.101.225002.

D. Neely, P. Foster, A. Robinson, F. Lindau, O. Lundh, A. Persson, C.-G. Wahlström, and P. McKenna. Enhanced proton beams from ultrathin targets driven by high contrast laser pulses. *Appl. Phys. Lett.*, 89:021502, 2006. doi: 10.1063/1.2220011.

K. Nemoto, A. Maksimchuk, S. Banerjee, K. Flippo, G. Mourou, D. Umstadter, and V. Y. Bychenkov. Laser-triggered ion acceleration and table top isotope production. *Appl. Phys. Lett.*, 78:595–597, 2001. doi: 10.1063/1.1343845.

F. Nuernberg, M. Schollmeier, E. Brambrink, A. Blažević, D. C. Carroll, K. Flippo, D. C. Gautier, M. Geissel, K. Harres, B. M. Hegelich, O. Lundh, K. Markey, P. McKenna, D. Neely, J. Schreiber, and M. Roth. Radiochromic film imaging spectroscopy of laser-accelerated proton beams. *Rev. Sci. Instrum.*, 80:033301, 2009. doi: 10.1063/1.3086424.

K. Ogura, M. Nishiuchi, A. S. Pirozhkov, T. Tanimoto, A. Sagisaka, T. Z. Esirkepov, M. Kando, T. Shizuma, T. Hayakawa, H. Kiriyama, T. Shimomura, S. Kondo, S. Kanazawa, Y. Nakai, H. Sasao, F. Sasao, Y. Fukuda, H. Sakaki, M. Kanasaki, A. Yogo, S. V. Bulanov, P. R. Bolton, and K. Kondo. Proton acceleration to 40 MeV using a high intensity, high contrast optical parametric chirped-pulse amplification/Ti:sapphire hybrid laser system. *Opt. Lett.*, 37:2868–2870, 2012. doi: 10.1364/OL.37.002868.

T. M. Ostermayr, D. Haffa, P. Hilz, V. Pauw, K. Allinger, K.-U. Bamberg, P. Böhl, C. Bömer, P. R. Bolton, F. Deutschmann, T. Ditmire, M. E. Donovan, G. Dyer, E. Gaul, J. Gordon, B. M. Hegelich, D. Kiefer, C. Klier, C. Kreuzer, M. Martinez, E. McCary, A. R. Meadows, N. Moschüring, T. Rösch, H. Ruhl, M. Spinks, C. Wagner, and J. Schreiber. Proton acceleration by irradiation of isolated spheres with an intense laser pulse. *Phys. Rev. E*, 94:033208, 2016. doi: 10.1103/PhysRevE.94.033208.

S. Palaniyappan, C. Huang, D. C. Gautier, C. E. Hamilton, M. A. Santiago, C. Kreuzer, A. B. Sefkow, R. C. Shah, and J. C. Fernández. Efficient quasi-monoenergetic ion beams from laser-driven relativistic plasmas. *Nat. Comm.*, 6, 2015. doi: 10.1038/ncomms10170.

C. A. J. Palmer, N. P. Dover, I. Pogorelsky, M. Babzien, G. I. Dudnikova, M. Ispiriyan, M. N. Polyanskiy, J. Schreiber, P. Shkolnikov, V. Yakimenko, and Z. Najmudin. Monoenergetic proton beams accelerated by a radiation pressure driven shock. *Phys. Rev. Lett.*, 106: 014801, 2011. doi: 10.1103/PhysRevLett.106.014801.

C. A. J. Palmer, J. Schreiber, S. R. Nagel, N. P. Dover, C. Bellei, F. N. Beg, S. Bott, R. J. Clarke, A. E. Dangor, S. M. Hassan, P. Hilz, D. Jung, S. Kneip, S. P. D. Mangles, K. L. Lancaster, A. Rehman, A. P. L. Robinson, C. Spindloe, J. Szerypo, M. Tatarakis, M. Yeung, M. Zepf, and Z. Najmudin. Rayleigh-Taylor instability of an ultrathin foil accelerated by the radiation pressure of an intense laser. *Phys. Rev. Lett.*, 108:225002, 2012. doi: 10.1103/PhysRevLett.108.225002.

M. Passoni, A. Sgattoni, I. Prencipe, L. Fedeli, D. Dellasega, L. Cialfi, I. W. Choi, I. J. Kim, K. A. Janulewicz, H. W. Lee, J. H. Sung, S. K. Lee, and C. H. Nam. Toward high-energy laser-driven ion beams: Nanostructured double-layer targets. *Phys. Rev. Accel. Beams*, 19: 061301, 2016. doi: 10.1103/PhysRevAccelBeams.19.061301.

M. Passoni, A. Zani, A. Sgattoni, D. Dellasega, A. Macchi, I. Prencipe, V. Floquet, P. Martin, T. V. Liseykina, and T. Ceccotti. Energetic ions at moderate laser intensities using foam-based multi-layered targets. *Plasma Phys. Controll. Fusion*, 56:045001, 2014. doi: 10.1088/0741-3335/56/4/045001.

P. K. Patel, A. J. Mackinnon, M. H. Key, T. E. Cowan, M. E. Foord, M. Allen, D. F. Price, H. Ruhl, P. T. Springer, and R. Stephens. Isochoric heating of solid-density matter with an ultrafast proton beam. *Phys. Rev. Lett.*, 91:125004, 2003. doi: 10.1103/Phys-RevLett.91.125004.

S. M. Pfotenhauer, O. Jckel, A. Sachtleben, J. Polz, W. Ziegler, H.-P. Schlenvoigt, K.-U. Amthor, M. C. Kaluza, K. W. D. Ledingham, R. Sauerbrey, P. Gibbon, A. P. L. Robinson, and H. Schwoerer. Spectral shaping of laser generated proton beams. *N. J. Phys.*, 10: 033034, 2008. doi: 10.1088/1367-2630/10/3/033034.

I. V. Pogorelsky, M. Babzien, I. Ben-Zvi, J. Skaritka, and M. N. Polyanskiy. BESTIA the next generation ultra-fast CO_2 laser for advanced accelerator research. *Nucl. Inst. Meth. Phys. Res. A*, 829:432–437, 2016. doi: 10.1016/j.nima.2015.11.126.

H. W. Powell, M. King, R. J. Gray, D. A. MacLellan, B. Gonzalez-Izquierdo, L. C. Stockhausen, G. Hicks, N. P. Dover, D. R. Rusby, D. C. Carroll, H. Padda, R. Torres, S. Kar, R. J. Clarke, I. O. Musgrave, Z. Najmudin, M. Borghesi, D. Neely, and P. McKenna. Proton acceleration enhanced by a plasma jet in expanding foils undergoing relativistic transparency. *N. J. Phys.*, 17:103033, 2015. doi: 10.1088/1367-2630/17/10/103033.

I. Prencipe, A. Sgattoni, D. Dellasega, L. Fedeli, L. Cialfi, I. W. Choi, I. J. Kim, K. A. Janulewicz, K. F. Kakolee, H. W. Lee, J. H. Sung, S. K. Lee, C. H. Nam, and M. Passoni. Development of foam-based layered targets for laser-driven ion beam production. *Plasma Phys. Contr. Fusion*, 58:034019, 2016. doi: 10.1088/0741-3335/58/3/034019.

B. Qiao, S. Kar, M. Geissler, P. Gibbon, M. Zepf, and M. Borghesi. Dominance of radiation pressure in ion acceleration with linearly polarized pulses at intensities of 10^{21} W cm^{-2}. *Phys. Rev. Lett.*, 108:115002, 2012. doi: 10.1103/PhysRevLett.108.115002.

B. Qiao, M. Zepf, M. Borghesi, B. Dromey, M. Geissler, A. Karmakar, and P. Gibbon. Radiation-pressure acceleration of ion beams from nanofoil targets: The leaky light-sail regime. *Phys. Rev. Lett.*, 105:155002, 2010. doi: 10.1103/PhysRevLett.105.155002.

K. Quinn, P. A. Wilson, C. A. Cecchetti, B. Ramakrishna, L. Romagnani, G. Sarri, L. Lancia, J. Fuchs, A. Pipahl, T. Toncian, O. Willi, R. J. Clarke, D. Neely, M. Notley, P. Gallegos, D. C. Carroll, M. N. Quinn, X. H. Yuan, P. McKenna, T. V. Liseykina, A. Macchi, and M. Borghesi. Laser-driven ultrafast field propagation on solid surfaces. *Phys. Rev. Lett.*, 102:194801, 2009. doi: 10.1103/PhysRevLett.102.194801.

A. P. L. Robinson. Production of high energy protons with hole-boring radiation pressure acceleration. *Phys. Plasmas*, 18:056701, 2011. doi: 10.1063/1.3562551.

A. P. L. Robinson, P. Gibbon, M. Zepf, S. Kar, R. G. Evans, and C. Bellei. Relativistically correct hole-boring and ion acceleration by circularly polarized laser pulses. *Plasma Phys. Contr. Fusion*, 51:024004, 2009, doi: 10.1088/0741-3335/51/2/024004

A. P. L. Robinson, R. M. G. M. Trines, N. P. Dover, and Z. Najmudin. Hole-boring radiation pressure acceleration as a basis for producing high-energy proton bunches. *Plasma Phys. Contr. Fusion*, 54:115001, 2012. doi: 10.1088/0741-3335/54/11/115001.

A. P. L. Robinson, M. Zepf, S. Kar, R. G. Evans, and C. Bellei. Radiation pressure acceleration of thin foils with circularly polarized laser pulses. *N. J. Phys.*, 10:013021, 2008. doi: 10.1088/1367-2630/10/1/013021.

L. Robson, P. T. Simpson, R. J. Clarke, K. W. D. Ledingham, F. Lindau, O. Lundh, T. McCanny, P. Mora, D. Neely, C.-G. Wahlstrom, M. Zepf, and P. McKenna. Scaling of proton acceleration driven by petawatt-laser–plasma interactions. *Nat. Phys.*, 3:58–62, 2007. doi: 10.1038/nphys476.

L. Romagnani, J. Fuchs, M. Borghesi, P. Antici, P. Audebert, F. Ceccherini, T. Cowan, T. Grismayer, S. Kar, A. Macchi, P. Mora, G. Pretzler, A. Schiavi, T. Toncian, and O. Willi. Dynamics of electric fields driving the laser acceleration of multi-MeV protons. *Phys. Rev. Lett.*, 95:195001, 2005. doi: 10.1103/PhysRevLett.95.195001.

M. Roth, A. Blazevic, M. Geissel, T. Schlegel, T. E. Cowan, M. Allen, J.-C. Gauthier, P. Audebert, J. Fuchs, M. ter Vehn, M. Hegelich, S. Karsch, and A. Pukhov. Energetic ions generated by laser pulses: A detailed study on target properties. *Phys. Rev. ST Accel. Beams*, 5:061301, 2002. doi: 10.1103/PhysRevSTAB.5.061301.

M. Roth, T. E. Cowan, M. H. Key, S. P. Hatchett, C. Brown, W. Fountain, J. Johnson, D. M. Pennington, R. A. Snavely, S. C. Wilks, K. Yasuike, H. Ruhl, F. Pegoraro, S. V. Bulanov, E. M. Campbell, M. D. Perry, and H. Powell. Fast ignition by intense laser-accelerated proton beams. *Phys. Rev. Lett.*, 86:436–439, 2001. doi: 10.1103/PhysRevLett.86.436.

H. Ruhl, S. V. Bulanov, T. E. Cowan, T. V. Liseikina, P. Nickles, F. Pegoraro, M. Roth, and W. Sandner. Computer simulation of the three-dimensional regime of proton acceleration in the interaction of laser radiation with a thin spherical target. *Plasma Phys. Rep.*, 27, 2001.

M. I. K. Santala, M. Zepf, F. N. Beg, E. L. Clark, A. E. Dangor, K. Krushelnick, M. Tatarakis, I. Watts, K. W. D. Ledingham, T. McCanny, I. Spencer, A. C. Machacek, R. Allott, R. J. Clarke, and P. A. Norreys. Production of radioactive nuclides by energetic protons generated from intense laser–plasma interactions. *Appl. Phys. Lett.*, 78:19–21, 2001. doi: 10.1063/1.1335849.

J. Schreiber, F. Bell, F. Grüner, U. Schramm, M. Geissler, M. Schnürer, S. Ter-Avetisyan, B. M. Hegelich, J. Cobble, E. Brambrink, J. Fuchs, P. Audebert, and D. Habs. Analytical model for ion acceleration by high-intensity laser pulses. *Phys. Rev. Lett.*, 97:045005, Jul 2006. doi: 10.1103/PhysRevLett.97.045005.

J. Schreiber, P. R. Bolton, and K. Parodi. Invited review article: "Hands-on" laser-driven ion acceleration: A primer for laser-driven source development and potential applications. *Rev. Sci. Instrum.*, 87:071101, 2016. doi: 10.1063/1.4959198.

H. Schwoerer, S. Pfotenhauer, O. Jaeckel, K. U. Amthor, B. Liesfeld, W. Ziegler, R. Sauerbrey, K. W. D. Ledingham, and T. Esirkepov. Laser–plasma acceleration of quasi-monoenergetic protons from microstructured targets. *Nature*, 439:445, 2006. doi: 10.1038/nature04492.

C. Scullion, D. Doria, L. Romagnani, A. Sgattoni, K. Naughton, D. R. Symes, P. McKenna, A. Macchi, M. Zepf, S. Kar, and M. Borghesi. Polarization dependence of bulk ion acceleration from ultra-thin foils irradiated by high-intensity ultrashort laser pulses. *Phys. Rev. Lett.*, 119:054801, 2017. doi: 10.1103/PhysRevLett.119.054801.

M. Seimetz, P. Bellido, R. Lera, A. R. de la Cruz, P. Mur, I. Sánchez, M. Galán, F. Sánchez, L. Roso, and J. M. Benlloch. Proton acceleration with a table-top TW laser. *J. Inst.*, 11: C11012, 2016. doi: 10.1088/1748-0221/11/11/C11012.

A. Sgattoni, S. Sinigardi, and A. Macchi. High energy gain in three-dimensional simulations of light sail acceleration. *Appl. Phys. Lett.*, 105:084105, 2014. doi: 10.1063/1.4894092.

R. A. Snavely, M. H. Key, S. P. Hatchett, T. E. Cowan, M. Roth, T. W. Phillips, M. A. Stoyer, E. A. Henry, T. C. Sangster, M. S. Singh, S. C. Wilks, A. MacKinnon, A. Offenberger, D. M. Pennington, K. Yasuike, A. B. Langdon, B. F. Lasinski, J. Johnson, M. D. Perry, and E. M. Campbell. Intense high-energy proton beams from petawatt-laser irradiation of solids. *Phys. Rev. Lett.*, 85:2945–2948, 2000. doi: 10.1103/PhysRevLett.85.2945.

T. Sokollik, M. Schnürer, S. Steinke, P. V. Nickles, W. Sandner, M. Amin, T. Toncian, O. Willi, and A. A. Andreev. Directional laser-driven ion acceleration from microspheres. *Phys. Rev. Lett.*, 103:135003, 2009. doi: 10.1103/PhysRevLett.103.135003.

I. Spencer, K. W. D. Ledingham, R. P. Singhal, T. McCanny, P. McKenna, E. L. Clark, K. Krushelnick, M. Zepf, F. N. Beg, M. Tatarakis, A. E. Dangor, P. A. Norreys, R. J. Clarke, R. M. Allott, and I. N. Ross. Laser generation of proton beams for the production of short-lived positron emitting radio-isotopes. *Nucl. Inst. Meth. Phys. Res. B*, 183:449–458, 2001. doi: 10.1016/S0168-583X(01)00771-6.

S. Steinke, P. Hilz, M. Schnürer, G. Priebe, J. Bränzel, F. Abicht, D. Kiefer, C. Kreuzer, T. Ostermayr, J. Schreiber, A. A. Andreev, T. P. Yu, A. Pukhov, and W. Sandner. Stable laser-ion acceleration in the light sail regime. *Phys. Rev. ST Accel. Beams*, 16:011303, 2013. doi: 10.1103/PhysRevSTAB.16.011303.

A. Stockem Novo, M. C. Kaluza, R. A. Fonseca, and L. O. Silva. Optimizing laser-driven proton acceleration from overdense targets. *Scient. Rep.*, 6:29402, 2016. doi: 10.1038/srep29402.

F. Sylla, M. Veltcheva, S. Kahaly, A. Flacco, and V. Malka. Development and characterization of very dense submillimetric gas jets for laser–plasma interaction. *Rev. Sci. Instrum.*, 83:033507, 2012. doi: 10.1063/1.3697859.

M. Tabak, J. Hammer, M. E. Glinsky, W. L. Kruer, S. C. Wilks, J. Woodworth, E. M. Campbell, M. D. Perry, and R. J. Mason. Ignition and high gain with ultrapowerful lasers. *Phys. Plasmas*, 1:1626–1634, 1994. doi: 10.1063/1.870664.

S. Ter-Avetisyan, M. Schnürer, P. V. Nickles, M. Kalashnikov, E. Risse, T. Sokollik, W. Sandner, A. Andreev, and V. Tikhonchuk. Quasimonoenergetic deuteron bursts produced by ultraintense laser pulses. *Phys. Rev. Lett.*, 96:145006, 2006. doi: 10.1103/Phys-RevLett.96.145006.

S. Ter-Avetisyan, B. Ramakrishna, R. Prasad, M. Borghesi, P. V. Nickles, S. Steinke, M. Schnrer, K. I. Popov, L. Ramunno, N. V. Zmitrenko, and V. Yu. Bychenkov. Generation of a quasi-monoergetic proton beam from laser-irradiated sub-micron droplets. *Phys. Plasmas*, 19:073112, 2012. doi: 10.1063/1.4731712.

C. Thaury, F. Quere, J.-P. Geindre, A. Levy, T. Ceccotti, P. Monot, M. Bougeard, F. Reau, P. d'Oliveira, P. Audebert, R. Marjoribanks, and P. Martin. Plasma mirrors for ultrahigh-intensity optics. *Nat. Phys.*, 3:424–429, 2007. doi: 10.1038/nphys595.

D. A. Tidman and N. A. Krall. *Shock Waves in Collisionless Plasmas*, chapter 6. Wiley/Interscience, New York, 1971.

V. T. Tikhonchuk, A. A. Andreev, S. G. Bochkarev, and V. Y. Bychenkov. Ion acceleration in short-laser-pulse interaction with solid foils. *Plasma Phys. Contr. Fusion*, 47:B869, 2005. doi: 10.1088/0741-3335/47/12B/S69.

S. Tokita, S. Sakabe, T. Nagashima, M. Hashida, and S. Inoue. Strong sub-terahertz surface waves generated on a metal wire by high-intensity laser pulses. *Sci. Rep.*, 5:8268, 2015. doi: 10.1038/srep08268.

T. Toncian, M. Borghesi, J. Fuchs, E. d'Humières, P. Antici, P. Audebert, E. Brambrink, C. A. Cecchetti, A. Pipahl, L. Romagnani, and O. Willi. Ultrafast laser driven microlens to focus and energy-select mega-electron volt protons. *Science*, 312:410, 2006. doi: 10.1126/science.1124412.

O. Tresca, N. P. Dover, N. Cook, C. Maharjan, M. N. Polyanskiy, Z. Najmudin, P. Shkolnikov, and I. Pogorelsky. Spectral modification of shock accelerated ions using a hydro-dynamically shaped gas target. *Phys. Rev. Lett.*, 115:094802, 2015. doi: 10.1103/Phys-RevLett.115.094802.

V. Veksler. The principle of coherent acceleration of charged particles. *At. Energy*, 2:525–528, 1957.

D. von der Linde and H. Schüler. Breakdown threshold and plasma formation in femtosecond laser–solid interaction. *J. Opt. Soc. Am. B*, 13:216–222, 1996. doi: 10.1364/JOSAB.13.000216.

V. A. Vshivkov, N. M. Naumova, F. Pegoraro, and S. V. Bulanov. Nonlinear electrodynamics of the interaction of ultra-intense laser pulses with a thin foil. *Phys. Plasmas*, 5:2727–2741, 1998. doi: 10.1063/1.872961.

F. Wagner, S. Bedacht, V. Bagnoud, O. Deppert, S. Geschwind, R. Jaeger, A. Ortner, A. Tebartz, B. Zielbauer, D. H. H. Hoffmann, and M. Roth. Simultaneous observation of angularly separated laser-driven proton beams accelerated via two different mechanisms. *Phys. Plasmas*, 22:063110, 2015. doi: 10.1063/1.4922661.

F. Wagner, O. Deppert, C. Brabetz, P. Fiala, A. Kleinschmidt, P. Poth, V. A. Schanz, A. Tebartz, B. Zielbauer, M. Roth, T. Stöhlker, and V. Bagnoud. Maximum proton energy above 85 MeV from the relativistic interaction of laser pulses with micrometer thick CH_2 targets. *Phys. Rev. Lett.*, 116:205002, 2016. doi: 10.1103/PhysRevLett.116.205002.

S. C. Wilks, A. B. Langdon, T. E. Cowan, M. Roth, M. Singh, S. Hatchett, M. H. Key, D. Pennington, A. MacKinnon, and R. A. Snavely. Energetic proton generation in ultra-intense laser-solid interactions. *Phys. Plasmas*, 8:542, 2001. doi: 10.1063/1.1333697.

L. Willingale, S. P. D. Mangles, P. M. Nilson, R. J. Clarke, A. E. Dangor, M. C. Kaluza, S. Karsch, K. L. Lancaster, W. B. Mori, Z. Najmudin, J. Schreiber, A. G. R. Thomas, M. S. Wei, and K. Krushelnick. Collimated multi-MeV ion beams from high-intensity laser interactions with underdense plasma. *Phys. Rev. Lett.*, 96:245002, 2006. doi: 10.1103/Phys-RevLett.96.245002.

L. Willingale, S. R. Nagel, A. G. R. Thomas, C. Bellei, R. J. Clarke, A. E. Dangor, R. Heathcote, M. C. Kaluza, C. Kamperidis, S. Kneip, K. Krushelnick, N. Lopes, S. P. D. Mangles, W. Nazarov, P. M. Nilson, and Z. Najmudin. Characterization of high-intensity laser propagation in the relativistic transparent regime through measurements of energetic proton beams. *Phys. Rev. Lett.*, 102:125002, 2009. doi: 10.1103/PhysRevLett.102.125002.

A. Zani, D. Dellasega, V. Russo, and M. Passoni. Ultra-low density carbon foams produced by pulsed laser deposition. *Carbon*, 56:358–365, 2013. doi: 10.1016/j.carbon.2013.01.029.

K. Zeil, S. D. Kraft, S. Bock, M. Bussmann, T. E. Cowan, T. Kluge, J. Metzkes, T. Richter, R. Sauerbrey, and U. Schramm. The scaling of proton energies in ultrashort pulse laser plasma acceleration. *N. J. Phys.*, 12:045015, 2010. doi: 10.1088/1367-2630/12/4/045015.

K. Zeil, J. Metzkes, T. Kluge, M. Bussmann, T. Cowan, S. Kraft, R. Sauerbrey, and U. Schramm. Direct observation of prompt pre-thermal laser ion sheath acceleration. *Nat. Comm.*, 3:874, 2012. doi: 10.1038/ncomms1883.

K. Zeil, J. Metzkes, T. Kluge, M. Bussmann, T. E. Cowan, S. D. Kraft, R. Sauerbrey, B. Schmidt, M. Zier, and U. Schramm. Robust energy enhancement of ultrashort pulse laser accelerated protons from reduced mass targets. *Plasma Physics and Controlled Fusion*, 56:084004, 2014. doi: 10.1088/0741-3335/56/8/084004.

X. Zhang, B. Shen, X. Li, Z. Jin, and F. Wang. Multistaged acceleration of ions by circularly polarized laser pulse: Monoenergetic ion beam generation. *Phys. Plasmas*, 14:073101, 2007. doi: 10.1063/1.2746810.

J. Ziegler, J. Biersack, and M. Ziegler. *Stopping and range of ions in matter*. SRIM Co., 2008. ISBN 9780965420716.

6

Neutron Generation

Markus Roth

6.1 Introduction to Laser-Driven Neutron Generation

Ultra-intense lasers have demonstrated the capability of accelerating short and intense bursts of ions [Snavely 2000, Borghesi 2006]. These ion bunches have been used to generate short bursts of neutrons by irradiating a converter in close proximity to the source, making this scheme a very compact and bright source of neutrons of up to more than 100 MeV in energy [Roth 2013]. Using novel laser ion acceleration mechanisms, directed bunches of neutrons can be generated, which increases the brightness of these sources when compared to previous attempts. We review the recent research and present experimental data using a mechanism based on the relativistic transparency to drive the most intense laser-driven neutron sources and use them for initial applications.

Neutrons in general can be released from nuclei using energetic radiation. Electromagnetic radiation can excite collective modes that yield to neutron evaporation. The impact by energetic electrons or ions can release neutrons from the nucleus, or a high velocity ion can break up in the Coulomb field of another atom releasing one or more neutrons. In addition, neutrons can be released as a consequence of fusion reactions, as with present inertial confinement fusion experiments or by using high-energy proton beams for spallation of heavy target nuclei. The promise of future ultra-intense lasers as drivers for brilliant, compact and highly efficient particle accelerators for electrons and ions (potentially replacing in some cases much larger conventional accelerators) also ushers prospects for driving next generation neutron sources. While laser-driven sources at present certainly cannot compete with accelerator-driven sources or high flux reactors in terms of average neutron flux (we view peak flux as: # of particles/bunch duration and average flux as # of particles/second, regardless of temporal substructure) for medium size neutron sources, as well as for high peak neutron sources, they can complement large-scale facilities and be easily attached to secondary drivers, such as accelerators, lasers or pulsed power machines.

6.2 Previous Research

In the following, we do not address the production of neutrons by thermonuclear fusion reactions, as this usually requires MJ (megajoule) laser systems the size of a 'football stadium' (NIF is roughly 100 × 170 m. [Hurricane 2014]). Instead we will focus on neutron production using compact, high-powered short pulse lasers.

6.2.1 Photon-Induced Reactions

Early experiments used excitation of giant resonances in nuclei to determine laser on-target intensities [Cowan 2000, Guenther 2012]. This research was motivated by the ultimate question "How much laser intensity actually reaches the critical target surface?". The interaction of the laser is driving a hot electron component into the target due to the ponderomotive force. The hot electron energy distribution function can be estimated according to Wilks [Wilks 1992] to scale with the laser strength parameter $a_0 = eE_L/m_e\omega_L c$, resulting in a temperature equivalent of $k_B T_h = m_e c^2((1+ a_0^2)^{1/2}-1$. The laser strength parameter is linked to the laser intensity by $a_0 \cong 0.85 \times 10^{-9}\ \lambda_L$ [μm] $(I_L[W/cm^2])^{1/2}$ and becomes of order of one at high intensities near $10^{18} W/cm^2$.

The Bremsstrahlung photons from this high-energy (i.e. hot) electron component therefore directly scales with the laser intensity; the higher the intensity, the hotter the electron component. The range of the Bremsstrahlung spectrum thereby can exceed the threshold of photon-induced nuclear reactions. Diagnostics based on photoneutron disintegration reactions have been used to determine the laser intensity and reactions up to $(\gamma, 7n)$ in gold have been observed indicating the presence of 'hard' photons above 60 MeV [Cowan 2000], as can be seen in Figure 6.1.

Recently, photo-induced neutrons have gained new attention as new short pulse lasers have produced very short neutron bursts using low density targets in a pitcher–catcher geometry [Pomerantz 2014].

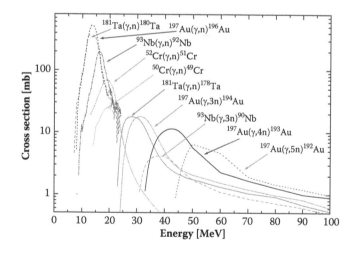

FIGURE 6.1 Energy-dependent cross section for photonuclear (γ, xn) reactions (1mb = $10^{-27}cm^2$) for different isotopes. (Reproduced with permission from Guenther, M.M. et al., 2011. "A novel nuclear pyrometry for the characterization of high-energy bremsstrahlung and electrons produced in relativistic laser plasma interactions" *Phys. of Plasmas* 18: 083102.)

6.2.2 Cluster Fusion

In the late 1990s, the production of neutrons from short pulse lasers reached a new level using cluster targets. The interaction of a short, intense laser pulse with deuteron clusters of an optimum size of around 50 Angstrom resulted in cluster coulomb explosion and the subsequent generation of d-d fusion neutrons from a compact source. In the early experiments in 1999, about 10^5 n per shot were produced [Ditmire 1999]. This method has been improved, resulting in a yield of 10^7 n/shot at the Texas Petawatt Laser [Bang 2013] that can deliver 120 J in a 170 fs pulse duration. The source size in these experiments is of order only a few tens of microns (which is the focal region in the cluster volume).

6.2.3 Ion Beam-Driven Fusion

Due to the large cross section of neutron producing reactions, ion-induced neutron sources have been widely used with conventional accelerators (see Figure 6.3 and a further description of conventional neutron sources in Chapter 20). The advent of laser-driven ion bunches spurred the idea of compact neutron sources. The first experiments in 2000 resulted in 4×10^8 n/shot using 15 J laser pulses provided by the laser system at Laboratoire pour l'Utilisation des Lasers Intenses (LULI). In these first experiments that used a deuterated plastic target (the 'pitcher') and a titanium 'catcher' highly doped with deuterons, the 2.45 MeV neutron signal from d-d- fusion reactions was clearly observed [Roth 2002].

During the following years, neutrons have been produced by laser-driven ion bunches using the TNSA mechanism (for a description of TNSA, see Chapter 5), which typically results in energetic proton bunches, as this mechanism preferentially accelerates surface contaminants according to their charge-to-mass ratio. The first experiments explored the possible neutron yield for applications. These experiments did not address the applications directly, but rather tried to optimize the neutron conversion efficiency in order to demonstrate the principle concept of laser-driven neutrons. In 2004, Lancaster et al. used the $^7Li(p,n)^7Be$ reaction for fast neutron radiography applications [Lancaster 2004]; and in 2010, Higginson et al. increased the neutron yield to 2×10^9 n/shot using the Titan laser at LLNL [Higginson 2010]. The result indicates the prospect for neutron resonance spectroscopy. A year later, using short laser pulses of 360 J energy, the neutron yield was further increased and the maximum neutron energy was increased to 18 MeV [Higginson 2011]. This time, not only was the yield increased to 8×10^9 n/shot, but also a pronounced directed neutron emission was observed. The basic reaction in all the above cases used the pitcher–catcher configuration and the conversion of proton bunches into neutrons using light catcher elements (Li, Be) (see Figure 6.2).

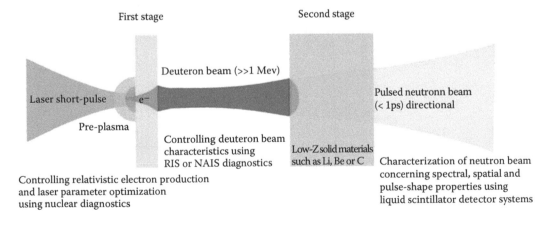

FIGURE 6.2 Typical pitcher–catcher geometry for laser-driven neutron production using ion bunches.

6.2.4 Relativistic Transparency

The fundamental problem of the prior experiments was the limit in ion energy as the cross sections, and therefore the yield for neutron production strongly increase with energy (see Figure 6.3). Moreover, accelerating ions other than protons requires special target treatment or the controlled application of the ion species of interest as the rear side surface contaminant [Krygier 2015].

In recent years the contrast of high-energy short laser pulses has largely improved, allowing the irradiation of solid targets with thicknesses well below one micron [Wagner 2014]. At the same time, the laser intensity at the target surface has increased to above 10^{20} W/cm^2. In 2007, a new ion acceleration regime was found in first 3D Particle In Cell (PIC) simulations, opening the onset of relativistic transparency of solids with large impact to laser-driven ion beams [Yin 2006, Yin 2007]. In this regime, the foil thickness is small, and during the rise time of the short laser pulse, the electrons in the target start to deplete. At the same time, the increasing intensity causes the quiver motion (i.e. mean oscillation velocity) of the electrons to become relativistic. The relativistic mass increase results in an increase of the critical electron density and a commensurate reduction of the plasma frequency to levels below the laser frequency, thus allowing the laser pulse to penetrate through the otherwise subrelativistically overdense target. The experimental confirmation of this theoretical prediction is known as the breakout afterburner (BOA) mechanism [Jung 2013, Hegelich 2011] and leads to ion energies in excess of 100 MeV for the first time, as well as the predicted change from a surface acceleration to the acceleration of the foil bulk material.

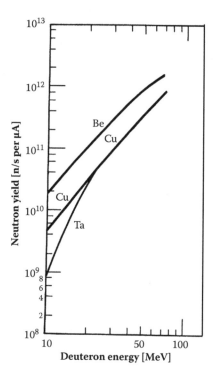

FIGURE 6.3 Neutron yield for deuteron-induced reaction in an infinite catcher geometry. (Reproduced with permission from Rindi, A., 1975. "Neutron Production from Heavy Ion Interaction: Some Very Empirical Considerations" http://escholarship.org.uc)

6.3 Neutron Production

Ion-induced neutron production can be separated by ion kinetic energy: a high-energy regime with ion (mostly proton) energies from hundreds of MeV up to several GeV, and a low energy regime starting at a few MeV; for the high energy regime neutron spallation is favoured by the use of heavy nuclei allowing for multiple neutron emissions per ion impact. The advantage of neutron multiplication comes at the cost of a long beam range in the converter and initially high neutron energies. While resulting in a high total neutron number, this regime is, at present, typically limited to very large accelerator facilities (e.g. the Los Alamos Neutron Science Center, LANSCE) [Lisowski 2006] and large converter installations. More discussion of conventional (i.e. not laser-driven) neutron sources can be found in Chapter 20. The low energy regime can be addressed with compact accelerators (e.g. cyclotrons) and, most recently, short pulse laser systems. For this regime, low-Z converter materials offer the best ion-to-neutron conversion efficiency.

For the efficiency simulations in Figure 6.4, we have assumed a typical TNSA proton spectrum with a cut-off energy near 25 MeV (see Figure 6.5).

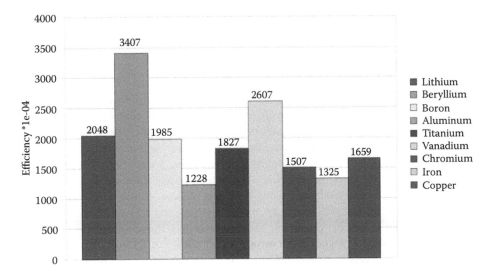

FIGURE 6.4 Simulated neutron production efficiencies per ion for different catcher isotopes, assuming a typical, laser-driven ion energy spectrum.

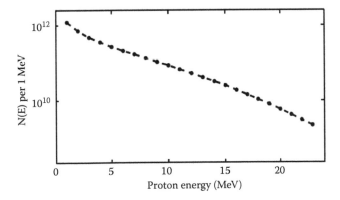

FIGURE 6.5 Typical TNSA proton spectrum driven by an ultra-intense laser pulse of 0.5 ps and 80 J energy.

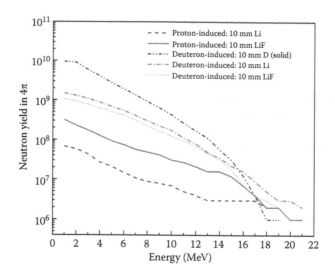

FIGURE 6.6　Energy-dependent proton and deuteron-induced neutron production using 10^{13} ions from a typical laser source at a cut-off energy near 25 MeV/u. (As seen in Figure 6.5.)

As efficient acceleration of the bulk material allows for the use of more efficient reaction channels for neutron production, the focus has changed from a proton-induced reaction for neutron production to deuteron-driven neutron production. The acceleration of deuterium offers two important advantages: First, every accelerated deuteron carries a neutron, which is loosely bound with a separation energy of only a few MeV; and second, following separation, the remaining proton can still evaporate a further neutron on impact with the converter material. Assuming a purely Coulomb breakup, the deuteron with initial energy E_D decelerates as it approaches the target nucleus. At a certain point R_s (the distance from the target nucleus) the deuteron breaks up, with the proton and neutron each emerging with half of the initially available energy, E_D. On leaving the target nucleus, the proton gets reaccelerated in the Coulomb field but the neutron keeps its given energy, E_n, where $E_n = \frac{1}{2}\,(E_D - Ze^2/R_S - 2.22)$ MeV (2.22 is the Q-value of the reaction [Oppenheimer 1935]).

Figure 6.6 reveals a distinction between p and d induced neutron reactions in light converter materials, assuming an input spectrum that is typical for and ion bunch with a maximum cut-off energy near 25 MeV accelerated by a single laser pulse with 70 J energy, as can be seen in Figure 6.5.

As a further benefit, because neutrons from the deuteron breakup follow the initial deuteron trajectory, the neutron angular distribution becomes non-isotropic, acquiring a distinct forward directionality. This results in higher neutron flux density and reduced shielding requirements compared to the more isotropic thermal sources.

6.4　Recent Experimental Results

With the use of solid target foils of submicron thickness and high contrast laser systems, recently the neutron yield has increased by orders of magnitude compared to the TNSA–based earlier attempts. The parameters of the recently developed neutron source are summarized in Table 6.1. In 2012, experiments at the Trident laser at Los Alamos National Laboratory resulted in a yield of 10^{10} n shot (i.e. per laser pulse) and a neutron energy in excess of 100 MeV using 80 J pulses of a 600-fs duration incident on a 3 μm diameter spot [Roth 2013, Jung 2013a]. The neutron distribution also revealed a clear directionality. This was accomplished using a thin, deuterated plastic target (pitcher) and a sealed Be catcher. In this work, neutrons had been used for the first time to perform for static neutron single bunch radiographs with fast neutrons. In 2013, the non-isotropic forward yield was further optimized with a cylindrical

TABLE 6.1   Demonstrated Laser-Driven Neutron Source Properties

Parameter/Item	Parameter Range/Comment
• Spectrum:	Energy range from thermal to >100 MeV
	Energy spread 100%
• Neutrons per bunch:	>10^{10}/sr (up to 2×10^{11} in 4π [Roth 2013])
• Spatial distribution:	Isotropic component and fourfold enhanced forward beam
• Opening angle:	Around 90°
• Forward neutrons:	2×10^{10}/sr
• Temporal structure:	Bunch duration initially sub-nanosecond
	Repetition rate: equals laser driver rate
• Neutron generation scheme:	Pitcher (deuterized plastic) + catcher (Be) + W reflector
	Deuteron-to-neutron conversion Efficiency: >10^{-2}
	Co-moving particles: e- (MeV), Gamma (MeV)
	Ions (depending on thickness of converter)
	Target: thin, free standing deuterated foil (200 to 800 nm thick)
• Laser parameters:	
Energy:	80 Joules @ 1053 nm, single shot
Pulse length:	500 femtoseconds
Contrast:	>10^{10}
Spot size:	>5 μm
Incident angle:	normal (10°)
	[Roth 2013]

tungsten reflector around the Beryllium converter. In 2015, experiments at the PHELIX laser system at the Helmholtzzentrum für Schwerionenforschung – GSI in Darmstadt [Neumayer 2005] lead to a neutron yield of 2×10^{11} n/shot with incident laser pulses of 80 J energy and 450 fs duration. In both cases, the neutron emission was strongly peaked in the forward direction. The catcher material and geometry had been improved to match the several centimeter range of the ion bunch. Consistent with the predictions from relativistic transparency, there was a strong dependence on the target thickness with a distinct optimum. The neutron pulse duration is basically determined by the time the ion bunch needs to pass the moderator. Given the fact that the ion bunch has an initial duration of only a few ps, and taking into account the temporal dispersion and increasing bunch duration commensurate with its energy spread (i.e. ion debunching), the neutron bunch can be of sub-nanosecond or nanosecond duration. This novel, short bunch capability enables various applications, as described in Figure 6.7.

6.5 Applications

The applications for a laser-driven neutron source are numerous. Because the laser driver can be made compact and mobile, for the first time a neutron source can be brought to users for various applications. The performance of laser-driven neutron sources has to be compared to the existing sources that are not laser-driven, listed in Table 6.2. At present, most of the laser systems used in neutron production are large-scale systems designed for basic science; whereas, future systems can be designed for compactness and transportability by a vehicle (truck or trailer). Because the infrared laser and the optical transport lines are non-hazardous with respect to radiation safety, only a small target area needs to be shielded, and only during operation. Thus, the entire system can become safe for transport and operation.

Moreover the short bunch duration affords excellent neutron energy resolution in time-of-flight measurements. High peak brightness allows for gated detector systems that can improve the signal-to-noise ratio and facilitate monitoring lower total neutron numbers and physical dimensions in applications.

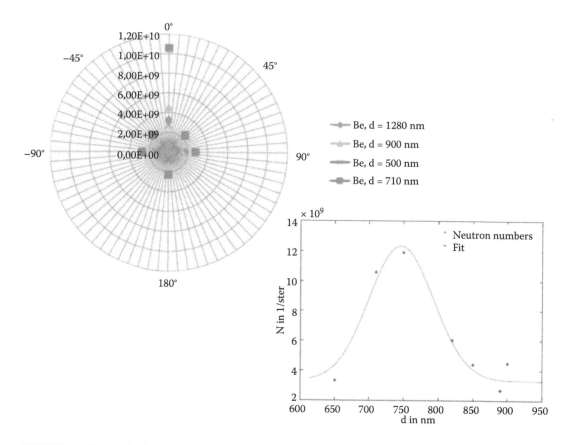

FIGURE 6.7 Target thickness sensitivity and directionality of the neutron emission based on ion bunches accelerated in the relativistic transparency regime. Numbers represent # of neutrons per sr.

Presented in what follows are a few application examples for laser-driven neutrons that are currently on the verge of technical readiness.

6.5.1 Fast Neutron Radiography

This technique has already been demonstrated using the LANL TRIDENT laser in 2012 [Roth 2013]. MeV neutrons from the laser-driven source were used to obtain radiographs of a test object made with tungsten blocks of different thicknesses (see Figure 6.8). Because neutron interactions with materials are distinct from X-ray interactions, they both can probe an object and provide complementary information. With gated detection, one can choose the energy range of the neutrons for observation, taking advantage of the energy-dependent differences in neutron scattering and absorption in different elements. Thus, the materials can be identified. An example is given in Jung [2013a] for neutrons between 2.5 and 15 MeV.

Besides its compactness, another advantage of a laser-driven neutron source is the directionality of the energetic neutrons. This also minimizes the need for shielding during the imaging.

6.5.2 Active Interrogation of Sensitive Nuclear Material

One of the most feared threats for modern societies is the uncontrolled proliferation of sensitive nuclear material and the possibility of a nuclear terrorist attack. Laser-driven neutron beams have been demonstrated to penetrate deeply into shielded containers and excite nuclear reactions in fissible material, resulting in detectable fission neutrons. Both prompt and delayed neutron emission has been measured at the LANL

TABLE 6.2 Conventional Neutron Sources (Not Laser-Driven)

Parameter for Source Type	Description
• Neutron source type:	High energy spallation driver (LANSCE)
• Spectrum:	Thermal (moderated) up to 800 MeV
• Neutrons per bunch:	10^{14} per bunch @ 20 Hz => 10^{15}/s
• Temporal structure:	200 nanosecond initial pulse width (recently 20 ns achieved)
	[Lisowski 2006, Lisowski 1990]
• Neutron source type:	Low energy (4MeV) RFQ (LLNL)
• Spectrum:	2.5–6.54 MeV
• Neutrons per bunch:	5×10^{9} n/s for 100 mA current (exit of collimator)
• Temporal structure:	CW
	[Hall 2007]
• Neutron source type:	D-T plasma tube
• Spectrum:	2.5 MeV or 14 MeV (DD or DT fusion)
• Neutrons per bunch:	Simple units limited to 10^{10} n/s (DT)
	10^{8} n/s(DD)
	(e.g. A-3062: 10^{8} n/s [up to 5 KHz])
• Temporal structure:	Microsecond long bunches
	[Buffler 2010]
• Neutron source type:	Dense plasma focus (DPF)
• Spectrum:	2.5 MeV or 14 MeV dependent on DD or DT
• Neutrons per bunch:	10^{13} at 3 MA current (Gemini) with DT
• Temporal structure:	A few nanoseconds, but jitter and multiple bunches
	[Vinogradov 2014]

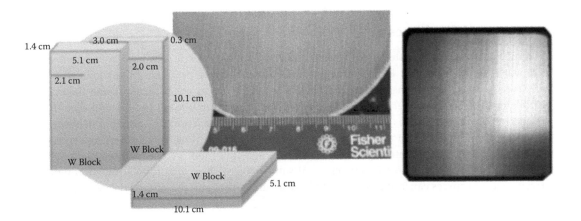

FIGURE 6.8 Example of laser-driven fast neutron radiography. The tungsten blocks in front of the detector result in absorption and scattering of the neutrons shown in intensity variations in the scintillating fibre detector. (Adapted with permission from Roth, M. et al., 2013. "A bright, laser-neutron source based on relativistic transparency of solids" *Phys. Rev. Lett.* 110: 044802; this image was taken without neutron energy selection [in contrast to Roth 2013].)

TRIDENT laser and the signal-to-noise ratio was high enough to allow for a large stand-off distance (distance between the source, the container to be probed and the detector assembly) and good threat discrimination.

6.5.3 Neutron Resonance Spectroscopy

Moderated epithermal or thermal neutrons can be used to identify isotopes based on their characteristic absorption cross sections. In order to use laser or accelerator-driven neutrons, they have to be moderated, i.e. slowed down by collisions in a low-Z material. Shorter neutron bursts mean better temporal resolution of detection and better energy resolution that facilitates material identification. Sub-nanosecond neutron bunches and a compact moderator system allow for high quality neutron spectroscopy [Pomerantz 2014]. An extension of this resonance technique is the determination of the absorption line width in an energy spectrum. The Doppler broadened absorption width is directly correlated to the ion Brownian motion, and therefore the sample's temperature.

If an absorption line can be detected in a single shot experiment (recently demonstrated, but as yet unpublished) the neutron bunch duration imposes a limit for this temperature measurement. This technique could then be used to measure for the first time the bulk temperature of a transient sample, e.g. in high energy density experiments. The shorter the initial neutron bunch duration, the more precise the energy distribution can be measured by time-of-flight methods. The resolution is a combination of distance (which increases the resolution, but decreases the flux density), detector temporal resolution and initial neutron bunch duration. Laser-driven neutron bunches are more than an order of magnitude shorter than accelerator-based neutron bunches. If the energy resolution becomes better than the Doppler width of a neutron resonance absorption line of a specific element, a temperature can be derived.

6.5.4 On-Site Neutron-Based System Tests

Neutrons are used in a variety of technical applications, e.g. in the oil and drilling industries [Schultz 1985], or in identifying structural defects. One example is testing turbine blade defects for aircraft engines [Thornton 2003]. A compact, laser-driven neutron source that might fit into a vehicle (as with mobile laser peening laser systems), could be envisioned as a mobile testing facility for jet engines. Colleagues in Japan have proposed testing the internal metal structure of bridges nationwide for corrosion using a mobile (laser-driven) neutron source to determine the need and urgency for repair [Seki 2014].

6.5.5 Material Sciences

We anticipate future laser systems that will operate at the multi-Hz level. Advanced laser systems such as the DIPOLE laser (100 J at 10 Hz) [Banerjee 2012] are being developed and can be modified for use with short pulse laser systems. Given the numbers already achieved experimentally and the neutron yield from an 80 J, 600 fs laser pulse is only three orders of magnitude lower than that of state-of-the-art neutron spallation sources (e.g. LANSCE). With future high repetition-rate laser systems delivering even higher intensities on target with high contrast, there is the prospect of achieving comparable average neutron fluxes for material science experiments and applications while maintaining the above-mentioned, advantageous short bunch temporal substructure. Thus, long-term irradiation by laser-driven neutron sources can lead to a widening in structural defect analysis in materials with neutrons.

6.5.6 DPA and Annealing Research

Neutron impact in material is known to cause lattice damage by replacing and shifting atoms from their initial positions. This phenomenon is expressed as DPA (displacement per atom) and is a crucial measure of the lifetime and stability of material exposed to neutron fluxes (such as materials in fission reactors and fusion devices in the future). However, it is known that the majority of neutron-induced defects are not detectable after long-term exposures and the annealing of defects plays an important role in real material behaviour. The underlying physics, timescales and dependences on initial conditions (stress, temperature) are poorly understood. One challenge in the research is to observe the damage and recovery on timescales shorter than the ones set by the diffusion of the displaced atoms (i.e. within a few ns). An intense, sub-ns neutron source coupled to high-resolution X-ray diffraction diagnostics (e.g. an X-ray free-electron laser) could, for the first time, explore the mechanisms leading to neutron-induced damage and annealing.

6.6 Academic Access

Many disciplines could benefit from a compact neutron source. Examples include biology, material sciences, geology and – of course – physics. Only a few universities in the world can have direct access to a large-scale spallation facility or an on-campus nuclear reactor, and even then the access is limited and involves complicated procedures to avoid contamination or exposure to radiation. The benefit of a laser-driven neutron source is the safety of the laser system itself; although, a class 4 laser system is operating a near infrared beam that must propagate up to the entrance of a compact target chamber has no radiation hazard. Also, fast, safe access is possible after experiments because the intense neutron pulse is only present during an experiment in operation that does not include sample activation. Moreover, the directionality of the neutron beam allows for beam guiding to the experiment that avoids unnecessary exposure to radiation. An affordable, compact short pulse laser is within the budget of most universities and the number of such systems is growing (e.g. throughout Europe). So, compact laser-driven neutron sources could broaden academic access and therefore boost neutron-related sciences in academia.

6.7 Summary

Lasers have been used to produce neutron beams for many decades. Apart from those experiments aimed at pursuing thermonuclear fusion, it has become clear that the development of modern short pulse lasers has been a breakthrough in this endeavour. First using cluster targets, then intense electron beams and after the discovery of laser-driven ion beams, the use of (p,n) reactions demonstrate that short pulse lasers have become an ever more efficient tool to produce neutron bunches. With the enhancement of the laser contrast and the exploration of the relativistic transparency regime, laser-driven sources now have entered a per-pulse yield level that is already suitable for real-world applications. The recent jump in neutron numbers mark only the beginning of what laser-driven sources might be able to do in the near future. Higher laser pulse energy, paired with higher wall plug efficiency (due to diode pumping) and a higher repetition rate, can at least add an order of magnitude to the average neutron flux. The very short initial neutron bunch length allows for very precise energy resolution. Based on the current experimental findings alone, a 200 J/10 Hz laser with high contrast could already provide around 10^{13} n/s. Due to the compact size of the system, the neutron flux density close to the converter already now exceeds 10^{20} n/s/cm^2. With the commissioning of the most modern systems, such as BELLA, ELI or Apollon [Zou 2015] for neutron experiments, new research can be envisioned that makes use of the excellent beam properties. As can be seen in Figure 6.9, the yield of laser driven neutrons per pulse has been increasing over the last few years and with the upcoming facilities can close the gap to large scale spallation sources. Finally, as smaller systems can be fielded at universities, laser-driven neutron sources can serve as a breeding ground for the next generation of young scientists in neutron science.

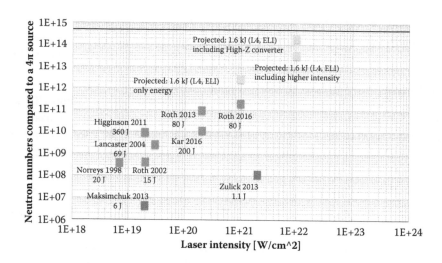

FIGURE 6.9 Laser-driven neutron numbers versus laser intensity. The neutron numbers of the directed neutron beam per laser pulse are compared to a conventional isotropic (into 4π) neutron source. Results in red indicate short pulse lasers below 100 fs, blue are high energy lasers with around 0.5 ps pulse duration and the yellow marks indicate the prospect for upcoming systems currently under construction. The top black line indicates the single, 250 ns pulse and neutron yield at the converter target, produced at the LANL LANSCE facility spallation source.

References for Chapter 6

Banerjee, S. et al. 2012. "High-efficiency 10 J diode pumped cryogenic gas cooled Yb: YAG multislab amplifier" *Optics Lett.* **37**: 2175.

Bang, W. et al. 2013. "Experimental study of fusion neutron and proton yields produced by petawatt-laser-irradiated D2-3He or CD4-3He clustering gases" *Phys. Rev. E* **87**: 023106.

Borghesi, M. et al. 2006. "Fast Ion Generation by High-Intensity Laser Irradiation of Solid Targets and Applications" *Fusion Science and Technology* **Vol. 49**: [3] 412.

Buffler, A. and Tickner, J. 2010. "Detecting contraband using neutrons: Challenges and future directions" *Radiation Measurements* **45**: 1186.

Cowan, T.E. et al. 2000. "Photonuclear Fission from High Energy Electrons from Ultraintense Laser-Solid Interactions" *Phys. Rev. Lett.* **84**: 903.

Ditmire, T. et al. 1999. "Nuclear fusion from explosions of femtosecond laser-heated deuterium clusters" *Nature* **398**: 489–492.

Guenther, M.M. et al. 2011. "A novel nuclear pyrometry for the characterization of high-energy bremsstrahlung and electrons produced in relativistic laser plasma interactions" *Phys. of Plasmas* **18**: 083102.

Guenther, M.M. et al. 2012. "Development of High-Power Laser Based Nuclear Applications" *Fusion Science and Technology* **Vol. 61**: [1] 231–236.

Hall, J.M. et al. 2007. "The Nuclear Car Wash: Neutron interrogation of cargo containers to detect hidden SNM" *Nucl. Instr. and Meth. B* **261**: 337.

Hegelich, B.M. et al. 2011. "Experimental demonstration of particle energy, conversion efficiency and spectral shape required for ion-based fast ignition" *Nucl. Fusion* **51**: 083011.

Higginson, D.P. et al. 2010. "Laser generated neutron source for neutron resonance spectroscopy" *Phys. Plasmas* **17**: 100701.

Higginson, D.P. et al. 2011. "Production of neutrons up to 18 MeV in high-intensity, short-pulse laser matter interactions" *Phys. Plasmas* **18**: 100703.

Hurricane, O. et al. 2014. "Fuel gain exceeding unity in an inertially confined fusion implosion" *Nature* **506**: 343–348.

Jung, D. et al. 2013. "Efficient carbon ion beam generation from laser-driven volume acceleration" *New Journal of Physics* **15**: 023007.

Jung, D. et al. 2013a. "Characterization of a novel, short pulse laser-driven neutron source" *Phys. Plasmas* **20**: 056706.

Kar, S. et al. 2016. "Beamed neutron emissiondriven by laser accelerated light ions" *New Journal of Physics* **18**: 053002.

Krygier, A.G. et al. 2015. "Selective deuterium ion acceleration using the Vulcan petawatt laser" *Phys. Plasmas* **22**: 053102.

Lancaster, K. et al. 2004. "Characterization of ^{7}Li (p,n)^7 Be neutron yields from laser produced ion beams for fast neutron radiography" *Phys. Plasmas* **Vol. 11**: 3404.

Lisowski, P.W. et al. 1990. "The Los Alamos National Laboratory Spallation Neutron Sources" *Nucl. Science and Engineering* **106**: 208.

Lisowski, P.W. and Schoenberg, K.F. 2006. "The Los Alamos Neutron Science Center" *Nucl. Instrum. and Meth. A* **562**: 910–914.

Maksimchuk, A. et al. 2013. "Dominant deuteron acceleration with a high-intensity laser for isotope production and neutron generation" *Appl. Phys. Lett.* **102**: 191117.

Neumayer, P. et al. 2005. "Status of PHELIX laser and first experiments" *Laser and Part. Beams* **Vol. 23**: [3] 385.

Norreys, P.A. et al. 1998. "Neutron production from picosecond laser irradiation of deuterated targets at intensities of 10^{19} W/cm^2" *Plasma Phys. and Control. Fusion* **40**: 175–182.

Oppenheimer, J.R. and Phillips, M. 1935. "Note on the Transmutation Function for Deuterons" *Phys. Rev.* **48**: 500.

Pomerantz, I. et al. 2014. "Ultrashort Pulsed Neutron Source" *Phys. Rev. Lett.* **113**: 184801.

Rindi, A. 1975. "Neutron Production from Heavy Ion Interaction: Some Very Empirical Considerations" http://escholarship.org.uc

Roth, M. et al. 2002. "Energetic ions generated by laser pulses: A detailed study on target properties" *Phys. Rev. ST-AB* **Vol. 5**: 061301.

Roth, M. et al. 2013. "A bright, laser-neutron source based on relativistic transparency of solids" *Phys. Rev. Lett.* **110**: 044802.

Seki, Y. et al. 2014. "A Simulation Study of Fast Neutron Imaging for Large-scale Concrete Structures" *Physics Procedia* **60**: 324.

Schultz, W.E. et al. 1985. "A borehole corrected pulsed neutron well logging system" *Nucl. Instrum and Methods B* **10–11**: 1023.

Snavely, R.A. et al. 2000. "Intense High-Energy Proton Beams from Petawatt-Laser Irradiation of Solids" *Phys. Rev. Lett.* **85**: 2945.

Thornton, J. et al. 2003. "The Detection and sizing of flaws in components from the hot-end of Gas Turbines using phase-contrast radiography with neutrons: A feasibility study" *NDT&E International* **36**: 289.

Vinogradov, V.P. et al. 2014. "Development and study of a portable plasma focus neutron source" *Plasma Physics Reports* **Vol. 40**: [2]; 146–159.

Wagner, F. et al. 2014. "Pre-plasma formation in experiments using petawatt lasers" *Optics Express* **Vol. 22**: 29505.

Wilks, S.C. et al. 1992. "Absorption of Ultra-Intense Laser Pulses" *Phys. Rev. Lett.* **69**: 1383.

Yin, L. et al. 2006. "GeV laser ion acceleration from ultrathin targets: The laser break-out afterburner" *Laser and Part. Beams* **Vol. 24**: [2]; 291–298.

Yin, L. et al. 2007. "Monoenergetic and GeV ion acceleration from the laser breakout afterburner using ultrathin targets" *Phys. Plasmas* **14**: 056706.

Zou, J.P. et al. 2015. "High Power Laser Science and Engineering Design and current progress of the Apollon 10 PW project" *High Power Laser Science and Engineering* **Vol. 3**: e2.

Zulick, C. et al. 2013. "Energetic neutron beams generated from femtosecond laser plasma interactions" *Appl. Phys. Lett.* **102**: 124101.

II

Applications of Laser-Driven Particle Acceleration

Preamble to Part II

As the main focus of this book, Part II consists of Chapters 7 to 23 and is organized according to the subject matter of envisioned applications. In each chapter, relevant experts describe how a laser-driven source of energetic particles or associated photons can be considered in their application. This is done either directly though explicit reference to laser-driven source requirements and/or capability, or indirectly through an adequate description of the application that can help the reader to infer laser-driven requirements, aided by the material in Part I. Of particular significance are those cases for which unique features of laser-accelerated particles and photon sources might be especially relevant and even exploited. In general, the reader can compare application requirements indicated in Part II to current and near-term future laser-driven capabilities presented in Part I. Some applications (particularly far-term ones) can require laser-driven source parameters that have not yet been achieved in practice. Nonetheless, the visionary applications mindset can provide guidance and motivation for the continued development of laser-driven sources.

Part II presents a variety of application possibilities which we consider to be an exemplary subset of what is possible and therefore worth targeting. Most of those included here are for ions. Although component chapters can be read in any order, the reader can also consider them according to the five topical groups listed below:

Chapters 7, 8, 10, 11, 12 and 14: Fundamental radiobiology (including space radiation effects), radiochemistry/radiolysis, laser-driven ion beam radiotherapy and medical imaging (it can be helpful to read some of Chapters 7, 8, 10 and 14 before Chapters 11 and 12)

Chapters 9, 15, 16, 17 and 18: Interactions with materials (such as semiconductors and Li-ion batteries) and related processing

Chapters 19 and 20 (and 6): Neutron source developments for novel applications (where we have included final brief sections in Chapter 6 of Part I)

Chapters 13, 21 and 22: Novel nuclear reactions in a plasma environment and radioisotope production
Chapters 17, 18 and 23: X-ray and ion (including proton) probes for materials, elemental assay and
  Li-ion battery performance

It is important to consider a variety of applications at this early stage to credibly assess key technical
challenges, critical constraints, as well as targeted strategies and timescales that can be relevant to laser-
driven particle acceleration. Also of significance in Part II is the implementation rationale and potential impact (or merit) of laser-driven particle acceleration to different research fields. Part II therefore
exposes the practical foundation for the laser-driven case, which is achieving the necessary scientific
and technical maturity that will inspire and foster continued development of a wide range of meaningful and doable applications.

7

New Tools for Facing New Challenges in Radiation Chemistry

Uli Schmidhammer

Jun Ma

Mehran Mostafavi

7.1 Introduction

Developing the study of radiation-condensed matter interactions on different scales of time and space, with advanced detection techniques, to understand real-time defect formation and oxidation processes under irradiation, is very challenging, and is the treated subject of Chapter 7. After a short overview to radiation chemistry, a few examples of research studies, which need new tools to be faced, will be presented. We describe the state of the art of the pulse radiolysis set-ups used currently in radiation chemistry, and discuss the performance and capabilities of the new pulse radiolysis installations, based on laser-driven electron and ion bunches.

7.2 Contribution of Radiation Chemistry

Radiation chemistry is a part of physical chemistry like photochemistry, plasma-chemistry, sonochemistry, etc. It is a mature branch of radiation science which evolves continually and finds wider applications [Spotheim-Maurizot 2008]. This is particularly apparent in the study of the roles of free radicals in biology generally, and radiation biology particularly [Virender 2013]. Ionizing radiation, such as highly energetic photons and accelerated charged particles, produces abundant secondary electrons and radicals which can induce a complex chemistry at the spot of energy deposit. In general, the interaction of one highly energetic particle with liquid or solid matter implies a multitude of primary and secondary ionization and excitation events. Their distribution along the trajectory of the initial particle is strongly inhomogeneous and depends on charge and velocity of the particle. The initial physical radiolytic event taking place on the attosecond scale is followed by decomposition and relaxation processes such as the thermalization of the kinetic energy of secondary particles, the reorganization of solvent molecules around them (solvation process) and spatial homogenization [Hatano 2010]. This interplay with the surrounding molecules takes places on several scales in time and space, and is decisive for chemical reaction steps. The main method to study this complex chemistry with a variety of transient species is to use the pulse radiolysis technique. A pulsed source of ionizing radiation (X-rays, accelerated electrons, protons, heavy ions) produces short-lived excited molecules within a very short time, radicals and ions with an initial concentration high enough to be directly detected by time-resolved techniques (optical absorption or emission, electron spin resonance, Raman spectroscopy, conductivity, etc.). The obtained time-dependent signals reflect the build-up or decay of the transient species, and the quantitative and spectral analysis of these signals permits the identification of the involved intermediates by establishing their emission or absorption spectrum, molar absorption coefficient, equivalent conductivity, equilibrium constant, reduction chemical potential, etc. These transient intermediates are of central importance in a variety of processes in chemistry, biology and nanoscience, as well as the environmental, polymer and material sciences. The ability of the time-resolved technique to produce data and to explain mechanisms in chemistry or biochemistry may be assessed by visiting the various data banks in the literature or online [Buxton 1988]. In addition to the chemical mechanisms, pulse radiolysis also enables us to understand the specific action of ionizing radiation on matter quantitatively and in detail. The importance of this understanding for the operation of nuclear reactors, nuclear waste treatment, storage and disposal, cosmic ray damage in space and in atmospheric chemistry, and for radiobiology and radiotherapy can scarcely be overstated (see also Chapter 10 by Shikazono, Moribayashi and Bolton). In addition, ionizing radiation is routinely used in the industry (for example) to cure polymers, to sterilize pharmaceuticals and food, to dope semiconductors and to remove pollutants from waste water (Figure 7.1). An understanding and control of the chemical changes induced by the radiation is central to the proper regulation of these industries. To face

Radiation chemistry and
application fields

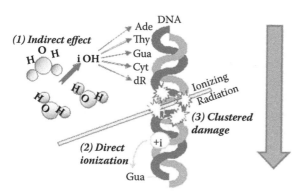
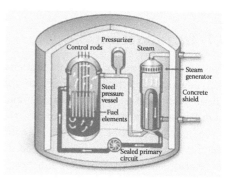

• Radiobiology, radiation protection

• Chemical mechanisms, biochemical, electrochemical, ...

• Chemistry of reactors, radioactive waste disposal

• Space chemistry

• Radiochemical synthesis

• Polymers and nanomaterials formation

• Non-homogenous and homogenous kinetics

FIGURE 7.1 Radiation chemistry and pulse radiolysis contribution to different fields.

these challenges, since the 1960s, many laboratories have made major investments in the pulse radiolysis technique, developing an increasingly short pulse resolution and various fast time-resolved detection systems.

Boag (Gray Laboratory) and Keene (Manchester) in the United Kingdom and Matheson and Dorfman, (Argonne) in the United States, developed pulse radiolysis in the early 1960s independently of each other for the study of the elementary steps of fast chemical reaction mechanisms [Kroh 1989]. Despite their rather high cost, several pulse radiolysis facilities were soon developed. They were generally implanted in laboratories with a long-standing expertise in radiation chemistry, such as in the United States, Canada, Japan, Israel and in Europe (i.e. the United Kingdom, France, Germany, Poland, Sweden, the Netherlands, Denmark, Italy and Hungary).

Radiation chemical techniques and especially pulse radiolysis, which enables the study of short-lived intermediates, are essential tools in the forefront of scientific development. They have been used for the elucidation of the mechanisms of many key processes in particular in irradiated water (Figure 7.2). The advantage of pulse radiolysis stems from the fact that ionizing radiation is absorbed mainly in the solvent and not in specific solutes: the radicals are generated in a calibrated way which may be used to produce secondary radicals quantitatively, particularly radical anions and cations, and to initiate a wide variety of chemical mechanisms. These radicals are key transients in a variety of chemical, biological, environmental and catalytic processes.

The properties of many transient species have been elucidated, which resulted in enriching our understanding of fundamental mechanisms: for example, a huge number of oxidation, reduction and proton transfer processes in water and non-aqueous solvents, fast electron solvation processes as a model of ion solvation mechanisms and properties of the strongly reducing solvated electron. In biological systems, essential radical processes include the synthesis of deoxyribose, many enzymatic processes, DNA damage, reactions involving radicals such as OH$^\bullet$ and NO$^\bullet$, and the destruction of bacteria by powerful oxidants. In addition, oxidative stress and the oxidation of proteins, induced by radical processes, are believed to cause over 70 diseases. Transition metal complexes with uncommon oxidation states are involved in a large variety of catalytic electron transfer processes, specifically in many enzymatic processes. In the environment, radicals are involved in the destruction of the ozone layer, in a variety

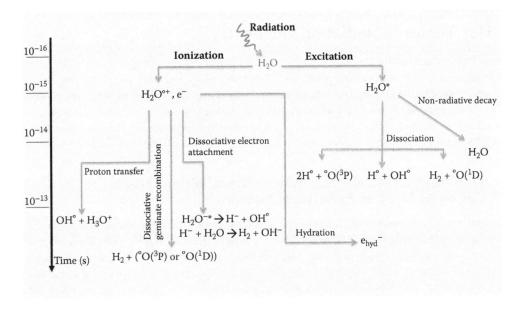

FIGURE 7.2 General scheme of reactions of transient species produced during the first picosecond after a radiolytic event in bulk water.

of processes in rain droplets leading to the formation of acid rain, in combustion, in the dehalogenation of halo-organic pollutants, in the degradation of dyes and organic pollutants, etc. Radicals catalyze many oxidations by oxygen and by peroxides, polymerization processes, electrochemical processes, etc. Nucleation and autocatalytic growth of metal and semiconductor nanoclusters and photographic mechanisms were also explained using pulse radiolysis [Belloni 2001].

Understanding radiation-induced processes is also of critical significance when exploited for the control and the optimization of protocols in diverse industrial applied fields, such as waste remediation in environmental clean-up, microlithography for chip production in electronics, synthesis of metal and semiconductor nanoclusters, radiation-induced doping of semiconductors, radiation processing of polymers, food sterilization, medical diagnosis and therapy, catalysis of chemical reactions, environmentally benign synthesis and nuclear energy production. Therefore, radiation chemistry is an irreplaceable and uniquely valuable tool for solving fundamental problems in chemistry, material and biological sciences, as well as providing quantitative information for technological and medical applications.

At present, the community of the radiation chemistry needs new tools to go beyond the state of the art. New studies to solve outstanding problems not yet understood could be undertaken with the help of advanced facilities. For example, there is particular need for an understanding of the structure and ultrafast dynamics of radiation 'tracks' produced by ionizing radiation as a central issue in the field. This is especially true for practically important, but poorly understood, non-standard environments such as interfacial systems, supercritical media (i.e. solvents at very high temperature and high pressure) and heterogeneous systems, which require short time-resolution and improvements in detection methodology. The reactivity of ultra-short-lived species such as pre-solvated electrons, precursor of the $OH^{\bullet}$ radical and that of $H^{\bullet}$ atom are also very important in bulk and at interface. The ultrafast kinetics studies of radical reactivity at high temperatures need also better time resolution than the state of the art. These goals necessitate the development and application of complementary programs of experiment and theory.

In the following sections, we illustrate with some examples of how current scientific topics of radiation chemistry are limited by the actual technology. We present the state of the art of the facilities used by radiation chemists and explain why the future needs the development of appropriated tools such as laser-driven particle bunches with short duration coupled to various optical spectroscopy approaches. Eventually, we will identify source features that are needed to undertake these studies.

7.3 Hot Topics in Radiation Chemistry

The reactions induced in the low linear energy transfer (LET) regime are now well studied mainly by electron pulse radiolysis with ps time resolution. Time resolved studies on (heavy) ion radiation are very scarce and limited to the nanosecond or microsecond scale. Among the different interesting topics here we give the example of the reactivity of the presolvated electron, precursor of H atom and precursor of $OH^{\bullet}$ radical, the ultrafast reactivity in water at high temperature and finally the spur reactions in the high LET regime.

7.3.1 Probing the Radical Precursors in Water: Water Radical Cation ($H_2O^{\bullet+}$) and Presolvated Electron

The ionization of liquid water, which has been extensively investigated over decades, involves the ejection of an electron, while leaving behind a positive charged radical, $H_2O^{\bullet+}$. The excess electrons are subsequently thermalized and hydrated by water molecules in their equilibrated state, which was first observed in 1960s using pulse radiolysis [Hart 1962]. With the sufficiently long lifetime ($\sim \mu s$) traced by many time-resolved techniques, the hydrated electron has been an object of central interest in radiation chemistry since its discovery [Buxton 1988]. However, the water radical cation, $H_2O^{\bullet+}$ in liquid water, which is assumed to react very rapidly (~100 femtoseconds) by transferring a proton along its hydrogen bond to a neighbouring water molecule, is still not experimentally identified yet and its delocalization remains unexplored due to the time resolution barrier in currently available methods. In fact, acting as a strong oxidizing intermediate, the basic knowledge of its nature and chemical role are particularly important in radiation physics, chemistry and biology. For instance, it is known that a large number of hydrating water molecules are in direct contact with DNA. When ionizing radiation is applied, part of the radiation energy is absorbed directly by DNA and breaks in sugars, phosphates and nucleobases units, while some is also absorbed by the water adjacent to the DNA. In that case, the generated $H_2O^{\bullet+}$ radical cations may induce a chemistry and some biological consequences different from $OH^{\bullet}$ radicals.

Transient absorption spectroscopy with an ultrafast pump pulse of duration below 30 fs has the potential ability to meet the challenges of directly observing this short-lived species, which is previously predicted by simulation to display a time-dependent absorption band in 400–500 nm region [Siefermann 2010]. Meanwhile, its chemical reactivity towards the biomolecules and many others could be continuously investigated using picosecond or sub-picosecond pulse radiolysis and time-resolved electron paramagnetic resonance to probe the formed secondary radicals. The combination of time-resolved Raman spectroscopy with pulse radiolysis (using either heavy ions or electrons) would also be very helpful in the near future to allow us to measure the process in a heterogeneous or interfacial system where $H_2O^{\bullet+}$ cation is involved (Figure 7.3).

The precursor of the solvated electron is usually probed by laser photolysis measurements. However, its reactivity in the radiolytic spurs is very interesting because it is unexpectedly high.

Moreover, an electron being trapped at the liquid water/vacuum interface is a state that has been recently suggested based on results of ultrafast photoelectron spectroscopy with micro-liquid jet technology in a vacuum [Turi 2012]. This surface-bound electron, with respect to its lower vertical binding energy (1.6 eV) compared to the interior state (3.3–3.6 eV), has a long lifetime (>100 ps) and its possible existence immediately turned into a very hot debate in radiation chemistry [Turi 2012]. The evidence for electron surface states was found in large water clusters and ice surfaces already before, but the finding of electrons near a liquid surface is strongly dependent on the 'surface sensitivity' of the probe pulse used to produce photoelectrons. The first observation in neat water was performed with a femtosecond extreme-ultraviolet (EUV) probe pulse (38.7 eV) from high-harmonic generation, has not been confirmed by a UV (4.8 eV) probe in a similar condition. Therefore, the 'inelastic mean free path' of photoelectrons relating to the probe depth describing how far on average the photoelectron can travel through a liquid before losing energy constitutes a key factor for detecting the hydrated electron at the surface. In addition, the distribution or relative abundance of electron sites localized in the interfacial and interior region cannot be determined accurately yet.

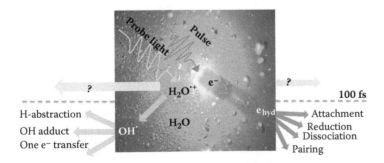

FIGURE 7.3 The reactivity study of the presolvated electron and the precursor of $OH^{\bullet}$ radical ($H_2O^{\bullet+}$) calls for an ultrafast electron bunch-laser probe setup with improved time resolution.

Even if the understanding and further identification of this transient species at an interface require many efforts, the radiation-induced interfacial processes have been a long-term issue in radiation chemistry. It would not be surprising if the reactivity of short-lived e_{hyd^-}, OH·, H· etc. differ at the liquid–liquid or liquid–solid interface from that in the homogenous phase. These modified physical and chemical processes undoubtedly have strong implications for different fields, for example nuclear spent fuel cycling, nuclear waste storage and radiotherapy.

7.3.2 Ultrafast Kinetics in High Temperature High Pressure Water

At elevated temperatures, the rate constants were only measured in diluted solutions. Knowledge of the absorption band of a hydrated electron as one of the main radiation-induced species is essential for determining the rate constant and radical yield at elevated temperatures. With picosecond pulse radiolysis, aqueous solutions were studied at elevated temperatures. However, due to the ultrafast reaction of radicals' species at elevated temperatures, the data on the rate constant were very scarce (Figure 7.4).

Knowledge of the rate of radical reaction at high temperatures and specifically in supercritical water is of great importance for the nuclear industry and for understanding the mechanism in water. For a simple acid solution with a pH of 1 at 350°C, the decay of the solvated electron is expected to occur within 1 picosecond. With the current pulse radiolysis setups, observation of the decay of the solvated electron in this condition is not accessible. Moreover, the solvation of electrons occurs within a very short time interval. Up to now, due to the absence of a short electron pulse facility, there is no data on the dynamics of electron solvation at high temperature.

7.3.3 Spurs Kinetics with High Linear Energy Transfer

Radiations of high LET have long been known to have greater biological effectiveness per unit dose than those of low LET for a wide variety of biological effects. The differences between high and low LET

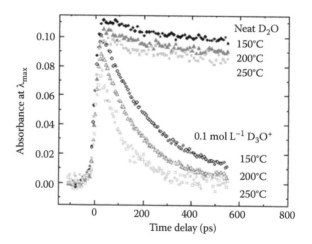

FIGURE 7.4 Example of the decay of hydrated electrons, observed at the maximum of the absorption band at different temperatures in neat D_2O (solid circle) and in 0.1 mol L^{-1} acid D_2O solution (open circle). With the state of the art pulse radiolysis setups, the observation of the reactions becomes limited at very high temperature and even not possible in supercritical water. (Adapted from Ma, J. et al., *Phys. Chem. Chem. Phys.* 17: 22934–22939, 2015.)

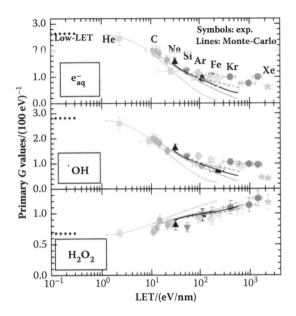

FIGURE 7.5 Primary radiolytic yield (obtained 100 ns after dose deposition) versus LET for hydrated electron (top), OH• radical (middle) and H_2O_2 (bottom). (Adapted from Shinichi, Y. et al., *Radia. Phys. Chem.* 77: 1224–1229, 2008.)

radiations may be due to factors related to radiation track structure. The spur reactions with high LET (from protons, α particles and other heavy ions) have been studied so far by time-resolved techniques with resolution limited to the nanosecond range. Usually the indirect method of measuring the amount of a stable product issued from the radical (scavenging method) is used to measure the LET effect on the yields of radicals and molecular products of water ionization and the G-values (i.e. the number of a species produced per 100 eV of energy loss of the ionizing particle). Still, knowledge about the initial distribution of energy and radicals is missing and only molecular simulations reproducing the results obtained by scavenging methods are typically used to understand the LET effect (Figure 7.5).

7.4 State of the Art of Pulsed Radiolysis Facilities

The predominant approach for performing time-resolved radiolysis is attributed to electron accelerator schemes. So, the results described in previous sections (except 7.3.3) are based on this approach to a high degree. Often, the development and advancement of accelerator technologies and of adapted transient detection schemes were motivated by the research in radiation physics and chemistry. For instance, the first realisation of pump-probe spectroscopy can be ascribed to an electron accelerator based approach with a time resolution better than 30 ps and a time window of 350 ps; developed in the late 1960s, it used the Cerenkov light generated by the electron pulse as an analyzing light flash that was delayed by moveable mirrors [Bronskill 1970]. The detection range and reliability was soon improved [Jonah 1975]. Nowadays, several accelerator facilities dedicated to pulsed radiolysis render technologies available to respond on different time scales to scientific and/or industrial needs. New nanosecond linacs for electron radiolysis were recently installed at Notre Dame University in the United States, the CEA of Saclay in France, the University of Pune in India and the University of Manchester in the United Kingdom.

Here, we focus on the forefront of time-resolved techniques in the different regimes of interaction between highly energetic radiation and matter that are arranged according to the LET properties of the particle radiation sources.

7.4.1 Picosecond Electron Radiolysis in the Low LET Regime

The state of the art for electron sources for picosecond radiolysis is based on linear accelerators (linacs) using radiofrequency (RF) photogun technology. An ultrashort light pulse is used to generate electrons from a photocathode surface that are in turn accelerated by at least one external RF field. Compared to the first generation of linacs that entered the low picosecond scale with thermionic cathodes as the initial electron source, the photogun approach brought important progress by reducing the construction costs (now a few million dollars including the shielding precautions) and the complexity of operation mainly with the distinctly increased beam quality and particularly its low emittance. The duration of the electron pulse is directly on the low picosecond scale thanks to the almost instantaneous electron bunch generation by the ultrashort laser via the photoeffect; no subsequent compression via pre-bunching is necessary, in contrast to thermionic emission with nanosecond duration. In addition, the intrinsic synchronization with the ultra-short laser pulse introduced a further milestone for time-resolved radiolysis that opened new possibilities for optical detection, particularly for electron bunch pump optical pulse probe spectroscopy.

This synchronization is a central part of the electron photogun technology. A master clock, such as a MHz quartz oscillator, is used to control the laser repetition rate and the microwave amplifier seed frequency (see also Figure 7.6). Commercially available phase lock electronics can control the cavity length of the laser oscillator with jitter less than 200 femtoseconds. However, a larger timing jitter and drift at the photocathode can occur due to thermal expansion of microwave system components of the electron gun and due to the beam transport of the laser with possible fluctuations and drifts in the air and of the optomechanical installation. Non-invasive single shot electron bunch diagnostics can monitor and optimize the length (and therefore temporal duration) and jitter of the bunches for experiments [De Waele 2009, Schmidhammer 2009].

The linac RF photogun technology has been adapted to and is now routinely available for picosecond radiolysis at a few facilities worldwide: the LEAF facility at the Brookhaven National Laboratory (BNL) [Wishart 1998], the ELYSE facility at the University of Paris-Sud [Belloni 2005], Osaka University [Kozawa 1999], Sumitomo Heavy Industries (SHI, Tokyo), Waseda University (Tokyo) [Sakaue 2008] and the NERL at the University of Tokyo [Muroya 2005]. For the past several years, facilities have been under construction at the Technical University of Delft for sub-picosecond, 4.5 MeV electron bunches and at the Bhabha Atomic Research Center in Mumbai. The facility at Waseda University is also able to generate soft X-rays by inverse Compton scattering [Sakaue 2008]. A recent overview of the specific technical parameters can be found in [Wishart 2010]. In summary, these facilities apply light pulses of wavelengths near 260 nm with pulse energies at the tens of µJ level and sub-picosecond to picosecond duration to the photocathode surface. The cathode material used is often a layer of the semiconductor material Cs_2Te. These highly oxygen-sensitive cathodes require time intensive *in situ* preparation in a special chamber that is (vacuum-) linked to the photogun. However, they exhibit more than one order of magnitude higher quantum efficiency (0.25%) compared to metal cathodes such as Cu or Mg. Once prepared and positioned, Cs_2Te cathodes can be operated as efficient electron sources for several months without significant degradation.

In the photogun (and in an additional booster at some facilities), the photoelectrons are accelerated by the RF fields to final relativistic energies. Several installations, such as those at BNL, University of Paris-Sud, Osaka University and Waseda University, operate at energies below 10 MeV in order to avoid radioactivation of the samples and experimental elements. In contrast, a higher kinetic energy implies a higher penetration depth and less scattering of the electrons in a medium. Both are factors that can deteriorate time-resolution and dose over the depth of the sample. Operation at 9 MeV is a good trade-off to

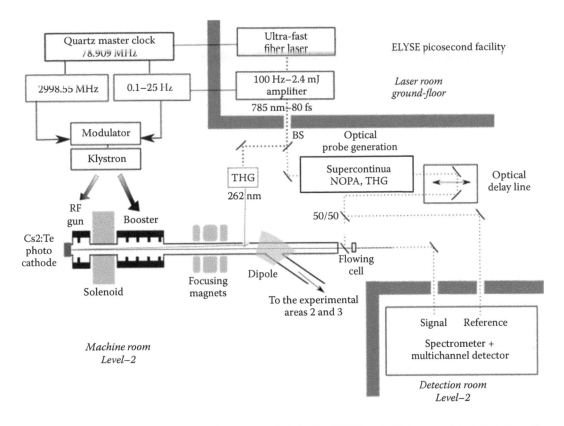

FIGURE 7.6 Scheme of the picosecond electron radiolysis facility ELYSE at the University of Paris Sud. Here, the pump-probe installation at experimental area 1 with several spectral options is given in more detail.

guarantee a sufficiently high beam quality over a sample thickness of some millimetres without the need for a specially equipped facility [Muroya 2005]. The energy deposit of such electrons in a 5 mm thick sample is about 10% of the incident particle kinetic energy.

Typical bunch durations that are obtained at the end of the linac are 5–10 ps with a bunch charge of several nanocoulombs (nC) and a repetition rate of 10 Hz. As discussed below, the effective time-resolution under experimental conditions is typically 10–20 ps depending on charge and sample thickness. At Osaka University, important and continuous effort was made towards femtosecond radiolysis. The limit of generation of electron bunches for radiolysis was successfully reduced to a duration of 200 fs for single 0.4 nC bunches by rotating the bunch in longitudinal phase space with a magnetic bunch compressor; under these conditions, the resolution in transient absorption measurements of the solvated electron was found to be 1.3 ps for 1 mm optical path [Yang 2011]. For this experiment, the temperature fluctuations of both the photocathode RF gun and the linac were reduced to +/– 0.05 °C using a fast feedback circuit in the cooling system. Also motivated by a reduction of jitter between the electron pump and the optical probe as well as that between the laser and RF pulses, the phase and amplitude of the RF system were stabilized with feedback.

Even if the described second generation of picosecond pulse radiolysis facilities is much less complex and more performant than its precursors, a technical team for maintenance and daily operation is very helpful. At the ELYSE facility, up to four specialized technicians work together with scientists to make the technology readily available and to benefit best from its potential.

Against the background of this complexity of the installation, the lack of access during operation and the expensive beam time, the time-resolved techniques to detect transient radiolytic species and to record their kinetics must also be particularly efficient and easy to run. So, in the LEAF, ELYSE and Waseda facilities, the electron beam can be bent by magnets to different beam lines and experimental areas that are equipped with various detection systems optimized for different tasks. For instance, at ELYSE, absorbance detection based on a Xenon flash lamp and a streak camera provides high flexibility for the temporal probe window. Broadband spectral evolution in the UV–Vis region can be recorded from picoseconds to microseconds from a single electron bunch pump. In a second experimental area, pump-probe set-ups realized in a modular approach allow highest sensitivity on the picosecond to nano-second time scale in different spectral regions. At LEAF, single wavelength detection with fast detectors coupled to a transient digitizer allows high spectral flexibility including the NIR with 150 picosoecond time resolution. Very recently, time-resolved infrared detection with ~40 nanosecond resolution for highly sensitive vibrational spectroscopy was coupled to pulsed radiolysis enabling improved structural characterization of intermediates [Grills 2015].

The potential of time-resolved emission spectroscopy was shown early [Sauer 1988]. The emitting excited states can be produced directly by the particle radiation, by geminate recombination, by other ground state reactions or by excited state energy transfer processes. With streak cameras, single shot acquisition of kinetics is possible with a typical instrument-limited time resolution of 5 picoseconds. On larger time scales, electron paramagnetic resonance is used to study the transient radicals [Trifunac 1980, Lund 2014]. The real part of time-resolved microwave conductivity probes the microwave absorp-tion influenced by mobility and concentration of radiation induced transients with nanosecond reso-lution, while the knowledge of the imaginary part of the conductivity gives access to the product of yield and the change in molecular polarizability [Bird 2014, Prins 2007]. Electrochemical detection was recently coupled to the spectroscopic analysis to probe the concentration of the redox intermediates with high sensitivity and a time resolution of 100 µs [Shahdo 2013].

However, the highest temporal resolution on the ultrafast time scale can be reached with pump-probe spectroscopy. The method of choice is transient absorption, particularly in the UV–Vis and NIR regions, to detect radiolytic species via electronic transitions from their ground state; this tech-nique is also sensitive to excited state absorption and stimulated emission. Broadband approaches use a supercontinuum synchronously generated by the ultrafast laser source of the accelerator, and a polychromator coupled to multichannel detection [Saeki 2006, Schmidhammer 2010, Schmidhammer 2014]. Entire transient spectra are recorded from one laser pulse and consequently independently of the shot-to-shot fluctuations and possible long-term drifts of the electron source. The evolution of several transient species can be followed simultaneously under the same experimental conditions. Often, the absorbance of the solvated electron in the visible or NIR is used as a reference for the local absorbed dose. With the idea to follow different transient species at the same time, multi-wavelength probing with the harmonics of the ultrashort laser pulse at 780 nm was applied to detect directly the hydroxyl radical OH$^\bullet$ in water via its low absorbance in the mid–UV region. With the absorbance of the hydrated electron as *in situ* dosimeter, the yield of the OH$^\bullet$ radical at 10 ps has been determined [El Omar 2011]. In a new optical configuration at ELYSE, the single wavelengths of the laser harmonics can be coupled to different types of supercontinua optimized for UV–Vis or NIR application [Schmidhammer 2010, Schmidhammer 2014].

Single shot approaches of pump-probe spectroscopy acquire efficiently the kinetics at selected wave-lengths on the ultrafast scale. Single-shot picosecond pulse radiolysis was realized with a bundle of fibres of different lengths that delay different parts of a light pulse to different extents; the exit of the fibre bundle is imaged to the pumped sample volume and, in turn, to a camera where the intensity of light from each fibre is measured simultaneously [Andrew 2009]. The probe wavelength can be tuned by optical parametric amplification. This technique is particularly useful for samples that are precious or for which it is difficult to exchange the irradiated volume during the experiment in order to avoid accumulation of radiolytic products.

Single shot ultrafast techniques can also circumvent one of the limitations of the time resolution of electron pump optical probe experiments. The 'time zero' for kinetics can be corrected before averaging over several shots to eliminate blurring due to the timing jitter between electron pump and optical probe. For the same motivation, double-decker pulse radiolysis was developed generating two electron bunches in a common photocathode radio-frequency gun. One of the electron bunches is converted to a probe that is inherently synchronized with the second electron bunch acting as the pump [Kan 2012].

In general, the overall time resolution, δt_T of an electron pump optical probe detection system can approximately be expressed as:

$$\delta t_T \approx \delta t_{GVM} + \sqrt{\delta t_E^2 + \delta t_L^2 + \delta t_{Sync}^2} \qquad (7.1)$$

where δt_{GVM} represents the group velocity mismatch between the electron bunch and probe pulse through the sample, δt_E and δt_L the durations of the electron bunch and light pulse, and δt_{Sync} the jitter between the electron bunches and optical probe pulses. For most state-of-the-art radiolysis facilities, the resolution is to a high degree determined by the electron bunch duration of several picoseconds. However, as the study on the solvated electron at the sub-picosecond electron pulse facility at Osaka University shows, jitter and group velocity mismatch of pump and probe become predominant factors if shorter electron pump bunches are available. So, while the relativistic electrons in a sample propagate almost with the speed of light (c), in contrast, the velocity of the optical probe pulse is slowed down in first order to c/n where n is the refractive index at the probe wavelength. This results typically in a velocity difference with differential delay of more than 1ps/mm. A trade-off between the transient signal amplitude and reduction of sample thickness must be found. Alternatively, non-collinear propagation of the electron pump to adapt its effective velocity to the velocity of light and tilting its bunch front to be parallel to the wave front of the light were proposed [Yang 2009]. This approach calls for precise control of the electron beam as well as the geometrical configuration of pump and probe size and direction in the sample. In general, for optimized ultrafast pulse radiolysis, parameters such as electron bunch charge and length, sample thickness and the necessary signal-to-noise ratio must be carefully considered to obtain best time resolution.

7.4.2 Nanosecond Heavy Ion Radiolysis in the High LET Regime

Proton and heavy ion sources are largely used in radiation therapy due to their capability to deposit high doses over a short, well-defined depth in the body while preserving healthy tissue along their path. In contrast, facilities that offer pulsed ion sources for scientific time-resolved studies are very rare. This fact is mainly due to the high instrumental effort associated with the cyclotron or synchrotron technology used for particle acceleration. This technology can deliver protons and heavy ion with energies of few ten of MeV up to few GeV and beam currents of µA that are necessary to penetrate samples sufficiently (some millimetres) and to generate detectable concentrations of radiolytic species. In order to access transient phenomena with an adapted temporal detection window, pulse control is needed to extract ion bunches from the bunch train of the cyclotron that is used as pump source. Among the facilities with the skill to resolve transient phenomena after proton and/or heavy ion radiation are the following:

- GANIL (Grand Accélérateur National d'Ions Lourds) of the Commissariat à l'Energie Atomique et aux Energies Alternatives (CEA) and of the Centre National de Recherche Scientific CNRS in Caen, France
- HIMAC (Heavy Ion Medical Accelerator in Chiba) of the National Institute of Radiological Science (NIRS), Japan, with a synchrotron stage
- TIARA (Takasaki Ion Accelerators for Advanced Radiation Application) of the Japan Atomic Energy Agency, Japan

All of these huge installations are shared by different applications in science, industry and/or medicine where the capability for time-resolved measurements is not necessary for some of them. This fact in combination with the expensive beam time renders the realization of experiments or even fixed installations for radiation chemistry difficult and costly. Nevertheless, motivated by the high impact of the results, important work on detection systems and their application to fundamental scientific topics have been performed. An overview of results of pulse radiolysis in water with heavy-ion beams is given in [Baldacchino 2008] and for GANIL in [Baldacchino 2015]. In general, the studies concentrate on the water medium. The yields of radiolytic species such as the solvated electron and the OH• radicals are studied on the nanosecond and microsecond scales for different ion species and energies (Figure 7.5). Similar to electron pulse radiolysis studies on the long several nanoseconds to microsecond time scale, the chemical scavenging method was often used particularly in the early studies. This indirect method measures the time-dependent concentration of the stable product of a reaction with the (initial) radiolytic species. For some initial radiolytic species, with the knowledge of the corresponding reaction rate and the extinction/emission coefficient of the scavenger, the yield of the radiolytic radical can be determined as a function of time. The approach benefits from the high transient (absorption) signal of a scavenger in an easy to access spectral range; it is however limited to rather long-time scales as it requires, on the nanosecond time scale, very high concentrations at which the direct ionization of the solute becomes non-negligible. Multiple reactions with different radiolytic species are possible.

Compared to the precursor technology based on van de Graaf accelerators that already reached the nanosecond scale [Rice 1982], the circular accelerator schemes brought important progress in particle energy and flexibility of ion type species in the 1990s. In contrast, no groundbreaking technical development on time-resolution has been achieved for a long time pointing out that the fundamental limitation of cyclotron acceleration seems to be reached [Baldacchino 2003, Noda 2005, Tetsuya 2008]. The detection schemes are adapted more and more to the nanosecond regime limited by the duration of a single ion bunch.

Optical sensing schemes are applied for probing without mechanical contact. As probe light for absorption spectroscopy, a continuous or quasi-continuous light source is most commonly used; laser diodes, HeNe lasers or pulsed xenon lamps that cover the UV to the near infrared. An absorption change sensitivity better than 10^{-4} can be reached allowing direct detection of the radical species of low extinction and low concentration in the high LET cases [Taguchi 2009]. In addition to the approach that follows the kinetics at a single wavelength, very recent work allows multichannel detection of transient absorption spectra with high sensitivity of 0.1% to 0.3% in absorbance with acquisition times of a few minutes [Iwamatsu 2016]. This approach might open the way for the direct determination of the OH• radical yield after heavy ion radiolysis. Alternatively, detection of emission can be in some cases indirect but sensitive in order to probe the concentrations and lifetimes of transient species in the spurs. A fluorescent state of the solvent or an added solute can be quenched by radiolytic species. Systems with a streak camera and single photon counting were used for time-resolved studies [West 1977, Laverne 1996, Taguchi 1997].

In the high LET regime associated with heavy ion radiolysis, special attention must be paid to the design of the sample and the geometrical configuration of probing. The penetration depth of the beam and the energy deposit along its track with the Bragg peak at its end must be considered. Before experiments, the linear energy transfer is therefore often simulated to quantify the local energy deposit in the probed area. This knowledge of the dose is central to determine transient radiolytic yields. The windows of the sample cell should be shielded against radiation or their possible transient signal should be considered during the analysis of the experimental data. Recently, a temperature- and pressure-regulated optical cell was used to study the OH• radical yield at elevated temperature, towards supercritical solvents [Balcerzyk 2014].

However, the number of time-resolved studies and in consequence the sophistication of our knowledge of the radiation track chemistry in the high LET regime is far from that of the low LET regime. A sound understanding of the spatial dose distribution, the character of the species initially induced by ionization or excitation and the diffusion and reactivity of these species and their successors calls for a number of systematic transient studies over the relevant time scales. To this aim, new installations dedicated to scientific studies and entering the picosecond scale would be very helpful (see Chapter 5). Such installations are also discussed for the optimization of cancer therapy [Getoff 2016].

7.5 Challenges of Future Laser-Driven Particle Acceleration for Femtosecond Radiolysis

The actual limitations and the scientific topics described in the previous sections point to the main challenges for the next generation of time-resolved radiolysis. The evident task is the improvement of time resolution, both for low and high LET. In the latter case, entering the ultrafast regime is going to be a real leap in technology with an improvement of more than two orders of magnitude. Laser-driven particle acceleration has the potential to realize this leap in technology by the design of accelerators with ultrashort proton/ion bunch durations. In the case of new pulsed electron sources for low LET the progress is less tremendous, but nevertheless important. Here the femtosecond scale should be reached in an efficient manner that allows the detection of the initial radiolytic species with medium or low extinction coefficient (<1000 L/mol*cm). So, the essential task is not only to provide an ultrashort source of electron/ion bunches, but also to implement a pump-probe experiment with sufficiently high signal-to-noise ratio and stability to realize time-resolved measurements in a reliable and 'day-to-day' routine (at least on the horizon of some years).

Figure 7.7 summarizes two important parameters of the particle sources for the state of the art of time-resolved radiolysis: particle bunch duration and (kinetic) energy. For the first ultrafast experiments of proton or heavy ion radiolysis a resolution of tens of picoseconds would be already great progress. In this case, the degradation of time resolution due to the velocity mismatch between pump and probe in the sample can be accepted in favour of a reasonable transient signal recorded over several millimetres of optical path. In consequence, the heavy ions should exhibit energies of at least 20 MeV per nucleon in order to penetrate samples such as water over several millimetres. Velocity matching could be achieved for example, by probing an aqueous sample in the visible region and pumping with proton bunches of about 500 MeV kinetic energy. If the probe signal is integrated over the depth of the ion track, particle energy and sample thickness should be adapted for homogenous and known LET over the probed area. Resolving the track structure including the Bragg peak by lateral imaging with a time-resolved detection system would provide particular important insights on the initial conditions of heavy ion radiolysis with high impact on fundamental research and applications such as radiotherapy.

The next wave of ultrashort electron radiolysis experiments calls for optimization of essential electron beam parameters, the configuration in the sample and the optical detection scheme to enter the sub-picosecond scale. As indicated in Figure 7.7, the currently best effective time resolution is distinctly longer than the realized electron bunch duration. Laser-driven particle acceleration overcomes one of

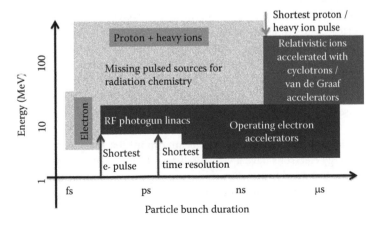

FIGURE 7.7 Particle sources for time-resolved radiolysis arranged according to pulse duration and energy; available regions are depicted in green, missing zones are shaded in blue.

the limitations of the RF photogun accelerator technology, the intrinsic timing jitter of the electron source. For laser-driven particle acceleration, the electron pump bunch and optical probe pulse are intrinsically synchronized because no external accelerating RF fields are used. Therefore, the time resolution depends only on the cross-correlation of pump and probe in the sample, which is determined by the pump and probe durations and their propagation in the sample (see Equation 7.1). The sample thickness should therefore be limited distinctly to less than 1 mm to keep the effect of velocity mismatch below 1 ps. In order to obtain short electron bunches the charge must be restricted to reduce pulse broadening due to coulombic repulsion (i.e. minimizing space-charge effect). Of course, both sample thickness and bunch charge are parameters that affect also the transient signal of radiolytic species to be recorded. In the case of a transient absorbance measurement

$$A(t, \lambda) = \varepsilon(\lambda) \, G(t) \, D \, \rho \, l \qquad (7.2)$$

with $\varepsilon(\lambda)$ is the extinction coefficient of the chemical species at the probe wavelength, $G(t)$ its time-dependent radiolytic yield, D the absorbed dose per particle bunch, ρ the sample density and l the optical path. The latter might be slightly increased with the proposed non-collinear pump-probe scheme [Yang 2009] (see Section 7.1) if the spatial beam overlap is carefully considered as a function of sample depth. For the absorbed dose, the electron fluence of the bunch is decisive; it should be on the scale of several 100 pC per mm^2 to obtain a dose of several Gy.

The example of recent detection of the hydroxyl radical in water with picosecond radiolysis based on the RF photogun technology can provide an estimate for the relevant experimental parameters [El Omar 2011]. For a sample thickness of 5 mm, its absorption, within the temporal resolution of ~10 ps, was found to be about 0.5% for an absorbed dose of about 23 Gy per pulse. Its initial yield was determined to be 4.80 ($\pm$0.12) $\times$ 10^{-7} mol/J with the optical extinction of 460 L mol/cm of OH$^{\bullet}$ at 263 nm. The electron bunches of 7 MeV had a beam diameter near 3 mm in the sample and an overall charge of 4 nC. The given absorbed dose is the peak value probed at the centre of the electron beam with the optical probe near a 0.1 mm diameter. Forty kinetic scans each with about 300 delay steps and for each delay an average of 10 electron bunch pump shots were recorded resulting in a noise of the absorbance change measurement of order of 10^{-4}. At a 10 Hz repetition rate, the overall measurement time was more than 6 hours.

The general feasibility of such an ultrafast pulse radiolysis transient absorption spectroscopy based on laser-driven acceleration has been shown at the Terawatt Ultrafast High Field Facility (TUHF) at the Argonne National Laboratory [Dmitri 2007]. In this proof of principle experiment, the solvated electron was monitored by its high absorption at the laser wavelength of 800 nm with a maximal time resolution of few picoseconds in a perpendicular pump-probe configuration. Shot-to-shot normalization of the transient absorption signal enabled exploitation of the data. The absorption amplitude for an optical probe path of 2 mm was about 0.3%. This study confirmed that it is in principal possible to generate laser-accelerated electron bunches at a repetition rate of 10 Hz with a charge of few nC in the MeV regime in a way sufficiently stable to realize pump-probe experiments. However, it reveals some limitations associated with this recent technology concerning particularly the electron beam quality.

Energetic and spatial dispersion of the electron beam can inhibit sub-picosecond resolution even for thin samples. Low energy electrons have a velocity distribution in matter distinctly below the speed of light leading to a spread in arrival times; they also lose energy and consequently are slowed down in the sample increasing the bunch scattering. For energies of several MeV, these effects are negligible in thin samples below 1 mm. Laser-driven accelerators exhibit polychromatic energy distribution. The energy spread of highly energetic electron bunches does not imply significant change of speed and energy deposit in the sample [Hooker 2013, Malka 2006]. In contrast, lower energy electrons should be removed using an adapted filter that minimizes the bunch scattering. More recently, based on laser-acceleration, the velocity dispersion of the polychromatic electron bunch was reduced to about 400 fs at the probe position [Gaudual 2010].

These first applications of laser-driven acceleration for radiolysis on the low picosecond scale use transient absorption that is also the very predominant probing approach at the operating RF photogun facilities. Entering the femtosecond scale can enable us to resolve and to understand the ultrafast deactivation processes of the radiolytically generated excited states that take place in parallel to the physicochemical mechanisms in the ground state. In addition to the possible excited state absorption and stimulated emission accessible by transient absorption measurements, fluorescing probing by optical Kerr gating can help to understand the expected complex transient signals.

A promising alternative for the repetitive pump-probe measurement with scanning of an optical delay line can be single-shot detection [Shin 2014] (see Chapter 9). This fastest possible approach could increase the efficiency and signal-to-noise ratio of radiolysis experiments with laser sources operating at a repetition rate above 10 Hz. Most notably, single-shot radiolysis enables the application of laser-driven accelerator schemes to operate at lower repetition rate, 0.03–1 Hz; it can therefore increase significantly the field of laser-driven plasma accelerators to be used for femtosecond radiolysis. A third important motivation for single-shot, pump-probe spectroscopy is the current limitation of laser-driven plasma accelerators concerning the stability of the bunch. Currently, parameters such as charge, beam pointing, and/or energy spread exhibit large shot-to-shot jitter [Hooker 2013], that can impact the absorbed dose in the optically probed area. Single-shot optical detection allows the normalization of each kinetic trace by the actual dose.

In conclusion, this efficient optical probing could help to accelerate the application of the laser-driven accelerator technology with its potential for highest time resolution for electron radiolysis. A second generation of tabletop femtosecond radiolysis with improved operation parameters could implement spectrally broadband probing in a repetitive manner. In the high LET regime, the use of state-of-the-art electronic detection schemes based on streak cameras or transient digitizers enables sufficient time resolution to launch a new age of radiation physics and chemistry. The realization of dedicated installations for heavy ion radiolysis provides not only access to experimentally unexplored time scales but opens also the route for systematic studies of the complex conditions in the tracks. The design of future ultrafast radiolysis laboratories should also consider, for the long term, the possibility to adapt the experimental conditions. Temperature and pressure should be tuneable, preferentially up to the supercritical regime, in regard to applicative studies for operation of high pressure nuclear reactors. Such laboratories would open the possibility for answering many fundamental scientific questions in physics, chemistry and biology that are often directly linked to applications such as radiotherapy, nuclear engineering and waste management.

References for Chapter 7

Andrew, C. and Shen, Y.Z. 2009. "Optical fiber-based single-shot picosecond transient absorption spectroscopy" *Rev. Sci. Instrum.* **80**: 073106 1–7.

Balcerzyk, A., Boughattas, I., Pin, S., Balanzat, E. and Baldacchino, G. 2014. "First observation of OH· reactivity in water under high energy ions at elevated temperature" *Phys. Chem. Chem. Phys.* **43**: 23975–23984.

Baldacchino, G. 2008. "Pulse radiolysis in water with heavy-ion beams. A short review" *Radia. Phys. Chem.* **77**: 1218–1223.

Baldacchino, G. 2015. "Water radiolysis with heavy-ion beams at Ganil. back to 20 years of investigations" *J. Phys: Confe Ser.* **629**: 012009 1–7.

Baldacchino, G., Vigneron, G., Renault, J.P., Pin, S., Remita, S., Abedinzadeh, Z., Deycard, S., Balanzat, E., Bouffard, S., Gardes-Albert, M., Hickel, B. and Mialocq, J.C. 2003. "A nanosecond pulse radiolysis study of the hydrated electron with high energy carbon ions" *Nucl. Instr. Meth. Phys. Res.* **209**: 219–223.

Belloni, J., Monard, H., Gobert, F., Larbre, J.-P., Demarque, A., De Waele, V., Lampre, I., Marignier, J.-L., Mostafavi, M., Bourdon, J.-C., Bernard, M., Borie, H., Garvey, T., Jacquemard, B., Leblond, B., Lepercq, P., Omeich, M., Roch, M., Rodier, J. and Roux, R. 2005. "ELYSE—A picosecond electron accelerator for pulse radiolysis research." *Nucl. Instrum. Meth. A.* **539**: 527–539.

Belloni, J. and Mostafavi, M. 2001. Metal and Semiconductor Clusters, *Studies in Physical and Theoretical Chemistry 87. Radiation Chemistry: Present Status and Future Trends,* Eds. Jonah C.D., Rao M. Elsevier. 411–450.

Bird, M.J., Reid, O.G., Cook, A.R., Asaoka, S., Shibano, Y., Imahori, H., Rumbles, G. and Miller, J.R. 2014. "Mobility of holes in oligo- and poly fluorenes of defined lengths." *J. Phys. Chem. C.* **118**: 6100–6109.

Bronskill, M.J., Wolff, R.K. and Hunt, J.W. 1970. "Picosecond pulse radiolysis studies. I. the solvated electron in aqueous and alcohol solutions." *J. Chem. Phys.* **53**: 4201–4210.

Buxton, G.V., Greenstock, C.L., Helman, W.P. and Ross, A.B. 1988. "Critical-Review of rate constants for reactions of hydrated electrons, hydrogen-atoms and hydroxyl radicals ($\cdot$OH/$\cdot$O$^-$) in aqueous-solution." *J. Phys. Chem. Ref. Data.* **17**: 513–886.

De Waele, V., Schmidhammer, U., Jean-Raphael, M., Monard, H., Jean-Philippe, L., Nicolas, B. and Mostafavi, M. 2009. "Non-invasive single bunch monitoring for ps pulse radiolysis." *Radiat. Phys. Chem.* **78**: 1099–1101.

Dmitri, A.O., Robert, A., Gosztola, D.J., Shkrob, I.A., Korovyanko, O.J. and Rey-de-Castro, R.C. 2007. "Ultrafast pulse radiolysis using a terawatt laser wakefield accelerator." *J. Appl. Phys.* **101**: 053102 1–9.

El Omar, A.K., Schmidhammer, U., Jeunesse, P., Larbre, J.-P., Lin, M., Muroya, Y., Katsumura, Y., Pernot, P. and Mostafavi, M. 2011. "Time-dependent radiolytic yield of OH radical studied by picosecond pulse radiolysis." *J. Phys. Chem. A.* **114**: 12212–12216.

Gauduel, Y.A., Glinec, Y., Rousseau, J.-P. et al. 2010. "High energy radiation femtochemistry of water molecules: Early electron-radical pairs processes." *Eur. Phys. J. D.* **60**: 121–135.

Getoff, N. 2016. "Pulse radiolysis using very-high-energy ions for optimizing cancer therapy." *In Vivo.* **30**: 118–120.

Grills, D.C., Farrington, J.A., Layne, B.H., Preses, J.M., Bernstein, H.J. and Wishart, J.F. 2015. "Development of nanosecond time-resolved infrared detection at the LEAF pulse radiolysis facility." *Rev. Sci. Instrum.* **86**: 044102 1–10.

Hart, E.J. and Boag, J. 1962. "Absorption spectrum of the hydrated electron in water and in aqueous solutions." *J. Am. Chem. Soc.* **84**: 4090–4095.

Hatano, Y., Katsumura, Y. and Mozumder, A. 2010. *Charged Particle and Photon Interactions with Matter: Recent Advances, Applications, and Interfaces.* CRC Press.

Hooker, M.D. 2013. "Developments in laser-driven plasma accelerators." *Nat. Photon.,* **10**: 775–782.

Iwamatsu, K., Muroya, Y., Yamashita, S., Kimura, A., Taguchi, M. and Katsumura, Y. 2016. "Quick measurement of continuous absorption spectrum in ion beam pulse radiolysis: Application of optical multi-channel detector into transient species observation." *Radia. Phys. Chem.* **119**: 213–217.

Jonah, C.D. 1975. "A wide-time range pulse radiolysis system of picosecond time evolution." *Rev. Sci. Instrum.* **46**: 62–66.

Kan, K., Kondoh, T., Yang, J., Ogata, A., Norizawa, K. and Yoshida, Y. 2012. "Development of double-decker pulse radiolysis." *Rev. Sci. Instrum.* **83**: 073302 1–5.

Kozawa, T., Mizutani, Y., Yokoyama, K., Okuda, S., Yoshida, Y. and Tagawa, S. 1999. "Measurement of far-infrared subpicosecond coherent radiation for pulse radiolysis." *Nucl. Instrum. Meth. A.* **429**: 471–475.

Kroh, J. 1989. "Early Developments in radiation chemistry." *Royal Society of Chemistry.* **513** 1st Edition.

Laverne, J.A. 1996. "Fluorescence in the heavy ion radiolysis of benzene." *J. Phys. Chem.* **100**: 18757–18763.

Lund, A. and Shiotani, M. 2014. "Applications of EPR in radiation research." *Springer.* **396**.

Ma, J., Yamashita, S., Muroya, Y., Katsumura, Y. and Mostafavi, M. 2015. "Deciphering the reaction between a hydrated electron and a hydronium ion at elevated temperatures" *Phys. Chem. Chem. Phys.* 17: 22931–22939.

Malka, V., Lifschitz, A., Faure, J. and Glinec, Y. 2006. "Staged concept of laser-plasma acceleration toward multi-GeV electron beams" *Phys. Rev. Accel. Beams.* 9: 091301 1–7.

Muroya, Y., Lin, M., Han, Z., Kumagai, Y., Sakumi, A., Ueda, T. and Katsumura, Y. 2008. "Ultrafast pulse radiolysis: A review of the recent system progress and its application to study on initial yields and solvation processes of solvated electrons in various kinds of alcohols" *Radiat. Phys. Chem.* 77: 1176–1182.

Muroya, Y., Lin, M., Iijima, H., Ueda, T. and Katsumura, Y. 2005. "Current status of the ultra-fast pulse radiolysis system at NERL, the University of Tokyo" *Res. Chem. Intermediat.* 31: 261–272.

Noda, K., Tann, D., Uesugi, T., Shibuya, S., Honma, T. and Hashimoto, Y. 2005. "Production of short-pulsed beam for ion-beam pulse radiolysis" *Nucl. Instr. Meth. Phys. Res.* 240: 18–21.

Prins, P., Ferdinand, C., Galbrecht, F.G., Ullrich Scherf, U. and Siebbeles, L.D.A. 2007. "Charge transport along coiled conjugated polymer chains" *J. Phys. Chem. C.* 111: 11104–11112.

Rice, S.A., Playford, V.J., Burns, W.G. and Buxton, G.V. 1982. "Nanosecond proton pulse radiolysis" *J. Phys. E: Sci Instrum.* 15: 1240–1243.

Saeki, A., Kozawa, T. and Tagawa, S. 2006. "Picosecond pulse radiolysis using femtosecond white light with a high S/N spectrum acquisition system in one beam shot" *Nucl. Instrum. Meth. A.* 556: 391–396.

Sakaue, K., Gowa, T., Hayano, H., Kamiya, Y., Kashiwagi, S., Kuroda, R., Masuda, A., Moriyama, R., Urakawa, J., Ushida, K., Wang, X. and Washio, M. 2008. "Recent progress of a soft X-ray generation system based on inverse compton scattering at Waseda University" *Radiat. Phys. Chem.* 77: 1136–1141.

Schmidhammer, U., de Waele, V., Marques, J.-R., Bourgeois, N. and Mostafavi, M. 2009. "Single shot linear detection of 0.01-10 THz electromagnetic fields" *Appl. Phys. B.* 94: 95–101.

Schmidhammer, U., Jeunesse, P., Stresing, G. and Mostafavi, M. 2014. "A broadband ultrafast transient absorption spectrometer covering the range from near-Infrared (NIR) down to green" *Appl. Spectrosc.* 68: 1137–1147.

Sauer Jr., M.C., Jonah, C.D., Le Motais, B.C. and Chernovitz, A.C. 1988. Sources of excited cyclohexane in the radiolysis of cyclohexance. *J. Phys. Chem.* 92: 4099–4103.

Schmidhammer, U., Pernot, P., de Waele, V., Jeunesse, P., Demarque, A., Murata, S. and Mostafavi, M. 2010. "Distance dependence of the reaction rate for the reduction of metal cations by solvated electrons: A picosecond pulse radiolysis" *J. Phys. Chem. A.* 114: 12042–12051.

Shahdo, M., Emmanuel, A., Daniel, M., Demarque, A., Larbre, J.P., Marignier, J.L. and Mostafavi, M. 2013. "Concomitant transient electrochemical and spectroscopic detection with electron pulse radiolysis" *Electrochem. Commun.* 35: 149–151.

Shin, T., Wolfson, J.W., Teitelbaum, S.W. et al. 2014. "Dual echelon femtosecond single-shot spectroscopy" *Rev. Sci. Instrum.* 85: 083115 1–10.

Shinichi, Y., Katsumura, Y., Lin, M.Z., Muroya, Y., Maeyama, T. and Murakami, T. 2008. "Water radiolysis with heavy ions of energies up to 28 GeV—2: Extension of primary yield measurements to very high LET values" *Radia. Phys. Chem.* 77: 1224–1229.

Siefermann, K.R., Liu, Y., Lugovoy, E., Link, O., Faubel, M., Buck, U., Winter, B. and Abel, B. 2010. "Binding energies, lifetimes and implications of bulk and interface solvated electrons in water" *Nat. Chem.* 2: 274–279.

Spotheim-Maurizot, M., Mostafavi, M., Douki, T., Belloni, J., (Eds). 2008. *Radiation Chemistry: From Basics to Applications in Material and Life Sciences.* EDP Sciences. 307 Pages.

Taguchi, M., Aoki, Y., Namba, H., Watanabe, R., Matsumoto, Y. and Hiratsuka, H. 1997. "Fast fluorescence decay of naphthalene induced by Ar ion irradiation" *Nucl. Instr. Meth. Phys. Res.* 132: 135–143.

Taguchi, M., Baldacchino, G., Kurashima, S., Kimura, A., Sugo, Y., Katsumura, Y. and Hirota, K. 2009. "Transient absorption measurement system using pulsed energetic ion" *Radia. Phys. Chem.* **78**: 1169–1174.

Tetsuya, F., Soma, I., Shinji, S., Koji, N., Toshiyuki, S., Hiromu, T. and Akira, N. 2008. "Formation and fast extraction of a very short-bunched proton beam for the Investigation of free radicals" *Nucl Instrum Meth A.* **588**: 330–335.

The Radiation Chemistry Data Center of the Notre Dame Radiation Laboratory. Retrieved from "https://www3.Nd.Edu/~Ndrlrcdc/."

Trifunac, A.D. and Smith, J.P. 1980. "Optically detected time resolved EPR of radical ion pairs in pulse radiolysis of liquids" *Chem. Phys. Lett.* **73**: 94–97.

Turi, L., and Rossky, P.J. 2012. "Theoretical studies of spectroscopy and dynamics of hydrated electron" *Chem. Rev.* **112**: 5641–5674.

Virender, K. and Sharma. 2013. "Oxidation of amino acids, peptides, and proteins: Kinetics and mechanism" Wiley, Steven E. Rokita, 420 Pages.

West, M.L. 1977. "Quenching of benzene fluorescence in pulsed proton irradiation" *J. Phys. Chem.* **81**: 377–380.

Wishart, J.F. 1998. "Accelerators and other sources for the study of radiation chemistry" in *Photochemistry and Radiation Chemistry*, Eds. Wishart J.F., and D.G. Nocera. *Adv. Chem. Ser.* **254**: 35–50.

Wishart, J.F. and Rao, B.S. eds. 2010. "Recent trends in radiation chemistry" *World Scientific*, 636 pages.

Yang, J.F., Kondoh, T., Kan, K. and Yoichi Yoshida, Y. 2011. "Ultrafast pulse radiolysis" *Nucl. Instrum. Meth. A.* **629**: 6–10.

Yang, J.F., Kondoh, T., Norizawa, K., Yoshida, Y. and Tagawa, S. 2009. "Breaking time-resolution limits in pulse radiolysis" *Radia. Phys. Chem.* **78**: 1164–1168.

Yang, J., Kondoh, T., Yoshida, A. and Yoshida, Y. 2006. "Double-decker femtosecond electron beam accelerator for pulse radiolysis" *Rev. Sci. Instrum.* **77**: 043302 1–5.

8

Application of Laser-Driven Beams for Radiobiological Experiments

Anna A. Friedl

Thomas E. Schmid

8.1 Introduction

Laser-driven particle sources offer great potential for application in radiation biology and radiation therapy. If the promise of adequate size reduction of laser-driven accelerators is achieved, they may in future replace conventional accelerators as tools for fundamental radiobiological experiments. Presently, an increasing number of conventional accelerators are in danger of being shut down due to cost reasons or because the facilities are getting very old. In addition, due to their special characteristics, laser-driven beams allow performing new types of experiments, e.g. simultaneous exposure of cells to photons and/or different particle types and thus radiation qualities* to mimic the situation encountered by astronauts in space or other complex exposure situations. So far, the first proof-of-principle experiments performed with laser-driven particle beams have mainly concentrated on the question of whether their ultra-high dose rates influence the cellular response mechanisms and sensitivity. While this question is of high relevance for future therapeutic applications, other applications will certainly be added in future.

---

* The term 'radiation quality' refers to the different patterns of energy deposition of various radiation types with different energies, which lead to different biological effects at the same absorbed dose.

8.2 Importance of Radiobiology for Understanding Fundamental Processes, Optimization of Radiation Therapy and Determination of Risks Associated with Exposures

Radiobiology is the science of the response of living systems to ionizing radiation. This comprises the identification and quantification of lesions in the DNA and other cellular structures induced by irradiation and the characterization of the complex response mechanisms that cells possess in order to deal with the damage induced. Elaborate mechanisms ascertain that lesions are detected, signal pathways are activated, lesions are repaired, and – if this is not possible – cells stop division and possibly enter a controlled form of cell death [Goldstein 2014]. Elucidation of these pathways and the factors and genes involved has contributed to our general understanding of cellular activities and facilitates the identification of pathogenic phenotypes associated with enhanced radiation sensitivity, as well as the estimation of radiation-associated risk. It is expected that this will, in the future, allow for the optimization of radiation therapy by the personalization of treatment concepts [Shah 2012, Boss 2014, Ree 2015].

Recent years have seen an increasing number of studies dealing with quantitative and qualitative differences in the cellular response mechanisms to different radiation qualities. While most of our knowledge on the cellular response to radiation stems from experiments conducted with sparsely ionizing photons (X-rays or gamma rays), the effects of irradiation with particles, especially densely ionizing particles, need further attention. Interest in the biological effects of particles stems from the fact that protons and heavier ions make up a significant part of the radiation encountered in space, which makes them an important factor for long-term missions. In addition, exposure to alpha particle-emitting radioactive materials at the workplace and in the environment (e.g. through radon and its decay products) contributes significantly to the total radiation exposure of human beings. However, the main reason for the increased interest is that radiation therapy using particles (mainly protons and carbon ions) is increasingly being utilized [Jermann 2014], due to its superior depth dose profile, which makes it possible to reduce dose to normal tissue.

Photons and charged particles differ in the pattern of energy deposition, with more localized energy deposition occurring from irradiation with charged particles, which results in an increased linear energy transfer (LET) and the induction of more DNA lesions exhibiting poor reparability [Goodhead 2006]. This enhances the biological effectiveness of high LET irradiation, so that less energy dose is required to obtain a given biological effect compared to photon irradiation (for further discussion of biological effects of ion irradiation see also Chapter 10 by Shikazono, Moribayashi and Bolton). The ratio of doses necessary for two radiation types (one of them being photon radiation as a reference) to induce a given effect is termed relative biological effectiveness (RBE). The RBE of heavier ions is much greater than 1, with the value roughly depending on LET. The RBE of high-energy protons, however, should not be very different from that of photons due to the similar low LET values. Pragmatically, in therapy settings an RBE of 1.1 for protons is assumed; however, optimized treatment planning would require more accurate, energy- and position-dependent RBE values [Paganetti 2015].

There is a high degree of variation in published RBE–LET relations (reviewed by Tommasino [2015]), emphasizing the importance of deeper and more accurate knowledge of the cell and tissue response to proton radiation. Several differences in the biological response induced by protons compared to photons were reported in the last decades. Even at comparable LET, various biological effects and response mechanisms induced by protons may differ from those obtained after photon irradiation, e.g. the amount and complexity of lesions, reparability of lesions and repair pathway choice, epigenetic patterns, gene expression patterns, angiogenesis and cell migration [Girdhani 2013, Tommasino 2015]. Similarly, heavier ions appear to induce response reactions that differ in quality from those induced by low–LET photon or proton irradiation (reviewed by Durante [2014]). It should be stressed, however, that direct comparisons of effects of different radiation types in most experimental settings were hampered by the limited comparability of experimental parameters that resulted from the irradiations being performed at different facilities using different samples. Further experiments will be required to unequivocally demonstrate qualitative response differences towards different radiation qualities and to understand the pathways involved.

8.3 Can Laser-Driven Particle Sources Elicit Fundamentally Different Responses as Compared to Conventionally Accelerated Beams?

So far, most of the biological experiments with laser-driven particle bunches were performed with the intention of setting the stage for future applications in radiotherapy, especially radiotherapy with accelerated protons and heavier ions. For further discussion of laser-driven beam applications in radiotherapy, see Chapter 11 by Enghardt, Pawelke and Wilkens. Already early after the first description of laser-driven particle acceleration (to tens of MeV energies), its potential application for therapy was discussed [Bulanov 2002]. A variety of differences between laser-driven and conventionally* driven particle beams was anticipated, including the exponential energy distribution and ultra-high dose rate. During the early acceleration phase at the source (i.e. laser target) single ion bunches are of tens of femtosecond duration with bunch charge adequate to deliver several Gy doses. However, due to their intrinsically large energy spread (typically characterized by an exponential energy distribution or spectrum) and commensurate debunching during transport to a biological sample or tumour site, the delivered dose rates are expected to be about 10^9 Gy/s or 1 Gy per nanosecond [Dollinger 2009]. Experiments using X-rays from the 1960s and 1970s suggested that at ultra-high dose rates, the cellular effects could be reduced due to depletion of intracellular oxygen. The well-known oxygen effect, i.e. enhanced radiation effects in the presence of oxygen as compared to hypoxic or anoxic conditions, is explained by higher radiation-induced production of free radicals and faster fixation of damage in the presence of oxygen. A recent literature review [Wilson 2012] delineated the conditions under which oxygen depletion may affect cellular radiation effects: for X-ray and electron irradiation at instantaneous doses rates of 10^9 Gy/s and higher and at single-shot delivery of sufficiently high doses (5–10 Gy), oxygen depletion can occur if the cells already have low oxygen levels. However, no data are available at present on oxygen depletion caused by other charged particles.

Several authors have demonstrated that laser-driven bunches cause DNA damage, as seen by formation of radiation-induced foci or strand breakage, without aiming to compare the effects to those of conventional irradiation [Yogo 2009, Kraft 2010, Rigaud 2010].

Others compared the effects of the laser-driven bunches to conventionally generated X-rays or gamma rays. Thus, any difference observed may be due not only to the dose rate, but also to differences in radiation quality and energy distribution. Tillman et al. [1999] compared colony formation after irradiation of hamster cells with laser-produced X-rays to irradiation with conventional X-rays and gamma rays and did not observe a marked difference. Although the mean peak dose rate of 10^{10}–10^{11} Gy/minute was rather high, the delivered single pulse dose was very low (1/600 Gy), and thus oxygen depletion would not be expected. Shinohara et al. [2004] compared the effects of laser-generated X-rays to gamma rays on colony formation of mouse lymphocytic leukaemia cells. While their instantaneous dose rate of 10^{12}–10^{13} Gy/second at a pulse duration of less than 1 picosecond meets the conditions where oxygen depletion may occur, differences in the observed survival might also be explained by the different energy spectra of both radiation qualities. Yogo et al. [2011] compared colony formation of human cancer cells after irradiation with laser-driven protons and X-ray irradiation. At an instantaneous dose rate of 10^7 Gy/second (delivering 0.2 Gy per 20 nanosecond bunch) conditions of oxygen depletion would not be expected and the observed RBE of 1.20 ± 0.11 at 10% survival for proton irradiation is well in line with published RBE data on conventionally accelerated protons of similar energy [Folkard 1996]. A similar RBE of 1.4 ± 0.2

---

* In the context of this chapter, conventionally accelerated refers to any non-laser acceleration.

at 10% survival, as compared to 225 kVp X-rays, was observed by Doria et al. [2012]. These authors investigated colony formation of hamster cells irradiated with laser-driven protons (1–5 MeV) at a dose rate of 10^9 Gy/second and doses up to 5 Gy delivered in a single shot, thus using conditions where oxygen depletion is possible, depending on the oxygenation status of the cells during the experiment. Bin et al. [2012] irradiated human cancer cells with laser-driven quasi-monoenergetic protons (5.2 MeV), at a peak dose rate of 7×10^9 Gy/second, thus delivering several Gy per single nanosecond bunch. They determined the number of gamma-H2AX foci 30 minutes after irradiation, which is a well-characterized method to quantify DNA double-strand breaks [Bouquet 2006, Redon 2012]. For comparison, cells were irradiated with 200 kV X-rays. At 30 foci per cell an RBE 1.3 ± 0.3 was determined, which agrees well with the survival data described above. The impact of laser-driven electrons was determined by researchers from Dresden for various endpoints. Laschinsky et al. [2012] irradiated malignant and non-malignant human cell lines with laser-driven electrons at a peak dose rate of 2.4×10^9 Gy/second and a single bunch dose of 2.4×10^{-3} Gy. These authors also drew attention to the fact that laser-driven bunches, while characterized by ultra-high instantaneous peak dose rates during the bunches, can have rather low average dose rates because of their low repetition rates. For example, in the experiments described by Laschinsky et al., the average dose rate was 0.36 Gy/minute, so the irradiation time in that reported work was about 30 minutes. Under these conditions a significant fraction of the lesions may be repaired during the irradiation procedure, thus possibly reducing the yield of complex lesions as compared to irradiation with dose rate of several Gy/minute. As reference radiation, these authors used an electron beam from a conventional therapy LINAC. The authors undertook great efforts to level out any differences between the beams (apart from dose rate) as much as possible. The authors state that their overall results show no significant differences in radiobiological response for *in vitro* cell experiments between laser-accelerated pulsed and clinically used electron beams, except for one endpoint where the laser-accelerated electrons appeared to exhibit an RBE less than 1.

Summarizing, the experiments published so far on the biological effects of laser-driven beams did not demonstrate ultra-high dose rate effects such as oxygen depletion, even for peak dose rates above 10^9 Gy/second and single bunch doses of order a few Gy. It should be noted, however, that even under these conditions cells grown in monolayers may not experience oxygen depletion because oxygen tension is too high. It is therefore important to also investigate the effects of laser irradiation in 3D culture systems or *in vivo* tumour models, where oxygen tension in the cells is expected to be lower. In this regard, the mouse ear tumour models described by Brüchner et al. [2014] and Oppelt et al. [2015] are important and relevant steps for near term applications. Human tumour cells were inoculated in mouse ears to form tumours and they were irradiated after a certain size was reached, under which conditions hypoxic regions were detectable. Irradiation delayed tumour growth to a similar extent after irradiation with laser-driven electrons and reference X-ray irradiation. However, it should be noted that the single bunch dose used in their work (of order 100 mGy) might be too low for oxygen depletion to occur.

In order to study the effects of ultra-high dose rate while eliminating differences in radiation quality, energy and energy distribution, irradiation set-up and other factors, researchers in Munich have developed a pulsed ion microirradiation set-up that allows delivering protons (20 MeV) with a single bunch dose of several Gy (of duration less than 1 nanosecond), thus obtaining peak dose rates greater than 10^9 Gy/second [Dollinger 2009, Greubel 2011]. A variety of radiobiological endpoints were investigated in a variety of biological systems, including 3D tissue and a tumour model. In only one case, a statistically significant difference was found between pulsed and continuous irradiation, which presumably is a chance event (see Table 8.1).

TABLE 8.1   RBE Estimation with Pulsed and Continuous Proton Beams at Munich Microbeam SNAKE

Endpoint	Biological System	RBE for Continuous	RBE for Pulsed	p-Value	Reference
Double-strand breaks (γH2AX)	HeLa cells	1.13 ± 0.21 1.16 ± 0.09	0.97 ± 0.19 1.13 ⊥ 0.09	n.s.	Zlobinskaya 2012
DNA strand damage (Comet)	HeLa cells	1.18 ± 0.21	1.11 ± 0.23	n.s.	unpublished data[b]
Micronuclei	HeLa cells	1.06 ± 0.07 1.05 ± 0.07	1.07 ± 0.07 1.09 ± 0.08	n.s.	Schmid 2009
Micronuclei	EpidermFT tissue	1.22 ± 0.15 1.13 ± 0.14	1.08 ± 0.20 1.00 ± 0.14	n.s.	Schmid 2010
Various chromosome aberrations	AL cells	0.78 ± 0.26 0.93 ± 0.04	0.73 ± 0.17 0.87 ± 0.05	n.s.[a]	Schmid 2011
Apoptosis induction	EpidermFT tissue	1.42 ± 0.12	1.24 ± 0.12	n.s.	unpublished data[b]
Colony formation	HeLa cells	0.91 ± 0.26	0.83 ± 0.33	n.s.	Auer 2012
Tumour growth delay	transplanted tumour	1.10 ± 0.18	1.22 ± 0.19	n.s.	Zlobinskaya 2014

[a] For all individual aberration types determined and for total aberrations, the difference was not significant (n.s.). For a combination of three types, a significant difference was obtained.

[b] T.E. Schmid et al., unpublished.

8.4 Laser-Driven Ion Sources as Workhorse for Radiobiological Experiments – State of the Art and Future Requirements

Today, the number of particle irradiation facilities accessible for researchers working with cells, tissues and animal models is rather limited. Furthermore, in a clinical setting access may be restricted concerning available time and the type of experiments that can be performed. On the other hand, only very few facilities are exclusively dedicated to experimental use, and their number appears to decrease rather than increase, in spite of the increasing need for basic radiobiological studies of particle irradiation [Durante 2014, Schlaff 2014]. Assuming that dose rate effects do not significantly affect the cellular response to irradiation, laser-driven particle sources may become an adequate complement to conventional beams in future experiments aiming at investigating the effects of particle irradiation. Given that laser-acceleration facilities are spreading rapidly, laser-driven sources can be envisaged to provide additional experimental options in the future. Beam requirements such as maximum energy, narrow energy spread and reasonable repetition rate will be less stringent in experimental use than is envisaged for the use of laser-driven beams in radiotherapy [Kitagawa 2010, Hofmann 2015]. Implementation of routine experiments at laser-driven accelerators is realistic in the near future and will foster the process of maturing this young technology. As a guide, Table 8.2 lists minimum requirements to enable radiobiological experiments.

So far, research has concentrated on protons and carbon ions, due to the fact that particle radiotherapy is mainly conducted with these ions. Future therapy concepts may, however, include other ions such as helium, lithium or oxygen ions [Burigo 2015, Grün 2015, Malinen 2015]. Therefore, broadening the spectrum of accelerated ions amenable to laser acceleration should be an important issue in future.

TABLE 8.2   Minimum Beam Requirements for Radiobiological Experiments

Type of Experiment	Radiation	Requirements
Cellular response to different radiation qualities (2D cell culture) Subtype: different fields irradiated with different radiation at same biological sample for microscopic comparison	Photons, e⁻, p⁺, carbon ions, other ions	- Range in air > several mm - Range in water > several tens of μm - Single pulse/bunch doses up to several Gy - Reasonably narrow energy spread - For certain types of experiments irradiation fields of several cm² to obtain sufficient quantity of cells - Combination of beam line with experimental set-up such as live-cell imaging microscopy or incubators allowing adjustment of oxygen tension - Comparable doses in both fields and ± simultaneous irradiation
Cellular response to different radiation qualities (3D cell culture/tissues/whole animals)	Photons, e⁻, p⁺, carbon ions, other ions	Same as above (2D experiments), but range in water/tissue extended to mm- or cm-scale
Simulation of space irradiation	Photons, e⁻, p⁺, carbon ions, other ions, as single or mixed beams	Same as above (2D/3D experiments), but broad exponential energy distribution
Effect of mixed irradiation fields	Mixed beams of photons plus particles	Same as above (2D/3D experiments)

8.5 Exploitation of Specific Characteristics of Laser-Driven Beams: Broad Energy Spectrum and Mixed Beams

Considering biological experiments with laser-driven particle bunches, emphasis in recent years was mainly on aiming to adjust their performance to that of conventionally accelerated beams. It may be wise, however, to exploit the specific characteristics of laser-driven sources. One example is the broad, more or less exponentially tapered energy distribution (i.e. spectrum), which has been noted to resemble the energy distribution of space radiation [Hidding 2011, Königstein 2012]. Most radiation occurring in space has an exponential or power-law energy distribution that cannot easily be reproduced with conventional accelerators, as these normally produce monoenergetic beams. On the other hand, there is a strong need for beam times to test electronic devices and materials that may be exposed to space radiation during missions (which can include exposure in the radiation belts of Earth and other planets). Since radiation exposure is regarded as a main problem in long-term manned space missions [Barcellos-Hoff 2015, Kim 2015], in a similar manner it might be worthwhile to investigate the biological effects of beams with broad energy distributions. More detail about laser-driven beam applications to exploring the effects of the space environment on biological systems and solar cells can be found in Chapters 14 by Reitz and Hellweg and 15 by Ohshima and Imaizumi, respectively. It should be noted, however, that dose rates in space are much lower than what is typically achieved with laser-driven particle sources.

Another important topic in risk estimation associated with space travel, but also in radiation therapy and the exposure of workers, is exposure to mixed beams [Staaf 2012a, Sotnik 2014, Kim 2015, Schneider 2015]. So far, for technical reasons, mixed beam exposure situations have mainly been simulated by subsequent exposures to different radiation qualities (i.e. not a simultaneous mix). Observed data were mainly interpreted in terms of adaptive response and bystander effects [Tseng 2013, Buonanno 2015]. Only a few investigations involved simultaneous exposure to different radiation qualities [Phoenix 2009, Staaf 2012a, Staaf 2012b, Staaf 2013]. Because some authors have observed synergistic effects of mixed beam irradiation [Staaf 2012a, Staaf 2012b, Staaf 2013], further experiments are required to elucidate cellular response mechanisms to simultaneous exposure to different radiation qualities and energies. Laser-driven sources provide a mixture of photons and particles present within the target and efforts have so far mainly concentrated on the preparation of single species as desired, for example, for radio-therapy applications [Hoffmann 2012]. Future studies should also encompass the generation of well-defined mixed beams for the simultaneous exposure of biological samples, together with the respective single component beams, for comparative analysis of their biological effects. In addition, for certain types of experiments (e.g. microscopic evaluation of damage response reactions in cells) it may be desir-able to select two beams of different radiation quality from the mixture and target them to different irra-diation fields on one biological sample. With such an approach, the response reactions and biological effects can be compared within one biological sample, thus minimizing variation caused by differential handling of samples.

8.6 Conclusion

Following a first wave of proof-of-principle experiments, biological research with laser-driven par-ticle bunches is expected to further mature into well-designed studies that specifically make use of the unique possibilities offered by this technology. First, concepts aiming at exploiting broad energy distribution and production of mixed beams have been proposed and new ideas will hopefully follow. The question of biological effects of ultra-high dose rates, while in the focus of recent research activi-ties, has not fully been resolved as of yet. Further experiments under conditions of limited oxygen supply are needed. Finally, there is still a strong need for basic radiobiological studies on the effects of irradiation by particles as compared to photons, which under the prevalent conditions of limited beam time are difficult to conduct with conventional accelerators. Laser-driven particle sources can offer complementary approaches, new technological solutions and also foster new ideas in the field of radiobiology.

References for Chapter 8

Auer, S., Hable, V., Greubel, C., Drexler, G.A., Schmid, T.E., Belka, C., Dollinger, G. and Friedl, A.A. 2011. "Survival of tumor cells after proton irradiation with ultra-high dose rates" *Radiat. Oncol.* **6**: 139.

Barcellos-Hoff, M.H., Blakely, E.A., Burma, S., Fornace, A.J. Jr., Gerson, S., Hlatky, L., Kirsch, D.G., Luderer, U., Shay, J., Wang, Y., Weil, M.M. 2015. "Concepts and challenges in cancer risk predic-tion for the space radiation environment" *Life Sci. Space Res. (Amst.)* **6**: 92–103.

Bin, J., Allinger, K., Assmann, W., Dollinger, G., Drexler, G.A., Friedl, A.A., Habs, D., Hilz, P., Hoerlein, R., Humble, N., Karsch, S., Khrennikov, K., Kiefer, D., Krausz, F., Ma, W., Molls, M., Michalski, D., Raith, S., Reinhardt, S., Schmid, T.E., Tajima, T., Wenz, J., Zlobinskaya, O., Schreiber, J. and Wilkens, J.J. 2012. "A laser-driven nanosecond proton source for radiobiological studies" *Appl. Phys. Lett.* **101**: 243701.

Boss, M.K., Bristow, R. and Dewhirst, M.W. 2014. "Linking the history of radiation biology to the hall-marks of cancer" *Radiat. Res.* **181**: 561–577.

Bouquet, F., Muller, C. and Salles, B. 2006. "The loss of gammaH2AX signal is a marker of DNA double strand breaks repair only at low levels of DNA damage" *Cell Cycle.* **5**: 1116–1122.

Buonanno, M., De Toledo, S.M., Howell, R.W. and Azzam, E.I. 2015. "Low-dose energetic protons induce adaptive and bystander effects that protect human cells against DNA damage caused by a subse-quent exposure to energetic iron ions" *J. Radiat. Res.* **56**: 502–508.

Burigo, L., Pshenichnov, I., Mishustin, I. and Bleicher, M. 2015. "Comparative study of dose distributions and cell survival fractions for 1H, 4He, 12C and 16O beams using Geant4 and Microdosimetric Kinetic model" *Phys. Med. Biol.* **60**: 3313–3331.

Brüchner, K., Beyreuther, E., Baumann, M., Krause, M., Oppelt, M. and Pawelke, J. 2014. "Establishment of a small animal tumour model for in vivo studies with low energy laser accelerated particles" *Radiat. Oncol.* **9**: 57.

Bulanov, S.V., Esirkepov, T.Zh., Khoroshkov, V.S., Kuznetsov, A.V. and Pegoraro, F. 2002. "Oncological hadrontherapy with laser ion accelerators" *Phys. Lett. A* **299**: 240–247.

Dollinger, G., Bergmaier, A., Hable, V., Hertenberger, R., Greubel, C., Hauptner, A. and Reichart, P. 2012. "Nanosecond pulsed proton microbeam" *Nucl. Instr. and Meth. B* **267**: 2008–2012.

Doria, D., Kakolee, K.F., Kar, S., Litt, S.K., Fiorini, F., Ahmed, H., Green, S., Jeynes, J.C.G., Kavanagh, J., Kirby, D., Kirkby, K.J., Lewis, C.L., Merchant, M.J., Nersisyan, G., Prasad, R., Prise, K.M., Schettino, G., Zepf, M. and Borghesi, M. 2012. "Biological effectiveness on live cells of laser driven protons at dose rates exceeding 10⁹ Gy/s" *AIP Advances* **2**: 011209.

Durante, M. 2014. "New challenges in high-energy particle radiobiology" *Br. J. Radiol.* **87**: 20130626.

Folkard, M., Prise, K.M., Vojnovic, B., Newman, H.C., Roper, M.J. and Michael, B.D. 1996. "Inactivation of V79 cells by low-energy protons, deuterons and helium-3 ions" *Int. J. Radiat. Biol.* **69**: 729–738.

Girdhani, S., Sachs, R. and Hlatky, L. 2013. "Biological effects of proton radiation: what we know and don't know" *Radiat. Res.* **179**: 257–272.

Goldstein, M. and Kastan, M.B. 2015. "The DNA damage response: implications for tumor responses to radiation and chemotherapy" *Annu. Rev. Med.* **66**: 129–143.

Goodhead, D.T. 2006. "Energy deposition stochastics and track structure: what about the target?" *Radiat. Prot. Dosimetry* **122**: 3–15.

Greubel, C., Assmann, W., Burgdorf, C., Dollinger, G., Du, G., Hable, V., Hapfelmeier, A., Hertenberger, R., Kneschaurek, P., Michalski, D., Molls, M., Reinhardt, S., Röper, B., Schell, S., Schmid, T.E., Siebenwirth, C., Wenzl, T., Zlobinskaya, O. and Wilkens, J.J. 2011. "Scanning irradiation device for mice in vivo with pulsed and continuous proton beams" *Radiat. Environ. Biophys.* **50**: 339–344.

Grün, R., Friedrich, T., Krämer, M., Zink, K., Durante, M., Engenhart-Cabillic, R. and Scholz, M. 2015. "Assessment of potential advantages of relevant ions for particle therapy: A model based study" *Med. Phys.* **42**: 1037–1047.

Hidding, B., Königstein, T., Willi, O., Rosenzweig, J.B., Nakajima, K. and Pretzler, G. 2011. "Laser-plasma-accelerators—A novel, versatile tool for space radiation studies" *Nucl. Instr. and Meth. A* **636**: 31–40.

Hofmann, K.M., Schell, S. and Wilkens, J.J. 2012. "Laser-driven beam lines for delivering intensity modulated radiation therapy with particle beams" *J. Biophotonics* **5**: 903–911.

Hofmann, K.M., Masood, U., Pawelke, J. and Wilkens, J.J. 2015. "A treatment planning study to assess the feasibility of laser-driven proton therapy using a compact gantry design." *Med. Phys.* **42**: 5120–5129.

Jermann, M. 2015. "Particle Therapy Statistics in 2014" *Int. J. Particle Ther.* **2**: 50–54.

Kim, M.H., Rusek, A and Cucinotta, F.A. 2015. "Issues for simulation of galactic cosmic ray exposures for radiobiological research at ground-based accelerators" *Front. Oncol.* **5**: 122.

Kitagawa, A., Fujita, T., Muramatsu, M., Biri, S. and Drentje, A.G. 2010. "Review on heavy ion radiotherapy facilities and related ion sources" *Rev. Sci. Instrum.* **81**: 02B909.

Königstein, T.K., Karger, O., Pretzler, G., Rosenzweig, J.B. and Hidding, B. 2012. "Design considerations for the use of laser-plasma accelerators for advanced space radiation studies" *J. Plasma Physics* **78**: 383–391.

Kraft, S.D., Richter, C., Zeil, K., Baumann, M., Beyreuther, E., Bock, S., Bussmann, M., Cowan, T.E., Dammene, Y., Enghardt, W., Helbig, U., Karsch, L., Kluge, T., Laschinsky, L., Lessmann, E., Metzkes, J., Naumburger, D., Sauerbrey, R., Schürer, M., Sobiella, M., Woithe, J., Schramm, U. and Pawelke, J. 2010. "Dose-dependent biological damage of tumour cells by laser-accelerated proton beams" *New J. Phys.* **12**: 085003.

Laschinsky, L., Baumann, M., Beyreuther, E., Enghardt, W., Kaluza, M., Karsch, L., Lessmann, E., Naumburger, D., Nicolai, M., Richter, C., Sauerbrey, R., Schlenvoigt, H.P. and Pawelke, J. 2012. "Radiobiological effectiveness of laser accelerated electrons in comparison to electron beams from a conventional linear accelerator" *J. Radiat. Res.* **53**: 395–403.

Ledingham, K.W.D., Bolton, P.R., Shikazono, N. and Ma, C.M. 2014. "Towards laser driven hadron cancer radiotherapy: Review of progress" *Appl. Sci.* **4**: 402–443.

Malinen, E. and Søvik, Å. 2015. "Dose or 'LET' painting—What is optimal in particle therapy of hypoxic tumors ?" *Acta. Oncol.* **54**: 1614–1622.

Oppelt, M., Baumann, M., Bergmann, R., Beyreuther, E., Brüchner, K., Hartmann, J., Karsch, L., Krause, M., Laschinsky, L., Leßmann, E., Nicolai, M., Reuter, M., Richter, C., Sävert, A., Schnell, M., Schürer, M., Woithe, J., Kaluza, M. and Pawelke, J. 2015. "Comparison study of in vivo dose response to laser-driven versus conventional electron beam" *Radiat. Environ. Biophys.* **54**: 155–166.

Paganetti, H. 2015. "Relating proton treatments to photon treatments via the relative biological effectiveness-should we revise current clinical practice?" *Int. J. Radiat. Oncol.* **91**: 892–894.

Phoenix, B., Green, S., Hill, M.A., Jones, B., Mill, A. and Stevens, D.L. 2009. "Do the various radiations present in BNCT act synergistically? Cell survival experiments in mixed alpha-particle and gamma-ray fields" *Appl. Radiat. Isot.* **67**: S318–320.

Redon, C.E., Weyemi, U., Parekh, P.R., Huang, D., Burrell, A.S. and Bonner, W.M. 2012. "γ-H2AX and other histone post-translational modifications in the clinic" *Biochim. Biophys. Acta.* **1819**: 743–756.

Ree, A.H. and Redalen, K.R. 2015. "Personalized radiotherapy: concepts, biomarkers and trial design" *Br. J. Radiol.* **88**: 20150009.

Rigaud, O., Fortunel, N.O., Vaigot, P., Cadio, E., Martin, M.T., Lundh, O., Faure, J., Rechatin, C., Malka, V. and Gauduel, Y.A. 2010. "Exploring ultrashort high-energy electron-induced damage in human carcinoma cells" *Cell Death Dis.* **1**: e73.

Schlaff, C.D., Krauze, A., Belard, A., O'Connell, J.J. and Camphausen, K.A. 2014. "Bringing the heavy: Carbon ion therapy in the radiobiological and clinical context" *Radiat. Oncol.* **9**: 88.

Schmid, T.E., Dollinger, G., Hauptner, A., Hable, V., Greubel, C., Auer, S., Friedl, A.A., Molls, M. and Röper, B. 2009. "No evidence for a different RBE between pulsed and continuous 20 MeV protons" *Radiat. Res.* **172**: 567–574.

Schmid, T.E., Dollinger, G., Hable, V., Greubel, C., Zlobinskaya, O., Michalski, D., Molls, M. and Röper, B. 2010. "Relative biological effectiveness of pulsed and continuous 20 MeV protons for micronucleus induction in 3D human reconstructed skin tissue" *Radiother. Oncol.* **95**: 66–72.

Schmid, T.E., Dollinger, G., Hable, V., Greubel, C., Zlobinskaya, O., Michalski, D., Auer, S., Friedl, A.A., Schmid, E., Molls, M. and Röper, B. 2011. "The effectiveness of 20 MeV protons at nanosecond pulse lengths in producing chromosome aberrations in human-hamster hybrid cells" *Radiat. Res.* **175**: 719–727.

Schneider, U. and Hälg, R. 2015. "The impact of neutrons in clinical proton therapy" *Front. Oncol.* **5**: 235.

Shah, D.J., Sachs, R.K., Wilson, D.J. 2012. "Radiation-induced cancer: a modern view" *Br. J. Radiol.* **85**: e1166–73.

Shinohara, K., Nakano, H., Miyazaki, N., Tago, M. and Kodama, R. 2004. "Effects of single-pulse (< or = 1 ps) X-rays from laser-produced plasmas on mammalian cells" *J. Radiat. Res.* **45**: 509–514.

Sotnik, N.V., Osovets, S.V., Scherthan, H., Azizova, T.V. 2014. "mFISH analysis of chromosome aberrations in workers occupationally exposed to mixed radiation" *Radiat. Environ. Biophys.* **53**: 347–354.

Staaf, E., Brehwens, K., Haghdoost, S., Nievaart, S., Pachnerova-Brabcova, K., Czub, J., Braziewicz, J. and Wojcik, A. 2012a. "Micronuclei in human peripheral blood lymphocytes exposed to mixed beams of X-rays and alpha particles" *Radiat. Environ. Biophys.* **51**: 283–293.

Staaf, E., Brehwens, K., Haghdoost, S., Czub, J. and Wojcik, A. 2012b. "Gamma-H2AX foci in cells exposed to a mixed beam of X-rays and alpha particles" *Genome Integr.* **3**: 8.

Staaf, E., Deperas-Kaminska, M., Brehwens, K., Haghdoost, S., Czub, J. and Wojcik, A. 2013. "Complex aberrations in lymphocytes exposed to mixed beams of (241) Am alpha particles and X-rays" *Mutat. Res.* **756**: 95–100.

Tillman, C., Grafström, G., Jonsson, A.C., Jönsson, B.A., Mercer, I., Mattsson, S., Strand, S.E. and Svanberg, S. 1999. "Survival of mammalian cells exposed to ultrahigh dose rates from a laser-produced plasma x-ray source" *Radiology* **213**: 860–865.

Tommasino, F. and Durante, M. 2015. "Proton radiobiology" *Cancers (Basel)* **7**: 353–381.

Tseng, B.P., Lan, M.L., Tran, K.K., Acharya, M.M., Giedzinski, E., Limoli, C.L. 2013. "Characterizing low dose and dose rate effects in rodent and human neural stem cells exposed to proton and gamma irradiation" *Redox Biol.* **1**: 153–162.

Wilson, P., Jones, B., Yokoi, T., Hill, M. and Vojnovic, B. 2012. "Revisiting the ultra-high dose rate effect: implications for charged particle radiotherapy using protons and light ions" *Br. J. Radiol.* **85**: e933–939.

Yogo, A., Sato, K., Nishikino, M., Mori, M., Teshima, T., Numasaki, H., Murakami, M., Demizu, Y., Akagi, S., Nagayama, S., Ogura, K., Sagisaka, A., Orimo, S., Nishiuchi, M., Pirozhkov, A.S., Ikegami, M., Tampo, M., Sakaki, H., Suzuki, M., Daito, I., Oishi, Y., Sugiyama, H., Kiriyama, H., Okada, H., Kanazawa, S., Kondo, S., Shimomura, T., Nakai, Y., Tanoue, M., Sasao, H., Wakai, D., Bolton, P.R. and Daido, H. 2009. "Application of laser-accelerated protons to the demonstration of DNA double-strand breaks in human cancer cells" *Appl. Phys. Lett.* **94**: 181502.

Yogo, A., Maeda, T., Hori, T., Sakaki, H., Ogura, K., Nishiuchi, M., Sagisaka, A., Kiriyama, H., Okada, H., Kanazawa, S., Shimomura, T., Nakai, Y., Tanoue, M., Sasao, F., Bolton, P.R., Murakami, M., Nomura, T., Kawanishi, S. and Kondo, K. 2011. "Measurement of relative biological effectiveness of protons in human cancer cells using a laser-driven quasimonoenergetic proton beamline" *Appl. Phys. Lett.* **98**: 053701.

Zlobinskaya, O., Dollinger, G., Michalski, D., Hable, V., Greubel, C., Du, G., Multhoff, G., Röper, B., Molls, M. and Schmid, T.E. 2012. "Induction and repair of DNA double-strand breaks assessed by gamma-H2AX foci after irradiation with pulsed or continuous proton beams" *Radiat. Environ. Biophys.* **51**: 23–32.

Zlobinskaya, O., Siebenwirth, C., Greubel, C., Hable, V., Hertenberger, R., Humble, N., Reinhardt, S., Michalski, D., Röper, B., Multhoff, G., Dollinger, G., Wilkens, J.J. and Schmid, T.E. 2014. "The effects of ultra-high dose rate proton irradiation on growth delay in the treatment of human tumor xenografts in nude mice" *Radiat. Res.* **181**: 177–183.

9

Ultra-Fast Opacity in Transparent Dielectrics Induced by Picosecond Bursts of Laser-Driven Ions

Brendan Dromey

9.1 Introduction

Direct investigation of ion-induced dynamics in matter on picosecond (ps, 10^{-12} seconds) timescales has been precluded to date by the relatively long nanosecond-scale (ns, 10^{-9} seconds) ion bunches (throughout this book the term, 'bunch' is used for particles with mass and the term, 'pulse' is used for photons) typically provided by radiofrequency accelerators [Baldacchino 2003]. By contrast, laser-driven ion accelerators provide bursts of ps duration [Macchi 2013, Dromey 2016], but have yet to be applied to the study of ultrafast ion-induced transients in matter. We report on the evolution of an electron-hole plasma excited in borosilicate glass by such bursts. This is observed as an onset of opacity to synchronized optical probe radiation and is characterized by the 3.0 ± 0.8 ps ion pump rise-time [Dromey 2016]. The resulting decay time of 35 ± 3 ps for the electron-hole plasma is shown to be in excellent agreement with modelling and reveals the rapidly evolving electron temperature ($>10^3$K) and carrier number density ($>10^{17}$cm^{-3}). This result demonstrates that ps laser-accelerated ion bunches are directly applicable to investigating the ultra-fast response of matter to ion interactions and, in particular, to ultra-fast ion radiolysis of water [Boag 1963, Hart 1964, Baldacchino 2008], the radiolytic decompositions of which underpin biological cell damage and hadrontherapy for cancer treatment [Kraft 2003].

9.2 Background

While the dynamics of the early stages of ionizing radiation interactions with matter have been investigated for photons [Audebert 1994] and electrons [Schmidhammer 2012, Yang 2010] the ultra-fast interaction of high-energy ions with matter is fundamentally different. A significantly higher linear energy transfer (LET, or stopping power) for ions results in highly localized energy deposition along their trajectory in matter. The unambiguous characterization of the dynamics of a system immediately following such an interaction is of central importance to multiple fields in science and medicine such as hadrontherapy, the safety of astronauts [Kraft 2003], electronics for use in space [Srour 1988] and nuclear engineering [Bulanov 2001]. Therefore, the ability to generate sufficiently short bunches of ions is critical to advancing the understanding of ultra-fast characteristics of ion interactions by direct experimental investigation.

9.3 Ion Interactions in Matter

To date, ion–matter interactions have been studied using microsecond (μs, 10^{-6} seconds) [Chitose 2001] and ns [Baldacchino 2003, Baldacchino 2004] ion bunches and have almost exclusively focused on radiolytic yields during the pulsed radiolysis of aqueous solutions. Such studies are typically performed using chemical scavenging techniques [LaVerne 1993, Appleby 1969] which themselves have the potential to generate significant uncertainty at the concentrations required for high temporal resolution. In recent years, however, laser-driven ion sources have developed rapidly (see Chapter 5), and ion bunches with initial durations below one picosecond, high particle numbers (>10^{13} per bunch) and excellent beam quality have been reported. Such parameters provide an excellent basis for addressing the problem of time-resolved ion-matter interaction dynamics on ultra-fast time scales for both aqueous solutions and amorphous and crystalline solids.

Here, we consider the excitation of electrons across the optical gap of a borosilicate (BK7) glass sample by a ps laser-driven ion bunch. BK7 is a multicomponent derivative of SiO_2. The resulting electron-hole plasma gives rise to transient opacity (TO) to optical probe radiation [Audebert 1994, Schmidhammer 2012]. This TO is ideally suited to our study since it is expected to exhibit a growth time that is essentially instantaneous compared to our ion bunch duration for low doses [Schmidhammer 2012, Bulgakova 2007], to subsequently relax rapidly on a timescale of tens of picoseconds [Bulgakova 2007] and does not consist of multiple pathways for species formation and decay, such as are found in the radiolysis of aqueous solutions [Baldacchino 2003, Boag 1963, Hart 1964]. The resulting pump and decay profiles of the TO–generated for proton fluence of ~50 μm^{-2} are found to be in excellent agreement with a semi-empirical Drude-like model (i.e. the charge carriers are assumed to be free and no selection rules are applied due to the lack of a crystalline structure) for our estimated ion bunch characteristics. This approach will, in the future, allow the estimation of physical quantities that are otherwise extremely difficult to measure directly, such as the electron temperature and carrier density of the electron-hole plasma immediately after excitation.

9.4 Experiment and Methodology: Accessing Ultra-Fast Interactions Using TNSA Proton Bunches

For our experiment, we used the TARANIS multi-terawatt (TW) laser facility at Queen's University Belfast [Dzelzainis 2010] to accelerate protons (that form part of a surface layer on a thin (10 μm) gold target foil) to an endpoint energy of ~15 MeV via the target normal sheath acceleration (TNSA) mechanism. After spectral filtering in the irradiated sample we obtain a proton bunch of about a 3 picosecond (full width at half maximum) duration in the region of interest (0.9–1mm depth) within the sample. The short proton bunch duration is made possible by the duration of each energy slice (in our case up to 1 MeV bandwidth) ultimately being set by the initial ps duration of the proton bunch [Macchi 2013].

Substantial stretching of the initial broadband bunch as it propagates with a large velocity spread from the target towards the sample results in a bunch duration greater than 70 picoseconds (since the glass sample is located 5 mm behind the target). The ultimate duration, however, is a function of the depth of propagation into the target. This is due to spectral filtering resulting from continuous energy loss and stopping of lower energy protons in the target, so that near the end of their track (~1 mm into the target) the interaction essentially involves a few MeV proton bunches of the duration of a few picoseconds, which is limited by the original ultra-short pulse duration of the accelerating laser. The ion bunch temporal profile, energy bandwidth (spread) and flux at a given depth have been estimated from semi-empirical calculations (SRIM [Ziegler 2010]) of the ion stopping power in BK7 based on the experimentally observed H+ spectral shape. These calculations are also in good agreement with an *ab initio* estimate of the ion stopping power in a similar material (cristobalite with experimental density imposed) obtained from TDDFT (time-dependent density functional theory) calculations [Schleife 2012, Correa 2012].

We use the well-established 'pump-probe' technique to study the ultra-fast interaction of ions in matter. In this scheme, a 'pump' (here a bunch of protons) excites a material or induces a desired effect. This effect is subsequently analysed using a suitable 'probe'. In our experiment, the H+–glass interaction region was transversely probed using a synchronized, ultra-fast optical probe (1053 nm, 400 fs, <10 μJ cm^{-2}) from the main laser system to produce spatially resolved snapshots of the TO (Figure 9.1). The temporal evolution of the opacity is recorded by varying the relative time delay Δt between the probe and ion bunch. As shown in Figure 9.1, these snapshots can reveal the spatial profile of the proton bunch and also demonstrate the recovery of the initial state of the subject material as Δt is significantly increased, indicating that no permanent macroscopic damage persists after the interaction.

However, obtaining a consistent data set of the temporal evolution of the TO in this manner is limited by shot-to-shot variation in H+ flux (±12%) and spectral cut-off energy (±5%). To permit a more accurate measurement of the TO, the full temporal evolution was instead recorded on a single-shot basis. This was achieved by optically streaking [Benuzzi-Mounaix 1999, Prasad 2010] the region outlined by the red dashed box in Figure 9.1.

Schematic of experimental setup

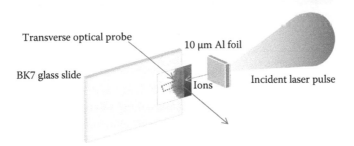

FIGURE 9.1 Probing of ultra-fast ion-induced opacity in a dielectric material. Figure 9.1 shows a schematic of the experimental setup. An intense laser pulse (>10^{19} Wcm^{-2}) is incident at an angle of 30° on to a 10 μm Au foil. An ion bunch produced via the TNSA mechanism (see Chapter 5) in the direction normal to the rear surface of the foil travels a distance of 5 mm to an optical quality transparent dielectric sample (1-mm thick in a horizontal dimension). The horizontal extent of the proton bunch is apertured to 100 ± 50 μm using a 1mm-thick Al collimating slit centred at the beam facing side of the sample. Low energy ions, electrons and X-rays were filtered using a 5 μm Al foil in front of the collimating slit (not shown). A synchronised probe laser pulse that can be either ~400 fs (compressed) or ~210 ps (chirped) is used to transversely image the pumped region (red line) onto an optical Andor CDD. The temporal evolution of the proton-induced TO along the red dashed region in Figure 9.1 is obtained on a single-shot basis. This region (100 ± 20 μm in vertical height) was imaged onto the entrance slit of a 1m imaging spectrometer with the pulse duration set to ~210 ps. The temporal evolution of TO was then encoded as variation of the signal in the probe spectrum (bandwidth ~4 nm) with respect to depth in the target [Benuzzi-Mounaix 1999, Prasad 2010] and detected using an optical Andor CCD.

In this arrangement, the original ultra-fast probe is stretched in time (or chirped) to create a long pulse (>200 ps) with an approximately linear frequency sweep [Benuzzi-Mounaix 1999]. This pulse in turn permits direct mapping of the temporal evolution of the TO onto the spectrum of the probe pulse as it passes through the interaction region, i.e. each successive frequency interval encodes a different level of opacity in the target at successive delay times. The optical streak is then achieved by passing the pulse over a diffraction grating to separate the frequency components spatially and imaging the interaction region onto a CCD [Prasad 2010]. In essence, this allows the ultra-fast dynamics of transient ion induced opacity to be observed with respect to depth in the sample.

A typical optically streaked image of the TO for a TNSA proton bunch with ~15 MeV maximum energy is shown in Figure 9.2. For this data TO is defined as a reduction in signal from full transmission (1 ± 0.03) by greater than 5% to allow for small non-uniformities in the near field transverse profile of the probing laser pulse. The maximum energy present in each proton bunch can be extracted from the slope of the detectable onset of opacity (solid black line). A maximum energy of 15 ± 0.75 MeV was observed, confirming a sharp cut-off of 1 ± 0.5 MeV bandwidth in the TNSA H+ spectrum. This observation is in line with spectral measurements made using radiochromic film (RCF) stacks.

The principal feature of this optical streak is the reducing level and greatly differing recovery time of the TO induced in the target with respect to depth. A complete discussion of this varying level of opacity is beyond the scope of this chapter, and can serve as basis for future investigations exploiting the benefit of laser-driven proton bunches. Here we limit our discussion to depths in BK7 >0.9 mm which can be reached only by the most energetic protons within ~1 MeV bandwidth of the incident proton spectrum.

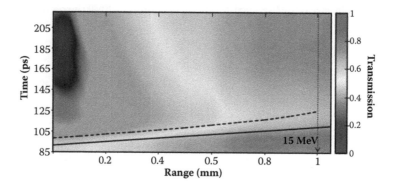

FIGURE 9.2 Optical streaking of ultra-fast ion induced kinetics in dielectric material. In this optical streak, the maximum proton energy is 15 ± 1 MeV. The vertical axis represents elapsed time since the (sudden) laser-acceleration of protons in the Au target while the horizontal axis represents depth of penetration of protons into the BK7 target with respect to the TNSA surface at 0 mm. The BK7 sample is placed ~5 mm behind the Au target. The TO occurs due to the excitation of electrons into the conduction band of the BK7 leading to increased free-free absorption. The dominant feature of the streaked image is the observation of a very sharp excitation front extending into the target (the black solid line represents the path of a 15 MeV proton with no stopping included). This is very clear evidence of a sharp cut-off in the TNSA spectrum with a (slice) bandwidth of <1 ± 0.5 MeV. The dashed horizontal line shows the expected spatio-temporal evolution of a 13 MeV proton front taking into account energy loss in the medium as calculated using *ab initio* TDDFT calculations [Schleife 2012, Correa 2012] and SRIM [Ziegler 2010]. It is this property that leads to ultra-short proton bunch generation. While H+ with incident energy of 15 MeV will penetrate further than 1 mm (vertical dotted line), H+ with incident energy of 13 ± 0.3 MeV will terminate fully due to energy loss in the medium (curved dashed line). H+ with incident energy less than 12.7 MeV will not reach a 1 mm depth. This intrinsically formed picosecond interaction permits a new regime of linear energy transfer to be explored at deeper depths in the target ultra-fast ion interactions with matter and the resultant short-lived species (in this case, an electron-hole plasma).

Another important feature is that the rise time of the TO is faster for traces taken at greater depths into the sample, corresponding to the progressively decreasing local duration of the broadband H^+ pulse due to the spectral filtering that narrows with increasing depth into the sample. The temporal evolution of the data in Figure 9.2 at a depth of 1 mm (dashed vertical arrow) is examined in detail in Figure 9.3.

Figure 9.3 shows how transmission at a 1 mm depth evolves as a result of a 3 ps (full width at half maximum [FWHM]) impulse of H^+ ions (black trace Figure 9.3) delivering an estimated total dose of 60 ± 15 Gy in the 1-mm depth region studied. Transmission reduces for the entire duration of the proton pump (~10 ps which includes the temporal portion of the bunch beyond the FWHM region). After this the medium is observed to rapidly recover with a time constant of 35 ± 3 ps (back to the value of maximum opacity at 1-mm depth). It is worth noting that due to their higher LET, protons create significantly more electron-hole pairs than an electron bunch of similar energy, dose (or fluence) and bunch duration in glass.

The ion bunch duration, τ_p is calculated as follows: the Bragg-induced cut-off (~118 ps after the interaction of the laser with the Au target) in the bunch shape (labelled in Figure 9.3) relates to protons with initial energy of approximately 13.1 ± 0.7 MeV incident on the sample. Energies below this have already been stopped in the sample and do not contribute to the TO at this depth. The high energy cut-off (~108 ps after the interaction of the laser with the Au target) is provided by experimental observation showing a 15.0 ± 1.5 MeV fast proton front. This analysis demonstrates that the proton bunch duration driving the TO at 1-mm depth is of order 3.0 ± 0.8 ps at the half maximum point, and that the total pumping time (ultimate duration of the bunch) is of order 10.0 ± 2.5 ps for this data set. These define start and endpoints of the bunch. The bunch profile is calculated as a convolution of the on-axis spectral shape of the protons coupled with the continuous energy loss due to propagation in the glass (corrected for straggling) using SRIM calculations.

To investigate the microscopic mechanism underlying the observed TO, we developed a simple Drude-type two-temperature model [Conwell 1950, Haug 2009, Zhang 2006, Mott 1978, van Exter 1990, Kaganov 1957] for the relaxation of the electrons promoted to the conduction band after the passage of the H^+ ions in the glass (Figure 9.4).

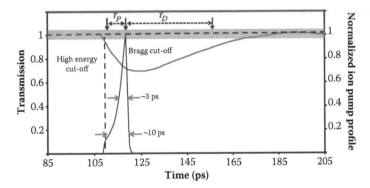

FIGURE 9.3 Temporal evolution of transient ion induced opacity induced in BK7 with picosecond scale proton bunches (part empirical, part analytically derived). Figure 9.3 shows lineouts of transmission at 1-mm depth (dotted vertical line figure 2). Here we define the onset of TO for a reduction in transmission by more than 5% due to fluctuations of ±3% on the intensity profile of the probe pulse. The thickness of the horizontal translucent red line shows the uncertainty in the transmission of the probe through the glass in the absence of the proton pump. It is critical to note that for shallower depths (< 850 μm) the opacity is heavily saturated, does not return to full transmission on the timescales under investigation and, as a result, does not provide reliable information on the transmission profile. The proton bunch profile is estimated from experimental results for ion bunch metrology in SiO_2 [Dromey 2016] giving an overall pump time at a depth of 1 mm, τ_p, of 3 ps FWHM with a total pump rise time of ~10 ps, while the decay constant for the induced opacity, τ_D, is estimated to be 35 ± 3 ps.

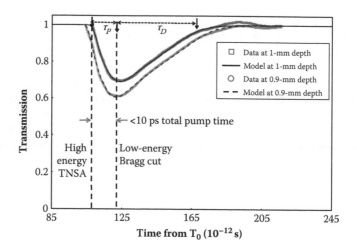

FIGURE 9.4 Drude-type model for electron-hole plasma excitation and relaxation fit to the experimental data at a time after the ion acceleration, t_0. Fits (black solid and dashed curves) to the data are shown for the evolution of transmission at 1 mm (dark red) and 900 μm (light red) penetration depths into the target (located 5 mm from the ion source). It is important to note that the model returns accurate values for ion stopping at depths >850 μm where the transmission returns to its original value. See Section 9.5 for full detail of the Drude-type two temperature model, fitting procedure and TDDFT [Schleife 2012, Correa 2012] calculations used to model the experimental data.

9.5 Drude-Type Two-Temperature Model

At room temperature, the thin glass appears transparent due to its vanishing small absorbance of photons whose energy range extends into the visible, below what is then called the *optical gap*, E_{opt}. However, an irradiated piece of glass can absorb light by means of: 1) the collective excitation of the electrons promoted to the conduction band and the holes created in the valence band by the irradiation, and 2) the excitation of the additional localised defect states. The second mechanism is negligible for low irradiation since the damage is also negligible.

To describe the collective behaviour of the electrons and holes created upon glass irradiation, we use a semi-empirical Drude-like model, i.e., the charge carriers are assumed to be free and no selection rule is applied due to the lack of a true crystalline structure in the glass. In addition, the carrier density is assumed to be thermal, being the electronic temperature, T_e, which is a time-dependent variable. A few parameters have been fit to reproduce the experimental findings, as explained below. Owing to its simplicity, the model provides a transparent microscopic interpretation of the observed transient opacity and gives access to physical quantities that are not directly assessed by the experiments, i.e. the electronic temperature, T_e, and the carrier density, n_e.

In our Drude-like model, the dielectric function is written as:

$$\varepsilon = \varepsilon_\infty - \frac{2\omega_{pl}^2}{\omega_L(\omega_L - i\gamma_s)} \tag{9.1}$$

where $\varepsilon_\infty = 2.25$ is the dielectric contribution at high frequency, ω_{pl} is the bulk plasma frequency, ω_L is the light (or laser angular) frequency and γ_s is the scattering rate, here assumed from ionized impurities [Conwell 1950]. In our model, the factor 2 in the numerator accounts for the equal contribution of electrons and holes to the conductivity. As in the standard Drude model of metals, the bulk plasma frequency is given by [Haug 2009]:

$$\omega_{pl} = \sqrt{\frac{n_e e^2}{\varepsilon_o m_e^*}} \tag{9.2}$$

where n_e is the carrier density (both electrons in the conduction band and holes in the valence band) and $m_e^* = 0.46\ m_e$ is their effective mass (both electrons and holes), while m_e is the free electron mass. From Equations 9.1 and 9.2, the absorbance is obtained as [Haug 2009]:

$$A(\omega) = \frac{\omega_L}{c} \frac{\varepsilon_2}{\sqrt{\varepsilon_1}} l \tag{9.3}$$

where $\omega_L = 1.18$ eV/$\hbar$ (corresponding to a probe central wavelength of 1053 nm) is the probe frequency, c is the speed of light, ε_1 and ε_2 are the real and imaginary components of the dielectric function defined in Equation 9.1, and $l = 0.5$ mm is the width of the collimated proton beam passing through the glass (see Figure 9.1). By neglecting the scattering and absorption inside the glass, the energy transmission of the probe pulse is simply given by exp $(-A(\omega_L))$.

In our physical picture, upon proton irradiation, secondary electrons are promoted to the conduction band and then quickly (within less than 0.1 ps) thermalized by electron–electron collisions. Note that this process (i.e. thermalization) does not change the total number of electrons (and holes). As carrier recombination takes place over a longer timescale [van Exter 1990], we also assume that the free carrier density, n_e, can be approximated by its instantaneous thermal value:

$$n_e(T_e) = 2\left(\frac{m_e^* k_B T_e}{2\pi\hbar^2}\right)^{3/2} e^{-E_{mob}/(2k_B T_e)} \tag{9.4}$$

Where $E_{mob} > E_{opt}$ and E_{mob} is the *mobility gap* of the glass. The difference between the mobility and optical gaps is due to a tail of localized (i.e. they do not contribute to the conduction) states right below (above) the conduction (valence) band.

As we assume the charge carriers to be thermalized at all times, the radiative and non-radiative relaxation processes determine the time evolution of the electronic temperature T_e. Through radiative recombination, n_e is decreased and the lattice is not heated, while by non-radiative relaxation, n_e is not changed and some energy is transferred to lattice vibration. We take as initial condition the thermal equilibrium between the electrons and the lattice, i.e. $T_e = T_L = 298$ K, where T_L is the lattice temperature. We then write down the following two-temperature model of energy transfer between excited carriers and lattice vibrations:

$$3k_B n_e \frac{\partial T_e}{\partial t} = -gk_B n_e(T_e - T_L) - n_e w(t)$$
$$3k_B n_L \frac{\partial T_L}{\partial t} = +gk_B n_e(T_e - T_L) \tag{9.5}$$

where n_L is the atomic density, g is the coupling term and $w(t)$ is meant to model the energy diffusion after the thermalization of the electrons knocked out by the protons. For simplicity, we describe this initial energy 'burst' as a finite 'sine-squared pulse':

$$w(t) \approx W \begin{cases} \dfrac{2}{t_{burst}} \sin^2\left(\pi \dfrac{t - t_{ini}}{t_{burst}}\right) & t \in [t_{ini}, t_{ini} + t_{burst}] \\ 0 & \text{elsewhere} \end{cases} \tag{9.6}$$

which starts at time t_{ini} and lasts for t_{burst} picoseconds (where burst duration equals bunch duration). Radiative processes are not accounted for in Equation 9.5, as they do not contribute to the lattice heating and they take place over a longer time scale. As very few electrons are ionized and (we have also verified this *a posteriori*) we always have $n_e \ll n_L$, the second line in Equation 9.5 can be safely ignored and the lattice temperature is set to $T_L = 298$ K at all times.

Finally, the scattering from ionized impurities in Equation 9.1 is modelled as:

$$\gamma = B(k_B T_e)^{-\frac{3}{2}} e^{-E_{opt}/(2k_B T_e)} \tag{9.7}$$

where the usual $T_e^{-3/2}$ dependence [Conwell 1950] is weighted with the probability to have an impurity ionized. Indeed, such impurities are expected to be activated in our high-quality borosilicate BK7 glass sample. Since they are energetically close to the centre of the gap, the energy difference between the highest occupied impurity states and the lowest (localized) unoccupied tail state is comparable to $E_{opt}/2$.

The parameters g, W, t_{ini}, t_{burst} and B – along with E_{mob} – have been fit to reproduce the experimental findings as shown in Figure 9.4. All six parameters were initially fit to reproduce the glass transient opacity at 1 mm. Next, only the last three parameters, W, t_{ini} and t_{burst}, have been adjusted to fit the transient opacity at 0.9 mm. Our fitting procedure is rather robust since essentially just the parameter W governing the intensity of the initial energy 'burst' needs to be adjusted.

By means of our Drude-type model, we can also compute the variation of free carrier energy density after the initial 'burst':

$$u_e = \left(E_{mob} + 3k_B T_e^{max}\right) n_e\left(T_e^{max}\right) - \left(E_{mob} + 3k_B T_L\right) n_e(T_L) \tag{9.8}$$

where T_e^{max} is the maximum achievable temperature, computed from Equation 9.5 for g = 0; u_e provides an estimate of the total energy deposited by proton stopping in the BK7 glass sample that can excite electrons and holes. In particular, we find that at the end of the proton propagation range ($\approx$1 mm depth for incident 10 MeV protons) $u_e = 4.84 \cdot 10^{-6}$ eV/Å^3. The measured density of proton tracks at this depth is $I \approx 50$ μm^{-2}, which implies that we can also estimate the electronic stopping power to be $S \approx u_e/I = 9.67$ eV/Å. As a further validation of our Drude-type model, this value of S is in good agreement with the independent estimates by SRIM [Ziegler 2010] and TDDFT *ab initio* calculations [Schleife 2012, Correa 2012].

9.6 Results and Discussion

For the interaction of ions in BK7 we assume that a uniform thermalized electron-hole plasma is quickly (during less than 0.1 ps) [Bulgakov 2007, van Exter 1990], generated inside the irradiated volume of the glass, the optical opacity of which provides a microscopic explanation of the observed macroscopic TO. For the ion bunch calculated in Figure 9.3, the model shows that approximately 5 picoseconds after the initial excitation the electron-hole plasma can be considered uniform in the irradiated region and the temperature can be described by the electron temperature [van Exter 1990]. As can be seen in Figure 9.4, this provides an excellent fit to the experimental results at 1 mm, and at 900 μm. The model suggests experimental peak carrier densities of 3–6 × 10^{17} cm^{-3} and peak electron temperatures of ~3000 K. In the future, it will be possible to use this modelling to investigate the evolution of these critical param-eters and examine the role of ion flux density in the ultimate recovery to equilibrium conditions. An independent indication of the quality of our simple Drude-type model is given by the agreement of the estimated maximum stopping power (~9.7 eV/Å) with the *ab initio* TDDFT modelling using the local adiabatic density approximation [Schleife 2012, Correa 2012] estimate of ~10 eV/Å, which is within the experimental uncertainty of ±10%.

While the model provides a good fit to the proton-induced opacity at depths greater than 850 μm, where the total dose is moderate, this is not true for shallower depths in the target (i.e. <850 μm where the transmission does not return to the initial high value). For the higher doses at those shallower depths (>100 Gy) initial investigations show that considering a radiative bottleneck to electron relaxation in our model increases the accuracy of the fit, but a study of electron-hole plasma evolution coupled with localized lattice heating and kinetic radiative bottleneck formation is beyond the scope of our study here and will form the basis of future work. In our discussion, we are exclusively concerned with interactions where the ion bunch is sufficiently short, and the dose sufficiently low such that lattice heating can be neglected.

9.7 Conclusion

The work described here offers a clear path towards real-time observation and metrology of hard-to-determine parameters, such as temperature and transient lifetimes of species excited in the immediate aftermath of ion stopping in matter. The intrinsic synchronicity of the laser-driven ion bunches and optical probe pulses ensures a high degree of reproducibility in the optical streaking technique used to investigate the early stages of such interactions. In the work presented here, the rise time of the transient opacity (excited electron population in the conduction band) is observed to be of order the pump (proton bunch) duration, while the decay constant of these electrons from the conduction band, with a time constant of 35 ± 3 ps, is consistent with our Drude-type model for electron-hole plasma dynamics. Overall, this technique demonstrates a method to directly estimate the picosecond time-dependent response of electronic temperature and carrier density in matter to ultra-fast ion bunches. This technique opens the way for picosecond pulsed ion radiolysis of aqueous solutions, and in particular that of pure water [Baldacchino 2003, Boag 1963, Hart 1964] which is of specific relevance to hadrontherapy [Kraft 2003] and neutron radiolysis for nuclear engineering [Bulanov 2001] to be performed at even small-scale laboratories.

Acknowledgements

The author would like to acknowledge direct contributions to the experimental and numerical data provided in this chapter: D. Adams[1], L. Stella[2], R. Prasad[1], K. F. Kakolee[1], R. Stefanuik[3], G. Nersisyan[1], G. Sarri[1], M. Yeung[1], H. Ahmed[1], D. Doria[1], T. Dzelzainis[1], D. Jung[1], S. Kar[1], D. Marlow[1], L. Romagnani[1+], A. A. Correa[4], P. Dunne[3], J. Kohanoff[2], A. Schleife[4], M. Borghesi[1], F. Currell[1], D. Riley[1], M. Zepf[1*] and C. L. S. Lewis[1] (1. Centre for Plasma Physics, Department of Physics and Astronomy, Queens University Belfast, BT7 1NN, UK; 2. Atomistic Simulation Centre, Department of Physics and Astronomy, Queens University Belfast, BT7 1NN, UK; 3. School of Physics, University College Dublin, Belfield, Dublin 4, Ireland; 4. Lawrence Livermore National Laboratory, Livermore, California, 94551, USA).

This work was supported by the Engineering and Physical Sciences Research Council (EPSRC) through both a Career Acceleration Fellowship (EP/H003592/1), standard grants EP/P016960/1 and EP/L02327X/1 and platform grant EP/I029206/1. R.S. was supported by Science Foundation Ireland (08/RFP/Phy1180). A.A.C. and A.S. performed work under the auspices of the US Department of Energy by Lawrence Livermore National Laboratory under Contract DE-AC52-07NA27344.

References for Chapter 9

Appleby, A. and Schwarzl, H. A. 1969. "Radical and Molecular Yields in Water Irradiated by X-Rays and Heavy Ions" *J. Phys. Chem.* **79**: 1937–1941.

Audebert, P. et al. 1994. "Space–time observation of an electron gas in SiO2" *Phys. Rev. Lett.* **73**: 1990–1993.

Baldacchino, G. 2008. "Pulse radiolysis in water with heavy-ion beams. A short review" *Rad. Phys. Chem.* **77**: 1218–1223.

Baldacchino, G. et al. 2003. "A nanosecond pulse radiolysis study of the hydrated electron with high energy carbon ions" *Nuc. Inst. and Meth. Phys. Res. B* **209**: 219–223.

Baldacchino, G. et al. 2004. "A nanosecond pulse radiolysis study of the hydrated electron with high energy ions with a narrow velocity distribution" *Chem. Phys. Lett.* **385**: 66–71.

Benuzzi-Mounaix, A. et al. 1999. "Chirped pulse reflectivity and frequency domain interferometry in laser driven shock experiments" *Phys. Rev. E.* **60**: R2488–R2491.

Boag, J. W. and Hart, E. J. 1963. "Absorption Spectra in Irradiated Water and Some Solutions: Absorption Spectra of 'Hydrated' Electron" *Nature* **197**: 45–47.

Bulanov, A. V. et al. 2001. "Protoka Experimental Stand for Studying High-Temperature Radiolysis of Water by Accelerated Protons" *Atomic Energy* **91**: 658–666.

Bulgakova, N. M., Burakov, I. M., Meshcheryakov, Y. P., Stoian, R., Rosenfeld, A., Hertel, I. V. 2007. "Theoretical models and qualitative interpretations of Fs laser material processing" *J. Lumin.* **2**: 76–86.

Chitose, N. et al. 2001. "Radiolysis of Aqueous Solutions with Pulsed Ion Beams. 4. Product Yields for Proton Beams in Solutions of Thiocyanate and Methyl Viologen/Formate" *J. Phys. Chem. A* **105**: 4902–4907.

Conwell, E. and Weisskopf, V. F. 1950. "Theory of Impurity Scattering in Semiconductors" *Phys. Rev.* **77**: 388–390.

Correa, A. A. et al. 2012. "Nonadiabatic Forces in Ion-Solid Interactions: The Initial Stages of Radiation Damage" *Phys. Rev. Lett.* **108**: 213201.

Dromey, B. et al. 2016. "Picosecond metrology of laser-driven proton bursts" *Nature Comms.* **7**: 10642.

Dzelzainis, T. et al. 2010. "The TARANIS laser: A multi-Terawatt system for laser-plasma investigations" *Las. Part. Beams* **28**: 451–461.

Hart, E. J., Gordon, S., Thomas, J. K. 1964. "Rate Constants of Hydrated Electron Reactions with Organic Compounds" *J. Phys. Chem.* **68**: 1271–1274.

Haug, H. and Koch, S.W. 2009. "Quantum Theory of the Optical and Electronic Properties of Semiconductors" World Scientific Publishing (Singapore), 5th edition.

Kaganov, M., Lifshitz, I. M. and Tanatarov, L. V. 1957. "Relaxation between electrons and crystalline lattices" *Sov. Phys. JETP* **4**: 173–178.

Kraft, G. 2003. "Radiobiological effects of highly charged ions" in *The Physics of Multiply and Highly Charged Ions: Vol. 1 Sources, Applications and Fundamental Processes*, Ed. F.J. Currell. Kluwer Academic Publishers, ISBN 1-4020-1565-8.

LaVerne, J. A. and Yoshida, J. 1993. "Production of the hydrated electron in the radiolysis of water with helium ions" *J. Phys. Chem.* **97**: 10720.

Macchi, A. et al. 2013. "Ion acceleration by superintense laser-plasma interaction" *Rev. Mod. Phys.* **85**: 0034–6861.

Mott, N. F. 1978. "Electrons in glass" *Rev. Mod. Phys.* **50**: 203–208.

Prasad, Y. B. S. R. et al. 2010. "Chirped pulse shadowgraphy for single shot time resolved plasma expansion measurements" *App. Phys. Lett.* **96**: 221503.

Schleife, A., Draeger, E. W., Kanai, Y. and Correa, A. A. 2012. "Plane-wave pseudopotential implementation of explicit integrators for time-dependent Kohn-Sham equations in large-scale simulations" *J. Chem. Phys.* **137**: 22A546.

Schmidhammer, U. 2012. "Transient absorption induced by a picosecond electron pulse in the fused silica windows of an optical cell" *Rad. Phys. and Chem.* **81**: 1715–1719.

Srour, J. R. and McGarrity, J. M. 1988. "Radiation Effects on Microelectronics in Space" *Proc. IEEE* **76**: 1443–1469.

van Exter, M. and Grischkowsky, D. 1990. "Carrier dynamics of electrons and holes in moderately doped silicon" *Phys. Rev. B* **41**: 12140.

Yang, J. et al. 2010. "Photocathode femtosecond electron beam applications: Femtosecond pulse radiolysis and femtosecond electron diffraction" *Proceedings of IPAC'10, Kyoto, Japan*, 4113.

Zhang, X. -G., Lu, Z. -Y. and Pantelides, S. T. 2006. "First-principles theory of tunneling currents in metal-oxide-semicondcutor structures" *Appl. Phys. Lett,* **89**: 032112.

Ziegler, J. F., Ziegler M. D. and Biersack J. P. 2010. "SRIM – The stopping and range of ions in matter" *Nucl. Instr. Meth. B* **268**: 1027–1036.

10

Using Laser-Driven Ion Sources to Study Fast Radiobiological Processes

Naoya Shikazono

Kengo Moribayashi

Paul R. Bolton

10.1 Introduction to Clustered DNA Damage

The high ionization and excitation densities produced by various types of ionizing radiation have long been related to biological effects. The idea was introduced nearly 60 years ago suggesting that a 'lethal hit' to the cell can be explained in terms of the number of ionization events within a given target on the scale of a few to tens of nanometers [Howard-Flanders 1958]. It was argued that the linear energy transfer (LET) effects of radiation account for this 'lethal hit'. The relevance and significance of the spatial distribution of DNA damage sites (lesions) to various biological effects, such as repair, lethality, induction of mutations or chromosome aberrations, were initially pointed out from the biophysical considerations by Goodhead as well as from the radiation chemical experiments by Ward in the 1990s [Goodhead 1994, Ward 1994]. Goodhead proposed that the spatial clustering of DNA damage is caused by dense ionization attributed to ionizing radiation, and that clustered DNA damage would retard repair and thus would become a significant impediment to cell survival. Ward also concluded that the spatial distribution of the DNA lesions is the key to describing the biological effectiveness of ionizing radiation. He took advantage of the fact that the types of DNA damage produced by ionizing radiation and by DNA oxidizing agents (such as H_2O_2), are considered to be chemically indistinguishable. He demonstrated that the biological effectiveness of ionizing radiation is much higher than that of H_2O_2, (which produces randomly spaced lesions) when identical numbers of DNA lesions were induced in cells.

Clustered DNA damage is defined as two or more lesions formed within 1–2 helical turns of DNA by a single radiation track. A double-strand break (DSB) is classified as bi-stranded clustered damage (Figure 10.1). DSBs are one of the most extensively studied types of clustered damage; probably because (1) DSBs break the DNA and reduce the molecular weight of the irradiated DNA, which makes them

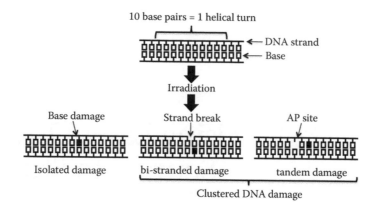

FIGURE 10.1 Types of DNA lesions defined from the spatial distributions. Irradiation of DNA leads to the generation of isolated and clustered DNA damage. Types of single DNA lesions induced by ionizing radiation are shown. Clustered DNA damage is further classified to two types of damage depending on the positions of comprising lesions: bi-stranded clustered damage and tandem damage. Examples of each type of damage are shown in the figure.

easier to detect and (2) a great deal of experimental evidence has accumulated over the years suggesting a close association between unrepaired DSB and induction of chromosome aberrations, lethality, etc. [Iliakis 1991]. Until recently, the occurrence and the processing of clustered DNA damage, especially of non–DSB types which are composed of base lesions, apurinic/apyrimidinic (AP) sites (which are locations in DNA that have lost bases) and single-strand breaks (SSBs) (that do not generate a DSB), by repair enzymes was not clearly demonstrated experimentally. Most, if not all, lesions (whether they are clustered or individual) are recognized and processed by the DNA repair machinery.

Significant progress has been made in recent years demonstrating that not only DSB but non–DSB clustered DNA damage is correlated to the biological effectiveness of the radiation. The lack of a simple yet sensitive method to detect base lesions and AP sites that are spatially in close proximity has long impeded a detailed study of non–DSB clustered DNA damage induced by ionizing radiation. During the end of 1990s and the beginning of 2000s, the use of DNA glycosylases and AP endonucleases, whose major substrates are oxidized purines/pyrimidines and AP sites, respectively, was established and has significantly impacted the detection of clustered damage [Hada 2008, Shikazono 2009]. As DNA glycosylases and AP endonucleases recognize base lesions and AP sites to create a SSB at the site of the damage through their AP lyase/endonuclease activity, clustered DNA damage in which there is one or more base lesion or AP site on each strand can be detected as a DSB after enzymatic treatment. The technique, in principle, transforms non–DSB clustered damage into a DSB, using a DNA repair enzyme as the probe (Figure 10.2). In cells, the amount of non–DSB clustered damage revealed by the treatments of different repair enzymes is at least 3–4 times larger than that of DSB after exposure to low LET radiation [Gulston 2002, Sutherland 2000, Sutherland 2002]. However, the enzymatic assay revealed that non–DSB clustered damage displays intrinsic limitations; one of which is that the capacity of repair enzymes to cleave at a lesion within a cluster could be reduced such that the damaged cluster does not lead to DSBs. Another limitation of the enzymatic assay is that detectable clustered DNA damage has the configuration with individual lesions on two strands (bi-stranded lesions). In other words, tandem damage (Figure 10.1), which is a clustered DNA damage of which lesions are only on one strand, would not be detected distinctively. These limitations can greatly hamper the accurate estimation of the yield of clustered lesions. Recently, an alternative method based on fluorescent resonance energy transfer (FRET), which enables the detection of clustered DNA damage without the use of DNA repair enzymes, has been developed [Akamatsu 2013]. FRET is a mechanism in which a donor fluorophore, initially in

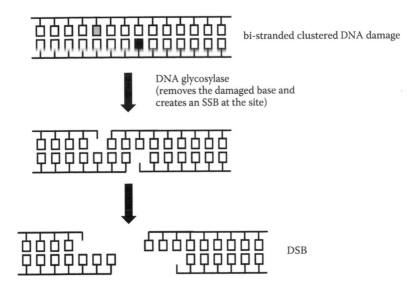

FIGURE 10.2 Detection of clustered DNA damage by enzymes. Clustered DNA damage comprised of two base lesions located on each strand is illustrated. AP endonuclease recognizes AP sites and creates a SSB at the site. When SSBs are in close proximity (< 10 base pairs), a DSB will be formed.

its electronic excited state, transfers energy to an acceptor fluorophore through non-radiative dipole–dipole coupling [Lakowicz 2006]. A fluorophore is a chemical compound which re-emits light upon light excitation (fluorescence). Since FRET usually occurs between fluorophores that are less than 10 nm apart, the clustered fraction of AP sites labelled with fluorescent dyes could be measured. Extensive clustering of AP sites was found due to carbon ion irradiation (with an LET of 760 keV/μm) [Akamatsu 2015].

How clustered DNA damage is processed and to what extent has also long remained unknown, most likely because the radiation-induced DNA lesions represent a random assortment of isolated and clustered DNA lesions. As the random nature of radiation-induced changes made it a challenge to prove the importance of clustered DNA damage, we and others have instead used chemically-synthesized clusters, for which the lesion type, number and relative positions could be specified. This approach afforded examination of the biological effect of DNA damage. Our current understanding of the biological effect of clustered DNA damage is largely based on the studies that analyzed the lethal or mutagenic potential of synthetic model clusters [Eccles 2011, Sage 2011, Sage 2017, Shikazono 2009]. Studies of similarly synthesized bi-stranded damage sites indicate that: (1) excision of a lesion on one strand often results in retardation of the repair of a base lesion on the other strand; (2) there appears to be a preferential order for excising different lesions; (3) when the repair of base damage is retarded, some of the clusters remain partly unrepaired by the time of replication; (4) one of the strands is lost or a replication fork is collapsed at the site of unrepaired clustered base damage and (5) unlike bi-stranded base clusters, some bi-stranded clustered AP sites and bi-stranded clusters comprised of an AP and an SSB are processed simultaneously and lead to deleterious DSB. These studies confirm a retardation and hierarchy for repair, and depending on the configuration of the cluster, deleterious DSBs can be formed during repair. The repair of the lesions within the cluster can be impaired to various extents and at various steps. The types and configurations of lesions within a cluster as well as the interplay between DNA repair proteins strongly affect the biological outcome of the clustered damage. As we have seen in this section, it is now widely accepted that clustered DNA damage is more lethal or mutagenic than isolated single lesions, and therefore one of the main causes and determining factors of the biological effect of radiation.

10.2 Unknown/Unsolved Issues Associated with Clustered DNA Damage: Fast Processes After Ion Irradiation

Despite extensive effort, it is still unknown how a given type of radiation causes clustered DNA damage. Although the clustered damage is considered to be more abundant with higher LET radiation, the nature of very fast processes immediately following such radiation remains largely unknown. The sequence of physical events for an isolated cell along a single particle track that leads to biological effects begins within femtoseconds after irradiation. Because a cell comprises mostly water (approximately 70%), we mainly focus on water radiolysis in what follows (for a detailed discussion of applications of laser acceleration to quantitative water radiolysis, see Chapter 7).

10.2.1 Simulation Study

We have developed a simulation model that allows us to treat physical phenomena following ion irradiation as realistically as possible [Moribayashi, 2011, Moribayashi 2013], aiming to acquire further insight into the generation of clustered DNA damage. Because there is a known correlation between the amount of DNA damage and dose [Shimobayashi 2013], we incorporate into our simulation a model for radial dose, which is the dose as a function of perpendicular distance (r) from the incident ion path [Moribayashi 2014, Moribayashi 2015].

Figure 10.3 shows the main difference between conventional simulation models [Katz 1969, Wang 2014] and our model [Moribayashi 2011, Moribayashi 2013] for the ion irradiation study. The composite electric field formed from molecular ions, which are created from the incident ion impact ionization, has not been considered in the conventional models. According to conventional models, newly freed electrons generated from the ion impact move away from the path (see Figure 10.3a). The free electron behaviour distinguishes our model [Moribayashi 2011, Moribayashi 2013] where the electric field pattern is different and acts on the electrons with a force given by:

$$\vec{F} = m_e \frac{d\vec{v}_e^{(i)}}{dt} = -\sum_{j \neq i} \frac{e^2 \vec{R}_{ij}}{4\pi\varepsilon_0 R_{ij}^3} + \sum_l \frac{q_l e \vec{R}_{il}}{4\pi\varepsilon_0 R_{il}^3}, \tag{10.1}$$

where ε_0, m_e, $\vec{v}_e^{(i)}$, and q_l, are the dielectric constant in vacuum, the mass of an electron, the velocity of the i'th electron, and the charge number of the l'th ion, respectively. $\vec{R}_{ij}$ and $\vec{R}_{il}$ are the distances from

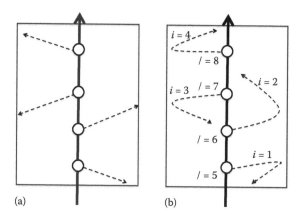

(a)   (b)

FIGURE 10.3 Main difference between (a) conventional simulation models and (b) our model: the ion and the electron paths are shown by → and --->, respectively. The water molecular ions are shown by ○. The number $i = 1 - 4$ and $l = 5 - 8$ are named for the electrons and the ions that appear due to an ion irradiation.

the *i*'th electron to the *j*'th electron and to the *l*'th ion, respectively. This force (or field) can trap some free electrons near the ion path (see Figure 10.3b). In order to explain the numbers *i* and *l* simply, we number $i = 1 - 4$ and $l = 5 - 8$ for the electrons and the ions that appear due to an ion irradiation in Figure 10.3b, respectively. In Figure 10.3b, we consider the electron of $i = 1, j = 2 - 4$.

The conceptual scenario for generating radial dose is as follows [Moribayashi 2014, Moribayashi 2015]: (1) a single incident ion interacts with water molecules along its path, (2) consequently free electrons are produced and (3) these electrons transfer energy to the target via the following electron impact processes k: ($k = $ [I]) electronic excitation ($e^- + H_2O \rightarrow e^- + H_2O^*$) [Pritchard 1990] and ($k = $ [II]) ionization of a water molecule ($e^- + H_2O$ (or H_2O^*) $\rightarrow e^- + H_2O^+ + e^-$) [Orient 1987]. The radial dose (D_r) can be expressed as:

$$D_r(r) = \frac{\displaystyle\sum_{k=(I)}^{(II)} \Delta E_k N_{ek}(r - \Delta r, r + \Delta r)}{\pi\left\{(r + \Delta r)^2 - (r - \Delta r)^2\right\}ld}, \tag{10.2}$$

where r, d, l, Δr, ΔE_k and N_{ek} are the perpendicular distance from the incident ion path, the target mass density, the incident ion migration length, the incremental change of r, the energy transfer during process k, and the number of times that process k occurs in the region between $r - \Delta r$ and $r + \Delta r$, respectively. In our simulations, we use the values of $d = 1$ g/cc, $l = 30$ nm, $\Delta r = 0.3$ nm (which corresponds to the average interval distance between water molecules), $\Delta E_k = 10.24$ and 12.6 eV for processes (I) and (II), respectively and N_{ek} is solved from our simulations. In order to treat the movement of the electrons at the top ($l = 0$ nm) and at the bottom ($l = 30$ nm) as accurately as possible, we move the secondary electrons that leave the bottom ($l = 0$ nm) or top area ($l = 30$ nm) of the target to the top or bottom area, respectively [Moribayashi 2011]. Our simulations spend too much memory and time which increase with increasing l. Since it takes several days to simulate radial dose at a single ion irradiation when we take $l = 30$ nm, we judge that $l = 30$ nm is the limiting value. We only consider the energy transfer of secondary electrons to the target for the radial dose simulations according to the conventional radial dose distribution model [Chatterjee 1976]. Namely, we exclude ionization energies of molecules due to the incident ion impact ionization processes in the radial dose calculations. These processes occur only at $r \sim 0$. Using ion impact ionization cross sections, we easily estimate the energy transfer to the target along the ion path ($r \sim 0$), however, for $r \neq 0$, it is difficult to estimate it. We need to consider the energy transfer from the secondary electrons for the dose calculation. Then, the large-scale simulations of the movement of the secondary electrons are required. We show the results averaged over the radial dose produced from the irradiation of a few hundred ions. We set up different positions of H_2O according to incident ions. We have described this simulation model in detail and the verification of our model in reported results [Moribayashi 2011, Moribayashi 2013, Moribayashi 2014, Moribayashi 2015].

10.2.2 Characteristic Physical Processes After Ion Irradiation

Figure 10.4A shows an example of the locations of water molecular ions created from the carbon ion impact ionization with $E_{ion} = 5, 10, 20$ and 30 MeV/u, where E_{ion} is the energy of the incident ion. We confirmed that the average intervals between molecular ions, produced from impact with the incident ions, agree well with the mean free path (τ) given by [Moribayashi 2011, Moribayashi 2013]:

$$\tau = \frac{1}{n_m \sigma_{ion}}, \tag{10.3}$$

where n_m and σ_{ion} are the number density of water molecules ($\sim 3 \times 10^{22}$/cm^3) and the ion impact ionization cross section, respectively. Figure 10.4B shows $E_f(r)$ as a function of r, where $E_f(r)$ is the electric

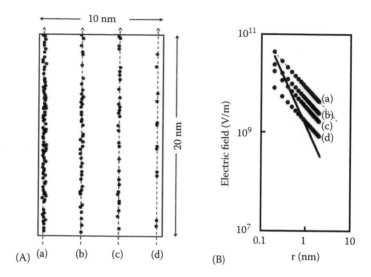

FIGURE 10.4 (A) Examples of positions (●) of water molecular ions created from the impact ionization of an incident carbon ion. The arrows with dashed lines show the motion of the incident carbon ions. In order to lead us to image the expanse of the incident ion path in the vertical direction, the vertical length is taken to be 10 nm. We can roughly estimate the interval distances between molecular ions produced here from Equation 10.3. (B) $E_f(r)$ formed from ions shown in (A) vs. r. the electric field produced from the point charge is shown by solid line. Dashed line shows the scaling of $1/r$. E_{ion} = (a) 5, (b) 10, (c) 20, (d) 30 MeV/u for both (A) and (B). Solid back line shows the electric field formed from a point charge of scaling of $1/r^2$.

field formed from molecular ions, which are shown in Figure 10.4A, toward the r direction. We show an example of one heavy ion irradiation in Figure 10.4A. On the other hand, we show the results averaged over $E_f(r)$ induced from the irradiation of a few hundred ions in Figure 10.4B with Equation 10.1. The electric field due to a point charge is also shown. We have found that the electric fields formed from a point charge and $E_f(r)$ decrease according to $\dfrac{1}{r^2}$ and $\dfrac{1}{r}$, respectively. These molecular ions produced from ion irradiation are located on a line along the incident ion path at almost regular intervals and the tendency of $E_f(r)$ is expected to be understood from

$$E_f(r) \propto \int_{-\infty}^{\infty} \frac{1}{\rho^2} \frac{r}{\rho} dz = \frac{2}{r}, \tag{10.4}$$

where z is the coordinate parallel to the incident ion path and $\rho = \sqrt{r^2 + z^2}$ is the distance from the position on the line of $r = 0$ for a given z. In this equation, $\dfrac{1}{\rho^2}$ and $\dfrac{r}{\rho}$ correspond to the strength and the r direction component of the electric field, respectively. The electric field induced from ion irradiation acts on longer ranges than that from point charges. Further, we have found $E_f(r) \propto \sigma_{ion}$ from Figure 10.4B, and therefore, $E_f(r) \propto \dfrac{\sigma_{ion}}{r}$. Figure 10.5 shows the rapid time variation of 'existence ratios' of the secondary electrons within $r = 1$ nm at an ionization site after the carbon ion with E_{ion} = 5, 10, 20 and 30 MeV/u irradiates the target (this ratio is 100% at r = 0 and delay = 0). The secondary electron is the electron emitted from the incident carbon ion impact ionization. In Figure 10.5 it can be seen that at E_{ion} = 5 MeV/u, 40% of secondary electrons are trapped within $r = 1$ nm within the first 100 femtoseconds following irradiation. The existence ratio decreases with increasing E_{ion} therefore lengthening τ

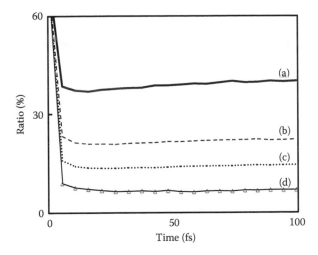

FIGURE 10.5 The existence ratio of secondary electrons within $r = 1$ nm as a function of time. For (a), (b), (c) and (d), the same ion energies are used as those in Figure 10.4. (existence ratio is 100% at $r = 0$ and delay = 0).

as shown in the figure and references [Moribayashi 2011, Moribayashi 2013]. This is because a shorter τ establishes a stronger electric field which traps a larger number of the electrons. The simulation results shown in Figure 10.5 approximately correspond to the electron densities of 2×10^{21}, 5×10^{20}, 2×10^{20} and 6×10^{19} (1/cm³) for $E_{ion} = 5$, 10, 20 and 30 MeV/u, respectively at the time of 100 fs.

Figure 10.6 shows D_r, defined in Equation 10.2, as a function of r for a single C^{6+} ion with $E_{ion} = 5$, 10, 20 and 30 MeV/u. In this figure, we show D_r simulated by including and excluding the effect of the electric field (as in Equations 10.1 and 10.2). In the region that extends more than 1 nm from the ion path ($r > 1$ nm), this field is observed to have no significant effect on D_r. This is because the effect of electric field weakens for $r > 1$ nm. The induced electric field $E_f(r)$ becomes stronger with decreasing E_{ion} [Moribayashi 2014, Moribayashi 2015]. Therefore, for $r < 1$ nm, we expect the effect of this field on D_r to become more significant with decreasing E_{ion}. For example, Figure 10.6 indicates that the doses differ by an order of magnitude at $r < 0.5$ nm with $E_{ion} = 5$ MeV/u. In order to widely spread our radial dose simulation results,

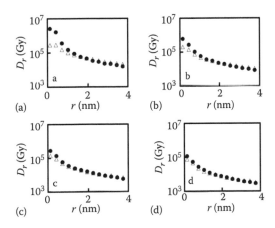

FIGURE 10.6 D_r vs. r for the irradiation of a C^{6+} ion with the energy of (a) 5, (b) 10, (c) 20 and (d) 30 MeV/u: in the case including (●) and excluding (△) the electric field. As can be seen in the figure, most of the energies of secondary electrons are deposited in the vicinity of an ion track and the local dose could be extremely high.

we have just started to provide simple expressions such as functions that reproduce our simulation results [Moribayashi 2015].

Our results suggest that characteristic physical processes immediately after ion particle irradiation could be strongly relevant to its biological consequences. Miller [2007] suggested that brain tumours could be eliminated and killed by electric fields. According to his proposed mechanism, when strong electric fields are applied locally during cancer cell division, polar molecules are displaced in a cell so that the cancer cell can no longer survive. Our simulations show that the ion irradiation produces a lot of molecular ions locally (see Figure 10.4A) and that these ions create a local strong electric field (see Figure 10.4B) in water similar to that discussed by Miller [2007]. The effect of the strong local electric field attributed to ion irradiation on polar molecules may be crucial in understanding the biological effect of ion radiation. It is interesting to know if the movement of polar molecules can be related to the formation of cluster DNA damage. Another physical process that may have relevance to the biological effects of ions is the potential generation of a thermal spike along an ion track. The 'thermal spike model' in solids [Toulemonde 1993, Hayashi 2015] has been proposed as another effect of incident ions. We speculate that thermal spikes are also generated in liquids. Thermal spikes can further generate shock waves, which in turn could produce a large force to shear or damage DNA through fast movement of water molecules. Because thermal spikes are formed along and in close proximity to the ion track, the effect of shock waves [Stöhlker 2015, Bottländer 2015] could lead to the formation of clustered DNA damage [Surdotovich 2013]. It is clear that the region in very close proximity to an ion track can be quite significant within the ultra-short temporal (<100 femtoseconds) and nanometer spatial scale. It is of note that we ignored the movement of water molecules in this paper for the following reasons: it takes longer than a few picosecnds for water molecules to move in the electric field [Boyd 2003] and, as previously mentioned, we have only treated up to 100 fs after the irradiation here.

10.3 Advantage of Using Laser-Driven Particle Acceleration

Emergent from a laser–plasma source, laser-driven particle acceleration can facilitate so-called pump-probe experiment configurations in radiolysis for exploring the very early aftermath of ion irradiation with ultra-short time resolution and high spatial resolution (please note that the overview of principles, state of the art and perspectives of laser-driven ion acceleration are presented in Chapter 5, and that the treatment of laser-driven pump-probe radiolysis in Chapter 7 focuses on the radiolytic species induced by ioniziation and excitation). In these cases, the incident ion bunch is the pump and the probe is a synchronously generated short (or ultra-short) optical pulse that can include terahertz (THz), near-infrared (NIR) and X-ray frequencies. In these options, we include pulses of XFEL light (for example at the SACLA XFEL in Japan, a synchronously coupled 500 terawatt laser, has been established). Specifically, we suggest here a conceptual configuration that can be used to directly investigate the effect(s) of a strong local electric field and potential holes (i.e. localized evacuation regions) created by thermal spikes close to the ion track.

10.3.1 Proposed Experimental Setup to Apply Heavy
Ion Beams Driven by Intense Lasers

Figure 10.7 illustrates a generic pump-probe configuration with a THz probe to be used in time-domain THz spectroscopy. The THz photon energy is comparable to that of rotational states in water molecules. Therefore, THz light is suitable for measuring molecular rotation effects; in particular, the dielectric relaxation time. This technique aims to observe dielectric relaxation effects that, in principle, come from measurements of THz refraction and absorption [Rønne 1997]. Refractive and absorptive behaviours are attributed to the real and the imaginary parts of the dielectric constant, respectively. The effect of a heavy ion-induced strong electric field on the movement of polar molecular ions can be investigated in this pump-probe application. A molecular ion along the incident ion path can draw polar molecules

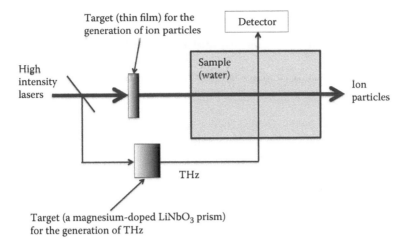

FIGURE 10.7 Proposed experimental set-up using ion beams driven by high intensity lasers with a synchronous THz probe.

toward itself thereby creating a 'cluster molecule' (a bigger sized molecule). Dielectric relaxation effects, which depend on molecular size, also depend on the number density of ions [Kondoh 2014]. Picosecond time resolution is needed because this relaxation is typically very fast in a liquid. Using THz spectroscopy in aqueous solutions such as LiCl and NaCl, Kondoh found that the clustering effect of Li^+ and Na^+ ions which form the uniform electric field can be observed for an ion number density greater than 1 mol/L (i.e. above 6×10^{20} ion/cm³) [Kondoh 2014]. In their experiments, they used a 1 kHz laser pulse train spectrally centred at 800 nm with a pulse duration of 120 fs and a pulse energy of 800 μJ from a titanium sapphire regenerative amplifier. Each ultrashort laser pulse generated a THz waveform (pulse) of energy of ~150 nJ with a corresponding photon number of ~2×10^{14} [Hebling 2008] by the well-established optical rectification technique in a magnesium-doped $LiNbO_3$ prism with wavefront control. In Figure 10.7, in place of Li^+ and Na^+, positive water ions are produced from the irradiation by heavy ions such as carbon, xenon and gold. The threshold ion number density of 1 mol/L corresponds to about 2% ionization of water molecules along the ion path. Therefore, we may determine the minimum required number (N_{ii}) of incident laser-generated ions as follows. We consider the situation where an ion passes a thin film with the surface area of 1 cm² and the thickness of Δl (see Figure 10.8). The number of molecules in this film N_t is $n_m V_s$, where n_m and V_s are the number density of the molecules and the volume of this film, respectively. On the other hand, the produced number (N_{ion}) of molecular water ions attributed to N_{ii} laser-generated ions is, $N_{ion} = N_{ii} n_m V_{ion} = N_{ii} n_m \sigma_{ion} \Delta l$ as seen in Figure 10.8, where, V_{ion}

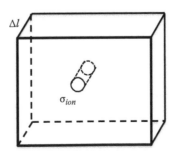

FIGURE 10.8 A thin water film with the width of Δl and σ_{ion}.

is the volume in which molecular ions are produced (set by the ionization cross section, σ_{ion}). For the 2% ionization fraction, we then have:

$$\frac{N_{ion}}{N_t} = \frac{N_{ii}n_m\sigma_{ion}\Delta l}{n_m S\Delta l} = N_{ii}\frac{\sigma_{ion}}{S} = 0.02,$$ (10.5)

where S and n_m are the spot size of the laser-driven ion bunch and the number density of molecules, respectively. We may control S by adjusting this spot-size using magnetic fields (i.e. ion focusing) but the diameter of S should exceed the wavelength of the THz probe (300 µm). For carbon ions with E_{ion} ≤ 5 MeV/u and $S = 1$ mm^2, $\sigma_{ion} \geq 10^{-15}$ cm$^2 = 10^{-13}$ mm^2 and $N_{ii} \leq 2 \times 10^{11}$, this is much larger than the number of ions typically obtained from high-intensity laser experiments [Schreiber 2016].

For such measurements, ion energies should be less than 5 MeV/u but the number of ions should be as high as possible. [Schreiber 2016] is a valuable review that provides a practical assessment of current laser acceleration capability where maximum ion kinetic energies are given along with differential spectral amplitudes for a range of laser pulse energies (for both TiS and Glass laser media). For example, using ultra-short, focused TiS laser pulses, maximum (spectral cut-off) proton and carbon energies of 5 MeV and 5 MeV/nucleon with differential spectral amplitudes of 10^6 to 10^7 and 10^5 to 10^6 (MeV msr × 1% energy) can be achieved for protons and carbon ions, respectively. The strength of the electric field induced from heavy ion irradiation has a longer- range force than that attributed to point charges (as shown in Figure 10.4B), and N_{ii} is expected to reduce with increasing distance, r from the ion track. The period of rotational motion of water molecules (T_{H2O}) is satisfied by the same equation to that of pendulums (T_p) given by $T_p = 2\pi\sqrt{\dfrac{l}{g}} = 2\pi\sqrt{\dfrac{lM}{F}}$, where l, g and M are the length of a pendulum, the gravitation constant, the mass, respectively, and $F = Mg$. Since the electric force acting on hydrogen atoms in H_2O becomes $2\alpha_H E_f(r)$, T_{H2O} is obtained as

$$T = 2\pi\sqrt{\frac{R_H M_H}{2\alpha_H E_f(r)}},$$ (10.6)

where M_H, R_H and α_H are the mass of a hydrogen atom (corresponding to M), the distance between O and H atoms (~ 0.95 Å) (corresponding to l) and the polarization (~0.325) of H in H_2O, respectively [Zhu 1991]. On the condition that $T \sim 1$ ps (which corresponds to 1 THz), r becomes approximately 70 nm for $E_f(r)$ produced from 5 MeV/u carbon ion irradiation (see Figure 10.4B), and approximately 3.5 nm for a point charge. Given that $E_f(r)$ has a longer range than the electric field from a point charge and that higher intensity THz sources with ~ 0.15 mJ (corresponding photon number of ~2 × 10^{17}) have been realized [Zhuo 2017], we consider that N_{ii} solved from Equation 10.5 is clearly an overestimation.

10.3.2 Measurement of Clustered Molecules or Holes Using SAXS

Figure 10.9 shows a second possible pump-probe scheme for picosecond time-resolved measurements of structural change in impacted material due to the laser-driven ion irradiation. The scenario depicted in this figure is as follows: (a) the heavy ions irradiate a solid target, (b) the ion passes through the target and excess heat deposited near the track can generate a transient hole, and (c) the hole is detected using the time-resolved x-ray small angle scattering technique (SAXS). We can apply this scheme to measure the clustered molecules discussed in Section 10.3.1 by using a water target.

Similar holes have been measured using electron microscopy [Leino 2014] or synchrotron-based SAXS in solids [Schauries 2015]. However, these measurements do not provide adequate time resolution (~ps) to isolate early time transient behaviour immediately subsequent to ion impact. In Figure 10.9a,

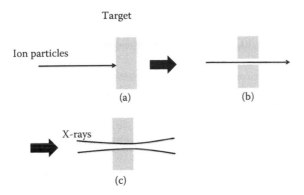

FIGURE 10.9 Setup for proposed resolution measurement of the structural change of the material due to irradiation by laser-accelerated ions: (a) the heavy ions irradiate a solid target, (b) the ion passes through the target and (c) the hole is detected using the time-resolved X-ray small angle scattering technique (SAXS).

where the solid target is shown, we can use as the probe either K-α X-rays that are synchronously produced with ions from a high intensity laser pulse [Galtier 2012)] or a synchronized XFEL light source.

For a 10 KeV XFEL probe, Tokuhisa [2012] have estimated in simulations that a spatial resolution of a few Å is achievable with an X-ray flux of 5×10^{21}/mm^2/pulse for some complex proteins such as lysozyme. Let us use this estimation as a standard because structures of the target required in our measurement are expected to be much simpler. We only measure the local density of water molecules or cluster molecules. With a maximum number of X-ray photons per pulse of 5×10^{11} (at the SACLA source), this X-ray flux can be achieved with focusing to ~100 nm^2 spot size. With the SACLA XFEL probe the measurement of the cluster molecules or a hole can require a spatial resolution of a few nanometres, according to the size of these molecules. This means that we can increase the spot size of XFEL pulse as the requirement of the spatial resolution lessens from a few Å to a few nm in our case. Therefore, it is adequate to focus the probe to an area of 10000 nm^2.

If a single heavy ion passes this area, we can measure the production of cluster molecules or holes, which means $N_{ii} = 1$ for S (spot size of ions) = 10000 nm^2. That is, $N_{ii} = 10^8$ when $S = 1$ mm^2. Even if S is taken to be slightly larger, we may be able to analyze the behaviour of the ions as follows. We can obtain a lot of measurement data proportional to the repetition rate (F_l) of high intensity lasers. For example, when $F_l = 1$ Hz, the number of measurement data can become $60 \times 60 \times 24 = 86400$ in a day. Even if ions are measured in a few % of shots, we can analyze the behaviour of the ions using more than one thousand samples. This indicates that the estimation of N_{ii} is not that critical in this measurement.

Table 10.1 lists N_{ii} estimated here for the THz and SAXS experiments and parameters of XFEL for the SAXS experiment. It should be noted that we tried to avoid underestimating N_{ii}, because this is a new proposal and we have little information for the estimation of N_{ii}. Although one may feel that N_{ii} is too large for these experiments, N_{ii} may be reducible according to our idea shown in this paper, and may even be further decreased depending on other new ideas and developments. We believe that the experiments proposed here have the possibility to advance the study of heavy ion beam science.

TABLE 10.1 Suggested Optical Probe Properties for use in Pump-Probe Experiments with Laser-Driven Heavy Ions

	N_{ii}	Photon Number	Wavelength	Pulse Length	Spot Size
THz	10^{11}	2×10^{14}	~300 μm	>1 ps	≥1 mm^2
SAXA	10^8	5×10^{11}	0.1 nm	10 fs	(100 nm)^2

10.4 Future Prospects

As discussed in previous sections, the very fast processes following ion particle irradiation of water or cells remains an important area to be elucidated in order to understand the underlying early stage mechanisms of ion particle-induced biological effects, particularly the formation process of clustered DNA damage. Our simulation studies which take into account all ejected electrons, molecular ions and polarized molecules have predicted that the electric field effect might play a fundamental role in determining the local dose near an ion track. Further consideration of the extremely high local dose also suggests the potential for shock wave generation and a subsequent transient cavitated region (i.e. hole) around an ion track. These hypothesized fast processes in the early aftermath of ion-particle irradiation are characteristic and specific for ion exposure, and thus usher a different picture of fast processes associated with low LET radiation. For example, the physical shearing force created with the generation of holes could produce DNA damage. Further, the predicted generation of electric fields (and holes) in early time following ion exposure can affect the behaviour of ions and radicals near the track, and thus impact the yield and spatial distribution of DNA lesions. We believe fast processes after ion particle irradiation would be one of the most exciting areas for radiolysis research.

Irradiation by laser-driven ions in combination with the synchronized probe light in a pump-probe configuration can be a valuable tool for verifying the occurrence of hypothesized fast processes that can immediately follow ion irradiation (see also Chapter 8 for other potential advantages of the use of laser-driven ion acceleration in radiobiology). It should be noted that the laser-driven case uniquely affords synchronous multi-beam experiments as have been briefly outlined here with optical probes. Applying laser-accelerated ions in such pump-probe schemes can advance our understanding of fast and localized processes and their impact on fundamental radiobiology and radiochemistry. The detailed knowledge of very fast processes, associated with characteristically ultra-high dose rates in the laser-driven case [Ledingham 2014], can be a key scientific basis for more sophisticated applications such as laser-driven ion beam radiotherapy (LIBRT).

References for Chapter 10

Akamatsu, K. and Shikazono, N. 2013. "A methodology for estimating localization of apurinic/apyrimidinic sites in DNA using fluorescence resonance energy transfer" *Anal Biochem* **433**: 171–180.

Akamatsu, K. et al. 2015. "Localization estimation of ionizing radiation-induced abasic sites in DNA in the solid state using fluorescence resonance energy transfer" *Radiat Res* **183**: 105–113.

Bin et al. 2015. Ion Acceleration "Using Relativistic Pulse Shaping in Near-Critical-Density Plasmas" *Phys. Rev. Lett.*, **115**: 064801/1–5.

Bottländer, D. et al. 2015. "Effect of swift-ion irradiation on DNA molecules: A molecular dynamics study using the REAX force field" *Nucl. Instru. Meth. Phys. Res. B*. **365**: 569–572.

Boyd, R.W. 2003. *Nonlinear Optics*, 2nd ed. Academic Press.

Cappello, C.D. et al. 2009. "Theoretical and experimental investigations of electron emission in C^{6+} + H_2O collisions" *Nucl. Instru. Meth. Phys. Res. B*. **267**: 781–790.

Chatterjee, A. and Schaefer, H.J. 1976. "Microdosimetric structure of heavy ion track in tissue" *Radiat. Environ. Biophys.* **13**: 215–227.

Eccles, L.J. et al. 2011. "Delayed repair of radiation induced clustered DNA damage: Friend or foe?" *Mutat Res* **711**: 134–141.

Galtier, E. et al. 2012. "High-resolution x-ray imaging of Ka volume radiation induced by high-intensity laser pulse interaction with a copper target" *J. Phys. B*. **45**: 025701/1–6.

Goodhead, D.T. 1994. "Initial events in the cellular effects of ionizing radiations: Clustered damage in DNA" *Int J Radiat Biol*. **65**: 7–17.

Gulston, M. et al. 2002. "Clustered DNA damage induced by gamma radiation in human fibroblasts (HF19), hamster (V79-4) cells and plasmid DNA is revealed as Fpg and Nth sensitive sites" *Nucleic Acids Res*. **30**: 3464–3472.

Hada, M. and Georgakilas, A.G. 2008. "Formation of Clustered DNA Damage after High-LET Irradiation: A Review" *J. Radiat. Res. (Tokyo)* **49**: 203–210.

Hayashi, H. et al. 2015. "Evaluation of local temperature around the impact points of fast ions" *Nucl. Instru. Meth. Phys. Res. B*. **365**: 569–572.

Hebling, J. et al. 2008, "Generation of high power terahertz pulses by tilted-pulse-front excitation and their application possibilities" *J. Opt. Soc. Am. B*, **25**: B6–B19.

Howard-Flanders, P. 1958. "Physical and chemical mechanisms in the injury of cells by ionizing radiations" *Adv. Biol. Med. Phys.* **6**: 553–603.

Iliakis, G. 1991. "The role of DNA double strand breaks in ionizing radiation-induced killing of eukaryotic cells" *Bioessays* **13**: 641–648.

Katz, R. et al. 1969. "Particle Tracks in Emulsion" *Phys. Rev.* **186**: 344–351.

Kondoh, M. et al. 2014. "Ion effects on the structure of water studied by terahertz time-domain spectroscopy" *Chem. Phys. Lett.* **591**: 317–322.

Lakowicz, J.R. 2006. *Principles of Fluorescence Spectroscopy*. Third Edition, Springer.

Ledingham, K.W.D. et al. 2014. "Towards Laser Driven Hadron Cancer Radiotherapy: A Review of Progress" *Applied Sciences-Basel* **4**: 402–443.

Leino, A.A. et al. 2014. "Swift Heavy Ion Shape Transformation of Au Nanocrystals Mediated by Molten Material Flow and Recrystallization" *Matter. Res. Lett.* **2**: 37–42.

Miller, J. 2007. "Electric fields have potential as a cancer treatment" *Physics Today*, **60**: 19–20.

Moribayashi, K. 2011. "Incorporation of the effect of the composite electric fields of molecular ions as a simulation tool for biological damage due to heavy-ion irradiation" *Phys. Rev. A*. **84**: 012702/1–7.

Moribayashi, K. 2013. "Movement of secondary electrons due to the irradiation of heavy ions: Role of the composite electric field formed from the polarization of molecules and molecular ions" *Rad. Phys. Chem.* **85**: 36–41.

Moribayashi, K. 2014. "Radial dose calculation due to the irradiation of a heavy ion: Role of composite electric field formed from the polarization of molecules and molecular ions" *Rad. Phys. Chem.* **96**; 211–216.

Moribayashi, K. 2015. "Development of the radial dose distribution function relevant to the treatment planning system for heavy particle cancer therapy" *Phys. Scr.* **90**: 054013/1–5.

Orient, O.J. et al. 1987. "Electron impact ionisation of H_2O, CO, CO_2 and CH_4" *J. Phys. B*. **20**: 3923–3936.

Pritchard, H.P. et al. 1990. "Electronic excitation of H_2O by low-energy electron impact" *Phys.Rev. A*. **41**: 546–549.

Rønne, C. et al. 1997. "Investigation of the temperature dependence of dielectric relaxation in liquid water by THz reflection spectroscopy and molecular dynamics simulation" *J. Chem. Phys.* **107**: 5319–5331.

Sage, E. and Harrison, L. 2011. "Clustered DNA lesion repair in eukaryotes: Relevance to mutagenesis and cell survival" *Mutat. Res.* **711**: 123–133.

Sage, E. and Shikazono, N. 2017. "Radiation-induced clustered DNA lesions: Repair and mutagenesis" *Free Radic Biol Med*. **107**: 125–135.

Schauries, D. et al. 2015. "Size characterization of ion tracks in PET and PTFE using SAXS" *Nucl. Instru. Meth. Phys. Res. B*. **365**: 573–577.

Schreiber, J. et al. 2016. "Invited Review Article: 'Hands-on' laser-driven ion acceleration: A primer for laser-driven source development and potential applications" *Rev. Sci. Inst.* **87**: 071101-1–071101-10.

Shikazono, N. et al. 2009. "The yield, processing, and biological consequences of clustered DNA damage induced by ionizing radiation" *J. Radiat. Res.* **50**: 27–36.

Shimobayashi S. F. et al. 2013. "Probability of double-strand breaks in genome-sized DNA by γ-ray decreases markedly as the DNA concentration increases" *J. Chem. Phys.* **138**: 174907/1–5.

Stöhlker, T. et al. 2015. "APPA at FAIR: From fundamental to applied research" *Nucl. Instru. Meth. Phys. Res. B*. **365**: 680–685.

Surdotovich, E. et al. 2013. "Biodamage via shock waves initiated by irradiation with ions" *Sci. Rep.* **3**: 1289.

Sutherland, B.M. et al. 2000. "Clustered DNA damages induced in isolated DNA and in human cells by low doses of ionizing radiation" *Proc. Natl. Acad. Sci. USA.* **97**: 103–108.

Sutherland, B.M. et al. 2002. "Clustered DNA damages induced by x rays in human cells" *Radiat. Res.* **157**: 611–616.

Tokuhisa et al. 2012. "Classifying and assembling two-dimensional X-ray laser diffraction patterns of a single particle to reconstruct the three-dimensional diffraction intensity function: Resolution limit due to the quantum noise" *Acta. Cryst.* **A68**: 366–381.

Toulemonde, M. et al. 1993. "Thermal spike model in the electronic stopping power regime" *Radiation Effects and Defects in Solids.***126.1–4**: 201–206.

Wang, H. et al. 2014. "Radial dose distributions from protons of therapeutic energies calculated with Geant4-DNA" *Phys Med Biol.* **59**: 3657–3668.

Ward, J.F. 1994. "The complexity of DNA damage: Relevance to biological consequences" *Int. J. Radiat. Biol.* **66**: 427–432.

Zhu, S.B. et al. 1991. "A Flexible/Polarizable simple point charge water model" *J. Phys. Chem.* **95**: 6211–6217.

Zhuo, H. B. et al. 2017. "Terahertz generation from laser-driven ultrafast current propagation along a wire target" *Phys. Rev. E* **95**: 013201/1–5.

11

Laser-Driven Ion Beam Radiotherapy (LIBRT)

Wolfgang Enghardt

Jörg Pawelke

Jan J. Wilkens

11.1 Introduction

Laser-driven particle acceleration may promise more compact and cost-effective heavy charged particle (proton and heavier ions) radiotherapy facilities and also potentially offers treatment benefits like better irradiation of moving tumours. In contrast to conventional beams, laser-driven ion beams are characterized by short pulses (bunches) of high particle flux, low repetition rate, broad energy spectrum, small emittance and large divergence. In addition to laser particle accelerator development, laser-driven ion beam radiotherapy (LIBRT) poses a new set of challenges, like the need for full characterization of radiobiological effects and development of new beam monitoring and dosimetry, as well as a beam line and treatment planning system for laser-driven bunches. Present status and further developments are reviewed, focusing on protons.

11.2 State of the Art in Radiotherapy

Due to the current developments in demography and improved prevention or treatment of heart and circulatory diseases in particular, the incidence of tumour-related diseases is increasing. At present in the developed world, approximately 50% of tumour patients undergo radiation therapy. Therefore, radiotherapy is, together with surgery and chemotherapy, one of the most important components of cancer therapy. Radiotherapy exploits ionizing radiation with the goal of devitalizing or inactivating tumour cells. The sensitive target of the cells with regard to ionizing radiation is primarily the DNA in the cell nucleus [Munro 1970]. To transport the radiation to the tumour and to deliver the prescribed dose, i.e. absorbed radiation energy per mass, different techniques are applied: external beam radiotherapy (EBRT) and brachytherapy, which are the methods in the discipline of radiotherapy, and radionuclide therapy as a therapeutic technique of nuclear medicine. The latter will not be further considered here; an overview and further comprehensive information on this topic is given by Hoefnagel [1998] and Speer [2010], respectively.

In external beam radiotherapy (sometimes called teletherapy or percutaneous therapy) the source of radiation is located outside of the patient. In order to deliver the dose prescribed to deep-seated tumours, the radiation species of a high penetration power has to be selected. Nowadays, beams of ultrahard X-rays, electrons and ions with atomic numbers ranging from 1 (hydrogen) to 6 (carbon) are primarily applied. The application of neutrons in therapy is very small. The usage of γ-rays from tele-cobalt irradiation units is steadily decreasing, since these devices were nearly completely taken out of operation in developed countries until the beginning of the 2000s. However, they still play an important role for radiotherapy in developing countries because of their robustness.

In brachytherapy, strong radioactive sources with millimetre dimensions are placed in close vicinity of the tumour, or even within the tumour volume. Depending on the activity of the sources, brachytherapy can be subdivided into high dose rate (HDR) brachytherapy, where the dose rate exceeds 12 Gy/h, and low dose rate (LDR) brachytherapy, with a dose rate below 2 Gy/h.

For radiation protection of the medical staff during HDR brachytherapy, the radioactive sources are moved to the desired location automatically under remote control from outside the treatment room by means of afterloading devices [Glasgow 1993]. Today, the radioisotope ^{192}Ir (half-life of 73.8 days, initial activity of 370 GBq and mean γ-ray energy of 0.355 MeV) and ^{60}Co (half-life of 5.27 years, initial activity of 74 GBq and mean γ-ray energy of 1.25 MeV) [Andrassy 2012] are predominantly used for HDR afterloaders. In HDR, the prescribed doses are delivered within 2 to 20 minutes, whereas LDR requires a temporary (for hours or days) or even permanent implantation of sources as needles, pellets or seeds (e.g. ^{125}I-seeds, half-life of 59.4 days, initial activity of 7–30 MBq and γ-ray energy of 0.0355 MeV for prostate cancer therapy). The application forms of brachytherapy are contact therapy for tumours growing superficially; intracavitary therapy, where the sources are brought to the tumour through orifices of the body (e.g. for treating gynaecological tumours); and interstitial brachytherapy, where the sources themselves or hollow needles or catheters for guiding afterloading sources are placed via a surgical intervention. Since the discovery of radium by Marie and Pierre Curie in 1898, brachytherapy has evolved to become a valuable component of radiotherapy. For several tumour species (e.g. prostate, cervical or breast cancer), brachytherapy is a highly efficient and cost-effective treatment modality. Nevertheless, in the last decade the use of brachytherapy has rapidly declined in developed countries mainly for the benefit of EBRT [Petereit 2015].

The 'workhorse' of EBRT is the compact, medical linear electron accelerator (LINAC) as shown in Figure 11.1, which is described in detail elsewhere (such as Podgorsak [2005]). To ensure sufficient radiotherapy care, five to six LINACs per one million inhabitants are required [Bentzen 2005, Grau 2014]. LINACs deliver pulsed electron beams with a typical pulse duration of 5 μs and a pulse repetition rate of 200 Hz (i.e. duty cycle of 1:1000). The electron energy varies between about 4 and 20 MeV corresponding to maximum ranges in water equivalent tissue between 2 and 10 cm. Utilizing optimized scatter and collimation systems, the primary electron pencil beam can be widened for irradiating fields up to a size of 25×25 cm^2. Only in about 10% to 20% of the clinical cases are electron beams utilized because of the limited range and considerable scatter of electrons in matter. In the majority of cases, ultrahard *bremsstrahlung* (usually named photons) is applied because of its capability to reach deep-seated tumours (Figure 11.2). Photons are produced by LINACs via the deceleration of electrons with kinetic energy between about 4 and 18 MeV in a radiator target of high atomic number, like tungsten and tantalum, of a few millimetres thickness. The irradiation fields can reach sizes up to 40×40 cm^2. State-of-the-art LINACs are equipped with dynamically controlled multileaf collimators (MLC) [Webb 2001, Ezzel 2003] and with instrumentation for image-guided radiation therapy (IGRT) [Verellen 2007], like planar kilovoltage or megavoltage radiography, kilovoltage or megavoltage cone beam computed tomography [Jaffray 2000] or in-room computed tomography (CT) on rails [Cheng 2003]. This equipment facilitates advanced, three-dimensional tumour conformal treatment applying the technique of intensity-modulated radiation

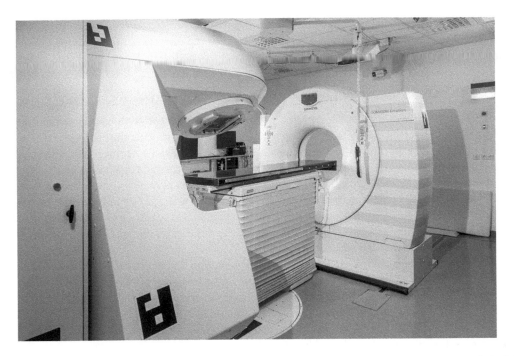

FIGURE 11.1 Radiotherapy room equipped with a LINAC (left) and with an in-room CT scanner on rails (right). The gantry of the LINAC can be rotated by 360° around a horizontal axis to facilitate the irradiation of patients lying on the treatment table (centre) from different directions. Photograph from Helmholtz-Zentrum Dresden-Rossendorf/Rainer Weisflog.

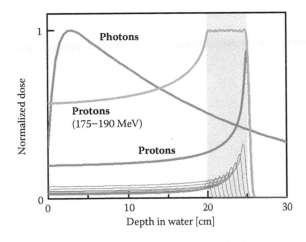

FIGURE 11.2 Normalized dose-depth profile of a photon beam versus a monoenergetic proton beam forming a pristine Bragg peak. For covering the whole target volume in depth, several pristine Bragg peaks have to be superimposed, which results in a spread out Bragg peak. Figure credit: Christian Richter, Technische Universität Dresden.

therapy (IMRT) [Webb 2001]. The main characteristic of IMRT is the spatially variable photon fluence of the impinging radiation field. These fluence maps are calculated by means of inverse treatment planning algorithms. The objective function of this optimization process considers specifications of the intended dose delivery, including its acceptable inhomogeneities in the tumour and normal tissue tolerances. With the introduction of MLCs, with up to 160 leaves of 5 mm or less thickness, the capability of IMRT for shaping tumour conformal dose distributions, even of concave geometry, has nearly reached its physical limit. Nowadays, further improvement of IMRT is sought in the direction of reducing treatment time, in order to minimize the influence of intrafractional movement on the dose distribution and to facilitate hypofractionated treatment schemes (i.e. the total dose is divided into larger dose fractions and delivered in fewer treatment sessions). One of these measures is the introduction of so-called flattening filter free LINAC beam deliveries [Georg 2011]. Conventionally, cone-shaped flattening filters are centred in the photon beam in order to compensate for the emission angle-dependent (with respect to the direction of the impinging electron beam) *bremsstrahlung* fluence and to obtain laterally constant dose profiles. However, MLCs allow for controlling the photon fluence across the field. Flattening filters, with their substantial photon attenuation, can be omitted and the dose rate during photon irradiation can be increased from about 4 Gy/min up to 20 Gy/min with a respective reduction of treatment time. A second technique for treatment time reduction is intensity modulated arc therapy (IMAT) [Yu 1995] or volumetric modulated arc therapy (VMAT) [Otto 2008]. In this technique, the LINAC gantry rotates around the patient while the beam is switched on (with modulated dose rate) and the leaves of the MLC are moving, changing size and shape of the open field, continuously.

The last development towards real-time motion compensation and in general adaptation of the treatment parameters is the development of magnetic resonance imaging (MRI) guided radiotherapy. In contrast to X-ray-based imaging, the MRI avoids ionizing radiation and can thus be performed during the whole treatment, offering the possibility of a real-time adapted radiotherapy. The first MRI–guided cobalt units [Saenz 2015] and MR LINACs [Lagendijk 2008] have been installed and the next years will be devoted to investigating their clinical benefit.

Over the last 20 years, the development of irradiation devices and techniques in ultrahard, photon-based radiotherapy brought remarkable progress primarily in tumour conformality of the dose distributions. However, all these technological developments cannot reduce the dose to normal tissue and must not exceed the tumour dose beyond the physical limits, which are given by the photon dose-depth profile with its exponential decline (Figure 11.2). Beams of heavy charged particles, like protons or heavier ions with atomic numbers $2 \leq Z \leq 8$, offer the potential for dose reduction in normal tissue (Figure 11.3) and

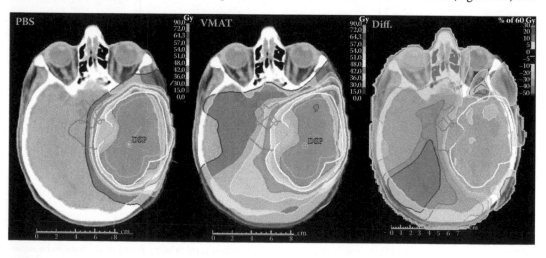

FIGURE 11.3 Illustration of the advantage of proton over photon therapy by comparing treatment plans (dose distributions superimposed onto X-ray-computed tomography images) for a patient with a high-grade glioma of the brain. Left: Proton dose delivery by means of pencil beam scanning (PBS); centre: Photon dose delivery by means of VMAT; right: Relative dose difference between PBS and VMAT. Obviously in proton therapy, large volumes of the brain are prevented from receiving dose and, furthermore, the dose to the radiosensitive structures brainstem and chiasma (contoured in pink) is substantially reduced. Colour version of figure can be accessed online. Treatment planning images from Christian Richter, Technische Universität Dresden.

dose escalation in the target volume. This is mainly due to the inverted dose-depth distribution (Figure 11.2), which is a consequence of the Coulomb interaction between a swift ion and the electrons of the target described by the Bethe Bloch formula (see for example Leo [1994]).

According to the statistics gathered by the Particle Therapy Co-Operative Group [PTCOG 2017] about 18,000 patients (which is much less than 1% of patients receiving radiotherapy) have been treated with particle beams worldwide, primarily with protons, at 65 facilities in operation in the year 2015. Further, 30 new particle therapy installations are under construction or in the planning phase. Proton beams alone are delivered at 55 facilities, and 10 facilities are capable of applying ^{12}C ion beams usually in combination with protons. The accelerator types operated for ion beam radiotherapy (IBRT) have to fulfil the following clinical requirements: the particle range in water has to reach at least 30 cm, corresponding to a particle energy of 220 MeV and 5.1 GeV for protons and carbon ions, respectively, and a dose rate of at least 1 Gy/min is requested for a reasonable therapeutic workflow. Considering a rather large treatment volume of $25 \times 25 \times 25$ cm^3 fluxes of about 6×10^9 protons/s (~1 nA average beam current) or 4×10^8 ^{12}C-ions/s (~0.4 nA) are required at the beam entrance into the patient. It should be noted that in this context, the average power of such beams is less than 1 W. This low beam power of therapeutic beams is, by the way, the rationale for the attempt to produce therapeutic ion beams by means of laser-driven particle acceleration. Considering the rather low proton energy of about 20 MeV, which is necessary for radionuclide production via nuclear reactions, it seems to be much easier to put into practice a laser-driven accelerator for that purpose (see Chapter 13 by Roso). However, the time-averaged beam power of accelerators for radionuclide production, usually compact cyclotrons, is of the order of 1 kW, which can hardly be reached with a laser-driven system.

Most proton centers are equipped with cyclotrons: 39 operate an isochronous cyclotron [Flanz 2011, Jongen 2011, Pearson 2016, Schippers 2012] with either normally conducting or superconducting magnets [Krischel 2012]. For producing beams of heavier ions (in particular ^{12}C) synchrotrons are exclusively applied. Typically, these machines reach diameters of 20 m or more [Eickhoff 2011]. Furthermore, 12 facilities operate pure proton synchrotrons. Since the beam rigidity of protons (2.26 Tesla meters, Tm at 220 MeV) is lower than that for carbon ions (6.57 Tm at 5.1 GeV), proton synchrotrons have diameters less than 10 m [Slater 1992, Smith 2009]. A rather new trend in proton therapy is the introduction of compact, superconducting synchrocyclotrons into clinical practice. Examples are the gantry mounted Monarch 250® of MEVION medical system or the floor-mounted S2C2® machine of ion beam applications (IBA) delivering its beam to a single gantry. The diameter of these machines is less than 2 m, their mass is less than 50 t, which is far less than the 'classical' therapy proton cyclotron Cyclone® of IBA weighting 220 t and having a diameter of 4.34 m [Pearson 2016]. The new synchrocyclotrons open prospects for including a rather compact proton facility with a single treatment room into large hospitals. It is expected that these compact proton treatment machines will improve the overall acceptance of proton therapy and will bring forward the necessary clinical studies for evaluating the benefit of proton therapy. It should be noted that up to now, such compact machines were not available for ions heavier than protons.

A particle therapy system is usually a modular system comprising three components (Figure 11.4):

1. The beam production system consists of the accelerator and a system for adjusting the beam energy in the range necessary for the particular treatment situation.

   For cyclotron-based systems, which deliver a beam of fixed energy, energy adjustment requires that the beam passes through an energy selection system: a degrader decelerates the protons and by a system of magnetic dipoles, quadrupoles and adjustable apertures the beam of desired energy is produced.

   For synchrotrons, an energy selection system is not necessary, since these machines provide beams of adjustable energy.

2. *The beam transport system* consists of a highly evacuated beam pipe surrounded by dipole magnets for bending and quadrupole magnets for focusing the beam. In particle therapy facilities with several treatment places, the transport beam line may reach a length of more than 100 m.

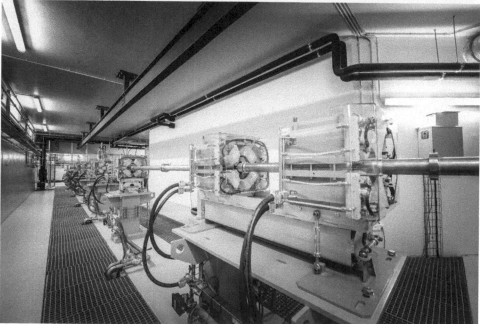

FIGURE 11.4   The beam production (above), beam transport (centre) and treatment beam delivery system (below) of the University Proton Therapy Dresden, Germany. Photographs from OncoRay/Andre Wirsig (above and centre) and University Hospital Dresden/Martin Förster (below).   (*Continued*)

3. *The treatment beam delivery system* laterally widens (either by pencil beam scanning or passive scattering), adjusts the lateral size and shape, modulates in range and analyzes the position and the current of the beam, in order to deliver the prescribed dose to the target volume in three dimensions. Most proton therapy facilities are equipped with isocentric gantries [Pedroni 1995] with a typical diameter and length of 10 m and a mass of about 100 t. Carbon ion facilities use fixed horizontal, vertical and even oblique beam lines and avoid gantries, because of their large extent and high mass. The only isocentric ion beam gantry in clinical use exists at the Heidelberg

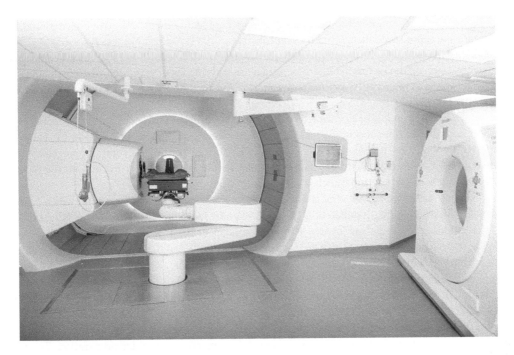

FIGURE 11.4 (CONTINUED)   The beam production (above), beam transport (centre) and treatment beam delivery system (below) of the University Proton Therapy Dresden, Germany. Photographs from OncoRay/Andre Wirsig (above and below) and University Hospital Dresden/Martin Förster (below).

Ion-Beam Therapy Center [Eickhoff 2011] and has a diameter of about 13 m, a length of 25 m and a mass of 600 t. Again, the reason for these huge dimensions of an ion beam gantry, in comparison to proton gantries, is the large magnetic rigidity of carbon ion beams, requiring a much longer travelling path for the ions between the deflection and the target points to obtain sufficiently large treatment fields of at least a 20 cm diameter.

The increased number of dedicated particle therapy facilities and the number of patients treated open possibilities for systematic clinical studies of the advantages of particle therapy and of improving the reliability and precision of the dose delivery by technological advancements. We must be reminded that, up to now, the advantages of particle therapy over modern photon therapy could only be shown for radioresistant tumours or tumours growing close to organs at risk (e.g. chordoma and chondrosarcoma at the skull base or the spinal cord, ocular melanoma and paediatric tumours, when the dose reduction in growing organs is highly desirable (Figure 11.3). For the majority of tumour species, in particular for the frequent ones, like lung, prostate or breast cancer, a superiority of particle therapy has yet to be demonstrated in well-designed clinical trials [Allen 2012, Baumann 2016, De Ruysscher 2012].

11.3 Present Status and Expected Progress in Laser-Driven Ion Beam Irradiation for Radiotherapy Application

When discussing laser-driven particle acceleration as a possible option to reduce the cost and size of ion beam radiotherapy (IBRT), it must be noted that a reduction in the size and cost of the ion accelerator alone will not be sufficient to establish the completely new laser-driven technology as a competitive alternative to conventional facilities, since the magnetic beam transport systems, including gantries for multidirectional irradiation of patients and the necessary surrounding radiation protection walls, contribute considerably to the large footprint and high costs of an IBRT facility (see Section 11.2).

Also, ongoing research and development aims for more compact and cost effective conventional IBRT accelerators like superconducting synchrocyclotrons.

Unlike quasi-continuous, monoenergetic, narrow beams from conventional therapy accelerators (see Section 11.2), laser-driven beams are characterized by short bunches of very high particle flux, low repetition rate, broad energy spectrum, large divergence and significant pulse-to-pulse fluctuation (see Chapter 5 by Macchi on laser-driven ion acceleration). These inherent properties make it very challenging to apply laser-driven ion beams directly for medical application with existing radiotherapy equipment and dose delivery approaches. In addition to laser and laser target development for generating therapeutically viable particle beams, such beams demand new optimized solutions for efficient and precise dose delivery to match clinical standards along with comprehensive research on the radiobiological consequences of radiation bunches with an ultra-high dose rate.

Over recent years, laser-driven irradiation technology was developed at such a rate that cell samples and small animals are being irradiated in the framework of radiobiological experiments (see Chapter 8 by Friedl and Schmid and also Chapter 10 by Shikazono, Moribayashi and Bolton). Extensive *in vitro* studies were performed with laser-driven electron [Rigaud 2010, Laschinsky 2012, Andreassi 2016] as well as proton [Yogo 2009, Kraft 2010, Yogo 2011, Bin 2012, Doria 2012, Zeil 2013, Raschke 2016] bunches using several tumour and normal tissue cell lines and investigating different biological endpoints. In a second translational step, also *in vivo* experiments were established [Brüchner 2014, Oppelt 2015]. A full-scale animal experiment was realized with laser-driven electrons and the radiation-induced tumour growth delay was determined for human head and neck squamous cell carcinoma (HNSCC) grown on nude mice [Oppelt 2015]. *In vivo* studies with laser-driven protons are under preparation. The necessary comprehensive experimental campaigns are challenging due to the rather poor beam stability and reproducibility as well as due to the very limited beam time of presently available laser accelerators with the capability to deliver minimal proton energies near 30 MeV to penetrate tumours of the size of few millimetres. Therefore, these studies await their realization.

So far, radiobiological experiments are performed at fixed horizontal beam lines with energy selection and irradiation field formation by magnetic steering systems. Doses up to about 10 Gy are delivered to the cells with proton bunches of nanosecond scale duration, energies below about 20 MeV and broad energy spectra (an energy full width at half maximum from about 10% to 100%) by using laser systems of about 100 TW power (peak or single pulse). In some experiments, the dose is fractionated and the average dose rate is comparable to that used in irradiation with conventional accelerators (a few Gy/min) [Yogo 2011, Zeil 2013]. In other cases, the dose is delivered by single-bunch irradiation and ultra-high dose rates up to about 10^{11} Gy/min have been estimated [Bin 2012, Doria 2012]. Bunch-to-bunch fluctuations of order 20% have been reported for the number of protons per pulse [Yogo 2009] and single bunch dose [Kraft 2010], respectively. The resulting uncertainty in delivery of a prescribed dose by single-bunch irradiation is acceptable for radiobiological cell experiments provided by the actual delivered dose is retrospectively determined at higher accuracy, but cannot be applied for patient irradiation. Apart from reducing the fluctuations, dose fractionation (multiple-bunch irradiation) in combination with beam bunch monitoring in real-time affords an increase in the accuracy of prescribed dose delivery.

It has to be noted that radiobiological experiments with electron bunches can be performed with laser systems of lower power (near 10 TW) [Laschinsky 2012, Oppelt 2015, Andreassi 2016], resulting in very high beam energy (100 MeV or more; see Chapter 2 by Malka) that exceeds the maximum energy of conventional clinical electron LINACs (see Section 11.2) by several times. Such very high energy electrons (VHEE) show reduced lateral scattering and an increased range (30 cm), allowing the direct use of electrons for treatment to extend from superficial to deep-seated tumours. However, clinical application of VHEE is complicated by radiation protection problems. Because of the rapidly increasing production of penetrating secondary (background) radiation (*bremsstrahlung* photons, neutrons) with increasing electron energy, special consideration must be given to the safe design, installation and utilization of a

VHEE accelerator, in order to protect patients, staff and the general public from unnecessary radiation risks

Radiotherapy application requires a precise knowledge of the absolute dose delivered by the ionizing radiation to the cells (tissue) which unfortunately cannot yet be measured directly. Dose is defined as absorbed radiation energy per unit mass and the energy transfer depends on the radiation (species, energy) and the irradiated material (elemental composition, density). National and international dosimetry protocols are established for radiotherapy with conventional beams (e.g. IAEA [2000]) and recommend the use of ionization chambers which are calibrated at a national standard (accredited) dosimetry laboratory for absolute dose absorbed by water. The ionization measured by the chamber (typically filled with air) is converted to absorbed dose by applying a dedicated chamber calibration factor. In addition, a set of chamber dependent correction factors must be applied to take into account any difference between reference (calibration) and experimental or clinical conditions of dose measurement at the hospital which affects the chamber signal, such as differences in beam quality (radiation species, energy, intensity and time structure), geometric arrangement, radiation field size and ambient conditions (air temperature and pressure). Besides an established absolute dosimetry, beam detection in real-time is necessary in clinical practice to confirm the accurate and consistent technical performance of the accelerator. Relative dosimetry is performed by monitoring the relative beam output with transmission ionization chambers to compensate for small beam intensity variations of the accelerator and to safely deliver the correct prescribed dose. For laser-driven beams, no national or international reference dosimetry protocol has been established so far and online beam monitoring detection is a challenge due to the peculiar, short duration and very high intensity of the generated particle bunches. A number of different types of established detectors and dosimeters have been investigated for dosimetric use in radiobiological experiments with laser-driven particles, but none of these fulfils all the requirements. They include nuclear track, film, image plate and luminescence detectors, which require post-irradiation processing, as well as online devices like ionization chambers, Faraday cups, scintillators, diamond and semiconductor pixel detectors [Karsch 2012, Bolton 2014, Nicolai 2014]. Most developed dosimetry protocols are based on sample-individual absolute dose measurement by radiochromic films [Richter 2011, Fiorini 2011, Bin 2012], whose response is independent of the dose rate and which can be arranged in such a way to obtain 3D dose maps, however providing retrospective dose information only. By combining film dosimetry with measurements by a second, independent absolute dosimeter and also including relative dose monitoring in real-time as well as determination of the energy spectrum of the beam, an overall uncertainty in delivery below 10% of a prescribed absolute dose to cell samples was already achieved [Zeil 2013] approaching the clinically established standard of 3–5%. Further developments aim to extend the use of ionization chambers in laser-driven acceleration, which is limited to relative dosimetry up to now. Absolute dose determination by reference dosimetry with ionization chambers requires the correction for recombination processes which cause incomplete collection of the charges released by the radiation-induced ionization of the chamber gas. The higher the dose rate, the higher the signal loss. Accurate correction methods are well established for the small corrections necessary for conventional clinical beams. Appropriate corrections for the ultra-high intensity of laser-accelerated particle bunches have recently been developed by experimental and theoretical investigations for commercial ionization chambers widely used in radiotherapy [Karsch 2011, Karsch 2016], but need verification at the prototype developments of future clinical laser accelerators. The challenges of dosimetry at laser-driven ion sources have triggered innovative detector developments like new detectors based on scintillator blocks in combination with optical tomography. As demonstrated by a prototype, such a transportable and cost-effective dosimetry system enables, for the first time, a very fast and direct measurement of 3D dose distributions which allows not only for monitoring the dose delivered by individual laser-driven particle bunches but is also of interest for conventional radiotherapy application [Kroll 2013].

11.4 Laser-Driven Ion Beam Delivery via Compact Gantry Systems Based on Pulse-Powered Magnets

Even though it is unlikely that laser-driven ion acceleration will result in beams that are comparable to beams deliverable by today's conventional accelerators, it is worth considering the beam transport and dose delivery offered by established methods for patient irradiation in conventional IBRT (see Section 11.2). This requires filtering out a monoenergetic, narrow (pencil-like) beam of variable energy and intensity from the divergent, intense, broad-energetic beam of the laser-driven accelerator. Although it may be possible in the future to adjust the beam parameters by the laser and laser target, a presently more realistic scenario is the adjustment in the beam line. Accordingly, the first attempt to design a gantry for laser-driven proton sources used a simple very low acceptance beam line with an aperture to limit the initial divergence for transport and a magnetic chicane approach for energy filtering of the laser-driven bunches [Luo 2005, Ma 2006]. However, such a system uses much less than 0.1% of all protons for dose delivery while depositing huge numbers of protons in beam dumps and producing a high level of secondary (background) radiation (like neutrons and photons), which has to be shielded by thick walls and makes the operation of LIBRT less economic. It is therefore advisable to follow a more sophisticated beam handling and dose delivery approach which exploits the particular properties of laser-driven bunches.

One of the main advantages of LIBRT would be the option of transporting laser light to the treatment room by compact optical lines with mirrors rather than transporting high energy ion bunches with large magnets, so that the high costs related to ion beam transport and radiation shielding on the way to the treatment room are reduced. This approach would benefit further from the development of a compact beam handling and delivery system for the treatment room as a replacement of the massive and costly gantries used in existing IBRT facilities. Here, another potential advantage comes from the short ion bunch duration in combination with the moderate pulse repetition rate (~ 10 Hz) of laser-driven acceleration. This time structure allows the use of high-field pulsed magnets for a compact gantry design. A higher magnetic field strength results in smaller beam bending radii which, in turn, determine the size of any gantry. The maximum magnetic field strength of conventional magnets is limited by saturation of the magnetization of the iron-core to a value of about 2 T. However, pulsed magnets are air-core designs and a much higher magnetic field strength (one order of magnitude or more) can be achieved, which is mainly limited by the peak current supplied by the power supply to the magnet and by the mechanical strength of the magnet structure needed to hold intense magnetic pressure. Also, due to the missing iron core, pulsed magnets are much lighter in weight. It has to be noted that superconducting magnets can generate maximum magnetic field strengths of about 10 T, exceeding normal magnets and compete to some extent with pulsed magnets, but are restricted to slow or small changes of the magnetic field.

In conventional IBRT, monoenergetic beams with decreasing energy and intensity are superimposed to deliver uniform doses to the tumour region via spread out Bragg peak (SOBP) irradiation (see Figure 11.2). A clinically relevant SOBP of certain width and at certain depth requires a broad energy range to cover the complete tumour in depth. For LIBRT, any efficient dose delivery scheme must make use of the large number and broad energy spread of the ions available in each bunch rather than filtering out narrow monoenergetic beams. This demands a magnetic beam handling system which can control, transport, shape and scan broad energetic ion bunches of variable spot size, regardless of the magnet type being used and independently of potential advanced dose delivery scheme (see Figure 11.5) being applied.

A lightweight, compact, isocentric, 360° rotatable gantry was recently designed for laser-driven proton bunches [Masood 2014]. This design has advanced features like a large acceptance particle capturing lens and an integrated bunch-to-bunch energy selection system to select and transport variable broad-energetic bunches (energy bandwidth range of 3–22%) in a robust gantry formation. Also, for an overall compact gantry design, high-field pulsed magnets have been introduced, which are much lighter in weight (by an order of magnitude) and smaller in size (by a factor of 2) than conventional iron-core

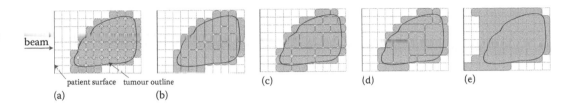

FIGURE 11.5 Potential dose delivery schemes for laser-driven proton bunches for a given beam direction. Dose is spread over increasingly larger areas ('clusters') in order to make efficient use of high proton numbers in a single bunch. (Reproduced with permission from K. Hofmann et al., 2012. "Laser-driven beam lines for delivering intensity modulated radiation therapy with particle beams" *J. Biophotonics* 5: 903–911 (© 2012 by WILEY-VCH Verlag GmbH & Co. KGaA, Weinheim.) (a) spot based, (b) lateral layer based, (c) axial layer based, (d) partial volume based and (e) target volume based.

electromagnets and can provide a magnetic field strength that is multiply higher and synchronized with traversing particle bunches. Moderate maximum field strengths of 9 T for dipole fields and 40 T for solenoidal fields as well as field gradients of 300 T/m for quadrupoles have been considered for calculating the gantry beam line. Due to the higher magnetic fields the size of the gantry can be reduced to about 2 m in radius and 3.5 m in length, which is about 2 times smaller in height and about 3 times shorter in length than most of the conventional isocentric gantries being deployed in conventional proton therapy (Section 11.2). The gantry design also makes use of the concept of a laser source outside the treatment room and a laser-target assembly is considered to be mounted inside the gantry where the laser-accelerated protons are generated.

For gantry realization, the research and development of pulsed magnets and pulsed power-supplies is ongoing. The feasibility of pulsed magnets as efficient capturing and focusing lenses (e.g. solenoids) for strongly divergent laser-driven ion bunches has already been demonstrated in several experiments, e.g. [Burris-Mog 2011, Busold 2014]. Also, pulsed dipole magnets for beam bending and energy selection as well as pulsed quadrupole magnets for beam shaping to desired field sizes are being developed [Schürer 2015]. The magnetic fields inside the pulsed magnets have rise times in the range of few hundred microseconds with pulse durations in the millisecond range, however, they can be considered uniform for the duration of the passage of the ion bunch, which can be up to few nanoseconds.

11.5 Tumour Conformal Dose Delivery Approaches and Treatment Planning for Laser-Driven Ion Beams

In order to derive the requirements for the parameters of laser-driven proton (or ion) beams for radiotherapy, it is worthwhile to specify potential dose delivery approaches and to perform treatment planning studies on real patient cases. This is important since many properties of laser-driven beams are quite different from 'conventional' beams, in particular the broad energy spectrum, the extremely pulsed nature (i.e. very short ion bunches) and the significant bunch-to-bunch fluctuation of laser-driven particle bunches.

One obvious approach would be to try to match conventional beam properties (e.g. a narrow energy spectrum) with the laser-driven beam, and to use the same dose delivery techniques (for state-of-the-art spot scanning, see Section 11.2) as in existing proton therapy centres. If the laser-generated spectra are broad, an energy selection system [Luo 2005] to select narrow energy spectra with variable nominal energy has to be applied. The additional requirement would be that the number of protons contained in a single bunch is high enough so that (for a given laser repetition rate) the treatment time becomes feasible (well below 15 min), and at the same time low enough so that the required degree of fluence modulation for each spot can be realized by a finite number of bunches per spot (since the required number of protons per spot as determined during treatment planning needs to be realized with an integer number of bunches for each spot). The disadvantages of this approach are a high number of secondary particles produced in the energy selection system (radiation protection issue) and the inefficiency of such systems, which may lead to the case that an extremely high number of laser pulses would be required and the treatment cannot be delivered within the mentioned time constraint (a repetition rate of > 10 Hz would be necessary in this case).

It is therefore advisable to follow a more sophisticated beam delivery approach, which utilizes the specific properties of laser-driven beams. For example, specific devices could be envisioned that allow for the delivery of a complete spread out Bragg peak in a single laser shot [Schell 2009]. Alternatively, delivery approaches which use, at least partially, a broader part of the energy spectrum (compared to conventional beams) are indicated. Such beams still show some kind of a (broadened) Bragg peak, and may be used in proximal parts of the target volume (see delivery schemes below). In the long run, it might even be possible to exploit the ultra-high dose rates of laser-driven beams for the treatment of moving targets (e.g. in a gating scenario or with a fast optical gantry for scanning or tracking; see Hofmann [2012]).

If we assume that the repetition rate of the laser and the corresponding proton bunches is 10 Hz (which is optimistic at this point), we have less than 10,000 proton bunches to deliver the dose to the target volume within an acceptable time. Since the number of beam spots in conventional proton therapy can easily be above that number, we would have basically one laser-driven bunch per spot position. This approach would only work if we can actively control the number of protons in each bunch, with a dynamic range of at least two (three is better) orders of magnitude. If this is not the case, we will need to spread the protons within one bunch to a much larger volume than a single spot position, either axially (i.e. using broad energy spectra) or laterally (i.e. using large spot size), or both at the same time [Schell 2010, Hofmann 2012]. Such delivery schemes are shown in Figure 11.5, starting from conventional 'spot-based' delivery over either lateral or axial spot clusters towards a combination of both (termed 'partial volume-based delivery') until the extreme case of delivering dose to the whole target volume in each proton bunch, similar to passive-scattering dose delivery in conventional proton therapy (see Section 11.2), which could be realized by 'spectral shaping' [Schell 2009] in the laser case.

To define more detailed requirements in terms of the required proton numbers per bunch and the required number of laser shots, treatment planning studies (based on the mentioned delivery schemes) have been performed for real patient cases, taking the properties of laser-driven beams into account [Schell 2010, Hofmann 2015]. These studies specifically addressed the fact that laser shots will be integer number (i.e. a 'half a shot' is not possible), and that the corresponding proton number per bunch can only be delivered as a whole. Based on a magnetic energy selection system and the axial clustering technique described above, Schell [2010] found clinical treatment plans with less than 10,000 laser shots, assuming 10^8 protons per bunch in a broad energy spectrum. Hofmann [2015] performed planning studies for a laser-driven gantry design with pulsed magnets [Masood 2014] (see also Section 11.4) and up to a 24% energy width of the transported spectra. They highlighted a clear dependence between the required number of laser shots and the number of protons contained in the initial polychromatic proton bunch. For initial proton numbers near 3×10^9 per bunch, a plan with less than 10,000 shots could be found. For higher proton numbers the plan quality was compromised (modulation problems due to the integer shot requirement), while for lower proton numbers treatment times would become too long. The situation could be improved if a certain degree of intensity modulation from shot to shot were possible [Hofmann 2015], either by the laser accelerator or by the beam delivery (gantry) system. Another potential problem is (random) shot-to-shot fluctuations of the proton numbers or the energy spectrum, which may cancel in some cases [Schell 2012, Hofmann 2015].

In summary, the key challenges in terms of laser-driven proton beam parameters (as already pointed out by Linz [2007] and Bolton [2016]) are beam energy and spectrum (up to 250 MeV and not too broad), number of protons per bunch (near 10^9), stability and reproducibility of energy and proton numbers (below a few percent), the repetition rate of the laser (at least 10 Hz) and intensity modulation from shot-to-shot (with a dynamic range of one or two orders of magnitude).

11.6 Summary and Conclusions

Radiation therapy is an important modality in cancer treatment and compact electron linear accelerators are widely used to provide electrons and photons for tumour irradiation in external beam radiotherapy. Due to their superior dose profile over photons and electrons, protons and heavier ions may provide higher tumour dose conformality and healthy tissue spearing. An increasing number of patients, but still a very low percentage of patients receiving radiotherapy, are treated with ions, mainly with protons. The high cost and huge size of existing facilities limits IBRT to few large centres. For the reduction of the overall size and cost of IBRT facilities, laser-driven particle acceleration may open the way not only to more compact ion accelerators but also to more compact beam transport systems including gantries for multidirectional irradiation of patients. However, laser-based particle accelerators deliver short, very intense particle bunches with low repetition rates, broad energy spectra and high divergences, i.e. very different beams than quasi-continuous, monoenergetic, pencil-like beams from conventional accelerators. In addition to laser particle accelerator development for providing beams of therapeutic quality, the particular features of laser-driven ion bunches demand new solutions for beam transport, dosimetric control and tumour conformal dose delivery along with research into the radiobiological consequences of short ion bunches with ultra-high pulse dose rate.

Over the last decade, research was mainly focused on laser-driven protons and important progress was achieved regarding all major aspects. Laser-driven bunches are already used at fixed horizontal beam lines for radiobiological experiments with cell samples and small animals. For translation to conformal irradiation of extended tumour volumes in patients, a compact lightweight gantry design based on pulsed, high-field magnets has been reported, enabling the capture and transport of divergent, broad energy proton bunches and including a novel beam shaping and dose delivery system [Masood 2017]. High quality treatment plans could be achieved based on the so-called axial and lateral clustering methods, deliverable via such a gantry and dose delivery system. Also, suggestions and solutions for precise dosimetric control [Gotz 2017], for example, and secondary radiation shielding [Faby 2015] have been reported. However, laser-driven proton bunch parameters are still far away from therapy requirements. Besides the necessary improvements in stability, reproducibility and control, a considerable energy increase of the laser-accelerated bunches is necessary which is expected from the ongoing efforts to develop high-repetition petawatt-class lasers.

Compact superconducting synchrocyclotrons have been introduced in clinical practice over recent years and compact conventional proton therapy facilities with a single treatment room are now commercially available [Contreras 2017]. In contrast to protons, progress towards more compact conventional facilities for heavier ions, which require synchrotrons, is almost non-existent. Therefore, further research and development of LIBRT should focus more on ions heavier than protons. Besides progress in the ongoing development of high power lasers and laser targets, but also of pulsed magnets for gantry realization, the need and possible success of LIBRT also relies on exploiting the special characteristics of laser-driven beams. One potential clinical advantage would be the dose concentration in short bunches for the irradiation of moving targets which is a major challenge for IBRT of many tumour entities. Also, the outcome of ongoing and future clinical trials comparing the clinical benefit of both ions versus photons and heavier ions over protons will affect the motivation to install cost-effective compact LIBRT facilities.

References for Chapter 11

Allen, A.M., Pawlicki, T., Dong, L., Fourkal, E., Buyyounouski, M., Cengel, K., Plastaras, J., Bucci, M.K., Yock, T.L., Bonilla, L., Price, R., Harris, E.E. and Konski, A.A. 2012. "An evidence based review of proton beam therapy: The report of ASTRO' emerging technology committee" *Radiotherapy and Oncology* **103**(1): 8–11.

Andrassy, M., Niatsetsky, Y. and Perez-Calatayud, J. 2012. "Co-60 versus Ir-192 in HDR brachytherapy: Scientific and technological comparison" *Revista de Fisica Médica* **13**(2): 125–130.

Andreassi, M.G., Borghini, A., Pulignani, S., Baffigi, F., Fulgentini, L., Koester, P., Cresci, M., Vecoli, C., Lamia, D., Russo, G., Panetta, D., Tripodi, M., Gizzi, L.A. and Labate, L. 2016. "Radiobiological effectiveness of ultrashort laser-driven electron bunches: micronucleus frequency, telomere shortening and cell viability" *Radiation Research* **186**(3): 245–253.

Baumann, M., Krause, M., Overgaard, J., Debus, J., Bentzen, S.M., Daartz, J., Richter, C., Zips, D. and Bortfeld, T. 2016. "Radiation oncology in the era of precision medicine" *Nature Reviews Cancer* **16**(4): 234–249.

Bentzen, S. M., Heeren, G., Cottier, B., Slotman, B., Glimelins, B., Lievens, Y. and van den Bogaert, W. 2005. "Towards evidence-based guidelines for radiotherapy infrastructure and staffing needs in Europe: the ESTRO QUARTS project" *Radiotherapy and Oncology* **75**(3): 355–365.

Bin, J., Allinger, K., Assmann, W., Dollinger, G., Drexler, G.A., Friedl, A.A., Habs, D., Hilz, P., Hoerlein, R., Humble, N., Karsch, S., Khrennikov, K., Kiefer, D., Krausz, F., Ma, W., Michalski, D., Molls, M., Raith, S., Reinhardt, S., Röper, B., Schmid, T.E., Tajima, T., Wenz, J., Zlobinskaya, O., Schreiber, J. and Wilkens, J.J. 2012. "A laser-driven nanosecond proton source for radiobiological studies" *Appl. Phys. Lett.* **101**: 243701.

Bolton, P. 2016. "The integrated laser-driven ion accelerator system and the laser-driven ion beam radiotherapy challenge" *Nuclear Instruments and Methods in Physics Research A* **809**: 149–55.

Bolton, P.R., Borghesi, M., Brenner, C., Carroll, D.C., de Martinis, C., Fiorini, F., Flacco, A., Floquet, V., Fuchs, J., Gallegos, P., Giove, D., Green, J.S., Green, S., Jones, B., Kirby, D., McKenna, P., Neely, D., Nuesslin, F., Prasad, R., Reinhardt, S., Roth, M., Schramm, U., Scott, G.G., Ter-Avetisyan, S., Tolley, M., Turchetti, G. and Wilkens, J.J. 2014. "Instrumentation for diagnostics and control of laser-accelerated proton (ion) beams" *Physica Medica* **30**(3): 255–270.

Brüchner, K., Beyreuther, E., Baumann, M., Krause, M., Oppelt, M. and Pawelke, J. 2014. "Establishment of a small animal tumour model for in vivo studies with low energy laser accelerated particles" *Radiation Oncology* **9**: 57.

Burris-Mog, T., Harres, K., Nürnberg, F., Busold, S., Bussmann, M., Deppert, O., Hoffmeister, G., Joost, M., Sobiella, M., Tauschwitz, A., Zielbauer, B., Bagnoud, V., Herrmannsdoerfer, T., Roth, M. and Cowan, T.E. 2011. "Laser accelerated protons captured and transported by a pulse power solenoid" *Physical Review Special Topics Accelerators and Beams* **14**: 121301.

Busold S., Schumacher, D., Deppert, O., Brabetz, C., Kroll, F., Blazevic, A., Bagnoud, V. and Roth, M. 2014. "Commissioning of a compact laser-based proton beam line for high intensity bunches around 10 MeV" *Physical Review Special Topics Accelerators and Beams* **17**: 031302.

Cheng, C. W., Wong, J., Grimm, L., Chow, M., Uematsu, M. and Fung, A. 2003. "Commissioning and clinical implementation of a sliding gantry CT scanner installed in an existing treatment room and early clinical experience for precise tumour localization" *American Journal of Clinical Oncology* **26**(3): e28–e36.

Contreras, J., Zhao, T., Perkins, S., Sun, B., Goddu, S., Mutic, S., Bottani, B., Endicott, S., Michalski, J., Robinson, C., Tsien, C., Huang, J., Fischer-Valuck, B.W., Hallahan, D., Klein, E. and Bradley, J. 2017. "The world's first single-room proton therapy facility: Two-year experience" *Practical Radiation Oncology* **7**: e71–e76.

De Ruysscher, D., Lodge, M.M., Jones, B., Brada, M., Munro, A., Jefferson, T. and Pijls-Johannesma, M. 2012. "Charged particles in radiotherapy: A 5-year update of a systematic review" *Radiotherapy and Oncology* **103**(1): 5–7.

Doria, D., Kakolee, K.F., Kar, S., Litt, S.K., Fiorini, F., Ahmed, H., Green, S., Jeynes, J.C.G., Kavanagh, J., Kirby, D., Kirkby, K.J., Lewis, C.L., Merchant, M.J., Nersisyan, G., Prasad, R., Prise, K.M., Schettino, G., Zepf, M. and Borghesi, M. 2012. "Biological effectiveness on live cells of laser driven protons at dose rates exceeding 10^9 Gy/s" *AIP Advances* **2**: 011209.

Eickhoff, H., Weinrich, U. and Alonso, J. 2011. "Design criteria for medical accelerators" Linz, U. (ed.) *Ion Beam Therapy*. Springer, Heidelberg, Dordrecht, London, New York, pp. 325–343.

Ezzell, G.A., Galvin, J.M., Low, D., Palta, J.R., Rosen, I., Sharpe, M.B., Xia, P., Xiao, Y., Xing, L. and Yu, C.X. 2003. "Guidance document on delivery, treatment planning, and clinical implementation of IMRT: Report of the IMRT subcommittee of the AAPM radiation therapy committee" *Medical Physics* **30**(8): 2089–2115.

Faby, S. and Wilkens, J.J. 2015. "Assessment of secondary radiation and radiation protection in laser-driven proton therapy" *Zeitschrift für Medizinische Physik* **25**: 112–122.

Fiorini, F., Kirby, D., Borghesi, M., Doria, D., Jeynes, J.C.G., Kakolee, K.F., Kar, S., Litt, S.K., Kirkby, K.J. and Merchant, M.J. 2011. "Dosimetry and spectral analysis of a radiobiological experiment using laser-driven proton beams" *Physics in Medicine and Biology* **56**(21): 6969–6982.

Flanz, J., Kooy, H. and DeLaney, T.F. 2011. "The Francis H. Burr proton therapy center" Linz, U. (ed.) *Ion Beam Therapy.* Springer, Heidelberg, Dordrecht, London, New York, 597–611.

Georg, D., Knöös, T. and McClean, B. 2011. "Current status and future perspective of flattening filter free photon beams" *Medical Physics* **38**(3): 1280–1293.

Glasgow, G.P., Bourland, J.D., Grigsby, P.W., Meli, J.A. and Weaver, K.A. 1993. "Remote afterloading technology" *AAPM Report No.* **41**: 1–107.

Gotz, M., Karsch, L. and Pawelke, J. 2017. "A new model for volume recombination in plane-parallel chambers in pulsed fields of high dose-per-pulse" *Physics in Medicine and Biology* **62**(22): 8634–8654.

Grau, C., Defourny, N., Malicki, J., Dunscombe, P., Borras, J.M., Coffey, M., Slotman, B., Bogusz, M., Gasparotto, C. and Lievens, Y. 2014. "Radiotherapy equipment and departments in the european countries: Final results from the ESTRO-HERO survey" *Radiotherapy and Oncology* **112**(2): 165–177.

Hoefnagel, C.A. 1998. "Radionuclide cancer therapy" *Annals of Nuclear Medicine* **12**(2): 61–70.

Hofmann, K.M., Masood, U., Pawelke, J. and Wilkens, J.J. 2015. "A treatment planning study to assess the feasibility of laser-driven proton therapy using a compact gantry design" *Medical Physics* **42**(9): 5120–5129.

Hofmann, K.M., Schell, S. and Wilkens, J.J. 2012. "Laser-driven beam lines for delivering intensity modulated radiation therapy with particle beams" *Journal of Biophotonics* **5**(11–12): 903–911.

IAEA (International Atomic Energy Agency) 2000. "Absorbed dose determination in external beam radiotherapy: An international code of practice for dosimetry based on standards of absorbed dose to water" *IAEA Technical Reports Series* **398**: 1–229.

Jaffray, D. and Siewerdsen, J. 2000. "Cone-beam computed tomography with a flat-panel imager: Initial performance characterization" *Medical Physics* **27**(6): 1311–1323.

Jongen, Y. 2011. "Commercial ion beam therapy systems" Linz, U. (ed.) *Ion Beam Therapy.* Springer, Heidelberg, Dordrecht, London, New York, 361–375.

Karsch, L. 2016. "Derivation of formulas to calculate the saturation correction of ionization chambers in pulsed beams of short, non-vanishing pulse durations" *Medical Physics* **43**(11): 6154–6165.

Karsch, L., Beyreuther, E., Burris-Mog, T., Kraft, S., Richter, C., Zeil, K. and Pawelke, J. 2012. "Dose rate dependence for different dosimeters and detectors: TLD, OSL, EBT films, and diamond detectors" *Medical Physics* **39**(5): 2447–2455.

Karsch, L., Richter, C. and Pawelke, J. 2011. "Experimental investigation of the collection efficiency of a PTW Roos ionization chamber irradiated with pulsed beams at high pulse dose with different pulse lengths" *Zeitschrift für Medizinische Physik* **21**(1): 4–10.

Kraft, S.D., Richter, C., Zeil, K., Baumann, M., Beyreuther, E., Bock, S., Bussmann, M., Cowan, T.E., Dammene, Y., Enghardt, W., Helbig, U., Karsch, L., Kluge, T., Laschinsky, L., Lessmann, E., Metzkes, J., Naumburger, D., Sauerbrey, R., Schürer, M., Sobiella, M., Woithe, J., Schramm, U. and Pawelke, J. 2010. "Dose-dependent biological damage of tumour cells by laser-accelerated proton beams" *New Journal of Physics* **12**: 085003.

Krischel, D. 2011. "Advantages and challenges of superconducting accelerators" Linz, U. (ed.) *Ion Beam Therapy.* Springer, Heidelberg, Dordrecht, London, New York, 377–396.

Kroll, F., Pawelke, J. and Karsch, L. 2013. "Preliminary investigations on the determination of three-dimensional dose distributions using scintillator blocks and optical tomography" *Medical Physics* **40**(8): 082104.

Lagendijk, J.J.W., Raaymakers, B.W., Raaijmakers, A.J.E., Overweg, J., Brown, K.J., Kerkhof, E.M., van der Put, R.W., Hardemark, B., van Vulpen, M. and van der Heide, U.A. 2008. "MRI/Linac integration" *Radiotherapy and Oncology* **86**(1): 25–29.

Laschinsky, L., Baumann, M., Beyreuther, E., Enghardt, W., Kaluza, M., Karsch, L., Lessmann, E., Naumburger, D., Nicolai, M., Richter, C., Sauerbrey, R., Schlenvoigt, H.P. and Pawelke, J. 2012. "Radiobiological effectiveness of laser accelerated electrons in comparison to electron beams from a conventional linear accelerator" *Journal of Radiation Research* **53**(3): 395–403.

Leo, W.R. 1994. *Techniques for Nuclear and Particle Physics Experiments.* Springer, Berlin, Heidelberg, New York.

Linz, U. and Alonso, J. 2007. "What will it take for laser driven proton accelerators to be applied to tumour therapy?" *Physical Review Special Topics Accelerators and Beams* **10**: 094801.

Luo, W., Fourkal, E., Li, J. and Ma, C.M. 2005. "Particle selection and beam collimation system for laser-accelerated proton beam therapy" *Medical Physics* **32**(3): 794–806.

Ma, C.M., Veltchev, I., Fourkal, E., Li, J.S., Luo, W., Fan, J., Lin, T. and Pollack, A. 2006. "Development of a laser-driven proton accelerator for cancer therapy" *Laser Physics* **16**(4): 639–646.

Masood, U., Bussmann, M., Cowan, T.E., Enghardt, W., Karsch, L., Kroll, F., Schramm, U. and Pawelke, J. 2014. "A compact solution for ion beam therapy with laser accelerated protons" *Applied Physics B* **117**(1): 41–52.

Masood, U., Cowan, T.E., Enghardt, W., Hofmann, K.M., Karsch, L., Kroll, F., Schramm, U., Wilkens, J.J. and Pawelke, J. 2017. "A light-weight compact proton gantry design with a novel dose delivery system for broad-energetic laser-accelerated beams" *Physics in Medicine and Biology* **62**(13): 5531–5555.

Munro, T.R. 1970. "The relative radiosensitivity of the nucleus and cytoplasm of Chinese hamster fibroblasts" *Radiation Research* **42**(3): 451–470.

Nicolai, M., Sävert, A., Reuter, M., Schnell, M., Polz, J., Jäckel, O., Karsch, L., Schürer, M., Oppelt, M., Pawelke, J. and Kaluza, M.C. 2014. "Realizing a laser-driven electron source applicable for radiobiological tumour irradiation" *Applied Physics B* **116**(3): 643–651.

Oppelt, M., Baumann, M., Bergmann, R., Beyreuther, E., Brüchner, K., Hartmann, J., Karsch, L., Krause, M., Laschinsky, L., Leßmann, E., Nicolai, M., Reuter, M., Richter, C., Sävert, A., Schnell, M., Schürer, M., Woithe, J., Kaluza, M. and Pawelke, J. 2015. "Comparison study of in vivo dose response to laser-driven versus conventional electron beam" *Radiation and Environmental Biophysics* **54**(2): 155–166.

Otto, K. 2008. "Volumetric modulated arc therapy: IMRT in a single gantry arc" *Medical Physics* **35**(1): 310–317.

Pearson, E., Kleeven, W., Nuttens, V., Zaremba, S., van de Walle, J., Forton, E., Choo, R. and Jongen, Y. 2016. "Development of cyclotrons for proton and particle therapy" in: Rath, A. K. and N. Sahoo (eds.) *Particle Radiotherapy.* Springer, New Delhi, Heidelberg, New York, Dordrecht, London.

Pedroni, E. 1995. "Concepts for Gantry Systems" Linz U. (ed.) *Ion Beams in Tumour Therapy.* Chapman & Hall, London, Glasgow, Weinheim, New York, Tokyo, Melbourne, Madras, 213–222.

Petereit, D.G., Frank, S.J., Viswanathan, A.N., Erickson, B., Eifel, P., Nguyen, P.L. and Wazer, D.E. 2015. "Brachytherapy: Where has it gone?" *Journal of Clinical Oncology* **33**(9): 980–983.

Podgorsak, E.B. 2005. *Radiation Oncology Physics: A Handbook for Teachers and Students.* International Atomic Energy Agency, Vienna.

PTCOG (Particle Therapy Co-Operative Group). 2017. "A non-profit organisation for those interested in proton, light ion and heavy charged particle radiotherapy" 25.07.2017. https://www.ptcog.ch/.

Raschke, S., Spickermann, S., Toncian, T., Swantusch, M., Boeker, J., Giesen, U., Iliakis, G., Willi, O. and Boege, F. 2016. "Ultra-short laser-accelerated proton pulses have similar DNA-damaging effectiveness but produce less immediate nitroxidative stress than conventional proton beams" *Scientific Reports* **6**: 32441.

Richter, C., Karsch, L., Dammene, Y., Kraft, S., Metzkes, J., Schramm, U., Schürer, M., Sobiella, M., Weber, A., Zeil, K. and Pawelke, J. 2011. "A dosimetric system for quantitative cell irradiation experiments with laser-accelerated protons" *Physics in Medicine and Biology* **56**(6): 1529–1543.

Rigaud, O., Fortunel, N.O., Vaigot, P., Cadio, E., Martin, M.T., Lundh, O., Faure, J., Rechatin, C., Malka, V. and Gauduel, Y.A. 2010. "Exploring ultrashort high-energy electron-induced damage in human carcinoma cells" *Cell Death and Disease* **1**: e73.

Saenz, D.L., Yan, Y., Christensen, N., Henzler, M.A., Forrest, L.J., Bayouth, J.E. and Paliwal, B.R. 2015. "Characterization of a 0.35 T MR system for phantom image quality stability and in vivo assessment of motion quantification" *Journal of Applied Clinical Medical Physics* **16**(6): 30–40.

Schell, S. and Wilkens, J.J. 2009. "Modifying proton fluence spectra to generate spread-out Bragg peaks with laser accelerated proton beams" *Physics in Medicine and Biology* **54**(19): N459–N466.

Schell, S. and Wilkens, J.J. 2010. "Advanced treatment planning methods for efficient radiation therapy with laser accelerated proton and ion beams" *Medical Physics* **37**(10): 5330–5340.

Schell, S. and Wilkens, J.J. 2012. "Dosimetric effects of energy spectrum uncertainties in radiation therapy with laser-driven particle beams" *Physics in Medicine and Biology* **57**(5): N47–N53.

Schippers, M. 2012. "Proton Accelerators" Paganetti, H. (ed.) *Proton Therapy Physics.* CRC Press, Boca Raton, London, New York, 61–102.

Schürer, M., Herrmannsdoerfer, T., Karsch, L., Kroll, F., Masood, U. and Pawelke, J. 2015. "Advanced short-pulsed high-field electromagnetic dipoles for laser-based proton therapy" *Biomedical Engineering/Biomedizinische Technik* **60**(Suppl.1): S284.

Slater, J.M., Archambeau, J.O., Miller, D.W., Notarus, M.I., Preston, W. and Slater, J.D. 1992. "The proton treatment center at Loma Linda university medical center: Rationale for and description of its development" *International Journal of Radiation Oncology Biology Physics* **22**(2): 383–389.

Smith, A., Gillin, M., Bues, M., Zhu, X.R., Suzuki, K., Mohan, R., Woo, S., Lee, A., Komaki, R., Cox, J., Hiramoto, K., Akiyama, H., Ishida, T., Sasaki, T. and Matsuda, K. 2009. "The M. D. Anderson proton therapy system" *Medical Physics* **36**(9): 4068–4083.

Speer, T.W. 2010. *Targeted Radionuclide Therapy.* Lippincott Williams & Wilkins, Philadelphia.

Verellen, D., de Ridder, M., Linthout, N., Tournel, K., Soete, G. and Storme, G. 2007. "Innovations in image-guided radiotherapy" *Nature Reviews Cancer* **7**: 949–960.

Webb, S. 2001. *Intensity-Modulated Radiation Therapy.* Institute of Physics Publishing, Series in Medical Science, Bristol and Philadelphia.

Yogo, A., Maeda, T., Hori, T., Sakaki, H., Ogura, K., Nishiuchi, M., Sagisaka, A., Kiriyama, H., Okada, H., Kanazawa, S., Shimomura, T., Nakai, Y., Tanoue, M., Sasao, F., Bolton, P.R., Murakami, M., Nomura, T., Kawanishi S. and Kondo, K. 2011. "Measurement of relative biological effectiveness of protons in human cancer cells using a laser-driven quasimonoenergetic proton beamline" *Applied Physics Letters* **98**: 053701.

Yogo, A., Sato, K., Nishikino, M., Mori, M., Teshima, T., Numasaki, H., Murakami, M., Demizu, Y., Akagi, S., Nagayama, S., Ogura, K., Sagisaka, A., Orimo, S., Nishiuchi, M., Pirozhkov, A.S., Ikegami, M., Tampo, M., Sakaki, H., Suzuki, M., Daito, I., Oishi, Y., Sugiyama, H., Kiriyama, H., Okada, H., Kanazawa, S., Kondo, S., Shimomura, T., Nakai, Y., Tanoue, M., Sasao, H., Wakai, D., Bolton, P.R. and Daido, H. 2009. "Application of laser-accelerated protons to the demonstration of DNA double-strand breaks in human cancer cells" *Applied Physics Letters* **94**: 181502.

Yu, C. X. 1995. "Intensity-modulated arc therapy with dynamic multileaf collimation: An alternative to tomotherapy" *Physics in Medicine and Biology* **40**(9): 1435–1449.

Zeil, K., Baumann, M., Beyreuther, E., Burris-Mog, T., Cowan, T.E., Enghardt, W., Karsch, L., Kraft, S.D., Laschinsky, L., Metzkes, J., Naumburger, D., Oppelt, M., Richter, C., Sauerbrey, R., Schürer, M., Schramm, U. and Pawelke, J. 2013. "Dose-controlled irradiation of cancer cells with laser-accelerated proton pulses" *Applied Physics B* **110**(4): 437–444.

12

Charged Particle Radiography and Tomographic Imaging

Reinhard Schulte

12.1 Introduction

Radiography and tomography with X-rays have a long history of medical applications, the former starting soon after the discovery of X-rays by Wilhelm Conrad Röntgen in 1895 and cumulating in the rapid development of X-ray computed tomography (CT) in the early 1970s. Charged particle radiography and tomography for medical applications, was off to a much slower start, mostly due to the large expense associated with classical particle accelerators as well as large and heavy gantries needed to deliver the beam from multiple directions. Another reason are problems with poor spatial resolution associated with multiple scattering of charged particles in matter. During early research in particle CT in the 1980s, it became obvious that particle radiography and CT, particularly when done on a particle-by-particle basis, are low-dose imaging modalities that provide more accurate values for material properties related to energy loss and the scattering of protons and ions used for therapy than X-ray radiography and CT. This explains the main focus of development in recent years, which is on treatment planning and image guidance associated with proton or ion therapy, improving their accuracy. With the increasing interest in developing imaging technology for particle therapy, there is also a need for developing suitable particle radiation sources for these imaging applications. Here, laser-driven particle accelerators may offer new opportunities as discussed in Section 12.4. Chapter 12 will give an overview of charged particle imaging principles and technological approaches, and the potential for using laser-driven particle accelerators for broader medical applications in charged particle therapy.

Charged particle therapy, which is radiation therapy with external beams of protons or ions heavier than protons, has undergone a rapid technological development since the early 1990s. Major technological drivers have been the impetus to develop compact accelerators and single-room installations that can be integrated into existing radiation therapy facilities [Owen 2016]. Since the 2000s, when existing technical solutions (i.e. synchrotrons or cyclotrons and gantries) were still relatively large and expensive, laser-accelerated particles are seen as the possible next step towards table-top delivery systems, similar in size to X-ray linacs [Linz 2016, Obceama 2016].

However, since then, compact, high-B-field, superconducting synchrocyclotrons, have been developed and are now offered by several vendors, making single-room installations and significant cost reduction a reality. On the other hand, the development of compact laser-based accelerators for particle therapy has been much slower than expected, as projections of the 2000s with respect to beam energy and intensity have not yet become a reality [Linz 2016].

Another technological revolution in particle beam therapy is the appearance of imaging systems in the treatment room for image-guided radiation therapy and also for adaptive particle therapy, allowing the radiation oncologist to make adjustments to the treatment plan when changes in anatomy or tumour size occur during a course of treatment. This new technology started in X-ray therapy rooms. Its major advantage is the potential for reducing the margins of planning treatment volumes, with the promise to reduce complications and for allowing higher doses to the tumour. Imaging capability was first two-dimensional (2D radiography), but then became 3D (volumetric) with cone beam CT installations and, more recently, 4D (including time) for moving targets such as lung and liver tumours. Lately, the first combined magnetic resonance imaging and linear accelerator (MRI–linac) systems have entered the market and promise another revolution in image guidance, offering online monitoring and correction of the treatment and utilizing the enhanced tissue contrast of MRI.

Image guidance in particle radiation therapy, despite being seen as critical for fully optimizing this form of radiation therapy, has been much slower in its development for particle therapy. It has now reached a stage were 3D cone beam CT is becoming common in particle treatment rooms. Some institutions also consider installing a compact diagnostic multi-slice CT scanner in the particle treatment room for image guidance. Even MRIs are now seen by some as an ultimate solution for image guidance in particle therapy.

There is, however, a problem with X-ray imaging in particle therapy, both for imaging studies done for treatment planning and for in-room image guidance, in that the range information, which is the most critical element of particle therapy, is inherently inaccurate due to the different dependence of X-ray attenuation on tissue composition (electron density and atomic number) and proton energy loss. Furthermore, MRIs cannot provide reliable stopping power values for particle therapy. In X-ray imaging, there are additional sources of inaccuracy due to beam hardening, which depends on the patient size, and due to high-density artefacts, which are typical for X-ray CT. These problems combined lead to an inherent inaccuracy of particle range prediction in treatment planning and pretreatment verification with current imaging modalities.

In X-ray CT treatment planning, the stopping power relative to water is converted from X-ray attenuation relative to water using the 'stoichiometric calibration' method [Schneider 1996]. In a recent report of a study measuring range errors in animal tissues, the authors stated that proton range can be calculated within about 1.5% plus 1.5 mm [Zheng 2016], excluding additional uncertainties due to, for example, metal artefacts. However, the range uncertainty in a real treatment situation can be much larger because of patient positioning errors, time-dependent changes in patient anatomy and internal organ motion.

The general solution to these problems is the development of particle imaging systems. The physical principles and current status of particle imaging, which has been pursued for many years now, will be summarized below. Based on this knowledge, we will then discuss whether there is a realistic role for laser-driven accelerators in particle imaging applications.

12.2 Principles of Particle Imaging and Image Reconstruction

12.2.1 Proton Radiography and Computed Tomography

Particle imaging in the context of this chapter is the common term of either one of two methods: The first method, called particle radiography, produces two-dimensional (2D) projection images of an object, called particle radiographs. The imaged signal detected across the 2D sensitive area of a particle radiography system can be the energy deposited, the particle fluence, the residual range of particles, or the 3D resolved integrated depth profile of a particle pencil beam that is scanned across the object.

The second particle imaging method, called particle computed tomography (CT), produces a 3D tomographic image set of the object consisting of a series of 2D slices usually stacked along the longitudinal axis of the human body (axial slices). For particle CT, the quantity imaged is usually the particle stopping power relative to water (RSP), but other proton interaction related quantities such as scattering power or attenuation due to nuclear interactions can also be reconstructed. Particle CT, being a 3D modality, requires a mathematical reconstruction algorithm that takes as input multiple projection images of the object from many directions covering an angle range of 180° or 360°. The standard reconstruction algorithm for X-ray CT is filtered back projection for a 2D fan beam arrangement or its 3D variant, the Feldkamp-Davis-Kress (FDK) algorithm for a 3D cone beam arrangement. Particle CT image reconstruction using data acquired in particle tracking mode (see below for a more detailed discussion), is usually done with an iterative reconstruction algorithm such as the algebraic reconstruction technique (ART).

12.2.2 Reconstruction of Relative Stopping Power

The basic method of tomographic image reconstruction of physical tissue properties is generally credited to Alan Cormack, who received the Nobel prize for his seminal work on tomographic image reconstruction in 1979. In his initial publication from the early 1960s [Cormack 1963], he described the idea of reconstructing both the linear attenuation coefficients of the object with X-rays and stopping power relative to water, RSP, with charged particles such as protons. Cormack was concerned about the poor spatial resolution of charged particle imaging due to multiple Coulomb scattering (MCS), which causes a deviation of the particle paths from straight lines, making it mathematically more challenging to reconstruct images with sufficient spatial resolution. In his 1979 Nobel speech, however, he was more optimistic and stressed the potential of low-dose imaging with protons [Cormack 1979].

The mean energy loss of charged particles is uniquely related to the line integral of RSP along their path, which is called water equivalent path length (WEPL); thus, particles that have identical energy loss can be assumed to have traversed an identical WEPL within the statistical uncertainty of energy straggling. It turns out that RSP, the stopping power relative to water, is mostly energy-independent and thus a characteristic quantity of tissues that is useful for dose-planning for proton and ion therapy.

12.2.3 Data Acquisition Modes for Particle Imaging

There are two principally different ways of acquiring data for particle radiography and CT [Parodi 2014]. The first is based on detecting the WEPL and reconstructing the track of *individual particles*, which has been called 'particle tracking' data acquisition. The particle trajectory before and after the object can be reconstructed using a pair of particle trackers on the entry and exit sides of the object, although with a scanned pencil beam the entry direction may be inferred and only one tracking plane is needed. The tracking data forms the input of a most likely path (MLP) algorithm [Williams 2004, Schulte 2008], which usually assumes that the object is water. The WEPL is determined in one of two ways: with a residual energy detector (calorimeter) or a stack or scintillating plates measuring the location of the Bragg peak of each proton (range telescope). The track information and WEPL of all particles is then fed into the reconstruction algorithm.

TABLE 12.1 Approximate Kinetic Energies Required for Tomographic Imaging of Head and Body Regions with Protons, Helium Ions and Carbon Ions

Body Region	Kinetic Particle Energy (MeV/amu)		
	Protons	Helium	Carbon
Head	200	200	380
Body	300	300	600

The second method of data acquisition does not detect individual particles but uses an *integration mode* summing the signal from many particles with a resulting value that depends on the water equivalent thickness (WET) of the region traversed by a charged particle pencil beam. There are different implementations of the integration mode of data acquisition scheme and examples for each approach will be given in Section 12.3. Using the integration mode, multiple 2D projection WET images can then be assembled during a full rotation of the scanned beam around the patient and reconstructed into a 3D tomographic image using, for example, a filtered back projection algorithm.

12.2.4 Particle Energy Required for Imaging

Since the human body contains a large fraction of water (60–75%), the range of charged particles in water is a good guide for estimating the charged particle energy that is needed to penetrate different parts of the human body to image them. Table 12.1 shows the recommended energy in MeV/amu for protons, helium ions and carbon ions that would be suitable for head tomographic imaging (including the neck region) and body imaging (chest, abdomen, pelvis) with these particles. The kinetic particle energy values required for the respective region were derived by assuming that the maximum encountered WET is 25 cm when imaging the head and neck of adult patients, and 50 cm when imaging the body. The maximum currently available energy for carbon ions is 430 MeV/amu, making carbon ions only suitable for head imaging. The maximum kinetic proton energy currently available from a commercial vendor of a clinical proton accelerator is 330 MeV, which will be sufficient to image all anatomical regions in most patients. Note that the energies stated are maximum energies for tomographic imaging. When variable energies are used during imaging with an actively scanned beam, this will be the maximum energy required for the region of interest while other pencil beams can have lower energy according to the WET of the subregion it covers. For radiographic imaging, the energy requirements are less stringent. For example, it was shown that radiography of an adult pelvis is possible with carbon ions of currently available clinical energies (430 MeV or less) when lateral beams at 90° angles are skipped (Rinaldi 2011).

12.3 Modern-Era Charged Particle Imaging Approaches

The first attempts to utilize charged particles for imaging were directed towards diagnostic imaging; over the last 20 years, the interest has shifted to developing particle imaging technology for improving treatment planning and verification of charged particle therapy. For a detailed review of the early approaches of imaging with protons and ions, the reader is referred to two recent comprehensive review articles on this subject [Parodi 2014, Poludniowski 2015]. Here, we will give an overview of contemporary technological approaches to particle imaging using particle tracking or the integration mode of data acquisition. All these approaches have benefited from the availability of Monte Carlo simulations; for coverage of this subject in the context of particle range verification, the reader is referred to the recent review by Kraan [2015].

12.3.1 First Proton Radiography System at PSI

The modern era of particle imaging technology for improving particle therapy began during the 1990s when researchers at the Paul Scherrer Institute (PSI) developed the first proton radiography

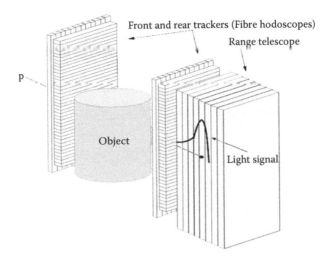

FIGURE 12.1 Schematic of the PSI proton radiography system, a prototype of a particle tracking imaging system. Entrance and exit coordinates are measured with two scintillating fibre hodoscopes and WEPL is determined with a stack of optically isolated plastic scintillator tiles (range telescope) as the difference of the WET from the entrance of the range detector to the scintillator with the highest signal (Bragg peak) and the average range of the particles.

system using the proton tracking acquisition mode, as explained in in Section 12.2.3. The design of this system, which served as a model for similar systems developed later, is shown in Figure 12.1. The entrance and the exit coordinates of tracked particles were measured with two identical scintillating fibre hodoscopes on either side of the patient, where each fibre was coupled to a channel of a photo-multiplier tube. The WEPL detector was a range telescope comprised of a stack of closely packed and optically isolated plastic scintillator tiles. The PSI proton radiography system was mounted on the gantry and produced images of head-size objects by scanning a proton pencil beam of reduced intensity (particle rates of 1 MHz) in one direction (y-direction) and the patient table in the orthogonal direction (x-direction). The PSI radiography system was used on dogs [Schneider 2005] but never on human patients.

12.3.2 First Proton CT System at the Harvard Cyclotron

In 2000, Zygmanski and colleagues described a cone beam proton CT system utilizing the 160 MeV proton beam at the Harvard Cyclotron [Zygmanski 2000]. This technique employed an integration mode of data acquisition. The schematic layout of this approach is shown in Figure 12.2. The cone beam was modified with a special modulator in the beam that produced a spatially dependent fluence signal that decreased linearly with WET of the traversed object. The detector system registering the fluence signal in a spatially resolved manner consisted of a Gd_2O_2S:Tb intensifying screen viewed by a cooled charge-coupled device (CCD) camera. The FDK cone beam reconstruction algorithm was used to reconstruct a proton stopping power CT image over a 16 cm^3 volume which was discretized into $512 \times 512 \times 512$ voxels.

The performance of the system was tested with a spatial and contrast resolution phantom. The resulting images had significant artefacts and its spatial resolution was severely degraded by MCS effects. The same fluence modulation approach with a screen–CCD detector was also investigated by Ryu et al. [Ryu 2008] in Korea with similar performance results. Overall, this approach based on the integration mode of data acquisition of a fluence modulated beam is less promising than particle tracking approaches for proton CT due spatial resolution limitations and the higher dose required. Still, it may be considered as a low-cost, and technologically less complex solution that could also work with laser-accelerated protons.

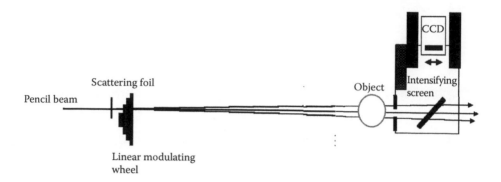

FIGURE 12.2 Schematic of the proton cone beam CT setup tested at the Harvard Cyclotron beam line (drawn approximately to scale). A monoenergetic 160 MeV proton pencil beam was scattered into a cone beam by a lead scatterer and then modulated by a linear modulating wheel. After traversing the CT phantom, the modulated proton beam was visualized in 2D by an intensifying screen CCD camera system. Various shielding materials were introduced to protect the camera from neutrons and scattered protons.

12.3.3 Heavy Ion CT Systems at HIMAC

Heavy ions are much less affected by MCS than protons. The fluence modulation approach to particle CT was also tested with carbon ions at the heavy ion medical accelerator (HIMAC) operated by the National Institute for Radiological Sciences (NIRS) in Japan. The first system reproduced the proton imaging system tested at the Harvard Cyclotron and achieved promising results with carbon ions of up to 400 MeV, in particular, the spatial resolution was better than 2 mm and the density resolution (which is equivalent to RSP resolution) was better than 7% [Abe 2002]. The imaging performance was further improved with successive upgrades of the system [Muraishi 2009, Muraishi 2016], eventually reaching a spatial resolution of 1 mm.

12.3.4 Heavy Ion Radiography and CT Systems Tested at HIT

In recent years, two teams of investigators in Germany built two different ion imaging systems, one an ion radiography system developed at the German Cancer Research Center (DKFZ) and the other an ion CT system developed by a team at Ludwig Maximilians Universität (LMU) which lead a collaboration between the Gesellschaft für Schwerionen Forschung (GSI) and Universitäts Klinikum Heidelberg (UKH). Both systems were tested with carbon ions at the Heidelberg Ion Therapy Center (HIT) and operated in the integration mode of data acquisition using an actively scanned carbon pencil beam. The DKFZ system employed a commercial amorphous silicon flat-panel detector originally designed for kV or MV portal imaging to obtain 2D WET maps for ion radiography [Telsemeyer 2012]. The energy deposited in the active layer of the detector depended on both the residual energy and intensity of the particles, which created some ambiguity. This was overcome with an ion range measurement technique called *energy scanning*: For each detector pixel, an energy signal curve was measured and the penetrated WET was correlated with the peak position of the signal. By actively varying the energy across a field of uniform intensity, one could thus obtain a spatially-resolved 2D WET map. The investigators tested this method with a PMMA phantom containing tissue-equivalent inserts of known RSP. For the investigated flat-panel detector with 0.8×0.8 mm^2 pixel size, an accuracy better than 0.5 mm WET was achieved, but only with a large dose of 8 Gy.

The system developed by the LMU–GSI–UKH collaboration featured a dedicated (non-commercial) range telescope comprising a stack of 61 parallel plate ionization chambers (PPIC) interleaved with 3-mm thick plastic (PMMA) absorber plates and a fast, multichannel electronics module for integration mode of data acquisition [Rinaldi 2013, Rinaldi 2014, Magallanes Hernandez 2017]. The spatially resolved WET of the traversed object was deduced from the measurement of the laterally integrated depth-dose profile (Bragg curve) in the range telescope. Using the measured depth profiles of multiple beams across the field and comparing it to theoretical profiles assuming a WET profile of the object in beam direction, the investigators reconstructed digital WET images that best fitted the measured profiles. Multiple 2D projected WET images were then assembled during a full rotation of the scanned beam around the patient and reconstructed as a 3D tomographic image using, for example, a filtered back projection algorithm.

In a recent publication [Meyer 2017], the ion CT of realistic human objects was compared for the integration and tracking mode of acquisition techniques using Monte Carlo simulations. This *in silico* study showed that using adequate post-processing strategies both strategies provided comparable RSP accuracy.

12.3.5 Recent Developments in Tracking Mode–Based Proton Imaging

The interest in proton CT and radiography increased rapidly after first experimental results had been published by Zygmanski et al. [2000]. Coincidental with the development of the large hadron collider at CERN and other scientific technology requiring position sensitive radiation detectors with single event resolution, particle tracking technology advanced to a state where millions of particles per second could be tracked individually, and as such their position and direction in 3D space could be determined with high efficiency and spatial resolution.

In early 2003, the proton CT (pCT) collaboration was formed during a two-day meeting at Brookhaven National Laboratory (BNL) and mapped out plans for further development of proton CT based on modern particle tracking and energy detector technology. Shortly thereafter, members of the collaboration defined design specifications based on clinical requirements and identified suitable detector technologies for a clinical proton CT system [Schulte 2004]. The agreement among the members of the group led to the conceptual design of a clinical proton CT scanner that is schematically shown in Figure 12.3. This design highlights the main components of a proton tracking–based design: an entrance tracking telescope with two pairs of x- and y-position sensitive detectors, an exit tracking telescope using the same design, and a calorimeter that measures the residual proton energy (alternatively the residual proton range).

The next steps in the development of preclinical pCT scanning were the further development of the most likely path (MLP) concept using the particle tracking information [Schulte 2008], and the development of iterative reconstruction techniques enhanced with the superiorization methodology for reconstruction of proton CT images with about 1-mm lateral spatial resolution [Penfold 2010].

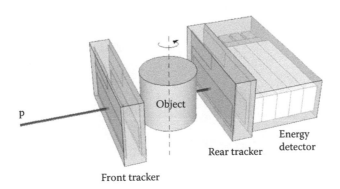

FIGURE 12.3 Schematic of the proton (or particle) CT design conceived by the pCT collaboration. The proton beam line is fixed and horizontal and scans the object across the sensitive tracker area. The object rotates relative to the fixed detectors. This design was eventually built (see Figure 12.4).

In 2011, the pCT collaboration received a four-year grant from the National Institute of Biomedical Imaging and Bioengineering (NIBIB) to build the detectors for a preclinical proton CT scanner (called the Phase II pCT scanner), able to track about 1 million protons per second and to acquire a complete 360° pCT acquisition of a head phantom in under 10 min using a low-intensity 200 MeV proton beam. The technical development of this scanner has been described recently [Bashkirov 2016b]. The system was installed and has been tested on the wobbled uniform scanning beam line at the Northwestern Medicine Chicago Proton Center, showing good resolution proton CT head scans with doses of the order of 1 mGy [Johnson 2017] (see Figure 12.4).

In the wake of the initial work by the pCT collaboration, there have been additional and more recent attempts to further improve or develop technological approaches to the detectors involved in proton tracking–based imaging. The Italian PRIMA project developed a scanning system similar to that of the pCT collaboration also using silicon strip detectors for proton tracking [Sipala 2013]. For the system built by the Northern Illinois University and Fermi National Laboratory (NIU–FNAL) collaboration, fibre trackers were used [Uzunyan 2014], and gaseous electron multipliers (GEM) detectors were used for the proton range radiography system PRR30 developed by the Italian TERA Foundation [Bucciantonio 2013]. For the energy/WEPL detector the pCT collaboration initially used a segmented CsI:Tl crystal calorimeter (Phase I), and then utilized a five-stage plastic scintillator (Phase II) [Bashkirov 2016a]; the PRIMA project used a calorimeter comprised of YAG:Ce crystals, and the NIU–FNAL project and TERA Foundation projects both used a range detector consisting of a stack of 3-mm plastic scintillators read by silicon photomultipliers (SiPMs).

During the last few years, we have seen the beginning of a new era of tracking detectors for proton imaging based on pixelated rather than strip detectors. Since 2013, the Proton Radiotherapy Verification and Dosimetry Applications (PRaVDA) consortium, funded by the UK Wellcome Trust, has developed a proton CT system in which the calorimeter was replaced by a pixelated, solid-state range telescope consisting of radiation-hard CMOS active pixel sensors (APS). The project utilizes trackers built for the ATLAS project at CERN. Due to the densely granulated pixel structure of the CMOS detector the system resolves multiple protons per frame time, whereas the strip-based detector of the pCT collaboration,

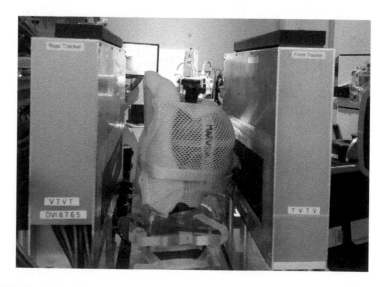

FIGURE 12.4 The Phase II pCT scanner completed by the pCT collaboration in 2014. The object shown between the front and rear trackers is an Alderson head phantom that was scanned to plan for a realistic head and neck cancer proton treatment in the presence of dental implants at the Northwestern Medicine Chicago Proton Treatment Center. A full treatment planning scan is acquired with a low-intensity scanning beam in 6 min and reconstructed in less than 5 min.

can only resolve one proton per frame; with about 500 frames per second, the PRaVDA system can handle proton rates of up to 1 MHz similar to the latest pCT collaboration scanner.

In parallel to the PRaVDA project, which benefited from the ATLAS project at CERN, other investigators also started taking advantage of modern pixelated tracking detectors and fast readout systems developed by CERN collaborations. Biegun et al. [2016] from the KVI-Center for Advanced Radiation Technology, Groningen, Netherlands, recently reported on a proton radiography system that uses two gas-based time projection chambers (TPCs) with GridPix technology for particle tracking and reading them out with the current Timepix chip designed by the Medipix2 collaboration [Biegun 2016]. The KVI team built a complete small prototype for proton radiography, featuring a sensitive tracker area of 2.8 cm × 3 cm and measuring the residual energy of 150 MeV protons traversing a tissue equivalent phantom object with a BaF_2 crystal. With this prototype system, the investigators could only read out a proton event at a rate of 100 Hz per tracker, but future implementations with the Timepix3 chip are expected to reach rates of up to 1 MHz.

A collaboration between the University of Bergen and two other universities in the Netherlands (Groningen and Utrecht), participating in the ALICE–FoCal collaboration at CERN, developed a proof of concept version of the digital tracking calorimeter (DTC) to be part of a proton CT system [Pettersen 2017]. This project has the goal of eventually acquiring proton data at a rate of up to 10 MHz, which makes acquisitions in under 1 minute possible.

Lastly, a novel tracking detector, called ProXY (for Protons X and Y) is currently under development by an INFN group, which is based on the so-called OrthoPix architecture, a readout scheme that was developed in particular for tracking particles over a large detector area at high rates [Mattiazzo 2015].

Describing these recent technology developments in further detail is beyond the scope of this chapter; the interested reader is referred to the recent review [Poludniowski 2015] that gives much more detail on this subject.

12.3.6 Recent Developments in Integration Mode Proton Imaging

There have been several approaches to acquiring data for proton imaging in the integration mode following the initial work by Zygmanski et al. [2000]. Only those approaches that resulted in experimental results will be summarized here in chronological order.

In 2010, Mumot et al. from PSI proposed a 'range probe' for *in vivo* range verification, using a multilayer parallel-plate ionization chamber to measure the integrated depth profiles (Bragg curves) of proton pencil beams scanned across and penetrating the patient in the direction of the treatment beam. For each pencil beam, the patient WET it traversed is verified by comparing the measured depth-dose profile to the X-ray and CT–based prediction of the Bragg peak depth-dose profile. In their initial publication [Mumot 2010], the PSI investigators used Monte Carlo simulations to prove this concept. Over the last few years, this concept has been tested by two groups. In 2015, investigators from the Heidelberg Collaboration for Image Processing reported a simulation study of the system they had previously built for ion radiography [Rinaldi 2013] as reported above for improved proton radiography in conjunction with an image processing method [Krah 2015]. With proton beam spot sizes between 8 and 10.3 mm FWHM, superior image quality and agreement with ground-truth digitally reconstructed radiographs (DRRs) showed a spatial resolution of 4–6 times improved over the beam spot size. In 2016, Farace et al. [2016] presented experimental results obtained with such a system for proton radiography using a commercially available multilayer ionization chamber. The range error maps obtained on the basis of proton radiographs compared to those predicted by the planning CT could be quite sensitive to small misalignment between the planning CT and the proton radiograph. An efficient algorithm to separate such misalignment from true range errors has recently been presented [Deffet 2017].

In 2011, Seco and Depauw at the Massachusetts General Hospital (MGH), presented experimental results of a radiography system that was based on integrating a proton beam signal with CMOS APS detectors [Seco 2011]. The detectors they employed had previously been developed and used for other proton therapy applications, i.e. for proton beam visualization and dosimetry. The investigators used these detectors to integrate the dE/dx signal of proton pencil beams traversing the object, and from this they calculated the average residual energy of the protons and inferred the average WET of the object at the location of the pencil beam. With this approach, the investigators demonstrated the capability of visualizing simple objects with relatively large density differences.

In 2013, Testa et al. at MGH presented a study of proton radiography and proton computed tomography based on a time-dependent signal produced by a modulator wheel (MW) in an object-penetrating, passively scattered proton beam [Testa 2013]. The signal was measured with a diode array detector located behind the object and capable of fast dose rate measurements (~100 ms resolution). The dose rate function (DRF) produced by the MW was periodic with the period of the wheel rotation (100 ms for their system) and had a unique pattern at each water-equivalent depth along the beam path that allowed spatially encoding the WET. An added benefit of this time-resolved method is that it can perform tumour tracking of a moving target. A disadvantage, however, is the need for a passively scattered beam with a range modulator, which is not compatible with active pencil beam scanning methods. At the time of the first publication, the method produced reasonable results with their prototype detector (diode array) at a dose of ~0.7 cGy (7 mGy) per radiograph. An update of this method with an improved detector using an X-ray flat-panel imaging system, was reported by Jee [2016]. For the updated version, a relatively slow rotation of a custom-designed MW (0.2 revolutions per second) and a fast image acquisition (30 frames per second, ~33 ms sampling) allowed sampling each DRF with sufficient temporal resolution at a dose rate of ~2,5 mGy per DRF period. The investigators reported that the RSP values of the inserts of a tissue surrogate phantom could be resolved within 2%.

Park and colleagues presented an integration mode proton imaging system that used a specifically designed dynamic water range shifter (DWRS) [Park 2017]. The DWRS was placed upstream of the object and the thickness of a water column was changed in small increments of 10 μm, stepping the proton Bragg peak of a passively scattered 230 MeV proton cone beam produced at the Korea National Cancer Center through the object. The dose signal behind the object was visualized with a scintillation screen that was viewed by a CCD camera during the scan. Two sets of proton radiographs with and without the scanned object were acquired from which the 2D distribution of the WET of the object was calculated. After optimization of the radiographic scan settings, a proton CT of a cylindrical phantom with multiple insert holes was obtained. The investigators successfully demonstrated the possibility of using the DWRS for proton radiography and CT.

To overcome the limitation of the described methods requiring the need for passive scattering proton beam lines, Bentefour and coinvestigators introduced and explored with simulations the concept of an energy resolved dose measurement (ERDM) to encode WET in a patient [Bentefour 2016]. The authors explored the WET range that could be covered for three clinical tumour scenarios (paediatric and adult head with AP and RL beams, two lung scenarios with AP beams and pelvis with AP beam) for typical (average-sized) patient anatomy and a maximum proton beam energy of 230 MeV. They concluded that the resulting projected dynamic ranges were wide enough to allow for the design of optimal tumour-specific proton imaging fields in order to produce proton radiographs covering the depth of the tumour site, awaiting further experimental verification.

12.3.7 Summary of Particle Imaging Approaches

Starting with the PSI particle tracking system and the MGH integration mode data acquisition system, Table 12.2 summarizes the particle imaging systems and approaches that led to the most recent approaches utilizing the technology developed for high-energy physics. Despite the higher complexity, the latter have reached preclinical status and are currently being tested in clinical proton facilities. The desirable particle rate of 10 million particles per second has yet to be demonstrated but may be reached within a few years. The development of compact technology provided by laser-accelerated

TABLE 12.2 Overview of Experimental Particle Imaging Systems Presented in this Section

First Author, Year, Institution	Beam Spread	Acquisition Mode	Particle Rate (s⁻¹)	Imaging Type	Imaging Dose
Pemler, 1999, PSI	Active[1]	Tracking	10^6	pRad	NA
Zygmanski, 2000, MGH	Passive[2]	Integration	NA	pCT	0.57 Gy
Ryu, 2008, KCCH	Passive	Integration	NA	pRad	0.45 mGy
Abe, 2002, NIRS	Passive	Integration	NA	iCT	40 Gy[3]
Murashi, 2009, 2016, NIRS	Passive	Integration	NA	iCT	40 Gy[3]
Telsemeyer, 2012, DKFZ	Active	Integration	NA	iRad	10 mGy
Rinaldi, 2013, LMU	Active	Integration	NA	iRad	0.5–1 mGy
Krah, 2015, HCIP	Active	Integration	NA	pRad	0.016 mGy
Magallanes Hernández, 2017, LMU	Active	Integration	NA	iCT	0.2–4 Gy[3]
Sipala, 2013, INFN, Catania	Passive	Tracking	10^6	pCT	NA
Johnson, 2017, UCSC	Passive or active	Tracking	10^6	pCT or pRad	1.5 mGy 0.015 mGy
Taylor, 2016, University of Liverpool	Passive	Tracking	10^6	pCT	NA
Biegun, 2016, KVI	Passive	Tracking	$10^{6\,[4]}$	pCT	NA
Pettersen, 2017, University of Bergen	Passive	Tracking	$10^{7\,[4]}$	pCT	NA

Abbreviations: DKFZ: Deutsches Krebsforschungszentrum; HCIP: Heidelberg Imaging Collaboratory; HIMAC: Heavy Ion Medical Accelerator; iCT: ion computed tomography; i-Rad: ion radiography; INFN: Istituto Nazionale di Fisica Nucleare; KCCH: Korea Cancer Center Hosptial; LMU: Ludwig Maximillians University; MGH: Massachusetts General Hospital; NA: not available/not applicable; NIRS: National Institute of Radiological Sciences; pCT: proton CT; pRad: proton radiography; PSI: Paul Scherrer Institute; UCSC: University of California Santa Cruz

Note: [1] Magnetically scanned beam, [2] passively scattered beam, [3] not optimized, [4] expected.

particle sources that may be combined with the latest detectors for particle imaging will be discussed in the final section of this chapter.

12.4 Laser-Accelerated Particles for Imaging

12.4.1 Motivation

Particle imaging (radiography and CT) is an attractive modality for improving the accuracy of planning particle treatments by providing more accurate stopping power values of the patient cross sections, and allowing a fast check of the correct patient position and 2D WET distribution before treatment with 2D radiographies. With particle imaging, treatment plans can also be verified prior to treatment and possibly adapted to the particular anatomical distribution of tumour and normal tissues.

Detectors for particle imaging have been developed and are expected to be implemented in clinical use within a few years from now. The attractive clinical applications in real patients would currently require to switch the particle beam sources (cyclotrons or synchrotrons) to higher energies than used for treatment, allowing the particles to penetrate the patient. The imaging beams would also be of much lower intensities than those required for treatment. Current beam delivery systems (gantries) are optimized for treatment but not designed to deliver beams during gantry rotation. While one could acquire individual tomographic projections at stationary gantry positions, this is impractical due to the large mass of the gantry, which requires several seconds to initiate and stop movement thereby, unduly prolonging the CT acquisition time. Implementing a continuous rotation during CT acquisition would require a change of the treatment control system, which is not an easy task and proton treatment system vendors may hesitate to do so.

Conventional ion sources for imaging deliver a more or less continuous ion beam, ideally to allow tracking every single ion. Laser-driven ions, however, are bunched, with at least 10^6 to 10^8 particles arriving within a bunch duration of order a nanosecond, i.e. with very high instantaneous dose rates. The subsequent bunch should follow within a few tens of ms (nowadays rather a few s). Therefore, the current detection strategy is not at all adapted to bunched laser-plasma based ion sources (see Chapter 5 for laser-driven ion acceleration) and would require fresh ideas, similarly to the case of ion therapy [Linz 2016]. Alternatively, one may think of alternative approaches such as dielectric laser acceleration of electrons, which could be adapted to ions in the future. The currently intrinsic problem of limited current for dielectric laser acceleration may be naturally well-adapted to the need of low intensity.

Replacing the particle source for particle imaging with an independent, compact laser-accelerated source is therefore an attractive option that would solve these problems: being compact and lightweight, these laser-driven accelerators could move with the gantry while providing low-intensity beams for imaging and would have an independent control system. They would then be moved out of the beam line at the time of treatment.

Proton therapy has matured and conventional proton accelerators have become much more compact. This makes it more difficult for laser-driven treatment accelerators to find a place in the commercial world of particle therapy. On the other hand, lasers built for particle imaging may provide a niche for lasers to enter the proton and maybe ion treatment rooms.

12.4.2 Energy, Energy Spread, Time Structure and Biological Effects

Beam energies for imaging are higher than those needed for therapy. For head imaging, a typical energy of 200 MeV/amu, and for body scans, a typical energy of 300 MeV/amu for protons and helium ions would be needed. Laser-driven proton acceleration can yield energies approaching 100 MeV (see Chapter 5 by Macchi), where typically 10^7–10^8 particles within a 1% energy spread are emitted into cone angles of ~100 msr. At a repetition-rate of 1 Hz, this would result in orders of magnitude more particles than needed for imaging. Hence, for particle imaging, it may be wise to look for alternative, low flux, more continuous acceleration schemes which may consume even less average power than today's petawatt lasers.

The energy spectra of laser-accelerated ions are typically peaked at low energies and fall off exponentially with increasing energy. Highly peaked (at higher energies) or 'quasi-monoenergetic' beams could be possible [Obceama 2016] but have yet to be demonstrated. Typically, the ions at higher energies needed for imaging represent only a small fraction of ions and most ions would stop in the patient and deposit unwanted dose. On the other hand, the higher than needed (for imaging) intensity of laser-accelerated ion beams makes it feasible to selectively filter high-energy ions upstream of the patient to obtain suitable lower intensity beams at higher energy. The polychromatic or possibly quasi-monoenergetic, high-energy, laser-driven charged particle beams could then be scanned across the patient, and would simultaneously deliver many different energies. From the energy spectrum of the outgoing ions, one can, in principle, deduce the WET along the path of the laser-accelerated pencil beam and thus produce WET radiographs or particle CT reconstructions of RSP.

The ultra-short nature of laser-accelerated ion bunches may have unexpected biological implications as recently noted [Obceama 2016]. It is believed that the initial ion bunch of the trail of bunches that come in very short wave trains could create a retardation effect (vicinage effect) on subsequent ion bunches at the very high instantaneous dose rates. The resulting increase of stopping power and thus biological effectiveness could mean an unwanted higher biological dose in imaging. This potential biological effect of ultra-short (ft/ps) laser-accelerated ion bunches is currently poorly understood and requires further investigation as it could affect imaging applications.

12.4.3 Size and Cost

The size and cost reduction of laser-driven particle accelerators have been cited as one of the major attractions for particle therapy. Conventional ion accelerators have a large footprint and are not optimized for imaging applications. Thus, adding a much more compact laser-driven accelerator is an attractive option that should be explored.

The advantage of the gantry-mounted laser system would be that the patient could be imaged in the treatment position without the need to be moved to a separate scanner location, and a different ion (e.g. helium) than the one for treatment could be used for imaging. Helium as the imaging beam source would have the advantage of lower multiple Coulomb scattering than protons and, therefore, a higher spatial resolution. The laser imaging beamline needs to be offset relative to the therapeutic beam, for example, by 90° or less in order not to interfere with the mechanical components of the treatment beam.

12.5 Conclusion

Imaging with charged particles for particle therapy applications has many advantages, but none of them have been clinically exploited. One reason for this is that medical accelerators and beam delivery systems have been designed and optimized for therapy and not for imaging. Therefore, imaging technology has matured around the parameter space provided by such existing sources. Imaging with charged particles requires several orders of magnitude lower beam intensities compared to therapy, especially when data are acquired in particle tracking mode. Laser-accelerated bunches in their current form are hence not suited to be adapted to existing solutions and approaches. If proton and helium sources could be made sufficiently compact to be mounted on existing gantries without increasing the size of the treatment vault, they could contribute to improved treatment planning accuracy and in-room image guidance, as well as treatment adaptation.

The laser requirements for such imaging systems will be determined by the combined requirements of higher particle energy and lower particle intensity. For example, near the high energy tail region of the particle spectrum and for a fixed required/desired number of particles, higher particle energy typically requires higher laser intensity; yet, at some fixed required or desired energy a reduction in the number of particles means lower laser intensity – in other words, imaging requirements can critically depend on the spectral shape of the particle yield, which requires a carefully planned trade-off between energy and intensity.

References for Chapter 12

Abe, S., Nishimura, K., Sato, H. et al. 2002. "Heavy ion CT system based on measurement of residual range distribution." *Igaku Butsuri* **22**: 39–47.

Bashkirov, V. A., Schulte R. W., Hurley R. F. et al. 2016a. "Novel scintillation detector design and performance for proton radiography and computed tomography." *Med. Phys.* **43**: 664–74.

Bashkirov, V. A., Johnson, R. P., Sadrozinski H. F. and Schulte, R.W. 2016b. "Development of proton computed tomography detectors for applications in hadron therapy." *Nucl. Instrum. Methods Phys. Res. A.* **809** [Feb 11]: 120–9.

Bentefour, E. H., Schnuerer, R. and Lu, H. M. 2016. "Concept of proton radiography using energy resolved dose measurement." *Phys. Med. Biol.* **61**: N386–93.

Biegun, A. K., Visser, J., Klaver, T. et al. 2016. "Proton radiography with timepix based time projection chambers" *IEEE Trans. Med. Imaging* **35**: 1099–105.

Bucciantonio, M., Amaldi, U., Kieffer, R., Sauli, F. and Watts, D. 2013. "Development of a fast proton range radiography system for quality assurance in hadrontherapy" *Nucl. Instrum. Methods Phys Res Sect A.* **732**: 564–7.

Cormack, A. M., 1979. "Nobel lecture: Early two-dimensional reconstruction and recent topics stemming from it" Nobelprize.org. Nobel Media AB 2014, http://nobelprize.org/nobel_prizes/medicine/laureates/1979/cormack-lecture.pdf

Cormack, A. M. 1963. "Representation of a function by its line integrals, with some radiological applications" *J. Appl. Phys.* **34**: 2722–27.

Deffet, S., Macq, B., Righetto, R., Vander Stappen, F. and Farace. P. 2017. "Registration of pencil beam proton radiography data with x-Ray CT" *Med Phys.* [Epub ahead of print].

Farace, P., Righetto, R. and Meijers, A. 2016. "Pencil beam proton radiography using a multilayer ionization chamber" *Phys. Med. Biol.* **61**: 4078–87.

Hegelich, B. M., Jung, D., Albright, B. J. et al. 2013. "160 MeV laser-accelerated protons from CH2 nanotargets for proton cancer therapy" arXiv:1310.8650v1 [physics.plasm-ph], https://arxiv.org/pdf/1310.8650.

Jee, K. W., Zhang, R., Bentefour, E. H. et al. 2017. "Investigation of time-resolved proton radiography using x-ray flat-panel imaging system" *Phys. Med. Biol.* **62**:1905–19.

Johnson, R. P., Bashkirov, V. A., Coutrakon, G. et al. 2017. "Results from a prototype proton-CT head scanner. Conference on the Application of Accelerators in Research and Industry" CAARI 2016, 30 October – 4 November 2016, Ft. Worth, TX, USA. arXiv:1707.01580v1 [physics.med-ph], https://arxiv.org/pdf/1707.01580.

Kraan, A. C. 2015. "Range verification methods in particle therapy: Underlying physics and Monte Carlo modeling" *Front Oncol.* **5**: 150.

Krah, N., Testa, M., Brons, S. et al. 2015. "An advanced image processing method to improve the spatial resolution of ion radiographies" *Phys. Med. Biol.* **60**: 8525–47.

Linz, U. and Alonso, J. 2016. "Laser-driven ion accelerators for tumor therapy revisited" *Phys. Rev. Accel. Beams* **19**: 124802.

Magallanes Hernández, L. 2017. "Low-dose ion-based transmission radiography and tomography for optimization of carbon ion-beam therapy" Dissertation, LMU München: Fakultät für Physik. https://edoc.ub.uni-muenchen.de/20468/13/Magallanes_Hernandez_Lorena.pdf.

Mattiazzo S., Bisello D., Giubilato P. et al. 2015. "Advanced proton imaging in computed tomography" *Radiat. Prot. Dosimetry* **166**: 388–92.

Meyer, S., Gianoli, C., Magallanes, L. et al. 2017. "Comparative Monte Carlo study on the performance of integration- and list-mode detector configurations for carbon ion computed tomography" *Phys. Med. Biol.* **62**: 1096–112.

Mumot, M., Algranati, C., Hartmann, M., Schippers, J. M., Hug, E. and Lomax, A. J. 2010. "Proton range verification using a range probe: Definition of concept and initial analysis" *Phys. Med. Biol.* **55**: 4771–82.

Muraishi, H., Hidetake, H., Abe, S. et al. 2016. "Experimental study of heavy-ion computed tomography using a scintillation screen and an electron-multiplying charged coupled device camera for human head imaging" *Jpn. J. Appl. Phys.* **55**: 036401.

Muraishi, H., Nishimura, K., Abe, S. et al. 2009. "Evaluation of spatial resolution for heavy ion CT system based on the measurement of residual range distribution with HIMAC" *IEEE Trans. Nucl. Sci.* **56**: 2714–21.

Obceama, C. 2016. "Potential clinical impact of laser-accelerated beams in cancer ion therapy" *Nucl. Instrum. Methods Phys. Res. A.* **829**: 149–52.

Owen, H., Lomax, A. and Jolly, S. 2016. "Current and future accelerator technologies for charged particle therapy" *Nucl. Instrum. Methods Phys. Res. A.* **809**: 96–104.

Park, S., Jeong, C., Lee, J. et al. 2017. "Proton radiography and computed tomography with dynamic water range shifter" *J. Instrum.* **12**: P 04004.

Parodi, K. 2014. "Heavy Ion radiography and tomography" *Phys. Med.* **30**: 539 43.

Pemler, P., Besserer, J., de Boer, J., Dellert, M., Gahn, C. Moosburger, M., Schneider, U., E. Pedroni, E. and Stäuble, H. 1999. "A detector system for proton radiography on the gantry of the Paul-Scherrer-Institute" *Nucl. Instrum. Methods Phys. Res. A.* **432**: 483–95.

Penfold, S. N., Schulte, R. W., Censor Y. and Rosenfeld, A. B. 2010. "Total variation superiorization schemes in proton computed tomography image reconstruction" *Med. Phys.* **37**: 5887–95.

Pettersen, H. E. S., Alme, J., Biegun, A. et al. 2017. "Proton tracking in a high-granularity Digital Tracking Calorimeter for proton CT purposes" *Nucl. Instrum. Methods Phys. Res. A.* **860**: 51–61.

Poludniowski, G., Allinson, N. M. and P. M. Evans. 2015. "Proton radiography and tomography with application to proton therapy" *Br. J. Radiol.* **88**: 20150134.

Rinaldi, I., Brons, S., Gordon, J. et al. 2013. "Experimental characterization of a prototype detector system for carbon ion radiography and tomography" *Phys. Med. Biol.* **58**: 413–27.

Rinaldi, I., Brons, S., Jäkel, O. et al. 2011. "Investigations on novel imaging techniques for ion beam therapy: Carbon ion radiography and tomography" 2011 *IEEE Nucl. Sci. Conf. R.* MIC11-3: 2805–10.

Rinaldi, I., Brons, S., Jäkel, O., Voss, B. and Parodi, K. 2014. "Experimental investigations on carbon ion scanning radiography using a range telescope" *Phys. Med. Biol.* **59**: 3041–57.

Ryu, H., Song, E., Lee, J. and Kim, J. 2008. "Density and spatial resolutions of proton radiography using a range modulation technique" *Phys. Med. Biol.* **53**: 5461–8.

Schneider, U., Pedroni, E., Lomax, A. 1996. "The calibration of CT Hounsfield units for radiotherapy treatment planning" *Phys. Med. Biol.* **41**: 111–24.

Schneider, U., Pemler, P., Besserer, J., Pedroni, E., Lomax, A. and B. Kaser-Hotz, B. 2005. "Patient specific optimization of the relation between CT-hounsfield units and proton stopping power with proton radiography" *Med. Phys.* **32**: 195–9.

Schulte, R., Bashkirov, V., Li, T. et al. 2004. "Design of a proton computed tomography system for applications in proton radiation therapy" *IEEE Trans. Nucl. Sci.* **51**: 866–72.

Schulte, R. W., Penfold, S. N., Tafas, J. T. and Schubert, K. E. 2008. "A maximum likelihood path formalism for application in proton computed tomography" *Med Phys.* **35**: 4849–56.

Seco J. and Depauw N. 2011. "Proof of principle study of the use of a CMOS active pixel sensor for proton radiography" *Med Phys.* **38**: 622–3.

Sipala, V., Bruzzi, M., Bucciolini, M. et al. 2013. "A proton computed tomography system for medical applications" *J. Instrum.* **8**: C02021.

Taylor, J. T., Poludniowski, G., Price, T. et al. 2016. "An experimental demonstration of a new type of proton computed tomography using a novel silicon tracking detector" *Med. Phys.* **43**: 6129–36.

Telsemeyer, J., Jäkel, O. and M. Martišíková. 2012. "Quantitative carbon ion beam radiography and tomography with a flat-panel detector" *Phys. Med. Biol.* **57**: 7957–71.

Testa, M., Verburg, J. M., Rose, M. et al. 2013. "Proton radiography and proton computed tomography based on time-resolved dose measurements" *Phys. Med. Biol.* **58**: 8215–33.

Uzunyan S. A., Blazey, G., Boi, S. et al. 2014. "A new proton CT scanner. 23rd Conference on Application of Accelerators in Research and Industry" CAARI 2014; 25–30 May 2014; San Antonio, TX, USA, arXiv:1409.0049v2 [physics.ins-det], https://arxiv.org/pdf/1409.0049.

Williams, D. C. 2004. "The most likely path of an energetic charged particle through a uniform medium" *Phys. Med. Biol.* **49** (13): 2899–911.

Zheng, Y., Kang, Y., Zeidan, O. and Schreuder, N. 2016. "An end-to-end assessment of range uncertainty in proton therapy using animal tissue". *Phys. Med. Biol.* **61**: 8010–24.

Zygmanski, P., Gall, K. P., Rabin, M. S. and Rosenthal, S. J. 2000. "The measurement of proton stopping power using proton-cone-beam computed tomography" *Phys. Med. Biol.* **45**: 511–28.

13

Radioisotope Production and Application

Luis Roso

13.1 Introduction

Radionuclides have a waste application in biomedical sciences [Kowalsky 2004, Ruth 2009] that resulted in a new branch of pharmacy: nuclear pharmacy [Saha 2010]. Unstable isotopes of medical interest have short lifetimes – from days to minutes – and need cyclotrons, linear accelerators or nuclear reactors to be produced. Most of them – if not all – are accessible via accelerated proton or deuteron collisions. Projectile energies are in the few MeV to a few tens of MeV range.

Radioisotope production is a branch of nuclear pharmacy that has been widely developed during many decades. A set of charged particle accelerators, such as cyclotrons, linacs and others, have been specifically designed to fulfil the needs of this sector. Although the fundamental concepts have not evolved very much during the last few decades, the technology of conventional accelerators has evolved with the introduction of superconductors that allow high magnetic fields and compact structures. However, this evolution has arrived to the limit allowed by those technological developments (in the absence of unexpected conceptual improvements). Laser accelerators have quite an interesting potential [Spencer 2001, Ledingham 2005] to become much simpler than conventional cyclotrons, at least in certain situations (as tailor-made, single-dose preparation, extremely short-lived isotopes). Moreover, lasers require much less shielding (since the laser itself is not radioactive) and can be used to prepare single-dose drugs at the hospital or a local PET radiology centre without the requirements of radioisotope transportation.

There are hundreds of radioisotopes that can be produced with charged particle accelerators, i.e. that are accessible through proton or deuteron collisions (or, less frequently, with ^{3}He or α particles). The cyclotron is the most commonly used accelerator for the isotopes accessible via reactions with projectiles at 10 MeV energies or less. Linacs (linear accelerators) are also frequently used. A few isotopes are obtained using nuclear reactors. An interesting comparison between different kinds of conventional accelerators can be found in Conrad [1994].

Lasers, in turn, are a very promising alternative for this application [Lefebvre 2006, Ma 2013]. Laser technology is quickly evolving in performance and cost reduction. However, laser technology is still not mature enough to match or exceed the capabilities of the cyclotron. A cyclotron is a robust, well-developed and reliable machine. Superconducting cyclotrons and mini-cyclotrons probably represent the limit of this accelerator technology. The very far end of this technology could be the isotron, a super-conducting cyclotron for radioisotope production with an extremely compact footprint [Michael 2013]. A dramatic reduction of the cyclotron size (i.e. less than one meter in diameter for a few MeV proton beams) is not foreseen. On the other hand, cyclotrons are machines able to deliver a very high current (hundreds of microampere) of accelerated particles. Also, linacs have been considered for the production of PET radionuclides. In all cases, a projectile (proton, deuteron, etc.) hits the precursor nucleus (contained within a gas, liquid or solid target, depending on the atom) with energies in the 3–20 MeV range for low-Z targets, and in the 10–70 MeV range for high-Z. Therefore laser-driven production of radioisotopes needs laser intensities high enough to achieve those projectile energies.

So, what could be the future role of a laser-driven accelerator? It is clear that to accelerate protons, deuterons or other ions at the energies relevant to radioisotope production, one needs [Macchi 2013] multi-TW laser systems (for a review of laser-driven ion acceleration see Chapter 5), probably beyond 50 TW (1 TW = 1 Terawatt = 10^{12} Watts), unless relevant breakthroughs appear in target development [d'Humières 2013]. Such systems now typically operate at repetition rates of 10 shots per second (10 Hz) or less. The required major increase in laser repetition rate in order to become competitive has been also quantitatively investigated. There are two perspectives: the first one is to wait for multi-hundred TW multi-kHz lasers. Those systems will eventually have both an adequate high peak (single pulse) power and repetition rate (therefore high average power). Recent developments with thick optical fibre amplifiers [Alam 2015, Jauregui 2013] are promising but not yet a reality. The avant-garde of this technology is represented by the extreme light infrastructure (ELI) [Gerstner 2007] in its three pillars [Gales 2016]. ELI is looking for extreme high intensities (focusing multi-PW lasers (1 PW = 1 Petawatt = 10^{15} Watts), too extreme for this purpose. Also, ELI is also looking for extremely high repetition rates [Amato 2016] and new developments on thin disk lasers are triggered by that project. The second perspective is to find innovative situations where a laser can introduce added value that overcomes the shortcomings of low average current yields.

On the path to the long-term perspectives pictured in the first option (high peak power and high average power lasers), it is worth concentrating on the second aspect (innovative situations that can benefit from the specific features laser accelerators offer) and determining scenarios where the employment of lasers can represent an advantage. One of the scenarios may be the production of very short-lived isotopes, monodose activation and extremely quick delivery. These scenarios will be explored after reviewing the basic concepts of radioisotope production.

13.2 Basic Concepts: Decay, Half-Life and Activity

Each unstable nucleus has a characteristic radioactive decay time, $t_{1/2}$, which is the time after which one-half of the initial population has decayed into some other state. This corresponds to a decay rate proportional to the number of unstable nuclei, N, that have not yet decayed. It is described by the well-known decay equation

$$dN/dt = -\lambda N = -(\ln 2 / t_{1/2})N, \tag{13.1}$$

ln2 (= 0.693) being the factor that links the half-life, $t_{1/2}$ with the exponential decay constant $\lambda = \ln 2/t_{1/2}$ (in the medical bibliography, the half-life $t_{1/2}$ is more frequently used than the exponential decay constant λ). If N_0 is the initial number of unstable nuclei, the time evolution of N(t) is thus simply given by

$$N(t) = N_0 \exp(-t\ln 2 / t_{1/2}) = N_0 \exp(-\lambda t). \tag{13.2}$$

However, the number of unstable nuclei is not the useful measure here. Instead, it is much more convenient to introduce the activity of an unstable sample. The activity A(t) is defined as the negative time derivative of N(t) and so

$$A(t) = \lambda N(t) = \ln 2 N(t)/t_{1/2}, \tag{13.3}$$

i.e. activity is defined as the number of decays per unit time. This is the more common parameter in radiopharmacy to determine the radioactive drug prescribed dose. Typical activities for diagnoses are 10^7 to 10^9 disintegrations per second (dps), with a high variability of situations.

The curie (Ci) is the historical unit for radioactivity. It was introduced originally as the number of dps of one gram of ^{226}Ra. This unit of activity was later set officially to the value of 3.7×10^{10} dps. Now the becquerel (Bq) is the SI unit of radioactivity and is defined as the activity that produces 1 dps, or 1 Bq = 1 dps. Therefore 1 Bq = 2.7×10^{-11} Ci, or inversely 1 Ci = 37 GBq. Typical doses are often measured in mCi (therefore, in terms of radioactivity level).

In addition, it is important to remember that an unstable nucleus can decay into one or more channels. If decay occurs through several channels, we refer to the branching ratio for one channel as the percentage of decays occurring that way. For example, one very long-lived isotope (with about 1.25 gigayears half-life) that we have in our body is ^{40}K. It decays in two different channels: one is electron capture (EC; i.e. capture of an inner electron by the nucleus), with a branching ratio of 10.7%, forming ^{40}Ar; and the other is β^- (electron) decay, yielding ^{40}Ca, with a branching ratio of 89.3%. For a radioisotope to be useful in medicine, we need to have a complete map of the different decay channels and branching ratios, up to the percent level or so. Some of the decay channels are applied for imaging or treatments. The rest of the decay channels represent dose that the patient receives without any benefit.

Most of the β^+ (positron) emitters have just this single decay channel available (in other words, the EC plus β^+ branching ratio is 100%). However, some of the newly developed radionuclides have other decay path besides the β^+ decay channel; an example is given by ^{64}Cu, with a half-life of 12.7 hrs, which decays via EC and β^+ channels 61.5% of the time, forming ^{64}Ni, and via the β^- channel 38.5% of the time, forming ^{64}Zn.

13.3 Nuclear Decay Emission Products

Many unstable isotopes are suitable for medical use as radiotracers or for radiotherapy. Radiotracers are unstable nuclei whose decay inside the body can be monitored. To this aim, their primary or secondary decays must be detectable. Radio-treatment implies that the product of the decay deposits ionizing energy into the target cells and destroys them. Decay products of an unstable nucleus can be:

- *A photon:* A 100 keV or MeV photon propagates quite deep inside the body, so its application for therapy is doubtful since the fractional energy that is absorbed is distributed over a long range inside the body and there is no possibility for precise deposition of energy. On the contrary, the long range of gamma rays makes them very interesting for diagnostics. Each nucleus has sharp characteristic gamma emission lines. We are natural emitters of gamma rays: the 1.46 MeV gamma rays from the ^{40}K in our body. (We naturally emit a few hundred of such gammas per second!) Applications must eventually consider such background radiation.

- *An electron:* Electrons released from a nucleus are not of much use for diagnostics because they are part of a β^- decay with energies in the few MeV or sub-MeV range, and thus with very short penetration lengths; too short to be useful since they cannot emerge out of the body. On the other hand, the body stopping power is not high enough to allow precise energy deposition [Turner 1995] and localized treatment. For example, the range in water is about 2 mm for 500 keV electrons and 2 cm for 4 MeV. So β^- emitters used in brachytherapy do not have very precise resolution for energy deposition. For that reason, emissions from β^- emitting radionuclides are effective mostly for large tumours. There is another source of electrons, indirectly emitted by the radiotracer.

It is well-known that certain radionuclide decay by electron capture or other mechanisms that end in the release of an Auger electron. Those low-energy electrons, and therefore well-localized electrons, have also application in brachytherapy [O'Donoghue 1996]. In this case the damage of the cells is very close to the location where the unstable nuclei decays [Knapp 2016]. Also, high-energy electrons can generate secondary radiation in the form of X-rays.

- *A positron:* Positrons are emitted with a range of kinetic energies due to the β^+ process, but there is a fundamental distinction from electrons. Our body fundamentally consists of atoms so that it is very likely that the positron annihilates with one electron. In fact, a one MeV positron just travels a distance of order one millimetre before being annihilated in a collision with an electron. So, positron emitters deposit the energy in a very precise way, within a few mm. Moreover, the annihilation leads to a pair of oppositely emitted gamma photons, each one with photon energy slightly more than 0.511 MeV (which is basically the rest mass of the electron plus the contribution of the kinetic energy of the positron).

- *A positive ion:* Conceptually any emitted ion (proton, deuteron, ^{3}H, ^{3}He, α particle) can be a candidate for therapy since its range inside the body is very small, sub-millimetre for sub-MeV energies (hence more precise than a positron with the same energy). Therefore, its energy is deposited within a short distance from the decay emission location. Direct detection is impossible, of course, but detection could be associated with secondary penetrating radiation, due to the deeper penetration of the secondary X-ray radiation.

- *A neutron:* Neutron emitters are not so interesting in this context because neutrons, due to their small interaction cross sections, travel far from the point of emission. However, released neutrons contribute to the risk of long-term activation of other molecules in the body and creation of a secondary cancer line far away from the point of emission of the neutron. Radiation-induced secondary cancer lines are extremely difficult to detect, because of the problems differentiating a naturally occurring second cancer line from an induced second line in a given patient.

13.4 Medical Applications

There are many possibilities to apply radioactive nuclei in medicine. We can differentiate between diagnostics and treatment as a preliminary classification, although in certain cases both applications are combined. In all cases, the nucleus obviously forms part of an atom, and this atom is part of a molecule.

Related to diagnostics, radiotracers allow a new branch of medical detection that is often called molecular imaging. They are systems not prepared to get a static picture of the body, anatomy such as X-ray based radiography, computer tomography (CT), etc. Conversely, they have been designed to measure the specific molecular activity in our body. Molecular imaging starts normally with the injection of a molecule marked with the radiotracer (ingestion or inhalation are also possible). This tracer molecule goes to a specific tissue, organ or other part of the body. The objective is to monitor how fast the molecule moves in the body and where it accumulates. Such molecular imaging can map the position of the molecule with the tracer inside the body. Of course, tracer decay from each molecule occurs only once, when its unstable nucleus decays. In order to detect such decays, different detector systems have been developed, tailored to the specific nuclear decays yielding photon emissions, as described in Section 13.3. Therefore, in this context, molecular imaging refers to the way to obtain images of the molecular activity inside the body.* There are different imaging techniques as indicated below.

Gamma cameras enable the simplest medical imaging technique involving radioisotopes. A gamma camera is just a flat scintillator. The resulting image is a 2D display resembling that obtained in radiography, except that the source of the radiation is inside the body. The medically injected, ingested or inhaled

---

* Note that molecular imaging has a totally different meaning in the ultra-fast laser applications community, where it refers to the femtosecond or sub-femtosecond evolution of electrons within a molecule using attosecond laser probes, and thus refers to imaging the binding electron motion inside the molecule.

radionuclides emit gamma rays that the scintillator detects. Because there is no optics for gamma rays, acquiring information about the direction of the incoming gamma rays requires the use of collimators.

Single photon emission computed tomography (SPECT) essentially represents the gamma ray analogue of a conventional CT, where the key distinction is that the gamma source is inside the body. Thus, it is a 3D imaging technique. As with the gamma camera, gamma photons emerge from a gamma emitter that is delivered to the patient normally through bloodstream injection. Its development is mostly with efficient gamma scintillators that can record and locate the origin of radiation coming from the radio-tracer with a proper system of collimators and detectors. In principle, multiple intermediate systems in between a flat camera and a real tomographic 3D system are possible. Examples of such systems are false 3D systems, or stereoscopic systems, simply combining two images at different angles, like our binocular vision, instead of the tomographic 3D.

Positron emission tomography (PET) [Valk 2003, Saha 2016] is another well-established technique for obtaining a 3D image that relies on coincidence detection of a pair of photons as an alternative strategy to achieve adequate signal-to-noise ratio and thereby reduce the time exposure within reasonable limits. In this case, the injected molecule has an unstable isotope that decays by emitting a positron. The positron annihilates in collision with an electron, yielding a pair of gamma photons for detection. Thus, the PET system is typically a ring of very sophisticated gamma-coincidence detectors (as illustrated in Figure 13.1) and collimators (gamma photon energies are slightly larger than 511 keV, due to the kinetic energy of the released positron).

Modern PET scanners can also incorporate very precise time-resolving systems (with precision from sub-ns down to 100 ps) that enables measuring the different arrival times of photons emitted in the opposing directions (time-of-flight PET or PET/TOF). On the nanosecond time scale this helps to improve the signal-to-noise ratio, by eliminating background radiation that is not in time coincidence and improving the PET image quality. This TOF information will be also used to determine the point of emission along the counter propagating gamma source line with high precision (note that 30 ps corresponds, at the speed of light, to a 1 cm propagation distance).

PET radiotracers are typically very short-lived isotopes, such as ^{11}C, ^{13}N, ^{15}O and ^{18}F, where the last one is the most widely used. Moreover, all of them must have the β^+ decay channel as the only or most likely decay channel (in most cases, the EC plus β^+ branching ratio is 100%). β^+ is conceptually represented as the transmutation of one proton to a neutron, releasing the positive charge as a positron. For example, the β^+ decay of ^{18}F can be depicted as, $^{18}F \rightarrow {}^{18}O + e^+ + \nu$. The positron is indicated by e^+ and ν is the neutrino. Globally this can be seen as a proton that disintegrates into a neutron, a positron and a neutrino. This reaction cannot happen for a free proton, but for the proton inside the ^{18}F nucleus, the energy comes from the binding field of the 18 nucleons. Schematically this is shown in Figure 13.2.

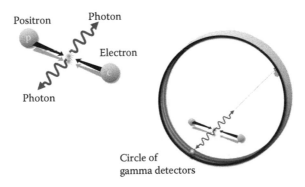

Positron   Photon

Electron

Photon

Circle of
gamma detectors

FIGURE 13.1 Schematic representation of positron emission tomography. An unstable nucleus emits a positron that quickly annihilates with its antiparticle, the electron, and creates two photons in coincidence. Correlated detection of the two photons reduces the noise background and determines the line containing the tracer source.

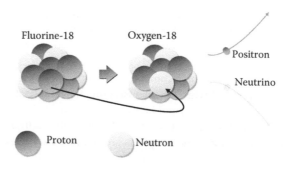

Fluorine-18　　　　Oxygen-18

Positron

Neutrino

Proton　　　Neutron

FIGURE 13.2 Schematic representation of the nuclear β^+ decay. Large, dark dots represent protons, and large, lighter dots represent neutrons. A proton is transmuted to a neutron and so one positron and one neutrino are released. The positron quickly annihilates with an electron inside the body. The neutrino is medically undetectable.

Although meaningless for our diagnostic application, the neutrino is basic for understanding the energy balance and originates a distribution of positron energies, instead of a well-defined, single energy value. The positron energy spectrum is relevant, because the mean free path of the positrons inside the body depends on their energy. Fortunately, the mean free path is very small (a few millimetres) so the blurring is acceptable for most tracers. In general, it is very difficult to resolve the internal organs in a PET image, and so it is desirable to combine it with conventional X-ray CT [Wolbarst 2006]. PET/CT systems are common. The patient goes through the PET ring and immediately after through the CT, all on the same motorized stretcher. Since recently the PET/MR combination is also possible. The most relevant characteristic of PET is that coincidence detection improves significantly the signal-to-noise ratio. Of course, PET, PET/CT and all other kinds of similar medical imaging techniques need a very specific expertise to provide a sharp interpretation [Fanti 2010, Conti 2016]. From a physicist's point of view, PET has the appealing feature of a medical use of antimatter.

As in all diagnostic applications, we must also consider the dose. The equivalent dose, i.e. the dose that accounts for the biological effects of the radiation is measured in sievert (Sv) units. The corresponding physical quantity is the gray (Gy). One gray (i.e. 1 Gy) corresponds to one joule of radiation energy deposited in one kilogram of tissue, without differentiation of the origin of the radiation. For photons, whatever its energy, the biological weighting factor is equal to one; so, 1 Gy of photons corresponds to 1 Sv. For protons, the biological weighting factor is 2, indicating that the biological effect is higher such that 1 Gy of protons represents 2 Sv. For neutrons, the factor is bigger than one and also depends on the energy of the neutrons. Natural radiation dose, i.e. the equivalent dose we typically receive on Earth, is about 2 mSv/year. The dose for a whole-body PET scanning depends on the tracer and also depends on whether or not it is combined with a CT. Typically it can be of the order of 10 mSv or more. This is a high value, comparable to a few years of natural radioactivity. Therefore, molecular imaging with radiotracers must be medically justified, particularly for whole-body images. Among other things, the activity of the radiotracer needed for a PET data acquisition depends on the photon sensitivity of the gamma detectors that form the image. Technology is evolving to increase this sensitivity, because it can result in a radiation-dose reduction. This aspect is relevant when considering laser-based approaches to radioisotope production.

Moreover, radioactive nuclei can also be used for treatment. Targeted radionuclide therapy uses molecules labelled with a radionuclide that decays into short-range ionizing radiation (Section 13.3). The molecule is therefore the carrier that brings the radiotracer to the specific organ or tissue inside the body. The distinction from the radiodiagnostic application is that the radionuclide delivers a toxic level of radiation. This radiation seeks the destruction of the targeted cell but, in turn, can kill other adjacent cells that are not marked. In certain cases, the destruction of peritumoural cells, hence avoiding microscopic disease spread, is an advantage of targeted radionuclides over other chemical molecular labels that just target and destroy the tumour cell. In addition, targeted radiotherapy sources can destroy not only the primary tumour but also metastatic cells while they spread though the body. So far, the diagnosis and treatment have been considered as distinct applications. Some advanced techniques start to consider new radiopharmaceuticals that combine both purposes. Labelled molecules attached to the tumour cells can be traced for diagnosis while they also deliver their energy and destroy the surrounding cells.

For the purpose of this book, the most relevant parameter is the half-life of the radiotracer or radiopharmaceutical [Welch 2003]. Storage is not possible for an unstable nucleus. Once generated, it intrinsically decays without any chance to modify/delay/manipulate its half-life.

Based on their half-lives, radiotracers can be classified into:

Long-lived tracers: In this context, this means unstable nuclei that can be easily transported from the production site to the patient. In the case of weeks or more this is straightforward. One example is ^{131}I (radioiodine), with a half-life of slightly more than 8 days that is used to destroy thyroid cells in hyperthyroidism cases.

Short-lived tracers: For this range of half-lives the transport has to be specifically prepared. One typical example is ^{18}F, with a 110-min half-life, that is probably the most widely used tracer for PET. Use of ^{18}F generation from a cyclotron (in many cases far from the hospital) in typical cases needs one half live for the transport and another in the hospital before administration to patient. The result is that only 2 of the 8 molecules (25%) are used as intended. To justify costs, ^{18}F has to be overproduced, which poses a grand challenge for the laser-based production of PET tracers as of today.

Very short-lived tracers: Some interesting tracers, such as ^{11}C and ^{15}O, have a half-life too short to realistically accommodate transport from one place to the other (the half-life of ^{11}C is 20 min, only 2 min for ^{15}O). So, they have to be generated very close to the PET facility where they will be used, and not many hospitals have a cyclotron on site. Therefore, very short-lived isotopes do not have many feasible applications. It is for these cases that laser-driven isotopes represent a very promising alternative. Moreover, a cyclotron has several constraints with space and radioprotection that can complicate its installation at a hospital. A multi-TW laser could become suitable for this purpose.

After the unstable isotope is generated and the organic molecule with this isotope is synthesized, some time is needed before it is introduced into the body of the patient. The organic molecule also needs some time to reach its target within the body, and after that time the PET data acquisition can start. Once the expert has recorded the necessary number of decay signals, the PET scan ends. But the tracer is still active in the body and the dose accumulation in the patient continues. This is schematically shown in Figure 13.3. Besides the production of the isotope, production of the tracer molecule (whenever

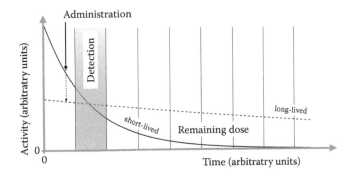

FIGURE 13.3 Schematic comparison of short- and long-lived radiotracers. There are three time zones during radiotracer use: administration, detection (shaded) and after-detection. Assuming the dose is proportional to the radioactivity, and that the area of both curves on the shaded region is the same (this area is proportional to the number of decays), longer half-life tracers can deliver more dose to the patient after acquisition of the diagnostic image.

needed) and transport to the PET facility, there are three consecutive steps in the radiotracer usage connected to dose delivery to the patient: administration, detection and after-detection. At the moment of administration, the radionuclide enters the body. Metabolization of the molecule requires time before the detection process starts. A certain number of decays are required to acquire an image (or a set of images). Then the medical PET imaging acquisition stops, but the patient has still a radioactive compound inside the body that continues decaying. Assuming the dose is proportional to the activity, longer half-life decays can deliver more dose to the patient following image data acquisition. If the long-lived tracer remains permanently in the body, for example incorporated in the bone tissue, the patient will be receiving a dose for a long time. In some other cases, the tracer is eliminated from the body by natural excretion and so patient irradiation ends. Obviously, it is preferable for the patient to receive tracers with the shortest possible half-life. This is schematically shown in Figure 13.3. Observe that in both cases (long- and short-lived) it has been considered that the number of decays to get proper imaging is the same. The two curves (continuous for the short-lived and dashed for the long-lived) have approximately the same area in the shaded region, meaning that the number of decays is the same in the two cases. Of course, this assumes that the image acquisition needs the same number of events in both cases, and this is quite realistic. The same area in the shaded region of Figure 13.3, or the same number of decays during the detection or data acquisition time interval correspond to a much higher after-detection dose for the long-lived tracer (areas of the two curves after the detection interval). Furthermore, the short-lived isotopes allow time evolution, or time resolution of metabolism. Disposal of short-lived isotopes must aim at minimizing the time required for production and transportation which are issues where the laser case could contribute. For applications in treatment, longer-lived isotopes are typically acceptable, for example ^{131}I (lifetime of 8 days) to treat thyroid cancer. For those cases, laser production has no obvious benefit yet.

13.5 Production of Radionuclides

In most cases there is a stable precursor nucleus that is at rest and the nuclear reaction is triggered by an accelerated projectile [IAEA 2008]. For example, the very popular fluorine-18 is obtained by the reaction of a proton on an oxygen-18 nucleus, yielding fluorine-18 and a neutron. This nuclear reaction is written as $^{18}O(p,n)^{18}F$. The measure of the probability of the reaction [Paetz 2014] is given by the reaction cross section [Qaim 2001]. An example is shown in Figure 13.4, which is an approximate representation of the

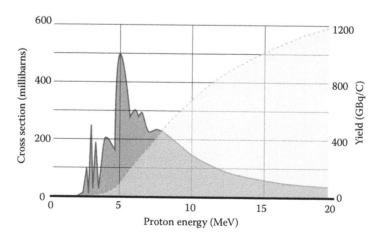

FIGURE 13.4 Reaction cross section in millibarns (continuous line plot with scale to the left) and radiotracer yield (dashed line plot with scale to the right) as a function of the incident proton energy. The figure corresponds to the $^{18}O(p,n)^{18}F$ reaction, but it is representative of many other population reactions.

values indicated in the literature [IAEA 2001]. The continuous line filled below in dark, and scale to the left, represents the energy dependence of the reaction cross section in millibarns (1 barn = 10^{-24} cm^2). In this typical plot two things are clear about the energy dependence of the cross section. Below 2 MeV the probability of reaction is negligibly small because the proton energy is not adequate to overcome the coulomb repulsion between the proton and the nucleus. After several local maxima, the cross section diminishes as energy increases. Consequently, there is no reason to accelerate protons beyond 20 MeV, since it takes a lot of effort without enhancing the reaction cross section. The dashed curve, and scale on the right, in Figure 13.4 indicate the yield, measured in GBq/C units. The GBq/C ratio refers to the electrical charge, in coulombs, needed to generate an amount of radioactive isotope with an activity of 1 GBq. This can be seen at the particle level. One coulomb of protons corresponds to 6.24×10^{18} particles. In Figure 13.4, protons of 15 MeV induce a yield of 1000 GBq/C. This indicates that 6.24×10^{18} protons of 15 MeV generate 1000 GBq of ^{18}F, approximately. Using Equation 13.3, we can calculate how many nuclei correspond to this activity level. For ^{18}F, $t_{1/2}$ = 110 min, and A(t) ~ N(t) 10^{-4} s^{-1}. This means that to have one decay per second, we need about 10^4 nuclei of ^{18}F. Thus, for 1000 GBq, we need 10^{16} nuclei. Now we can translate, for this reaction and for this incoming proton energy the yield, 1000 GBq/C indicates that we get 10^{16} nuclei of ^{18}F with one coulomb of protons, i.e. with 6.24×10^{18} protons, or, simplifying, we need more that 600 protons of 15 MeV energy to activate one ^{18}F nucleus. The cross section is quite peaked because there are resonant energies that enhance its magnitude. On the contrary, the total yield is an average over incident proton energy since the incoming proton reduces its energy in subsequent collisions. The yield is an increasing function. Figure 13.4 corresponds to the above-mentioned ^{18}O(p,n)^{18}F reaction, but it is quite representative of a wide set of common reactions.

Conventional cyclotrons operate at a nominal energy, in certain cases two are available. Protons are delivered at a given energy and its modification can be complicated. As a result, projectile energies are chosen high enough to ensure coverage of a desired set of possible reactions. Laser-acceleration of protons offers a different and potentially simpler way of energy modification via the choice of laser parameters. Moreover, depending on the specific acceleration mechanism, lasers produce a broad distribution of proton energies. In principle, assuming energies beyond the threshold, this well-established feature is not a problem for PET isotope production.

Radionuclide production in a cyclotron is achieved using a constant flux of protons with a given energy. We assume that they hit a target that initially is not activated. Therefore, while the proton beam is on, change in activation level over time can be described by an expression of the form:

$$\frac{dA}{dt} = g - \frac{\ln 2}{t_{1/2}} A, \tag{13.4}$$

g being a constant coefficient that accounts for the production/activation rate (i.e. a source term) which masks a much more complicated expression that accounts for the increase of activation (it depends on the cross section of the reaction as well as the energy distribution of the projectiles). Of course, it depends strongly on the proton beam energies and on the energy-dependent reaction cross sections. The decay is omnipresent, so as soon as the radionuclides are generated they start to decay. The integration of this differential equation is straightforward and the final expression, with the initial condition A(0) = 0, is

$$A(t) = g \frac{t_{1/2}}{\ln 2} \left(1 - \exp\left(-\frac{t}{t_{1/2}} \ln 2 \right) \right). \tag{13.5}$$

This activation curve is plotted in Figure 13.5, where it is clear that A saturates after many half-lives as indicated in the light grey region for a projectile beam from a continuous-wave source. There is

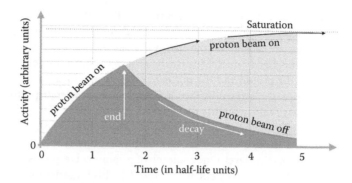

FIGURE 13.5 Typical plot of the time-variation of the activity during irradiation with a continuous (light grey) source and a pulsed (dark grey) source.

saturation at A = (g $t_{1/2}$/ln2). Here saturation means that regardless of how much time you keep the accelerated proton beam on, the activity is not going to increase significantly, as is shown in Figure 13.5, where the saturation limit is indicated by the horizontal dotted line. So, irradiating the ^{18}O target with protons for an extremely long time does not help due to saturation. In the saturation regime, unstable nuclei are generated at the same rate that they decay, so the activity level is constant (where this conclusion also assumes that we have an unlimited supply of the precursor ^{18}O). Approaching too much to the saturation regime is a wasted effort. For example, a reasonable balance in most cases is achieved irradiating for about one or two half-lives. The on/off cycle of the cyclotron is shown in the dark zone of Figure 13.5. There the vertical arrow indicates the point the proton beam is off. After that, g = 0, and A(t) is given by its half-life decay,

$$A(t)_{t>t_{off}} = A(t_{off})\exp\left(-\frac{t-t_{off}}{t_{1/2}}\ln 2\right), \tag{13.6}$$

which eventually goes to zero, as depicted in the decreasing part (dark grey) of Figure 13.5.

With a laser-driven source, one new element has to be taken into consideration. The laser is pulsed, and so is the activation of the radionuclides. Activation is in the form of very short bursts, so the g coefficient can be regarded as a step function with a very large value within the duration of the proton bunch, and zero as soon as the proton bunch ends.

Radioisotopes do not always come from just one simple reaction; there are some cases where they come from an intermittent unstable parent nucleus. One very important example is metastable ^{99m}Tc. It has a half-life of 6 hrs and it is straightforward to produce since it comes from the decay of ^{99}Mo (whose half-life is almost 66 hrs or 2.75 days). In this case we say that ^{99m}Tc comes from a precursor, and the problem is to produce the precursor. There are two common reactions for obtaining ^{99}Mo with proton interactions, which start from ^{100}Mo, found in nature. One is ^{100}Mo (p,pn) ^{99}Mo and the other is ^{100}Mo (p,2p) ^{99m}Nb, where the product ^{99m}Nb decays within about 15 sec to ^{99}Mo. There is a third reaction that complicates the production scenario, ^{100}Mo (p,2n) ^{99m}Tc. Optimization of one or another of these three channels depends on the energy of the proton projectile (i.e. associated with the reaction cross section energy dependence).

Radionuclides have to be produced in a useable chemical form [Long 2014]. Typically, a mini-laboratory for radiochemistry is needed. For the first proof of concept application of lasers to PET tracer production, it is not expected that this chemical step changes significantly compared to the cyclotron case. However, as laser technology will consolidate its role, significant changes in the chemical step are foreseen, particularly in tracers like ^{11}C, for example, that can be directly incorporated onto CO_2 or other gas molecules.

13.6 Some Examples of Radionuclides Useful for Medical Applications

The number of unstable atoms suitable for attachment to a molecular label that can go to a specific target is increasing very rapidly. In some cases, the molecule is fundamental and new molecules are developed routinely [Harada 2015].

Fluorine-18: By far ^{18}F is the most widely used tracer. It is so because its half-life, 110 min, represents a convenient compromise between decay time that is long enough to transport the tracer, yet short enough to have a measurable activity without too much remaining dose after scanning. There are other reasons, such as its compatibility with many biotracers. It is mainly produced by proton bombardment of ^{18}O enriched water (dense water) through the $^{18}O(p,n)^{18}F$ nuclear reaction, with proton energies between 2.5 and 20 MeV (the lower limit is a cut-off, while the upper limit is an asymptotic decrease) as indicated in Figure 13.4. Fluorine-18 is generally recovered as an aqueous fluoride solution and then is extracted. It is important to mention that proton bombardment is not the only way to produce fluorine-18. It can also be produced as a radioactive gas by deuteron bombardment through the $^{20}Ne(d, \alpha)^{18}F$ nuclear reaction with deuteron energies between 1.5 and 21 MeV. This production method, which is useful for electrophilic substitution, requires the addition into the target of fluorine-19 gas as carrier, and is currently seen as a less attractive method. The decay of ^{18}F is purely β^+. As stated already, the released positrons do not have a monoenergetic spectrum (remember the neutrino), the most probable energy is about 250 keV, and the endpoint energy (the maximum energy) is 635 keV, which corresponds to a range in water of about 2.4 mm. FDG (fluorodeoxyglucose, chemically 2-deoxy-2-(^{18}F)fluoro-D-glucose) is a glucose analogue, with the ^{18}F substituting an hydroxyl group. FDG is widely used in neuroimaging, in cancer diagnostics, and in other situations where the requirement is to determine the parts of the body that absorb such sugar. To get an idea of the radiation dose, we can say that typical ^{18}F-FDG PET/CT whole-body oncology imaging procedures involve the delivery of 5–8 mSv of effective dose from the PET scan, and an additional 8–10 mSv for the CT scan.

Carbon-11: Carbon by itself is the most relevant biological atom, so having an isotope of carbon as tracer opens a wide field of applications. In our bodies, we have a relevant amount of another unstable isotope, ^{14}C, (half-life 5730 years) which generates about 4 kBq activity for a 80 kg person. However, this is evenly distributed in the body, and so far it does not give useful medical information. ^{14}C decays to stable ^{14}N by β^- emission. On the contrary, the very-short lived ^{11}C, of half-life 20 min, is ideal for determining molecular activity. It is produced by proton bombardment of natural nitrogen through the $^{14}N(p, \alpha)^{11}C$ nuclear reaction. A target gas mixture of 2% of oxygen in nitrogen will produce radioactive carbon dioxide ($^{11}CO_2$), for example. Combined with a few per cent of hydrogen this reaction will finally end in methane ($^{11}CH_4$). There are many more compounds and possibilities [Rahman 2015]. From the point of view of this book, ^{11}C is a very interesting tracer because it has different populating possibilities [Paans 2000]. We analyze just a few ones that are very representative. There are two populating reactions involving protons: $^{14}N(p, \alpha)^{11}C$, with a threshold energy of 3.1 MeV and a cross section of 250 mb [Kane 1988]. The second one is $^{11}B(p,n)^{11}C$, with a threshold energy of 3.0 MeV and a cross section of 250 mb. There is another populating reaction involving deuterons, $^{10}B(d,n)^{11}C$, which is a reaction without a much lower threshold energy and with cross section of 250 mb [Firouzbakht 1998]. Deuterons present other problems, but the bombarding with deuterons has also to be accounted for within laser technology. The decay of ^{11}C is also purely β^+, the most likely energy of the positrons is 385 keV and the endpoint energy (the maximum energy) is 960 keV which corresponds to a range in water of 4.1 mm. This range accounts for the fundamental limit in the study of the PET image spatial resolution. In this context, it is necessary to mention that often the shorter lifetime is considered as a benefit but a lifetime that is too short indicates a nucleus too unstable that decays releasing high energy positrons. As a basic rule, more energetic positrons result in PET images with poorer spatial resolution. Among many biomolecules that can be used with ^{11}C, one is ^{11}C choline, widely used to follow certain tumours and their activity.

Nitrogen-13: ^{13}N (half-life 9.96 min), can be produced by proton bombardment of water through the $^{16}O(p, \alpha)^{13}N$ nuclear reaction. The decay of ^{13}N is also purely β^+, the mean energy of the positrons is 492 keV, and the endpoint energy (the maximum energy) is 1.19 MeV which corresponds to a range in water of about 5.4 mm.

Oxygen-15: ^{15}O (half-life 122 sec), is produced by deuteron bombardment of natural nitrogen through the $^{14}N(d,n)^{15}O$ nuclear reaction. As a result, unstable oxygen is produced as molecular oxygen ($^{15}O_2$). Alternatively, it can be directly produced as carbon dioxide ($C^{15}O_2$) by introducing a few per cent of natural carbon dioxide as a carrier. Carbon monoxide ($C^{15}O$) can also be produced by reduction of the dioxide. The decay of ^{15}O is also purely β^+, the mean energy of the positrons is 735 keV, and the endpoint energy (the maximum energy) is 1.72 MeV which corresponds to a range in water slightly above 8 mm. So, the large mean energy of the positron emitted by ^{15}O severely limits PET image resolution.

Copper-64: Another radiotracer produced by proton bombardment is ^{64}Cu. It is a very special radionuclide because it decays by positron emission (β^+ branching ratio is 17%) by electron emission (β^- branching ratio is 39%), as well as by electron capture (EC branching ratio is 44%) with the subsequent gamma emission. So, it can be used for PET imaging and at the same time have the therapeutic potential related to the three kinds of emissions. It has a half-life of 12.7 hrs and a great ability to be incorporated into complex molecules. Its half-life is not too short and can accommodate complex radiochemistry processes, and transport. The most common production reactions are $^{64}Ni(p,n)^{64}Cu$ and $^{64}Ni(d,2n)^{64}Cu$.

Technetium-99m: The metastable nuclear state ^{99m}Tc emits a single gamma photon so it is useful for 3D images (SPECT) as well as for 2D images (planar scintigraphy). The advantage of Technetium is that it is suitable for many applications, such as in medronic acid with ^{99m}Tc, a molecule that is often used to study bone abnormalities and metastases by imaging (gamma imaging) the gamma emission pattern. ^{99m}Tc is obtained by the β-decay of ^{99}Mo, as already indicated. The precursor/tracer radionuclide pair $^{99}Mo/^{99m}Tc$ is the radiotracer most widely used in the world.

Many more isotopes of medical applicability are now ready for use or under development [Velikyan 2014, Vallabhajosula 2009]. Particularly innovative are the combined drugs with one isotope for tracing and other unstable isotope on the same molecule for treatment. One example is the use of ^{177}Lu based radionuclide for tumour therapy in combination with other chemotherapy drugs.

13.7 Differences between Cyclotron Production and Laser Production

Going to a basic level there are three fundamental differences between a laser-based system and a cyclotron-based one: energy spread, time macrostructure (i.e. cw or pulsed) and angular divergence (i.e. directionality). Depending on the primary target the energy spectrum of laser-accelerated protons in a single bunch is much broader than that from a cyclotron. This is a major problem for specific applications, but not so relevant for activation of radiotracers (except where undesired reaction channels are allowed by the broadband spectrum). In the laser case, the proton flux is pulsed and is also less directional.

Moreover, laser-driven accelerators and conventional cyclotrons do have different requirements for shielding. A cyclotron has a very high ion current that must be shielded. All cyclotrons have to be inside bunkers with concrete walls of thickness typically more than one meter. When preparing the radionuclides in a separate facility out of or even far from the hospital, this is not a major problem. However, very short-lived isotopes must be prepared on site. Installing a cyclotron inside a hospital, particularly inside an existing hospital, has to consider these problems. The laser system itself, being infrared light, does not need radioactive shielding (although adequate light shielding will be essential). Only the region in the vicinity of the laser focus where the particle acceleration takes place has to be shielded. We can consider that radioprotection starts to become relevant with a near-infrared laser (in the 800–1200 nm wavelength range) only for intensities higher than 10^{18} W/cm^2. The high-intensity volume around the focal region is very small, and the target has to be precisely located at the focal spot. Most experiments with high power lasers today are equipped with very large vacuum chambers because there are many positioning and detection tools placed around the target for investigating fundamental physical processes. An optimized laser-target system, with basic metrology, can be designed to be much more compact. Radioactive shielding is just for the few litre volume around the focal spot (including primary target, accelerated ion yield and secondary target, as shown in Figure 13.6) plus the radiochemistry. So, shielding requirements can become compatible with installation restrictions in an existing hospital.

Cost is also an issue. Cyclotron technology is more or less stable, except for the relatively recent introduction of modern superconducting magnets that allow an important reduction in the cyclotron diameter. Price reductions are not foreseen, since a reduction in size can correspond to an increase of complexity. This situation is not yet reached for lasers. Laser technology is still evolving very rapidly and prices are reducing; in near term, with the introduction of diode pumped systems and later, in longer term by exploiting high power fibre laser technology. But the laser purchase price is not the only issue. Maintenance and failure rate are also very important. When the cyclotron is on, nobody can stay inside the shielded bunker. Even when the cyclotron is off, it is necessary to wait – for hours in some cases – for the deactivation of the area before entry [IAEA 2009]. When the laser is on, however, a technician can be in the laser room, so predictive maintenance and repair could be much simpler. When the laser is on the technician can be nearby (with protective goggles, of course). And when the laser is off, one can enter without any protection. Lasers are modular and can have redundant key components, so light can be set to propagate though one or some other amplifier chain (i.e. path) by simply adjusting one mirror, something impossible with conventional accelerators.

Laser-based systems have also their own unique issues and challenges. One big difference cited between conventionally-accelerated protons and laser-accelerated protons is the energy spectrum (or energy spread). With a conventional accelerator, the energy spectrum corresponds to a relatively narrow Gaussian velocity distribution. For a laser-driven proton accelerator the energy spectrum is much broader (this can be influenced greatly by the chosen acceleration mechanism(s)). In certain cases, it opens the possibility of having competing reactions. For example the foreseen ^{15}N(p,n)^{15}O reaction channel might not be the only one favoured by the incoming proton energies; ^{15}N(p,2p)^{14}C could be also possible. A lot of work remains to be done with laser-acceleration, with proton velocity distributions and

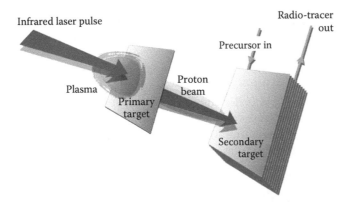

FIGURE 13.6 Schematic representation of the radionuclide generation process using laser-acceleration of protons. The laser pulse hits the primary target and generates by TNSA (target normal sheath acceleration) a proton bunch. Protons subsequently hit a secondary target where the nuclear reaction occurs. The distance between the primary and the secondary target can be reduced and also the geometry (not necessarily flat as seen in the figure) to use the complete uncollimated proton beam.

with the control of the acceleration mechanisms. Furthermore, the development of optimized high repetition targets is critically needed; particularly for high repetition-rate applications. Also target structures can affect the final energy spectrum of accelerated electrons that drive ion acceleration [Zigler 2013] and can significantly enhance it in certain cases.

To end this chapter, it is worth assessing the efficiency of the laser. To calculate the ratio of laser energy (i.e. laser pulse energy) to radiotracer production, we suggest a very simplistic calculation. Consider a laser delivering 1 J of pulse energy conveniently focused on target –this is a very conservative value, there are many lasers that can deliver 10 to 1000 J 'on target' [Danson 2015]. Consider also if 1% of this energy goes to 15 MeV protons. This proton yield efficiency is realistic and probably higher in a real case. Less realistic is the assumed zero energy spread. From this assumption, we can estimate the number of ^{18}F nuclei that are activated. A 15 MeV proton has 2.4×10^{-12} J of kinetic energy. So, 10 mJ of 15 MeV protons corresponds to about 4.2×10^9 protons from a 1 J laser pulse. On the other end, we have seen before (see Figure 13.4) that 10 MeV protons correspond to an activity yield of about 1000 GBq/C for the ^{18}O(p,n)^{18}F reaction. This means that one coulomb of protons produces ^{18}F with an activity of 1000 GBq (assuming approximately instantaneous irradiation and activity just at the end of irradiation). One coulomb of protons is 6.2×10^{18} protons, so an activity of 1 Bq requires 6.2×10^6 protons. This is an interesting number that relates the activity of the radioisotope produced with the number of protons necessary to produce it. Observe that this is a very inefficient result. For a half-life of 110 min, one Bq of activity corresponds (according to the relation previously indicated) to about 9500 nuclei of ^{18}F. This indicates that to generate one unstable ^{18}F nucleus we need about 700 protons accelerated to exactly 15 MeV. A typical dose is of order 10^8 Bq, so within this example this corresponds to 6.2×10^{14} protons. At 4.2×10^9 protons accelerated to 15 MeV per laser shot, this activity level requires more than 10^5 shots. At a 1 kHz laser repetition rate this number of shots can accumulate within 2 min. This result is acceptable for ^{18}F, because is only a fraction of the half-life. At a 1 Hz repetition rate the required number of shots and therefore the accumulation time is too long. This is the main drawback of laser acceleration. Of course, in this simplistic calculation we have not considered that the efficiency can be much higher and that the whole spectrum of protons accelerated to a few MeV or more will contribute to the nuclear reaction, but compared to the required increase of repetition rate from state of the art 1–10 Hz to at least 1 kHz, this is negligible. It is clear that laser repetition rates have to increase and acceleration mechanisms have to be optimized. Also, a novel concept is to coherently combine multiple laser beams [Mourou 2013] and this opens new possibilities for the target design, much of them unexplored so far.

Finally, it is important to point out that the first generation of laser-based radiotracers will, in some way, copy the conventional approach simply substituting the cyclotron for a laser plus a primary target. A beam of accelerated protons emerging from the primary target impacts the secondary target with the same geometry than in conventional systems (although the proton beam features are expected to be very different as mentioned above).

One important problem that needs to be addressed is the current lack of quantitative general models that describe laser-induced proton acceleration (including particle yields). The most widely accepted mechanism is TNSA, as indicated before, but its scaling laws and validity range are not completely understood. At the same time, other mechanisms have been observed and analyzed, such as BOA (breakout afterburner) and RPA (radiation pressure acceleration), The reader is again referred to Chapter 5 for details on laser-driven acceleration mechanisms. TNSA is interesting because the proton beam is emitted with certain order and within a divergent cone and therefore can be manipulated.

Moreover, laser systems are also amenable to other architectures, such as having the two targets together. For example, in a cyclotron the proton beam needs to be extracted from the cyclotron and delivered to the primary target. In the case of the laser, there is room for considering new designs avoiding proton transport and therefore simplifying the device.

Significant changes in the chemical post processing part can also be foreseen [Bengström 2003], such as ion chemistry and laser-assisted reactions. However, detailed considerations on this topic are likely to start only after laser-driven ion acceleration has proven to become more viable over medium term future.

Moreover, it is clear that a laser pulse couples much better to an electron than to a proton due to the different mass. Electrons are thus the ideal particles to be accelerated by lasers to significant energies (see Chapter 2 for discussion of laser-driven electron acceleration). However very few reactions triggered by electrons are interesting in the context of radioisotope production. There is an indirect possibility to use laser-accelerated electrons, to generate gammas [Lefebvre 2003] in the multi-MeV range. Those photons can trigger the production reactions. One such example is the reaction $^{12}C(\gamma, n)^{11}C$, with a threshold energy of 18.7 MeV but with a cross section of only 4 mb [Karpuz 2015]. This reaction requires multi-MeV photon sources, which are not yet routinely available, laser-driven betatron sources or a similar demonstrated technology potential (see also Chapter 4 for a review laser-driven photon sources).

Laser technology may simplify the production of short-lived PET radiotracers. This can in turn foster the development of new drugs (conventional drugs marked with radiotracers) [Lappin 2003]. Also, the whole target/projectile concept might require revision when considering the laser case. Conventional accelerators work with projectiles travelling very long distances (say metres) over linear, helical or circular trajectories and therefore need to be very well-controlled and collimated with a low-energy spread. All those concepts have to be very significantly revisited for considering competitiveness of laser production of radionuclides in the medical sector. Laser accelerators, or better laser systems that activate nuclei may have the projectile and the precursor at very short distances and use acceleration of projectiles with a large, angular spread. This applies uniquely to radiotracer production while other applications may require more controlled conditions closer to the conventional accelerator output parameters.

Cyclotrons were invented by E. O. Lawrence in 1934, so the technology matured over 80 years. Chirped pulse amplification, being the only driver for laser-driven particle acceleration (the RF source, so to say), was invented in 1985 by G. Mourou and D. Strickland, and the evolution of laser technology based on this idea is still in progress. Such lasers just now start to enter industrial applications. Low repetition rate systems, (i.e. about one shot per minute), present a high shot-to-shot instability because the amplified pulse is the result of slightly different conditions for each shot. In principle, newly designed 10 Hz or 100 Hz systems must have reduced fluctuations and the proposed kHz multi-TW systems must have a much better shot-to-shot stability, but again it is a long way for laser-driven particles sources to achieve the stability of conventional accelerators. To convey an optimistic message to the laser community it is important to remark that those instabilities, broad energy spectra and large angular divergence are acceptable for radionuclide production and specific designs can be foreseen.

13.8 Conclusions

High-powered lasers have potential for complementing conventional cyclotrons in radionuclide production in certain areas, particularly for very short-lived isotopes that need to be generated close to the patient. Modern microfluidic chemical units to synthesize the drug just before administration, can play a fundamental role [Benström 2003]. Availability of single-dose, very short-lived tracers can open new possibilities for personalized drugs and dynamical molecular imaging.

Currently, lasers cannot compete with cyclotron mass production. However, lasers represent potential advantages in final cost, technology evolution and radiation shielding which can make them relevant at small facilities in local hospitals and also for preclinical uses with animals [Levin 2007]. Laser-based systems may have a degree of versatility that is not possible with conventional sources and even not yet fully ascertainable. By considering the laser as a driver and enabling tool, there will be room to analyze complementary approaches and projectiles such as electrons or gamma photons within the same community that develops laser-driven particle acceleration.

References for Chapter 13

Alam, S.U. et al. 2015. "Current status of few mode fiber amplifiers for spatial division multiplexed transmission" *Journal of Optics* **45**: 275–284.

Amato, E. et al. 2016. "Study of the production yields of ^{18}F, ^{11}C, ^{13}N and ^{15}O positron emitters from plasma-laser proton sources at ELI-Beamlines for labeling of PET radiopharmaceuticals" *Nuclear Instruments and Methods in Physics Research Section A: Accelerators, Spectrometers, Detectors and Associated Equipment* **811**: 1–5.

Bergström, M. et al. 2003. "Positron emission tomography microdosing: A new concept with application in tracer and early clinical drug development" *European Journal of Clinical Pharmacology* **59**: 357–366.

Conard, E.M. 1994. "High intensity accelerator for a wide range of applications" *Nuclear Instruments and Methods in Physics Research Section A: Accelerators, Spectrometers, Detectors and Associated Equipment* **353**: 1–5.

Conti, S.P. and Kaushik, A. Eds. 2016. *PET-CT A Case-Based Approach.* Springer Berlin.

Danson, C. et al. 2015. "Petawatt class lasers worldwide" *High Power Laser Science and Engineering* **3**: 3 (14).

Ditmire, T. et al. 1999. "Nuclear fusion from explosions of femtosecond laser-heated deuterium clusters" *Nature* **398**: 489–492.

Fanti, S. et al. 2010. *PET-CT beyond FDG. A quick Guide to Image Interpretation.* Springer Berlin.

Firouzbakht, M.L. et al. 1998. "Yield measurements for the ^{11}B(p,n)^{11}C and the ^{10}B(d,n)^{11}C nuclear reactions" *Nuclear medicine and biology* **25**: 161–164.

Gales, S. 2015. "Laser driven nuclear science and applications: The need of high efficiency, high power and high repetition rate Laser beams" *The European Physical Journal Special Topics* **224**: 2631–2637.

Gerstner, E. 2007. "Laser physics: Extreme light" *Nature* **446**: 16–18.

d'Humières, E. et al. 2013. "Optimization of laser-target interaction for proton acceleration" *Physics of Plasmas* **20**: 023103 (8).

Harada, R. et al. "^{18}F-THK5351: A novel PET radiotracer for imaging neurofibrillary pathology in Alzheimer disease" *Journal of Nuclear Medicine* **57**: 208–214.

IAEA 2001. "Charged particle cross-section database for medical radioisotope production: Diagnostic radioisotopes and monitor reactions" TECDOC 1211. Vienna.

IAEA 2008. "Cyclotron Produced Radionuclides: Principles and Practice" Technical Report Series 465. Vienna.

IAEA 2009. "Cyclotron Produced Radionuclides: Guidelines for Setting Up a Facility" Technical Report Series 471. Vienna.

IAEA 2010. "Production of Long Lived Parent Radionuclides for Generators: ^{68}Ge, ^{82}Sr, ^{90}Sr and ^{188}W" Vienna.

Jauregui, C. et al. 2013. "High-power fiber lasers" *Nature Photonics* **7**: 861–867.

Kane, K.S. 1988 *Introductory Nuclear Physics.* John Wiley.

Karpuz, N. et al. 2015. "Calculation of Gamma Strength Functions for Photonucleon Reactions" *Acta Physica Polonica A* **128** B414–416.

Knapp, F.R. and Dash, A. 2016. *Radiopharmaceuticals for Therapy.* Springer New Delhi.

Kowalsky, R.J. and Falen, S.W. 2004. *Radiopharmaceuticals in Nuclear Pharmacy and Medicine.* American Pharmacists Asociation Washington D.C.

Lappin, G. and Garner, C. 2003. "Big Physics, small doses: The use of AMS and PET in human microdosing of development drugs" *Nature Reviews, Drug Discovery* **2**: 233.

Ledingham, K.W.D. 2005. "Laser induced nuclear physics and applications" *Nuclear Physics A* **752**: 633–644.

Lefebvre, E. et al. 2003. "Electron and photon production from relativistic laser–plasma interactions" *Nuclear Fusion* **43**: 629–633.

Lefebvre, E. et al. 2006. "Numerical simulation of isotope production for positron emission tomography with laser-accelerated ions" *Journal of Applied Physics* **100**: 113308 (5).

Levin, C.S. and Zaidi, H. 2007. "Current trends in preclinical PET system design" *PET Clinics* **2**: 125–160.

Long, N., Wong, W.T. and Immergut, E.H. 2014. *The Chemistry of Molecular Imaging.* John Wiley and Sons.

Ma, C.-M. Ch. and Lomax, T. 2013. *Proton and Carbon Iontherapy.* CRC Press, Boca Raton.

Macchi, A., Borghesi, M. and Passoni, M. 2013. "Ion acceleration by superintense laser-plasma interaction" *Reviews of Modern Physics* **85**: 751–793.

Michael, P.C. et al. 2013. "Test of a conduction-cooled, prototype, superconducting magnet for a compact cyclotron" *Applied Superconductivity, IEEE Transactions on*, **23**: 4100304 (25).

Mourou, G. et al. 2013. "The future is fiber accelerators" *Nature Photonics* **7**: 258–261.

O'Donoghue, J.A. and Wheldon, T.E. 1996. "Targeted radiotherapy using Auger electron emitters" *Physics in Medicine and Biology* **41**: 1973–1992.

Paans, A.M.J. and Vaalburg, W. 2000. "Positron emission tomography in drug development and drug evaluation" *Current Pharmaceutical Design* **6**: 1583–1591.

Paetz gen. Schieck, H. 2014. *Nuclear Reactions.* Springer Lecture Notes in Physics, Springer Berlin.

Qaim, S.M. 2001. "Nuclear data relevant to the production and application of diagnostic radionuclides" *Radiochimica Acta* **89**: 223–245.

Rahman, O. et al. 2015. "Synthesis of ((^{11}C) carbonyl) raclopride and a comparison with ((^{11}C) methyl) raclopride in a monkey PET study" *Nuclear Medicine and Biology* **42**: 893–898.

Ruth, T.J., 2009. "The use of Radiotracers in the life sciences" *Reports on Progress in Physics* **72**: 016701 (23).

Saha, G.B. 2010. *Fundamentals of Nuclear Pharmacy.* Springer New York.

Saha, G.B. 2016. *Basics of PET Imaging. Physics, Chemistry and Regulations.* Springer New York.

Spencer, I. et al. 2001. "Laser generation of proton beams for the production of short-lived positron emitting radioisotopes" *Nuclear Instruments and Methods in Physics Research Section B: Beam Interactions with Materials and Atoms* **183**: 449–458.

Turner, J.E. and Kelsey, C.A. 1995. *Atoms, Radiation, and Radiation Protection.* John Wiley and Sons New York.

Valk, P.E. et al. 2003. *Positron Emission Tomography: Basic Science and Clinical Practice.* Springer London.

Vallabhajosula, S. 2009. *Molecular Imaging. Radiopharmaceuticals for PET and SPECT.* Springer Berlin.

Velikyan, I. 2014. "Prospective of ^{68}Ga-Radiopharmaceutical Development" *Theranostics* **4**: 47–80.

Welch, M.J. and Redvanly C.S. 2003. *Handbook of Radiopharmaceuticals. Radiochemistry and Application.* John Wiley.

Wolbarst, A.B. and Hendee W.R. 2006. "Evolving and experimental technologies in medical imaging" *Radiology* **238**: 16–39.

Zigler, A. et al. 2013. "Enhanced proton acceleration by an ultrashort laser interaction with structured dynamic plasma targets" *Physical Review Letters* **110**: 215004 (5).

14

Space Radiation and Its Biological Effects

Günther Reitz

Christine E.
Hellweg

14.1 Introduction: Background and Driving Forces

Galactic cosmic rays (GCR) and solar cosmic rays (SCR) are the primary sources of the radiation field in space. GCR have their origin in previous cataclysmic astronomical events, such as supernova explosions. They contain all particles from hydrogen to uranium, all fully ionized, with energies up to 10^{11} GeV and with low fluxes of around a few particles $cm^{-2} s^{-1}$. SCR are solar energetic particles (SEP) originating from solar flares or shock waves driven by a coronal mass ejections (CME), as well as in corotating interaction regions in the interplanetary medium. Of most concern for human spaceflight are solar proton events (SPEs) a subgroup of SEP events. Such events consist mostly of protons, with a small percentage of heavy ions with energies up to several GeV. The duration of the events last from hours to days, in which fluences of up to 10^{11} particles cm^{-2}, can be reached.

GCR expose biological systems and humans to quite low mean dose rates not leading to acute radiation effects, but the exposure causes an additional risk of carcinogenesis, degenerative tissue effects, damages to the central nervous system (CNS) and accelerated aging. Exposures to SPE also contribute to these listed risks but, in addition, may cause acute effects, like performance degradation, acute radiation sickness or even death. To prevent exposures due to solar particles, spacecraft can be equipped with shelters, but shielding is not effective against GCR. While the effects of high doses are relatively well-investigated and reasonably understood, the biological effects caused by heavy ions are poorly understood. Mitigation of the effects of heavy ions is one of the most important challenges to be solved for the exploration of the solar system. This chapter describes the space radiation field, its biological effects and measures that are necessary to limit the exposures in space missions to acceptable levels. Laser-driven ion acceleration can provide an advanced tool to study heavy ion effects in order to close gaps of knowledge.

Humans leaving the Earth to visit other planets of our solar system are therefore exposed to the full spectrum of GCR and SCR. Especially critical are exposures during interplanetary travel and during extravehicular activities (EVA), as well as during excursions on planetary surfaces. Habitats on planets can be constructed finally as thick shelters providing sufficient shielding thickness to reduce radiation levels to those on Earth. In interplanetary travel, only limited shielding can be provided by the spacecraft structures; the same holds true for spacesuits. Since GCR consists of particles up to extremely high energies, their penetration ability ranges up to several hundred centimetres in aluminium. Penetrating the material, GCR loses energy by two ways, by the Coulomb interaction and by nuclear collisions. In Coulomb interaction, most of the energy is transferred to the electrons of the target atoms. In nuclear collisions, target and projectile fragments are produced. One particle can cause a whole cascade of secondary radiation. Having the same energy as the projectiles, projectile fragments penetrate even deeper. Excited target atoms may explode thereby producing low energy particles, mostly protons, neutrons and alpha particles, which deposit high amounts of energy in small volumes.

In contrast to GCR, short-term exposures by SCR particles may become so high, that life-threatening exposures become possible. The good message is that the energies of the particles are moderate and thus shelters can provide sufficient protection in interplanetary travel. Exposures especially during EVA and planetary excursions have to be prevented through adequate forecasting and mission planning.

Radiation effects to be taken into consideration are both early and late effects. Early radiation effects, such as acute radiation sickness, manifest in minutes to days. They occur only through elevated exposures by SCR and may cause performance degradation, life span reduction or even inflight death. GCR cannot cause early effects due to the low fluence rates, but increases the risk of late effects, from which carcinogenesis and damage to the CNS are of major concern. Late effects manifest within years or even decades.

While radiation effects at high doses are reasonably well understood, this is not the case for low doses of heavy ions. The exposure is not homogenous, since some cells in the human body are exposed and some are not. Heavy ions deposit a large amount of energy along their path with a core of about 50 nm around it followed by a very rapid decrease of the energy transfer with increasing distance from the path centre due to so-called delta electrons. One ion hitting the nucleus of a cell may cause cell death [Bücker 1974, Reitz 1995].

The main target is DNA in the cell nucleus, where a single heavy ion can cause complex damages consisting of single-strand breaks (SSBs), base damages and double-strand breaks (DSBs), while sparsely ionizing particles like protons mostly produce single-strand breaks. Whereas SSBs can be repaired easily by the cell, DSBs are difficult to repair and are potentially mutagenic and lethal (see Section 14.4.1). Although the flux of heavy ions is low, they represent the major contributor to radiation risk in space missions. Since the biological effects of heavy ions are poorly understood, they represent the main source of uncertainties in risk estimates.

Preparing a mission to Mars requires an extensive ground-based radiobiological program to achieve the reduction of uncertainties in radiation risk assessment and the development of appropriate countermeasures. The major facility is the NASA Space Radiation Laboratory (NSRL), established by NASA at the Brookhaven National Laboratory on Long Island, NY, followed by the Heavy Ion Medical Accelerator (HIMAC) in Chiba, Japan. The NASA Space Radiation Health Program is by far the most intense undertaking towards mitigating the radiobiological risk in explorative missions (see website http://humanresearchroadmap.nasa.gov/). A strong collaboration between Europe and NASA was envisaged in the IBER Program (see website https://www.gsi.de/work/forschung/biophysik/esa_iber .htm), the first European Accelerator-based Research Program (EARP) [Durante 2007] performed at GSI (GSI Helmholtzzentrum für Schwerionenforschung). Having supported the performance of 25 experiments, the program was terminated due to the ongoing construction of the new facility, Facility for Antiproton and ION Research (FAIR) at GSI. New beam times have become recently available as part of a 'FAIR phase 0' transitional program allowing to implement a new IBER research program for the period of 2018 until 2022. After FAIR becomes available, the IBER program will need to be continued and intensified with a facility which has the potential to become the leading facility worldwide for space radiobiology. A further facility, but only providing relatively low heavy ion energies up to 100 MeV/n, is the Grand Accélérateur National de Ions Lourds (GANIL) in Caen, France, with its biological facility Laboratoire d' Accueil et de Recherche avec les Ions Accélérés (LARIA). More facilities around the world exist, but most of them suffer either providing the appropriate particles and energies or from the provision of the infrastructures, such as biological laboratories and animal facilities. The accelerators listed are heavily used in nuclear physics research and medical treatment, therefore only limited beam time can be provided to address all open issues present for explorative missions. In space all particles and energies are available, therefore the simulation of the space radiation field at accelerators is a challenging time-consuming task [Norbury 2016].

Laser-driven ion beams represent an excellent opportunity to increase the available beam time for studies of the biological effects of heavy ions of energies up to 100 MeV/n. Laser facilities additionally have the potential advantage of being physically much smaller than particle accelerators. The broad distribution of laser-driven ion species and their kinetic energies, if properly controllable, could become a complementary feature to available accelerator sources.

A detailed description of the needed research topics can be also found in the HUMEX study (see website, http://www.dlr.de/me/PortalData/25/Resources/dokumente/publikationen/humex-summary.pdf) and the THESEUS study (see website, http://www.esf.org/fileadmin/Public_documents/Publications /Cluster3_web.pdf). The following paragraphs will give a more detailed description of the radiation field in space and its specifics followed by a description of biological effects and the studies needed for the radiation risk reduction in explorative missions.

14.2 Radiation Fields in Space

The following description includes only radiation which is relevant to the radiation exposure of humans in space. Such types of radiation need energies that are sufficiently high to enter into the human body and the ability to ionize the atoms or molecules of the body. The radiation field inside the solar system is dominated by our sun and consists of a complex mixture of particles of solar and galactic origin. All particles and energies are present in the field [Wilson 1978]. Three main sources can be identified: GCR originated outside the solar system, SCR emitted from the solar surface and trapped radiation (TR) caused by interaction of GCR and SCR with planetary magnetic fields.

There are long-term and short-term temporal variations caused by the activity of the sun. Although different cycles have been already identified [Braun 2005], the most important for the radiation field modulation is the Schwabe cycle, with a mean duration of approximately 11 years, in which the solar activity passes from one solar minimum activity period through a phase of maximum activity and back to the next minimum. One continuous measure (since 1755) that describes this activity is the Zürich sunspot number [Hathaway 2002].

The sun continuously emits particle radiation, primarily electrons and protons: the solar wind. Although the particle energies are so low that they would be absorbed within some micrometres of tissue, the solar wind is the major driver which determines the extent of the radiation exposure to GCR and TR. The solar wind fills out the complete solar system and carries a magnetic field which represents a heliocentric potential against which the GCR particles have to work when entering the heliosphere. During maximum solar activity, the GCR exposure becomes lower, since the lower energy particles are no longer capable of entering the solar system. The solar modulation of GCR causes a shift of the energy for maximum intensity from solar minimum to solar maximum by about 500 MeV, with a strong attenuation of the flux during solar maximum below energies of about 30 GeV/nucleon [Badhwar 1997]. GCR are composed mainly of protons to about 85% and alpha particles to about 14%, with the remainder of about 1% being heavier nuclei [Mewaldt 1988].

In addition, there are also episodes of extreme solar activity in which sudden releases of magnetic energy may occur in the corona previously stored in non-potential (stressed) fields, which result in coronal mass ejections (CMEs) and solar flares. CMEs are huge clouds of magnetized plasma expelled from the solar corona into interplanetary space [Chen 2001]. Solar flares manifest themselves as enhanced radiation across the whole electromagnetic spectrum due to heating and interaction of high-energetic particles with the solar atmosphere. Both CMEs and solar flares may be associated with SPEs, which pose a radiation hazard to astronauts. Such extreme SPEs may last from hours to several days. The high magnetic fields embedded in such events lead to a reduction of GCR, the so-called Forbush decreases. On the other hand, a huge amount of particles, mostly protons with a small varying amount of heavy ions with energies up to several GeV, can be released. There is no long-term forecast for such events. In the case of solar flares where the ejected particles immediately spiral around the interplanetary magnetic field lines, the travel time to Earth may be in the order of 30 min to 1 day, whereas in case of CME's the particles may reach the Earth after 1–4 days. The onset of a SPE can be recorded by satellites, but neither the flux nor the energy spectra can be predicted. Energy spectra of candidate SPEs demonstrate an enormous variability of energy spectra as well as for the range of intensities observed [Wilson 1997].

The third radiation source is the radiation belt around the Earth. The radiation belt is a product of the interaction of GCR and SCR with the Earth's magnetic field and the atmosphere [Allkofer 1975]. The radiation belts extend over a region from 200 km to about 75,000 km around the geomagnetic equator. The radiation belts consist of electrons and protons, and some heavier ions trapped in the magnetic field. Different processes contribute to the build-up of the radiation belts. The inner belt is mainly formed by protons and electrons as products from decaying neutrons, produced in interaction of cosmic particles with the atoms of the atmosphere. The outer belt is filled mainly by solar particles, which are injected during magnetic disturbances caused by particle events hitting the Earth. In each zone, the charged particles spiral around the geomagnetic field lines and are reflected back between the magnetic poles which act as mirrors. At the same time, because of their charge, electrons drift eastwards, while protons and heavy ions drift westward. The electrons reach energies up to 7 MeV, the protons up to 700 MeV with a fluence maximum near 100 MeV. As with GCR, the TR is modulated by the solar activity. With increasing activity, the proton flux decreases, while the electron flux increases. The proton flux decreases because the Earth's atmosphere expands leading to a loss of protons due to increased interaction with the molecules of the atmosphere. The electron contribution increases because more electrons are fed into the radiation belts through enhanced solar activity. The electron fluxes in the outer zone show diurnal variations up to a factor of 16, but short-term variations may raise the mean flux by 2 or 3 orders of magnitude. The flux in the centre of the inner belt is quite stable. However, at the lower edge of the belt where the International Space Station (ISS) is operating, electron and proton fluxes may vary by a factor of 5.

Charged particles from GCR and SCR have to penetrate the Earth's magnetic field to reach an orbiting spacecraft. For each point inside the magnetosphere, there exists a cut-off rigidity which is proportional to the magnetic field component perpendicular to the direction of the particle motion. To reach this point, the particle rigidity (particle momentum divided by its charge) must exceed the local geomagnetic rigidity cut-off. The rigidity is a function of the geomagnetic latitude and increases from high latitude towards the equator. At the poles, the cut-off rigidity is zero, so particles of any energy can enter. As a result, the GCR flux decreases from the poles to the equator.

14.3 Radiation Fields Inside Spacecraft, on Planetary Surfaces and in the Human Body

The radiation field inside a spacecraft or on planetary surfaces differs significantly from the primary field in space. In penetrating spacecraft walls, radiation is partly absorbed, and secondary radiation is produced by scattering and nuclear interaction. Due to the non-homogenous distribution of the equipment inside spacecraft, the internal radiation field depends on the location inside the spacecraft. On planetary surfaces, the field depends on the existence of a magnetic field, the thickness and the composition of the atmosphere and the surface material. This field is further modified when entering the human body. The analysis of the particle transport inside the body is a prerequisite when determining absorbed doses and the risk of early and late effects.

Several radiation transport codes are in place to allow the calculation of particle fluxes and dose rates behind defined shielding. The Boltzmann transport equations for atomic and nuclear collisions may be solved by numerical and analytic techniques [Wilson 1993] or by Monte Carlo techniques [Ferrari 2001, Agostinelli 2003, Allison 2006, Townsend 2005, Waters 2002, Niita 2006]. The calculation of doses inside the human body additionally needs to employ computational phantoms, which represent the anatomy of the human bodies or parts of them [ICRP 2009, Yucker 1990, ICRP 2013].

The codes have to be validated by measurements with area monitors inside and outside spacecraft at well-selected locations with known shielding distributions and by individual measurements with personal dosimeters. The organ-absorbed dose cannot be measured in the human body, therefore human phantoms equipped with radiation monitors are used to provide the essential confidence for radiation transport calculations [Reitz 2009].

Usually, the absorbed dose is the basic quantity to measure radiation exposure. The absorbed dose is the quotient of the energy deposited by ionizing radiation within an elemental volume to the mass of matter in that volume. The absorbed dose is measured in units of Gray (Gy) (1 Gy = 1 J/kg [= 100 rad]). Whereas different radiation qualities produce the same type of effect, the magnitude of the effect per unit of absorbed dose can be different. For radiation protection, the quality factor (Q) was introduced in order to account for the different relative biological efficiencies (RBEs) of different types of ionizing radiation. This factor depends not only on appropriate biological data, but primarily it reflects a judgement concerning the importance of the biological endpoints. Q is defined in dependence of the linear energy transfer (LET). It is set to 1 for LET < 10 keV/μm, in the LET range from 10–100 keV/μm to 0.32* LET–2.2 keV/μm and for LET > 100 keV/μm to 300/(LET)$^{0.5}$ [ICRP 1991]. The dose equivalent at a point is defined as the product of absorbed dose and Q. The quantity effective dose equivalent [ICRP 2013] is the sum of all organ doses which can be calculated as product of Q and absorbed dose by additionally applying tissue weighting factors as defined in ICRP103 [ICRP 2007]. The effective dose equivalent is given in units of Sievert (Sv) (1 Sv = 1 J/kg [=100 rem]).

During space missions, astronauts are constantly exposed to GCR. This chronic whole-body exposure with single energetic particles (electrons, protons, α-particles and heavy ions) results in an inhomogeneous dose distribution in the body. The flux is quite low and counts to about 4 protons, 0.4 Helium ions and 0.04 heavier particles per cm^{-2} s^{-1}. Therefore, the cells that are hit by a single energetic heavy ion are exposed to a high dose, and others that are not hit receive no dose at all. The traversing ions produce 'ionization channels' in the hit cells and the biological effect depends on the extent of damage to sensitive biomolecules. Assuming an ionization channel of 10 μm in diameter for iron as example doses can be as high as 100 kGy. The damage of DNA DSBs has been visualized by immunofluorescence in a human skin fibroblast exposed to 2 Gy of ionizing radiation (iron ions); DNA DSBs are located along a particle trajectory [Durante 2006]. Therefore, the radiation protection system which is based on mean absorbed dose is just an approximated surrogate and consequently providing the highest uncertainty in risk estimates.

In interplanetary missions, in addition to this chronic, in average low-dose exposure at low-dose rate, an acute whole-body exposure to a high radiation dose at a high dose rate can occur during a SPE. For high doses, the above-mentioned radiation protection quantities should not be used, instead the RBE weighted mean absorbed dose in the organ or tissue should be applied, which is given in Gray equivalents (Gy-Eq). The best estimates of RBE values for the different radiations are given in in ICRP Publication 58 [ICRP 1989]. For protons >2 MeV, a RBE of 1.5 is recommended.

In low earth orbit (LEO), where the ISS is operating, the SPE exposures are low to moderate due to a quite efficient geomagnetic shielding, while the contribution of the charged particles trapped in the Earth's magnetic field (Van Allen belts) to the total dose cannot be neglected [Facius 2006].

Radiation monitors were flown as part of all human missions and on numerous satellites. In the early times of spaceflight, passive detectors which integrate dose and flux over time dominated. Later on, active devices complemented the information on the radiation environment by adding temporal information. Good summaries of measurements can be found elsewhere [Badhwar 2002, Benton 2001, Reitz 2005, Reitz 2008, Casolino 2002, Dudkin 1995, Dudkin 1996]. A summary of effective dose equivalent rates for various missions, such as the Apollo, Skylab and Shuttle missions, MIR and early ISS can be find in a compilation by Cucinotta [2003]. The Apollo deep space missions (to the Moon) show mean dose rates up to 3 mSv/day. In these missions, the nuclei of the GCR dominate the exposure. Shuttle missions with inclination between 40° and 60° show very nicely the influence of the solar cycle and maximum effective equivalent dose rates up to 1 mSv/day. Up to 4 mSv/day were observed in low inclination missions (around 28° inclination) above 600 km altitude which are a result of the higher contribution of the protons from the radiation belt. The ISS has been operating since 1998 at an altitude around 400 km and an inclination of about 51.5°; the first long-duration human stay was from November 2000 to March 2001. Environmental measurements on board the ISS in June 2016 show equivalent dose rates up to 0.7 mSv/day. For comparison, the mean exposure rate on Earth is about 0.0066 mSv/day. Measurements on the ISS are permanently done by a suite of instruments; results can be found on the web unde http://wrmiss.org. Figure 14.1 shows particle count rates recently measured with a silicon detector of 6.93 cm² detection area on the ISS. The peaks represent proton counts when the ISS is passing through the South Atlantic Anomaly, a region where the radiation belt comes closer to the Earth surface. The other counts are GCR particles; the maximum flux is in the polar regions and the lowest is at the equator.

A summary of the measurements of the Mars Science Lab Radiation Assessment Detector (MSL–RAD) is given in Table 14.1. The table presents the particle flux, dose and dose equivalent rates with which we are faced in interplanetary travel or on the surface of Mars.

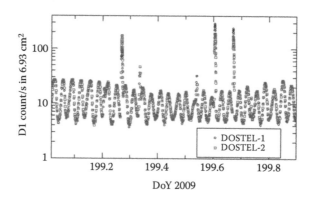

FIGURE 14.1 Count rate measurements onboard the ISS of the two DOSTEL telescopes as part of the DOSIS experiment (Principal Investigator: Dr. Guenther Reitz, German Aerospace Center). The GCR count rates are due to the changing magnetic cut-off conditions during the ISS orbit with dependence of latitude; GCR count rates are at a minimum at the magnetic equator and a maximum at the highest latitude position; the spikes occur when the ISS is crossing the South Atlantic Anomaly of the radiation belt; the peaks are mainly due to protons. (DoY = Day of Year)

TABLE 14.1 Flux, Dose and Dose Equivalent Rate Measurements of GCR During the Cruise to and on the Surface of Mars by the MSL–RAD

RAD Measurements	Mars Surface	Cruise	Units
Differential charged particle flux density	0.26	0.64	Particles cm^{-2} s^{-1} sr^{-1}
Dose Rate	0.21 ± 0.04	0.46 ± 0.06	mGy d^{-1}
Average Quality Factor <Q>	3.05 ± 0.26	3.82 ± 0.30	(dimensionless)
Dose Equivalent Rate	0.64 ± 0.06	1.84 ± 0.30	mSv d^{-1}

TABLE 14.2 Calculation of Dose Equivalent for Two NASA Reference Missions Based on MSL–RAD Measurements Compared to Calculation of Effective Dose Equivalent Taking into Account a Human Body in a Space Suit on the Mars Transfer Vehicle and on Mars

Mission Phase	Dose Equivalent/Effective Dose Equivalent (Sv)	Notes
Astronaut Career Limits[a]	~0.60–1.20	Depends on age, gender, etc.
Cruise to Mars (180 days)	~0.33/0.22	Near Solar Max
Mars Surface Mission (600 days)	~0.38/0.24	Thin habitat shielding
Mars Surface Mission (300 days)	~0.19/0.12	Thin habitat shielding
Return to Earth (180 days)	~0.33/0.22	Near Solar Max
Total Mission Dose Equivalent (300 days on Mars)	**~0.85/0.56**	
Total Mission Dose Equivalent (600 days on Mars)	**~1.04/0.68**	

[a] Lower and upper values for NASA career limits based on 3% excess risk of fatal cancer.

The dose equivalent rates during cruise, compared to that on ISS, are close to a factor of three higher. On the Mars surface, we experience about the same exposure as on the ISS; and the exposure is due to GCR only, as in the cruise, where on ISS the radiation field is a mixture of protons from the radiation belts and GCR.

The total number of particle hits per cell nucleus (diameter 11.3 μm, area 100 μm²) in the human body at average skin depth while exposed during a stay on Mars at the mean surface level during one year based on measurements of the Martian Radiation Environment experiment (MARIE) arrived at about 40 protons, 1 helium ion and 0.1 ions with charge greater than two [Saganti 2002].

Until now only doses of GCR including 'belt' protons in LEO are reported. The doses contributed by SPE in interplanetary space vary from zero to a couple of Gy depending on the fluence, the shape of the energy spectra and provided shielding thickness. Because the onset of the SPE can be monitored (no forecast of fluence or shape of energy spectra is possible), an inflight interplanetary crew can at least move to shelter. About 40 g/cm² of material (corresponding for example to 15 cm aluminum) is considered to be sufficient shielding. This shelter thickness allows the crew a safe cruise even at times of high solar activity. Excursions on planetary surfaces present of course a higher risk, especially on celestial bodies without atmospheres. The Mars atmosphere with a shielding thickness of 20 g/cm² of CO_2 provides already a complete protection against most SPEs. Housings adequately constructed with 2–3 m wall thickness of regolith or located deep enough in caves can guarantee exposure levels close to the exposure levels on the Earth surface.

Table 14.2 shows the calculation of dose equivalent and effective dose equivalent based on MSL–RAD measurements compared with calculations of GEANT4 transport model with a human phantom implemented in the MSL mission.

14.4 Effects in Humans

Astronauts can 'see' the exposure to GCR and trapped radiation in form of light flashes when they close their eyes [Pinsky 1975]. This phenomenon occurs after about a 15-min dark adaptation and is explained by highly energetic heavy ions that interact directly or indirectly (via Cherenkov radiation *in vitreous humour*) with the retina [Bidoli 2002], or possibly with the optic nerve or the visual centres in the brain, producing a visual sensation.

The extent, type and onset of radiation effects observed in humans depends on the dose, the dose rate, the radiation quality and the individual sensitivity of the exposed human or animal (see Figure 14.2). Acute radiation effects in humans appear quite soon after exposure to a high dose in a short period of

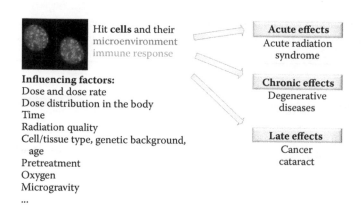

Hit **cells** and their
microenvironment
immune response

Acute effects
Acute radiation
syndrome

Chronic effects
Degenerative
diseases

Late effects
Cancer
cataract

Influencing factors:
Dose and dose rate
Dose distribution in the body
Time
Radiation quality
Cell/tissue type, genetic background,
  age
Pretreatment
Oxygen
Microgravity
...

FIGURE 14.2 Acute, chronic and late effects of radiation exposure. Exposure to ionizing radiation produces DNA double strand breaks (pink foci) in the cell nucleus (blue) that can be visualized by immunofluorescence staining of the phosphorylated histone H2AX (γH2AX). The outcome depends on many influencing factors, including the microenvironment of the hit cells and a local or systemic immune response.

time (minutes to a few days). Late effects, such as cancer, can manifest after years or even decades after exposure. Their probability for occurrence is proportional to the exposure level.

14.4.1 Basic Mechanisms: DNA Damage and Cellular Radiation Response

Ionizing radiation reacts with cellular macromolecules by direct ionization or indirectly via radiolysis of the cellular water. Thereby, generated oxygen radicals attack the DNA molecule and disrupt the ribose-phosphate backbone leading to single- and double-strand breaks, and induce base damage, loss of bases, or DNA–DNA and DNA-protein crosslinks. Such damage can disturb the information carried by the affected DNA molecule. DNA damage is regarded as the central element in cell killing by ionizing radiation, whereby DSB are considered to be the most cytotoxic damages [Jackson 2009]. These early events after radiation can pave the way to disease by activation of the cellular radiation response.

The cellular response to radiation is predominantly a DNA damage response (DDR) (see Figure 14.3) that detects lesions, signals their presence and promotes their repair [Jackson 2009]. This signal transduction pathway involves multiple sensors for different types of DNA lesions, transducer molecules and a variety of effector molecules and enzymes for repair. As radiation can simultaneously activate multiple pathways [Dent 2003], the DDR results in potentially cell-protective (cell cycle arrest, DNA repair, survival) or cell-altering (misdifferentiation, premature differentiation, senescence, mutations) or even destructive responses (different types of cell death) [Khanna 2001, NASA 2004, Ohnishi 2002].

14.4.2 Non-Targeted Effects

In addition to the direct effects on DNA as the main biological target, non-targeted effects such as the bystander effect and the adaptive response can influence the outcome after exposure to ionization radiation and are therefore important for space radiation risk assessment. Bystander cells are not directly exposed or traversed by radiation, but are in the neighbourhood of a cell that had been hit or are incubated with medium from irradiated cells and show responses [Mothersill 2004].

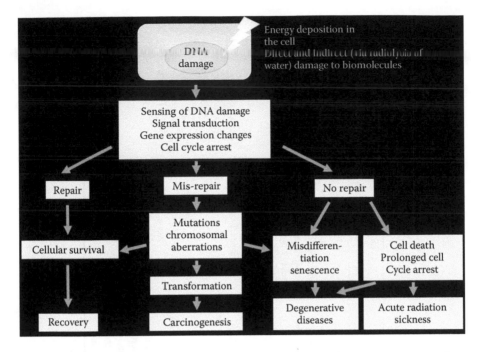

FIGURE 14.3 The DNA damage response as central element of the biological reaction to ionizing radiation exposure and the different outcomes.

14.4.3 Acute Effects

SPE events pose the risk of acute high doses and high dose rate exposures. A large SPE, such as the one that occurred in August 1972, might result in absorbed dose rates as high as 1.4 Gy/h [Parsons 2000]. Exposure to an acute single dose elicits the acute radiation syndrome (Table 14.3), also known as radiation sickness, with symptoms depending on absorbed total radiation dose, type of radiation, the dose distribution in the body and the individual radiation sensitivity [Cronkite 1964, Drouet 2010]. A rapid onset of nausea, vomiting and malaise (absorbed dose > 0.5 Gy) characterizes the prodromal stage, which is followed by a nearly symptom-free latent phase of weeks to days, depending on dose. The prodromal stage can be life threatening, if vomiting occurs while the astronaut performs an EVA in a space suit.

The acute radiation syndrome affects the tissues with rapid turnover in the first place. Cellular fate is determined by decision between cell death and survival. Accordingly, the effects on the individual result from depletion of already differentiated cells by cell death mechanisms, and from failing replacement by stem cells due to cell cycle block and cell death. Only cells overcoming the cell cycle block are able to replace radiation-damaged tissue to regain its normal function.

The bone marrow is one of the most radiosensitive organs in the body. In the manifest phase of radiation sickness, depression of its function (bone marrow or hematopoietic syndrome) appears in humans exposed to doses of 0.7–4 Gy. These doses are lower compared to those eliciting the gastrointestinal tract syndrome (5–12 Gy) or the CNS syndrome (>20 Gy). The lethal dose for 50% of the exposed human individuals within 30 days was estimated to be 3–4 Gy ($LD_{50/30}$).

Recent studies suggest that in all three subsyndromes, exacerbated innate immune responses play a major role in pathogenesis [Drouet 2010, Jacob 2010, van der Meeren 2005]. Epithelial and endothelial cells are suggested as sources of the proinflammatory cytokines in the acute radiation syndrome

TABLE 14.3 The Acute Radiation Syndrome

Dose Range (Gy)	Sub-Syndrome or Stage	Affected Cells/Tissues	Symptoms	Therapeutic Approach	References
>0.5	Prodromal stage	Neurons in the CNS, e.g. in the caudal medulla, *Nervus vagus*	Rapid onset of nausea, vomiting and malaise	Antiemetics	[Makale 1993, Tofilon 2000, Marquette 2003]
0.5–4	Hematopoietic	Bone marrow stem cells including megakary-oblasts	Progressive lymphopenia, immune system suppression, susceptibility to infections, thrombocytopenia with increased bleeding tendency, early granulocytosis followed by progressive granulocytopenia, anaemia	Antibiotics, antifungal drugs, isolation, electrolyte and platelet or blood transfusions, Granulocyte Colony-Stimulating Factor (G–CSF), allogenic bone marrow transplantation	[Cronkite 1964, Dainiak 2003, Chao 2007, Drouet 2010]
5–12	Gastrointestinal	Differentiating cell compartment – stem cells in crypts; endothelial cells, activation of the innate immune system	Diarrhoea, sepsis, multiple organ failure; >10–12 Gy: 100% lethality		[Drouet 2010, Jacob 2010, Gaugler 2005, Singer 2004, Gourlemon 2005]
>20–30	Neurovascular	Mature functioning cells: neurons and endothelial tissue	Loss of coordination, confusion, convulsions, eventually coma, and signs of the bone marrow and gastrointestinal syndromes, vomiting, dehydration, cerebral oedema, injuries to the nerves, death occurs within few days or hours	None	[Jacob 2010]

[van der Meeren 2005]. Immune suppression is the predominant feature of the bone marrow syndrome. Due to the longer lifespan of erythrocytes, anaemia develops later (within 2–6 weeks) than the lymphopenia. Death usually occurs from sepsis at 30–60 days after radiation exposure, if the patient cannot be carried through the critical period of the possibly reversible aplastic state of the bone marrow [Cronkite 1964]. Cytological abnormalities (multipolar mitosis, micronuclei, mitotic bridges, binucleated cells) and a reduced mitotic index were observed in human bone marrow cells (e.g. erythroblasts) during the first days after accidental sublethal whole body γ-radiation exposure, and cytological abnormalities persist at a lower frequency for years after the accident [Fliedner 1964].

The gastrointestinal tract syndrome appears after a short latent period after whole-body irradiation with 5–12 Gy. It is due to loss of the intestinal epithelium after massive cell death and lack of mitotic success in the intestinal epithelium crypts, and injury to the fine vasculature of the submucosa. Invading bacteria produce local and systemic inflammation and sepsis with multiple organ failure [Drouet 2010]. Enhanced by the concomitant immunosuppression, death occurs between 3 and 10 days post-exposure. If some crypt cells survive, they will regenerate functional crypts and repopulate the villi.

The onset of the CNS syndrome occurs after a very short latent period of several hours to days after exposure to very high acute doses (>20 Gy in humans). Cells with highest differentiation status and lowest reproductive capacity, the neurons, are affected. The prognosis is fatal.

In spaceflight, exposure to mostly protons during a large SPE could result in the acute radiation syndrome. Experiments with mice have shown, that 2 Gy of protons delivered within 36 hrs induce anaemia and immunosuppression with decreased numbers of erythrocytes, lymphocytes, monocytes and granulocytes, and a decreased relative spleen mass [Gridley 2008]. In mice exposed to 3 Gy of protons, a strong immune depression was observed already 12 hrs after exposure, with the nadir on day 4 [Kajioka 2000]. Increased susceptibility to infections, cancer induction by promotion of initiated cells and disorders of the immune system, such as autoimmunity or hypersensitivity, are possible consequences of the immunosuppression.

Therapeutic approaches for the acute radiation syndrome were tested in non-human primates; dogs, mice, rats and pigs [McVittie 2005, Donnadieu-Claraz 1999] and in accidently irradiated humans. The efficacy of other growth factors, such as keratinocyte growth factor (KGF) and combinations of different growth factors (stem cell factor [SCF], nerve growth factor [NGF], erythropoietin, pegylated growth factors), antioxidants (e.g. N-acetyl cysteine) and anti-inflammatory approaches (e.g. inhibitors of cyclooxygenase-2- COX-2-, anti-IL-antibodies, curcumin, Ghrelin) are currently under investigation [Drouet 2010, Jacob 2010, Neal 2003].

In case of high-dose SPEs, astronauts have to be protected by a warning system allowing movement to a radiation shelter.

14.4.4 Chronic and Late Effects

Delayed or chronic effects of radiation exposure include cancer and non-cancer effects such as degenerative diseases. Cancer induction is a highly relevant and life threatening late effect of radiation exposure by radiation accidents, in atomic bomb survivors, for human spaceflight or radiotherapy. Ionizing radiation can definitively provoke tumour initiation, due to its DNA damaging effects. The role of radiation in tumour promotion and progression is less clear.

In radiotherapy, induction of secondary tumours by low radiation doses to the tumour-surrounding tissue occurs stochastically. The probability of tumour induction in the dose range below 1 Gy was derived from the cancer incidence (solid tumours, leukaemia) in atomic bomb survivors who were exposed at high dose rates [Pierce 1996, Pierce 2000]. With increasing survival times of patients after cancer radiotherapy, there is growing concern for the risk of secondary cancer, especially in children who are inherently more radiosensitive [Baskar 2010]. These epidemiological data were derived almost entirely from low–LET radiation exposures.

Human data for cancer risk by GCR, especially high–LET heavy ions, are absent, and radiobiological data are limited. Currently, the Q for biological weighting of the GCR dose is governed by results derived from experiments with mice revealing a very high RBE for induction of tumours of the Harderian gland by heavy ions (LET > 100 keV/μm) [Fry 1983, Alpen 1994]. During a long-term space exploration mission, astronauts accumulate high exposures to GCR. A causal relationship between the frequency of chromosomal aberrations in peripheral blood lymphocytes and cancer allows estimation of cancer risk by appropriate analysis of the aberration frequency [Durante 2001]. A significant increase in chromosomal aberrations was observed in astronauts after long duration flights [George 2001]; data on cancer in astronauts are limited and lack statistical confidence, due to small group size [Longnecker 2004]. By the chromosomal aberration risk assessment technique, the cancer risk after a long-term mission on the former Russian space station Mir was estimated to be increased by 20–30% compared to a control group [Durante 2001]. For a Mars mission, the uncertainties in cancer risk projections were estimated to be 400–600% [Cucinotta 2001a]. The latter projection involved many biological and physical factors, each adding a different range of uncertainty attributed to limited data and knowledge to the overall uncertainty. Several factors contribute to the overall uncertainty from which radiation quality is the major contributor followed by dose rate effects, others are risk transfer across populations, dosimetry, errors in human data and microgravity [Durante 2008].

Different organs can be affected by radiation-induced degenerative diseases, e.g. the eye lens, and the central nervous, digestive, respiratory, endocrine, immune system or cardiovascular system. The eye lens is a quite radiosensitive tissue, as it has no mechanism for removal of dead or damaged cells and is not connected to the blood stream. During life, lens epithelial cells are continuously produced by mitosis in the germination zone of the lens and they differentiate into transparent lens fibres. Cataracts are detectable changes in the normally transparent lens of the eye; for example, resulting from disturbed differentiation into lens fibres. The threshold for cataract formation after protracted radiation exposure was as low as 2 Gy for sparsely ionizing radiation. Other investigators suggest even a threshold of 100–300 mGy for cataract formation, or dismiss any threshold [Worgul 1999]. Concluding from studies with rabbits exposed to neon or argon ions, Lett supposed that astronauts could experience late radiation effects one or more decades after a long-term space mission beyond LEO [Lett 1980]. In mice and rats exposed to energetic heavy ions, it was shown that high–LET ionizing radiation is especially effective in cataract formation, even at doses below 2 Gy [Worgul 1989, Worgul 1993, Hall 2006]. A lower threshold for cataracts of about 100 mGy was published recently by Blakely et al. [2010]. Late cataractogenesis in Rhesus monkeys (*Macaca mulatta* with median life span near 24 years) was observed about 20 years after the exposure to protons of different energies, and in rabbits after exposure to energetic iron ions [Cox 1992, Lett 1991]. Besides animal experiments, data from radiotherapy patients are used for space radiation risk assessment concerning cataract formation or cancer induction in astronauts [Lett 1994, Wu 1994, Blakely 1994]. Up to now, lens opacities are the only proven space radiation late effect in astronauts [Cucinotta 2001b, Rastegar 2002] and they occur with higher frequency in astronauts exposed to higher proportions of high–LET radiation [Cucinotta 2001b]. As cataract surgery can restore the vision, it is not warranted to consider cataractogenesis as a major critical health risk in short to medium-term spaceflight, in view of an overall mission death risk of a few percent. During long-term missions that can occur over several years, the expression time for cataracts has to be considered. During such missions, a cataract might develop before return to Earth and thereby it represents a mission risk.

During an interplanetary mission, astronauts can accumulate a considerable number of hits at critical sites of the brain [Craven 1994, Curtis 1998]. This gives rise to concern about CNS effects that could reduce the neurocognitive performance of the crew during the mission. Possible mechanisms are induction of DNA damage and cell death in neurons and neural precursor cells, disturbance of neurogenesis, electrophysiological activity, synaptic plasticity and neurotransmitters, neuroinflammation and oxidative stress [Rola 2005, Limoli 2007, Rola 2008, Vlkolinsky 2007, Machida 2010, Tseng 2014, Rivera 2013, DeCarolis 2014, Baulch 2015]. Animal experiments revealed learning, memory and executive function deficits after heavy ion exposure [Rabin 2000, Rabin 2005, Britten 2012, Lonart 2012].

In cancer therapy patients, an elevated risk for cardiovascular disease has been known for decades. Recent epidemiological data show that also low dose exposure can result in cardiovascular effects [Bhattacharya 2015] and has to be considered in space radiation risk assessment [Boerma 2015]. Ionization radiation can affect endothelial cells or cardiomyocytes, for example [Soucy 2011, Colemann 2015].

14.5 Ground-Based Research

The limited knowledge of the biological effects from exposure to heavy charged particles is an ongoing concern in human spaceflight [Cucinotta 2011]. Open questions in space radiation biology are the difference in mechanisms for high vs. low LET radiation, the extent of the dose rate effect and the radiation quality dependence, the extrapolation of cell culture and animal data to humans and the individual radiation sensitivity (radiosensitivity syndromes represent the extremes; polymorphisms), all contributing to the radiation risk assessment. The atomic bomb survivor data serve as the main reference for radiation risk estimates of space radiation, but the exposure conditions for the survivors and astronauts are quite different (Table 14.4). The exposure profile in space differs quite substantially from those which usually apply for reference experiments or data sources on Earth; especially where fluence rates and time scales are concerned.

Limited data are available for non-cancer late effects, e.g. cataracts and cardiovascular diseases. A quantitative risk assessment of CNS effects is completely missing [Cucinotta 2013].

Understanding acute effects requires data for whole body and partial body irradiation. Accidental exposures to SPEs result in inhomogeneous irradiation of the body, depending on the slope of the energy spectra and exposure situations. There are no data on dose rate dependency for acute effects. Modification of the radiation response through the spaceflight environment is totally neglected in all agency radiation protection programs.

Experimental data on molecular and cellular mechanisms in cell, tissue and organ cultures are needed, and animal studies are required for studies on cancerogenesis, degenerative diseases and countermeasure testing. It needs to be proven whether or not the prodromal syndromes are different in humans and animal systems. As astronauts in a space suit or in a planetary habitat live in an atmosphere different of that on Earth, effects of altered oxygen and carbon dioxide concentrations, leading to hypoxia or hyperoxia or hyperkapnia certainly cannot be neglected.

Furthermore, countermeasures have to be provided, such as shielding and biomedical approaches (diet and radioprotectors). Effective biomedical countermeasures beyond amifostine and antioxidants

TABLE 14.4 Differences in Exposure Conditions of the Atomic Bomb Survivors (Reference for Radiation Risks) and Astronauts in Space

Exposure Parameters	Atomic Bomb Exposure[a]	Space Exposure
Dose rate, duration	Instantaneous/acute, seconds	GCR: chronic/protracted, months–years; SPE: acute, hours–days
Radiation quality	Low–LET gamma rays	Most complex mixture of disparate radiation qualities conceivable
Body distribution	Essentially homogeneous	Depending on external shielding, inner organs largely homogeneous
Exposed population	Common age, gender, health status distribution of a city population	Selected for physical/psychic prowess, age, and health
Environmental conditions	Normal terrestrial	Microgravity, confined ecological system, artificial/technical components dominating

[a] Database: atomic bomb survivor epidemiology.

have to be developed by ground-based research with appropriate models. Most recently it was found that CDDO-Me (bardoxolone methyl) showed a dose reduction factor (DRF) of almost 2, which is very promising. Clinical tests indicate that it shows minimal toxicity in animal studies [Eskiocak 2010]. Ion sources are also needed for detector development and characterization to allow for ground and space borne intercomparisons. Improved detectors are needed to provide field dosimetry.

14.6 Summary

This chapter is intended to provide the major dosimetric features that are needed to draw a picture of which particle types, fluences and doses are required to perform reliable space radiation simulation. Because this picture is far from being complete, please refer for more detailed information to cited literature. The chapter also lists major biological questions that must be solved in order to reduce the uncertainties of risk estimates. The simulation of the radiation field in space is rather difficult due to the mixture of particles needed and by the rather long exposure times required to accumulate significant dose in the human body or the cell system under investigation. A recent paper describes in detail the needs for a GCR simulation [Norbury 2016]. The state of laser-driven heavy ion beams is currently far from providing all particle energies that are present in space. However, it is valuable to employ such laboratory sources providing a range of particles and kinetic energies for space research. The laser-driven case for energetic electron and ion production is in an early stage of development and we can expect useful advancements of this technology. We envisage protons and heavy ions with kinetic energies up to 100 MeV/n for this new laser-driven technique (see Chapters 2, 4 and 5 (Part I) for electron and ion acceleration status with high-powered lasers). For example, many biological investigations require a high LET, which means that energies of 20–50 MeV/n (with carbon for example) can be efficient for acquiring missing radiobiological data. Although laser-driven particle fluences can be very high and therefore suited to provide high doses in short time, space reference systems typically require low particle fluences continuously distributed over an extended period. Therefore, it can be a challenging task to reduce flux levels using collimator optics. A notable advantage of laser-driven ion sources is there emission characteristics, i.e. the broad angular divergence which reduces fluence quickly over a certain distance from the source and enables large field irradiation in a potentially compact setup. A second key advantage might be the intrinsically broad energy distributions of laser-driven ion sources – so, one can consider irradiating samples with a broad energy spectrum which comes closer to the space environment than an ion beam of well-defined energy at a conventional accelerator. A third key advantage might be the intrinsic capability of the laser-driver to generate multiple synchronous beams with different particles, again better simulating the real space environment if properly controllable. Those potential advantages usher optimism that laser-driven particle sources can expand the experimental toolbox for radiation biological studies. In particular, they can help to close the current gap for facility beam time for radiation biologists to answer the open questions and to reduce the extraordinarily large uncertainties in radiation biology for explorative missions.

Acknowledgements

This publication was supported by the Open Program for Research, Development and Education, MEYS, under the project 'CRREAT', "Reg. No. CZ.02.1.01/0.0/0.0/15_003/0000481."

References for Chapter 14

Agostinelli, S., Allison, J. and Amako, K. 2003. "GEANT4 - A simulation toolkit" *Nuclear Instruments and Methods in Physics Research A* **506**: 250–303.

Allison, J., Amako, K., Apostolakis, J. et al. 2006. "Geant4 developments and applications" *IEEE Transactions on Nuclear Science* **53**: 270–278.

Allkofer, O.C. 1975. "Introduction to Cosmic Radiation" Buchreihe der Atomenergie, Band 10, Karl Thiemig Verlag München.

Alpen, E.L., Powers-Risius, P., Curtis, S.B., DeGuzman, R. and Fry, R.J. 1994. "Fluence-based relative biological effectiveness for charged particle carcinogenesis in mouse Harderian gland" *Adv. Space Res.* **14**: 573–581.

Badhwar, G.D. 1997. "Deep space radiation sources, models, and environmental uncertainty" Wilson, J.W., Miller, J., Konradi, A. and Cucinotta, F.A., Eds. Shielding Strategies for Human Space Exploration. *NASA Conference Publication* **3360**: Chapter 2, 17–28, Washington DC, USA, NASA Headquarters.

Badhwar, G.D., Atwell, W., Reitz, G., Beaujean, R. and Heinrich, W. 2002. "Radiation measurements on the Mir Orbital Station" *Radiation Measurements* **35**: 393–422.

Baskar, R. 2010. "Emerging role of radiation induced bystander effects: Cell communications and carcinogenesis." *Genome Integr.* **1**:13.

Baulch, J.E., Craver B.M., Tran, K.K., Yu, L., Chmielewski, N., Allen, B.D. and Limoli, C.L. 2015. "Persistent oxidative stress in human neural stem cells exposed to low fluences of charged particles" *Redox Biol.* **5**: 24–32.

Benton, E.R. and Benton, E.V. 2001. "Space radiation dosimetry in low-Earth orbit and beyond" *Nuclear Instruments and methods in Physics Research B* **184**: 255–294.

Bhattacharya, S. and Asaithamby, A. 2016. "Ionizing radiation and heart risks" *Semin. Cell Dev. Biol.* **58**: 14–25.

Bidoli, V., Casolino, M., de Pascale, M.P. et al. 2002. "The Sileye-3/Alteino experiment for the study of light flashes, radiation environment and astronaut brain activity on board the International Space Station" *J Radiat. Res.* **43** Suppl: 47–52.

Blakely, E.A., Daftari, I.K., Meecham, W.J., Alonso, L.C., Collier, J.M., Kroll, S.M., Gillette, E.L., Lee, A.C., Lett, J.T., Cox, A.B., Castro, J.R. and Char, D.H. 1994. "Helium-ion-induced human cataractogenesis" *Adv. Space Res.* **14**: 501–505.

Blakely, E.A., Kleiman, N.J., Neriishi, K., Chodick, G., Chylack, L.T., Cucinotta, F.A., Minamoto, A., Nakashima, E., Kumagami, T., Kitaoka, T., Kanamoto, T., Kiuchi, Y., Chang, P., Fujii, N. and Shore, R.E. 2010. "Radiation cataractogenesis: epidemiology and biology" *Radiat. Res.* **173**: 709–717.

Boerma, M., Nelson, G.A., Sridharan, V., Mao, X.W., Koturbash, I. and Hauer-Jensen, M. 2015. "Space radiation and cardiovascular disease risk" *World J. Cardiol.* **7**: 882–888.

Braun, H., Christl, M., Rahmstorf, S., Ganopolski, A., Mangini, A., Kubatzki, C., Roth, K., Kromer, B. 2005. "Possible solar origin of the 1470-year glacial climate cycle demonstrated in a coupled model" *Nature* **438**: 208–211.

Britten, R.A., Davis, L.K., Johnson, A.M., Keeney, S., Siegel, A., Sanford, L.D., Singletary, S.J. and Lonart, G. 2012. "Low (20 cGy) doses of 1 GeV/u (56) Fe—Particle radiation lead to a persistent reduction in the spatial learning ability of rats" *Radiat. Res.* **177**: 146–151.

Bücker, H. 1974. "The Biostack Experiments I and II aboard Apollo 16 and 17, Life Sci." *Space Res.* **XII**: 43–50.

Casolino, M., Bidoli, V., Furano, G. et al. 2002. "The Sileye-3/Alteino experiment on board the International Space Station" *Nuclear Physics B (Proc. Suppl.)* **113**: 71–78.

Chao, N.J. 2007. "Accidental or intentional exposure to ionizing radiation: Biodosimetry and treatment options" *Exp. Hematol.* **35**: 24–27.

Chen, J. 2001. "Physics of Coronal Mass Ejections: A New Paradigm of Solar Eruptions" *Space Science Reviews* **95**: 165–190.

Coleman, M.A., Sasi, S.P., Onufrak, J., Natarajan, M., Manickam, K., Schwab, J., Muralidharan, S., Peterson, L.E., Alekseyev, Y.O., Yan, X. and Goukassian, D.A. 2015. "Low-dose radiation affects cardiac physiology: Gene networks and molecular signaling in cardiomyocytes" *Am. J. Physiol. Heart Circ. Physiol.* **309**: 1947–1963.

Cox, A.B., Lee, A.C., Williams, G.R. and Let, J.T. 1992. "Late cataractogenesis in primates and lagomorphs after exposure to particulate radiations" *Adv. Space Res.* **12**: 379–384.

Craven, P.A. and Rycroft, M.J. 1994. "Fluxes of galactic iron nuclei and associated HZE secondaries, and resulting radiation doses, in the brain of an astronaut" *Adv. Space Res.* **14**: 873–878.

Cronkite, E.P. 1964. "The Diagnosis, Treatment, and Prognosis of Human Radiation Injury from Whole-Body Exposure" *Ann. N. Y. Acad. Sci.* **114**: 341–355.

Cucinotta, F.A. and Chappell, L.J. 2011. "Updates to astronaut radiation limits: Radiation risks for never-smokers" *Radiat. Res.* **176**: 102–114.

Cucinotta, F.A., Kim, M.H., Chappell, L.J. and Huff, J.L. 2013. "How safe is safe enough? Radiation risk for a human mission to Mars" *PLoS One* **8**: e74988.

Cucinotta, F.A., Manuel, F.K., Jones, J., Iszard, G., Murrey, J., Djojonegro, B. and Wear, M. 2001b. "Space radiation and cataracts in astronauts" *Radiat. Res.* **156**: 460–466.

Cucinotta, F.A., Schimmerling, W., Wilson, J.W., Peterson, L.E., Badhwar, G.D., Saganti, P.B. and Dicello, J.F. 2001a. "Space radiation cancer risks and uncertainties for Mars missions" *Radiat. Res.* **156**: 682–688.

Cucinotta, F.A., Wu, H., Shavers, M.R. and George, K. 2003. "Radiation Dosimetry and Biophysical Models of Space Radiation Effects" *Gravitational Space Biology Bulletin* **16(2)**: 11–18.

Curtis, S.B., Vazquez, M.E., Wilson, J.W., Atwell, W., Kim, M. and Capala, J. 1998. "Cosmic ray hit frequencies in critical sites in the central nervous system" *Adv. Space Res.* **22**: 197–207.

Dainiak, N., Waselenko, J.K., Armitage, J.O., MacVittie, T.J. and Farese, A.M. 2003. "The hematologist and radiation casualties" *Hematology Am. Soc. Hematol. Educ. Program* 473–496. (https://www .ncbi.nlm.nih.gov/pubmed/14633795)

DeCarolis, N.A., Rivera, P.D., Ahn, F., Amaral, W.Z., LeBlanc, J.A., Malhotra, S., Shih, H.Y., Petrik, D., Melvin, N., Chen, B.P. and Eisch, A.J. 2014. "56Fe Particle Exposure Results in a Long-Lasting Increase in a Cellular Index of Genomic Instability and Transiently Suppresses Adult Hippocampal Neurogenesis in Vivo" *Life Sci. Space Res.* **2**: 70–79.

Dent, P., Yacoub, A., Contessa, J., Caron, R., Amorino, G., Valerie, K. and Hagan, M.P. 2003. "Grant S, Schmidt-Ullrich R. Stress and radiation-induced activation of multiple intracellular signaling pathways" *Radiat Res.* **159**: 283–300.

Donnadieu-Claraz, M., Benderitter, M., Joubert, C. and Voisin, P. 1999. "Biochemical indicators of whole-body gamma-radiation effects in the pig" *Int. J Radiat. Biol.* **75**: 165–174.

Drouet, M. and Herodin, F. 2010. "Radiation victim management and the haematologist in the future: time to revisit therapeutic guidelines?" *Int. J. Radiat. Biol.* **86**: 636–648.

Dudkin, V.E., Karpov, O.N., Potapov, Y., Akopova, A.B., Magradze, N.V., Moiseenko, A.A., Melkumyan, L.V. and Rshtuni, S. 1995. "Studying radiation environment on board STS-55 and STS-57 by the method of passive detectors" *Radiation Measurements* **25**: No* 1–4: 483–484.

Dudkin, V.E., Potapov, Y., Akopova, A.B., Melkumyan, L.V., Bogdanov, V.G., Zacharov, V.I., Plyuschev, V.A., Lobakov, A.P. and Lyagyshin, V.I. 1996. "Measurements of fast and intermediate neutron energy spectra on MIR space station in the second half of 1991" *Radiation Measurements* **26** [No. 1]: 535–539.

Durante, M. 2006. Courtesy of Marco Durante.

Durante, M., Bonassi, S., George, K. and Cucinotta, F.A. 2001. "Risk estimation based on chromosomal aberrations induced by radiation" *Radiat Res.* **156**: 662–667.

Durante, M. and Cucinotta, F.A. 2008. "Heavy ions carcinogenesis and human space exploration" *Natur. Reviews Cancer* **8**: 465–472.

Durante, M., Kraft, G., O'Neill, P., Reitz, G., Sabatier, L. and Schneider, U. 2007. "Preparatory study of a ground-based space radiobiology program in Europe" *Advances in Space Research* **39**: 1082–1086.

Eskiocak, U., Kim, S.B., Roig, A.I., Kitten, E., Batten, K., Cornelius, C., Zou, Y.S., Wright, W.E. and Shay, J.W. 2010. "CDDO-Me protects against space radiation-induced transformation of human colon epithelial cells" *Radiat Res.* **174**: 27–36.

Facius, R. and Reitz, G. 2006. "Space weather impacts on space radiation protection" Bothmer V., Daglis I.A., Eds. *Space Weather-Physics and Effects*. Heidelberg: Springer, 289–353. ISBN:3-540-23907-3.

Ferrari, A., Ranft, J. and Sala, P.R. 2001. "The FLUKA radiation transport code and its use for space problems" Ciria, R., Cucinotta, F.A., Durante, M. Eds. Proceedings of the 1st Intern. Workshop on Space Radiation Research and 11th Annual NASA Space Radiation Health Investigators' Workshop. *Physica Medica* **17** Suppl. 1: 72–81.

Fliedner, T.M., Andrews, G.A., Cronkite, E.P. and Bond, V.P. 1964. "Early and Late Cytologic Effects of Whole Body Irradiation on Human Marrow" *Blood* **23**: 471–487.

Fry, R.J., Powers-Risius, P., Alpen, E.L., Ainsworth, E.J. and Ullrich, R.L. 1983. "High-LET radiation carcinogenesis" *Adv. Space Res.* **3**: 241–248.

Gaugler, M.H., Vereycken-Holler, V., Squiban, C., Vandamme M., Vozenin-Brotons, M.C., and Benderitter, M. 2005. "Pravastatin limits endothelial activation after irradiation and decreases the resulting inflammatory and thrombotic responses" *Radiat. Res.* **163**: 479–487.

George, K., Durante, M., Wu, H., Willingham, V., Badhwar, G. and Cucinotta F.A. 2001. "Chromosome aberrations in the blood lymphocytes of astronauts after space flight" *Radiat. Res.* **156**: 731–738.

Gourmelon, P., Marquette, C., Agay, D., Mathieu, J. and Clarencon, D. 2005. "Involvement of the central nervous system in radiation-induced multi-organ dysfunction and/or failure" *BJR Suppl.* **27**: 62–68.

Gridley, D.S., Rizvi, A., Luo-Owen, X., Makinde, A.Y., Coutrakon, G.B., Koss, P., Slater, J.M. and Pecaut, M.J. 2008. "Variable hematopoietic responses to acute photons, protons and simulated solar particle event protons" *In Vivo* **22**: 159–169.

Hall, E.J., Worgul, B.V., Smilenov, L., Elliston, C.D. and Brenner, D.J. 2006. "The relative biological effectiveness of densely ionizing heavy-ion radiation for inducing ocular cataracts in wild type versus mice heterozygous for the ATM gene" *Radiat. Environ. Biophys.* **45**: 99–104.

Hathaway, D.H., Wilson, R.M. and Reichmann, E.J. 2002. "Group sunspot numbers: Sunspot cycle characteristics" *Solar Physics* **211**: 357–370.

ICRP, 1989. "RBE for deterministic effects" ICRP Publication 58. *Ann. ICRP* **20**(4).

ICRP, 1991. "The 1990 Recommendations of the International Commission on Radiological Protection" ICRP Publication 60 *Ann. ICRP* **21**.

ICRP, 2007. "The 2007 Recommendations of the International Commission on Radiological Protection" ICRP Publication 103 *Ann. ICRP* **37**.

ICRP, 2009. "Adult reference computational phantoms" ICRP Publication 110 *Ann. ICRP* **39** (2).

ICRP, 2013. "Assessment of radiation exposure of astronauts in space" ICRP Publication 123 *Ann. ICRP* **42**(4).

Jackson, S.P. and Bartek, J. 2009. "The DNA-damage response in human biology and disease" *Nature* **461**: 1071–1078.

Jacob, A., Shah, K.G., Wu, R. and Wang, P. 2010. "Ghrelin as a novel therapy for radiation combined injury" *Mol. Med.* **16**: 137–143.

Kajioka, E.H., Andres, M.L., Li, J., Mao, X.W., Moyers, M.F., Nelson, G.A., Slater, J.M. and Gridley, D.S. 2000. "Acute effects of whole-body proton irradiation on the immune system of the mouse" *Radiat. Res.* **153**: 587–594.

Khanna, K.K., Lavin, M.F., Jackson, S.P. and Mulhern, T.D. 2001. "ATM, a central controller of cellular responses to DNA damage" *Cell Death Differ.* **8**: 1052–1065.

Lett, J.T., Cox, A.B., Keng, P.C., Lee, A.C., Su, C.M. and Bergtold, D.S. 1980. "Late degeneration in rabbit tissues after irradiation by heavy ions" *Life Sci. Space Res.* **18**: 131–142.

Lett, J.T., Lee, A.C. and Cox, A.B. 1991. "Late cataractogenesis in rhesus monkeys irradiated with protons and radiogenic cataract in other species" *Radiat. Res.* **126**: 147–156.

Lett, J.T., Lee, A.C. and Cox, A.B. 1994. "Risks of radiation cataracts from interplanetary space missions" *Acta. Astronaut* **32**: 739–748.

Limoli, C.L., Giedzinski, E., Baure, J., Rola, R. and Fike, J.R. 2007. "Redox changes induced in hippocampal precursor cells by heavy ion irradiation" *Radiat. Environ. Biophys.* **46**: 167–72.

Lonart, G., Parris, B., Johnson, A.M., Miles, S., Sanford, L.D., Singletary, S.J. and Britten, R.A. 2012. "Executive function in rats is impaired by low (20 cGy) doses of 1 GeV/u (56)Fe particles" *Radiat. Res.* **178**: 289–294.

Longnecker, D.E., Manning, F.J. and Worth, M.H. 2004. "Committee on the Longitudinal Study of Astronaut Health – Review of NASA's Longitudinal Study of Astronaut Health" Washington, D.C., USA, The National Academic Press.

Machida, M., Lonart, G. and Britten, R.A. 2010. "Low (60 cGy) doses of (56)Fe HZE-particle radiation lead to a persistent reduction in the glutamatergic readily releasable pool in rat hippocampal synaptosomes" *Radiat. Res.* **174**: 618–623.

MacVittie, T.J., Farese, A.M. and Jackson, W. III. 2005. "Defining the full therapeutic potential of recombinant growth factors in the post radiation-accident environment: The effect of supportive care plus administration of G-CSF" *Health Phys.* **89**: 546–555.

Makale, M.T. and King, G.L. 1993. "Plasticity of autonomic control of emesis in irradiated ferrets" *Am. J Physiol.* **265**: 1092–1099.

Marquette, C., Linard, C., Galonnier, M., Van Uye, A., Mathieu, J., Gourmelon, P. and Clarencon, D. 2003. "IL-1beta, TNFalpha and IL-6 induction in the rat brain after partial-body irradiation: Role of vagal afferents" *Int. J Radiat. Biol.* **79**: 777–785.

McDonald, F.B. 1965. "Review of Galactic and Solar Cosmic Rays" Reetz, A. Jr., Ed. Second Symposium on Radiation Protection Against Radiation in Space, *NASA SP71*: 19–29.

Mewaldt, R.A. 1988. "Elemental Composition and Energy Spectra of Cosmic Rays", Feynmann, J. and Gabriel, S. B. Eds. Interplanetary Particle Environment. *JPL Publication 88–28*. Pasadena, CA, Jet Propulsion Laboratory.

Mothersill, C. and Seymour, C. 2004. "Radiation-induced bystander effects and adaptive responses—The Yin and Yang of low dose radiobiology?" *Mutat. Res.* **568**: 121–128.

NASA. National Aeronautics and Space Administration. 2004. "The Vision for Space Exploration" *NP-2004-01-334-HQ*. Washington DC, USA, NASA Headquarters.

Neal, R., Matthews, R.H., Lutz, P. and Ercal, N. 2003. "Antioxidant role of N-acetyl cysteine isomers following high dose irradiation" *Free Radic. Biol. Med.* **34**: 689–695.

Niita, K., Sato, T., Iwase, H., Nose, H., Nakashima, H. and Sihver, L. 2006. "PHITS-a particle and heavy ion transport code system" *Radiation Measurements* **41**: 1080–1090.

Norbury, J.W., Schimmerling, W., Slaba T.C. et al. 2016. "Galactic cosmic ray simulation at the NASA Space Radiation Laboratory" 2 *Life Sciences in Space Research* **8**: 38–51.

Ohnishi, T., Takahashi, A. and Ohnishi, K. 2002. "Studies about space radiation promote new fields in radiation biology" *J Radiat. Res.* (Tokyo) **43 Suppl**: 7–12.

Parsons, J.L. and Townsend, L.W. 2000. "Interplanetary crew dose rates for the August 1972 solar particle event" *Radiat. Res.* **153**: 729–733.

Pierce, D.A. and Preston, D.L. 1996. "Risks from low doses of radiation" *Science* **272**: 632–633.

Pierce, D.A. and Preston, D.L. 2000. "Radiation-related cancer risks at low doses among atomic bomb survivors" *Radiat. Res.* **154**: 178–86.

Pinsky, L.S., Osborne, W.Z., Hoffman, R.A., and Bailey, J.V. 1975. "Light flashes observed by astronauts on skylab 4" *Science* **188**: 928–930.

Rabin, B.M., Joseph, J.A., Shukitt-Hale, B. and McEwen, J. 2000. "Effects of exposure to heavy particles on a behavior mediated by the dopaminergic system" *Adv. Space Res.* **25**: 2065–2074.

Rabin, B.M., Shukitt-Hale, B., Joseph, J., and Todd, P. 2005. "Diet as a factor in behavioral radiation protection following exposure to heavy particles" *Gravit. Space Biol. Bull.* **18**: 71–77.

Rastegar, N., Eckar, P. and Mertz, M. 2002. "Radiation-induced cataract in astronauts and cosmonauts" *Graefes Arch Clin Exp Ophthalmol* **240**: 543–547.

Reitz, G. 2008. "Characteristic of the radiation field in low earth orbit and in deep space" *Z. Med. Phys.* **18** (4): 233–243.

Reitz, G., Beaujean, R., Benton, E., Burmeister, S., Dachev, T., Deme, S., Luszik-Bhadra, M. and Olko, P. 2005. "Space radiation measurements on-board ISS – The DOSMAP experiment" *Radiat. Prot. Dosim.* **116**: 374–379.

Reitz, G., Berger, T., Bilski, P. Facius, R. et al. 2009. "Astronaut's organ doses inferred from measurements in a human phantom outside the International Space Station" *Radiat. Res.* **171** (2): 225–235.

Reitz, G., Horneck, G., Facius, R. and Schäfer, M. 1995. "Results of Space experiments" *Radiat. Environ. Biophys.* **34**: 139–144.

Rivera, P.D., Shih, H.Y., LeBlanc, J.A., Cole, M.G., Amaral, W.Z., Mukherjee, S., Zhang, S., Lucero, M.J., DeCarolis, N.A., Chen, B.P. and Eisch, A.J. 2013. "Acute and fractionated exposure to high-LET (56)Fe HZE-particle radiation both result in similar long-term deficits in adult hippocampal neurogenesis" *Radiat. Res.* **180**: 658–667.

Rola, R., Fishman, K., Baure, J., Rosi, S., Lamborn, K.R., Obenaus, A., Nelson, G.A. and Fike, J.R. 2008. "Hippocampal neurogenesis and neuroinflammation after cranial irradiation with (56)Fe particles" *Radiat. Res.* **169**: 626–632.

Rola, R., Sarkissian, V., Obenaus, A., Nelson, G.A., Otsuka, S., Limoli, C.L., Fike, J.R. 2005. "High-LET radiation induces inflammation and persistent changes in markers of hippocampal neurogenesis" *Radiat. Res.* **164**: 556–560.

Saganti, P.B., Cucinotta, F.A., Wilson, J.W. and Schimmerling, W. 2002. "Visualization of particle flux in the human body on the surface of Mars" *J. Radiat. Res.* 43 Suppl., S119–S124.

Singer, M., de Santis, V., Vitale, D., and Jeffcoate, W. 2004. "Multiorgan failure is an adaptive, endocrine-mediated, metabolic response to overwhelming systemic inflammation" *The Lancet* **364**: 545–548.

Soucy, K.G., Lim, H.K., Kim, J.H., Oh, Y., Attarzadeh, D.O., Sevinc, B., Kuo, M.M., Shoukas, A.A., Vazquez, M.E., and Berkowitz, D.E. 2011. "HZE (5)(6)Fe-ion irradiation induces endothelial dysfunction in rat aorta: Role of xanthine oxidase" *Radiat. Res.* **176**: 474–485.

Tofilon, P.J. and Fike, J.R. 2000. "The radioresponse of the central nervous system: A dynamic process" *Radiat. Res.* **153**: 357–370.

Townsend, L.W., Miller, T.M., and Gabriel, T.A. 2005. "HETC radiation transport code development for cosmic ray shielding applications in space" *Radiat. Prot. Dosim.* **116**: 135–139.

Tseng, B.P., Giedzinski E., Izadi, A., Suarez, T., Lan, M.L., Tran, K.K., Acharya, M.M., Nelson, G.A., Raber, J., Parihar, V.K. and Limoli, C.L. 2014. "Functional consequences of radiation-induced oxidative stress in cultured neural stem cells and the brain exposed to charged particle irradiation" *Antioxid. Redox Signal* **20**: 1410–1422.

Van der Meeren, A., Monti P., Vandamme, M., Squiban, C., Wysocki, J. and Griffiths, N. 2005. "Abdominal radiation exposure elicits inflammatory responses and abscopal effects in the lungs of mice" *Radiat. Res.* **163**: 144–152.

Vlkolinsky, R., Krucker, T., Smith, A.L., Lamp, T.C., Nelson, G.A., and Obenaus, A. 2007. "Effects of lipopolysaccharide on 56Fe-particle radiation-induced impairment of synaptic plasticity in the mouse hippocampus" *Radiat. Res.* **168**: 462–470.

Waters, Ls. 2002. "MCNPX™ 2.4.0 Monte Carlo N-Particle Transport Code System for Multiparticle and High Energy Applications" *LA-CP-02-408* Edn. Oak Ridge: U.S. Department of Energy.

Wilson, J.W. 1978. "Environmental Geophysics and SPS Shielding" *LBL-8581*: 33–116. Berkeley, CA, Lawrence Berkeley Laboratory. Lawrence Berkeley Laboratory Report.

Wilson, J.W., Shinn, J.L., Simonsen, L.C., Cucinotta, F.A., Dubey, R.R., Jordan, W.R. et al. 1997. "Exposures to Solar Particle Events in Deep Space Missions" *NASA TP 3668*. Washington DC, USA, NASA Headquarters.

Wilson, J.W., Townsend, L.W., Schimmerling, W., Khandelwal, G.S., Khan, F., Nealy, J.E., Cucinotta, F.A., and Norbury, J.W. 1993. "Transport Methods and Interactions for Space Radiation" Swenberg, C.E., Horneck, G., Stassinopoulos, E.G. Eds. *Biological Effects and Physics of Solar and Galactic*

Cosmic Radiation, Part B. New York and London: NATO ASI Series **Volume 243B**, Plenum Press, New York and London, IBSN 0-306-44418-6.

Worgul, B.V., Brenner, D.J., Medvedovsky, C., Merriam, G.R., Jr. and Huang, Y. 1993. "Accelerated heavy particles and the lens. VII: The cataractogenic potential of 450 MeV/amu iron ions" *Invest Ophthalmol. Vis. Sci.* **34**: 184–193.

Worgul, B.V., Kundiev, Y., Chumak, V.V., Ruban, A., Parkhomenko, G., Vitte, P., Sergienko, N.M, Shore, R., Likhtarev, I.A., Medvedovsky, C., Junk, A.K. 1999. "The Ukranian/American Chernobyl Ocular Study" Junk, A.K., Kundiev, Y., Vitte, P., Worgul, B.V., Eds. Ocular Radiation Risk Assessment in Populations Exposed to Environmental Radiation Contamination. Dordrecht/Boston/London: Kluwer Academic Publishers, 1–12.

Worgul, B.V., Medvedovsky, C., Powers-Risius, P. and Alpen, E. 1989. "Accelerated heavy ions and the lens. IV. Biomicroscopic and cytopathological analyses of the lenses of mice irradiated with 600 MeV/amu 56Fe ions" *Radiat. Res.* **120**: 280–293.

Wu, B., Medvedovsky, C. and Worgul, B.V. 1994. "Non-subjective cataract analysis and its application in space radiation risk assessment" *Adv. Space Res.* **14**: 493–500.

Yucker, W.R. and Huston, SL. 1990. "The computerized anatomical female. Final report" *MDC-6107.* Huntington Beach, CA, McDonnell Douglas Company.

15

Space Irradiation Effects on Solar Cells

Takeshi Ohshima

Mitsuru Imaizumi

15.1 Radiation Effects on Semiconductors

When charged particles pass through materials, their energy decays due to ionizing and non-ionizing effects. As a result, the reliability and lifetime of semiconductor devices decrease. Single event effect (SEE), total ionizing dose (TID) effect and displacement damage effects are known as three major radiation effects in semiconductor devices. Since charges (electron-hole pairs) generated by even one incident charged particle can lead to malfunction of electronic devices, this effect is called SEE. A wide variety of SEE was reported since there are wide varieties of structures and functions of electronic devices. For example, the flip-flop of memory information in large-scale integrated circuits (LSIs) (single event upset and single event transient) [Dodd 2003, Reed 2002, Dodd 2004, Makino 2009] and destructive malfunction of electronic devices triggered by dense charges induced by ion incidence (single event latch-up, single event burnout and single event gate rupture) [Leavy 1969, Titus 2001, Wheatley 2001, Sexton 2003] are known. Charge accumulated in an insulator and/or at an interface between an insulator and a semiconductor of metal-insulator-semiconductor (MIS) structures, such as metal-oxide-semiconductor (MOS) devices, gradually change the characteristics of MIS devices, and this effect is referred to as the TID effect [McWhorter 1986, Ohshima 2001, Oldham 2003]. The change in device characteristics due to TID is larger with increasing radiation dose until, finally, the fatal malfunction of such devices occurs.

Charged particles introduce not only charge but also cause lattice damage in semiconductors, in which atoms at lattice sites are displaced into non-lattice sites (knock-on effects). Primary defects generated by particle irradiation thermally migrate and transform into more stable defects, and as a result, residual defects such as divacancies, vacancy clusters, and vacancy-impurity complexes are formed in semiconductors. Since such defects act as scattering/recombination centres for free carriers, electrical properties of semiconductors are degraded by the generated damage [Walters 1999, Yamaguchi 2001, Imaizumi 2005a, Matsuura 2006]. This is the mechanism of electrical characteristic degradation of semiconductor devices by the displacement damage effect. Since the defect density generated by irradiation increases with increasing particle fluence, the degradation of semiconductor devices due to the displacement damage effect is likewise enhanced, resulting in fatal malfunction.

In space, solar cells installed on satellites are subjected to high energy electrons and protons trapped in the Van Allen belt. As a result, the electrical performance of space solar cells is degraded by crystal damage due to proton and electron irradiation. Thus, displacement damage is the most dominant effect for space solar cells. Therefore, in order to clarify the decrease in power generation of solar cells during space missions, electron and proton irradiation experiments for solar cells using accelerators are conducted at ground-based facilities (the details will be described in the Sections 15.3.2 and 15.3.3).

Next, the degradation mechanism of solar cells due to irradiation is briefly described. Photons of energy higher than the bandgap of a semiconductor can generate photo-induced carriers (electron hole pairs) in the semiconductor. Solar cells generate electric power using such photo-induced carriers. For example, if a solar cell consists of an n-type top layer and a p-type base layer, photo-induced electrons in the base layer mainly contribute electric power generation when they reach the top layer (see Figure 15.1a). Electrons and holes in a p-type region are referred to as minority and majority carriers. The photo-induced electrons (minority carriers) migrate in the base layer by diffusion and recombine with holes (majority carriers) within a certain time. Therefore, the diffusion length of minority carriers is one of the most important parameters for solar cell performance. Since radiation-induced defects generally act as recombination centres, a decrease in minority carrier diffusion length due to irradiation is the most dominant effect in the degradation of solar cell performance (see Figure 15.1b). Tada et al. [Tada 1982] reported that the decrease in minority carrier diffusion length can be described by the following semi-empirical equation:

$$(1/L_\varphi)^2 = (1/L_0)^2 - K_L \varphi, \tag{15.1}$$

where L_0 and L_φ are the minority carrier diffusion length before and after irradiation, respectively. K_L and φ are the damage coefficient of the minority carrier and the particle fluence, respectively. The short circuit current (I_{SC}), which is defined as the value of current at zero voltage under light illumination, for silicon (Si) solar cells irradiated with protons at an energy of 10 MeV for a fluence up to 10^{12}/cm^2 is shown in Figure 15.2. The values of the remaining factor, which are obtained from I_{SC} for irradiated

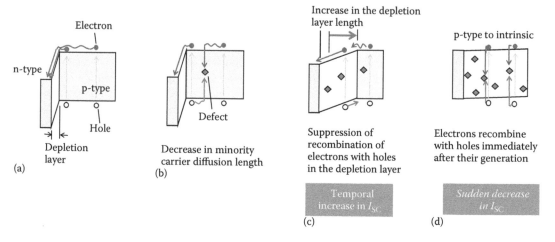

FIGURE 15.1 Schematic band diagram for a solar cell which consists of an n-type top layer and a p-type base layer: (a) before irradiation; (b) during low fluence irradiation; (c) and (d) during high fluence irradiation.

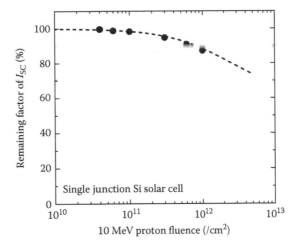

FIGURE 15.2 Remaining factor of short circuit current (I_{SC}) for single junction Si space solar cells as a function of 10 MeV proton fluence up to 10^{12}/cm^2 (circles). Fitting result calculated using the semi-empirical equation is also plotted in the figure (dashed line).

samples normalized by I_{SC} for the pristine sample, are plotted in the figure. The fitting line calculated using the semi-empirical equation well agrees with experimental results. Thus, since I_{SC} is known to be proportional to the minority carrier diffusion length, the degradation behaviour of I_{SC} also follows the degradation behaviour of the minority carrier diffusion length described in Equation 15.1.

A majority carrier removal effect appears in a high fluence region, in addition to the minority carrier diffusion length shortening effect (see Figure 15.1c). Thus, radiation-induced defects act as compensation centres for majority carriers, and as a result, the concentration of majority carriers decreases. In general, this effect can be observed when the concentration of radiation-induced defects is comparable to the concentration of majority carriers. The reduction of majority carriers can be empirically expressed as:

$$p_\varphi = p_0 \exp(-R_C \varphi / p_0), \tag{15.2}$$

where, p_0 and p_φ are majority carrier concentration (typically, order of 10^{15}–10^{17}/cm^3 for solar cells) before and after irradiation, respectively. R_C is the 'carrier removal rate' in units of/cm. The carrier removal effect is clearly observed when the carrier concentration in the base layer of solar cells decreases. Thus, the depletion layer length extends to the base layer by the decrease in majority carrier concentration in the base layer. This indicates that the number of minority carriers arriving at the base layer (top layer) from the top layer (base layer) increases because the recombination of electrons with holes is suppressed by the electric field in the depletion layer. As a result, the value of I_{SC} increases in high fluence regions. However, after this very unique increase in I_{SC}, the value of I_{SC} suddenly drops to almost zero upon further irradiation. This behaviour can be interpreted in terms of the base layer which becomes intrinsic (or type conversion), and as a result, no pn junction exists [Yamaguchi 1996, Ohshima 1996]. (see Figure 15.1d). Figure 15.3 shows the remaining factor of I_{SC} for a space Si solar cells irradiated with 10 MeV protons as a function of fluence up to about 10^{14}/cm^2. In fluence regions above 5×10^{13}/cm^2, the intermittent increase in I_{SC} before suddenly dropping down to 0 is observed. This behaviour can be attributed to the majority carrier removal effect described above.

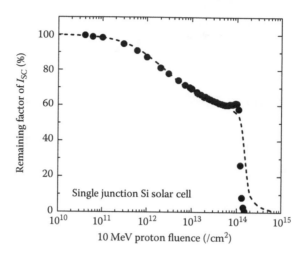

FIGURE 15.3 Remaining factor of short circuit current (I_{SC}) for single junction Si space solar cells as a function of 10 MeV proton fluence up to 10^{14}/cm² (circles). Fitting result in which majority carrier removal as well as decrease in minority carrier diffusion length are considered is also plotted in the figure (dashed line).

15.2 Space Solar Cells

Up to about two decades ago, the majority of space solar cells were silicon cells. However, in response to demands for higher electric power on spacecraft, multi-junction (MJ) solar cells using III–V compound semiconductor materials have become the major space solar cell type due to their higher conversion efficiency and radiation resistance compared to those of silicon solar cells. Recently, the efficiency of triple-junction (3J) space solar cells with a InGaP/GaAs/Ge 3-subcell stack structure could be increased to 30% [Strobl 2012, Fatemi 2013].

In space, particles exhibit an energy distribution. Their flux as well as the energy distribution depends on location in space. Because the performance of solar cells degrades in space due to the radiation exposure we can mitigate radiation degradation with a thin glass sheet (~ 100um) (which is called a cover glass) typically attached to the solar cell surface for radiation shielding. However, high energy protons and electrons still enter solar cells through the cover glass because their penetration depth is relatively high compared to other heavier radiation particles. As a result, solar cells operating in space still exhibit a significant decrease in output power. Therefore, we need to design space solar cells to be less sensitive to radiation damage and this is called radiation resistant design. Radiation resistance is an important representative feature of space solar cells.

MJ cells are composed of a number of series-connected p-n junction diodes. Therefore, the voltage output of multi-junction cells is the sum of the photovoltage of subcells, while the current output of multi-junction cells is limited by the lowest photocurrent among the subcells. Consequently, the operation output characteristics of multi-junction cells are determined by the combination of those of the individual subcells. InGaP/GaAs/Ge 3J solar cells are composed of InGaP, GaAs, and Ge diodes/subcells, and the radiation resistances of each of the three subcells differ from each other. Generally, the InGaP subcell is the most radiation resistant, while the GaAs subcell is the least resistant [Imaizumi 2005a]. Because of the different radiation responses of the three subcells in the 3J cells, changes in output characteristics and degradation of performance due to radiation exposure are more complicated.

Figure 15.4 illustrates a general current–voltage characteristic of a solar cell under dark (i.e. no) and light illumination. Output of a solar cell is expressed in the fourth quadrant. Three representative output characteristic values, short-circuit current (*Isc*), open-circuit voltage (*Voc*) and maximum power

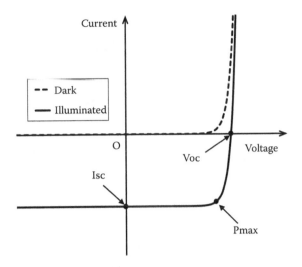

FIGURE 15.4 A general current–voltage characteristic of a solar cell under dark (i.e. no) and light illumination. Short circuit current (I_{SC}), open-circuit voltage (*Voc*) and maximum power (*Pmax*) points are indicated.

(*Pmax*), are indicated. The *Isc* is a current output value at zero voltage: the *Voc* is a voltage output value at zero current: and the *Pmax* is the maximum power output derived from multiplication of current and voltage values on the current–voltage curve in the fourth quadrant.

For single-junction solar cells, the degradation of the three solar cell output parameters, *Isc*, *Voc* and *Pmax* can be expressed in the following way:

$$X/X_0 = 1 - A\log(1 + \varphi/\varphi_0), \tag{15.3}$$

where X_0 and X are output parameter (*Isc*, *Voc* or *Pmax*) values before and after radiation exposure with a fluence of φ, respectively, with A and φ_0 as fitting parameters. For MJ cells, although *Voc* follows this equation, *Isc* does not always follow it because a current-limiting subcell can change at certain fluence. Figure 15.5 is an example of output parameter degradation of (a) Si and (b) InGaP/GaAs/Ge 3J solar cells due to 1 MeV electron irradiation. In the case of (a), all of *Isc*, *Voc* and *Pmax* follow the fitting curve, while in the case of (b), *Voc* follows the curve but *Isc* does not well. Power is principally the product of voltage and current, so, it does not follow the curve well, either.

There is another feature that has to be taken into account when discussing radiation degradation of multi-junction solar cells in space. As noted above, high energy electrons and protons contribute to solar cell degradation. The energy ranges we have to consider are typically 500 keV to 10 MeV for electrons and 50 keV to 10 MeV for protons. In general, the magnitude of radiation damage increases with energy in the case of electrons, while the damage decreases with an increase of proton energy according to the nonionizing energy loss (NIEL) calculation [Summers 1993], where the NIEL represents the ability of radiation defect creation by means of elastic collision. Electrons create a fairly uniform radiation defect distribution in a solar cell. However, protons, especially low-energy ones, create localized defects since they can stop in the middle of solar cell structure. Because multi-junction solar cells have a layered structure, effects of radiation damage strongly depend on proton energy. The approximate stopping depth of protons in a InGaP/GaAs/Ge 3J solar cell is depicted in Figure 15.6 [Sumita 2003].

The electric power output from solar panels on spacecraft has to be maintained above the desired power value until the end of its mission. Because the output power of solar cells decreases due to radiation damage, the prediction of solar cell power output in orbit is mandatory for utilizing solar cells in space.

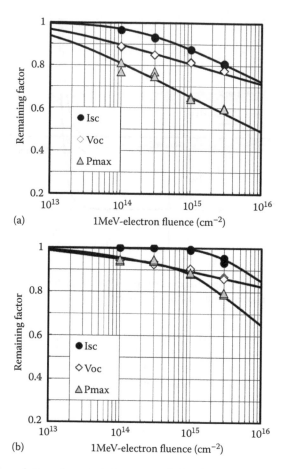

(a)

(b)

FIGURE 15.5 Typical degradation characteristic curves of (a) silicon single-junction space solar cell and (b) InGaP/GaAs/Ge 3J space solar cell. Remaining factors of short circuit current (I_{SC}), open-circuit voltage (Voc) and maximum power (Pmax) are plotted as a function of 1 MeV electron fluence. Fitting lines of all the three in (a) and only of Voc in (b) are obtained using Equation 14.3.

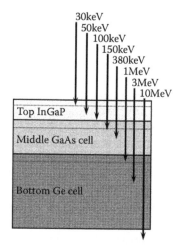

FIGURE 15.6 Approximate stopping depth of protons in an InGaP/GaAs/Ge 3J solar cell with a typical layered structure. The incident angle of protons is assumed to be normal to the surface of the solar cell.

15.3 Lifetime Prediction Method

15.3.1 Space Radiation Environments

A variety of high energy radiative particles with a certain density or flux density exist in the space environment. However, electrons and protons dominantly cause solar cell degradation. The effects of alpha particles and heavier charged particles are negligible as far as solar cell damage is concerned. Therefore, electrons and protons in space are the primary particles of interest here.

Electrons and protons with a wide range of energies characterize the space radiation environment. Their flux density is a function of energy, and the energy distribution (spectrum) depends on the time and location that is relevant to the orbit, since the spectrum is affected by solar activity such as solar flares. The electrons and protons come from the Sun and are trapped by Earth's magnetic field. They form the so-called Van Allen radiation belts. The inner portion of the belts consists mainly of protons, while the outer portion consists primarily of electrons. In addition, great increases in particle flux sometimes happen due to the sun's coronal mass ejections (i.e. solar flares) can occur. Figure 15.7 illustrates an example of flux–energy spectral data for protons collected by one of JAXA's satellites 'MDS-1' during its mission in orbit which passed through the Van Allen belts [Imaizumi 2005b].

Due to such instability and the variety of flux–energy spectra of electrons and protons, prediction tools have been developed to estimate electron and proton exposure for a given space mission (date, duration and orbit). The most widely accepted tools for its calculation are the AP9 (protons) and AE9 (electrons) codes [Ginet 2013] for the trapped particles. On the other hand, solar flares are modelled with other tools such as the JPL 91 code [Feynman 1993].

15.3.2 Irradiation Test Method

Currently ground-based testing is done by irradiating solar cells with electrons and protons from accelerators tuned to various specified energies with changing fluence. This test is to obtain degradation characteristics of solar cells due to accelerator-supplied electrons or protons, as shown in Figure 15.5. Usually, solar cells without any shielding material are used because we can calculate shielding effects in the actual space radiation environment.

To guarantee uniform exposure of the solar cell to irradiation particles, care must be taken to ensure that the beam intensity is uniform over the entire area of a solar cell to be irradiated. To achieve sufficient

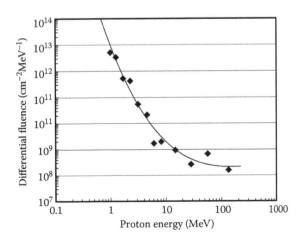

FIGURE 15.7 An example of the flux–energy spectrum data of protons collected by a JAXA's satellite 'MDS-1' on its orbit which passed through the Van Allen belt. (From Imaizumi 2005b.)

uniformity of particle flux in the irradiation area (deviations from uniformity are typically less than 10%), the particle beam is generally focused to a small area and then the focused beam is scanned horizontally and also vertically to expand the irradiation area. Another way to expand the irradiation area is to diffuse or defocus the beam using metal foils or magnets.

The sample solar cells are set in an environment at atmospheric air pressure or vacuum. Irradiation in an atmosphere of air can be acceptable for high-energy particles if scattering of the particle in air, which results in an energy distribution, is negligible. However, for lower energy particle beams, the energy distribution generally becomes significant and therefore tests must be done in vacuum. In addition, to avoid any side effects such as oxidization, keeping the solar cells in vacuum is preferable.

Particle irradiation can heat the sample solar cells and since heating the sample can affect the radiation-induced damage such as annealing effects, the irradiation temperature must be maintained as adequately low. This is typically achieved in two ways: (1) limiting the particle flux and (2) mounting the samples on a temperature-controlled plate. The detail of the irradiation testing methods is described in the ISO 23038 standard [ISO 2005].

15.3.3 JPL Method

In the following, the most widely used analyzing method for degradation prediction of space solar cells, the so-called 'JPL method' (JPL refers to the Jet Propulsion Laboratory in California) or 'relative damage coefficient method' is explained [Tada 1982, Messenger 2001].

In this method, the relative damage coefficients (RDC) for solar cells have to be derived from experimentally obtained radiation degradation curves for perpendicularly incident electrons/protons on unshielded solar cells. It is necessary to create degradation curves for the solar cell output parameters (*Isc*, *Voc*, and *Pmax*), as shown in Figure 15.5, over a necessary range of energies for both protons and electrons. Therefore, a significant number of degradation curves for both electrons and protons with sufficient variation of incident energies are needed to obtain RDC values.

In order to generate the degradation curves, irradiation tests with at least four electron energies (for example, 0.5, 1, 2 and 10 MeV) and eight proton energies (for example, 0.05, 0.1, 0.2, 0.3, 0.5, 1, 3 and 10 MeV) are recommended. Further, at least four fluence levels are required for each incident particle energy. Further still, at least two cells for each fluence level must be irradiated for cumulating good statistics. Consequently, several tens of solar cells, tens of irradiation tests, and hundreds of current–voltage measurements are required to obtain RDC values.

For each incident particle type, the fluence at which an output parameter is reduced to, generally, 90% (*Isc* and *Voc*) or 80% (*Pmax*) of its initial value (this is the so-called remaining factor) is determined from the respective degradation curve. These fluences can be obtained by fitting the experimental degradation results with equation 15.3. Then, the ratio of the fluence corresponding to the 90% or 80% remaining factors with 10 MeV protons, $\varphi_{10\text{MeV}}(90/80\%)$, to that for protons at other energies, E, $\varphi_E(90/80\%)$, defines the proton RDC value (RDC$_p$) according to the following expression:

$$\text{RDC}_p(E) = \varphi_{10\text{MeV}}(90/80\%)/\varphi_E(90/80\%), \tag{15.4}$$

RDCs for electrons (RDC$_e$) are obtained in the same manner, but in this case the reference energy is 1 MeV. Thus,

$$\text{RDC}_e(E) = \varphi_{1\text{MeV}}(90/80\%)/\varphi_E(90/80\%), \tag{15.5}$$

Figures 15.8a and 15.8b are examples of proton and electron RDCs of a 3J solar cell.

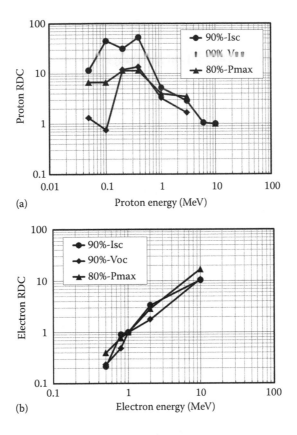

FIGURE 15.8 Examples of RDC for (a) protons and (b) electrons of a InGaP/GaAs/Ge 3J solar cell.

In addition, another parameter $C_{e/p}$, which is called the 'proton to electron damage equivalency ratio' can be calculated. $C_{e/p}$ converts the 10 MeV proton fluence to an equivalent 1 MeV electron fluence for the 90% or 80% remaining factors as in the following equation.

$$C_{e/p} = \varphi_{1MeVe} / \varphi_{10MeVp}, \tag{15.6}$$

Note that the φ_{1MeVe} and φ_{10MeVp} are fluences applying equivalent degradation to an output parameter (*Isc*, *Voc* or *Pmax*) of a solar cell. $C_{e/p}$ is different for each parameter and for the specified remaining factors.

The electron and proton environment for the mission under consideration is required in an integrated form for a mission duration as input to this degradation prediction calculation. The predicted radiation environment information for mission duration can be available from models such as AE9 and AP9, and electron and proton energy–fluence spectra can be obtained in differential form ($d\Phi_e(E)/dE$ and $d\Phi_p(E)/dE$). Then, integration of the product with the appropriate electron and proton RDCs ($RDC_e(E)$ and $RDC_p(E)$) determines the equivalent 1 MeV electron and 10 MeV proton fluences for the mission. These RDCs must be modified from the original RDCs (unshielded, normal incidence) to those corresponding to shielded, omnidirectional incidence as in actual space environment (for the modification calculation see Tada [1982]).

These integrals are shown in the following equations, for electrons and protons, respectively:

$$\phi_{1\,\mathrm{MeVe}}^{\mathrm{Eq}} = \int d\Phi_e(E)/dE \cdot RDC_e(E)dE, \tag{15.7a}$$

$$\phi_{10\,\mathrm{MeVp}}^{\mathrm{Eq}} = \int d\Phi_p(E)/dE \cdot RDC_p(E)dE, \tag{15.7b}$$

The total equivalent 1 MeV electron fluence in a mixed proton and electron environment is obtained by adding the results of Equations 15.7a and a scaled 15.7b, where the value represents the net fluence of 1 MeV electrons normally incident on bare cells that is expected to cause the equivalent damage to a spacecraft at the end of its mission. Then the total equivalent 1 MeV electron fluence can therefore be expressed as:

$$\phi_{\mathrm{total}\,1\,\mathrm{MeVe}}^{\mathrm{Eq}} = C_{e/p}\,\phi_{10\,\mathrm{MeVp}}^{\mathrm{Eq}} + \phi_{1\,\mathrm{MeVe}}^{\mathrm{Eq}} \tag{15.8}$$

For a given total equivalent 1 MeV electron fluence accumulated during the mission, the predicted degradation can be obtained for each solar cell output parameter from the experimentally determined 1 MeV electron degradation curves.

As described above, RDCs for both electrons and protons are necessary to perform degradation prediction calculation for space solar cells. However, a great number of accelerator-based irradiation tests are required to determine RDCs. That these tests consume significant cost and time is a serious issue for space solar cell development and utilization.

15.4 Physically Simulating the Space Environment with Ground-Based Laser-Driven Sources

As has been discussed above, one of the most important issues for solar cells in space is the degradation of their electrical performance due to irradiation by protons and electrons. Consequently, ground-based testing must be carried out to predict the lifetime of space solar cells as well as to develop new space solar cells. For the ground-based testing, it is very important to establish systems for spatial uniformity of beams, sample temperature control and accurate beam current measurement. These systems are steadily being developed [Tanaka 1980, Yuri 2007, Harris 2008, Yuri 2011]. However, because conventional accelerators are used for ground-based testing, solar cells must be irradiated separately with electrons and protons at any given single energy; although protons and electrons in space have known spectra. Therefore, single energy protons from tens keV to MeV and electrons from several hundred keV to MeV must be applied for ground-based testing at accelerators to cover these spectra ranges (JPL method described in 15.3.3) [Tada 1982]. A conventional accelerator cannot accelerate beams with a wide range of energy instantaneously nor irradiate a solar cell simultaneously with different types of particles. Even for a single particle type, two or more accelerators are required for ground-based testing in order to complete data collection for lifetime prediction and further development of space solar cells. State-of-the-art space solar cells are multi-junction solar cells, in which three or more subcells made of different materials are stacked into one solar cell. As a result, the degradation behaviour of multi-junction solar cells is not simple, and ground-based testing using electrons and protons in small energy steps should be done to clarify degradation of all subcells in multi-junction solar cells. Such a test increase is both cost and time-consuming.

Laser-driven ion/electron beams with broad energy spectra can solve this issue [Orimo 2007, Nishiuchi 2009, Imaizumi 2012]. We refer the interested reader to Chapter 2 by Malka and Chapter 5 by Macchi for a more detailed treatment of laser-driven electron and ion acceleration respectively. To apply laser-driven beams for ground-based testing of space solar cells, the broad energy spectrum affords adjustment (tuning) to match the actual space spectrum as shown in Figure 15.7. This can also be a key feature of relevance in studies of biological effects of space radiation for which the reader is referred to Chapter 14 by Reitz and Hellweg. Spatial uniformity of beam flux is also important to create spatially uniform damage in a solar cell. The size of solar cells for testing is usually 2 cm × 2 cm, and the uniformity should be within 90% or more as a preferred requirement. In addition, it is important to accumulate a specified beam flux for irradiation tests within a certain reasonable time interval. For example, fluences of electrons and protons up to 1×10^{16}/cm^2 and 1×10^{14}/cm^2, respectively, are required for predicting lifetimes of a few decades in GEO (geostationary Earth orbit). Furthermore, the total accumulated fluence strongly depends on the specific satellite mission. If a satellite is launched into an orbit in which it passes across the Van Allen radiation belt, such as GTO (geostationary transfer orbit), the required total fluence should be higher than that for GEO.

15.5 Prospects for Laser-Driven Beams for Evaluation of Space Solar Cells

As mentioned in Section 15.4, since laser-driven particle acceleration techniques can produce proton and electron beams with broad energy spectra, it is very attractive for us to apply these techniques to ground-based testing for space solar cell evaluation. To realize laser-driven sources as a tool for ground testing, we need to achieve the above-mentioned specifications (such as uniformity and fluence). Of course, it is also important to establish the evaluation technique of particle bunches. Thus, the measurement techniques spanning a large fluence range and energy range (from keV to MeV) must be established. Also, the irradiation should neither adversely heat the samples nor destroy them (by partial melting or/and sputtering) by particle bunches with high peak flux density. Thus, relatively low flux and high repetition rate are preferable to high flux and low repetition rate. Since these conditions strongly depend on materials and structures of solar cells, the optimum condition for each solar cell must be determined by experiments. In addition, the beam quality and purity are also important. Thus, if other particles, such as heavy ions, are contained in proton and electron beams, the accurate lifetime prediction for solar cells in space cannot be carried out, because heavy ions create greater damage than electrons and protons and, as a result, solar cell degradation in space is overestimated. Therefore, measurement of contamination, such as heavy ions in proton beams, and subsequent removal of them are also required.

In Chapter 15, space radiation effects on solar cells and necessary ground-based testing for their lifetime prediction have been explained. Prospects for laser-driven beams for testing solar cells were also described. In conclusion, laser-driven beams can usher significant innovation for ground-based irradiation of space solar cells because they can deliver broad and controlled energy spectra, features that can potentially complement studies at conventional accelerators.

References for Chapter 15

Dodd, P. E. and Massengill, L. W. 2003. "Basic mechanism and modeling of single-event upset in digital microelectronics." *IEEE Trans. Nucl. Sci.* **50**: 583–602.

Dodd, P. E., Shaneyfelt, M. R., Felix, J. A. and Schwank. J. R. 2004. "Production and propagation of single-event transients in high speed digital logic ICs." *IEEE Trans. Nucl. Sci.* **51**: 3278–3284.

Fatemi, N., Lyons, J. and Eskenazi, M. 2013. "Qualification and production of Emcore ZTJ solar panels for space missions." *Proc. 39th IEEE Photovoltaics Specialists Conf., Tampa*: 2793–2796.

Feynman, J., Spitale, G., Wang, J. and Gabriel, S. 1993. "Interplanetary proton fluence model: JPL 1991." *J. Geophys. Res.* **98**: 13281–13294.

Ginet, G. P., O'Brien, T. P., Huston, S. L., Johnston, W. R., Guild, T. B., Friedel, R., Lindstrom, C. D., Roth, C. J., Whelan, P., Quinn, R. A., Madden, D., Morley, S. and Su, Yi-Jiun. 2013. "AE9, AP9 and SPM: New Models for Specifying the Trapped Energetic Particle and Space Plasma Environment" *Space Sci. Rev.* **179**: 579–615.

Harris, R. D., Imaizumi, M., Walters, R. J., Lorentzen, J. R., Messenger, S. R., Tischler, J. G., Ohshima, T., Sato, S., Sharps, P. R. and Fatemi, N. S. 2008. "In situ irradiation and measurement of triple junction solar cells at low intensity, low temperature (LILT) conditions" *IEEE Trans. Nucl. Sci.* **55**: 3502–3507.

Hong, K-H., Noh, Y-C., Ko, D-K. and Lee, J-G. 2007. "Simultaneous proton and x-ray imaging with femtosecond intense laser driven plasma source" *Jpn. J. Appl. Phys.* **46**: 5853–5858.

Imaizumi, M., Matsuda, S., Kawakita, S., Sumita, T., Takamoto, T., Ohshima, T., Yamaguchi, M. 2005. "Activity and current status of R&D on space solar cells in Japan" *Prog. Photovolt. Res. Appl.* **13**: 529–543.

Imaizumi, M., Sumita, T., Kawakita, S., Aoyama, K., Anzawa, O., Aburaya, T., Hisamatsu, T., Matsuda, S. 2005. "Results of flight demonstration of terrestrial solar cells in space" *Prog. Photovolt. Res. Appl.* **13**: 93–102.

Imaizumi, M., Yuri, Y., Bolton, P. R., Sato S-I. and Ohshima, T. 2012. "Innovative technologies on proton irradiation ground tests for space solar cells" *Proc. of 38th IEEE Photovoltaic Specialists Conf., Austin, TX (UAS)*: 2831–2834.

ISO 23038. 2005. "*Space systems — Space solar cells — Electron and proton irradiation test methods*", ISO.

Leavy, J. F. and Poll, R. A. 1969. "Radiation-induced integrated circuit latchup", *IEEE Trans. Nucl. Sci.* **6**: 96–103.

Makino, T., Kobayashi, D., Hirose, K., Takahashi, D., Ishii, S., Kusano, M., Onoda S, Hirao, T. and Ohshima, T. 2009. "Soft-error rate in a logic LSI estimated from SET pulse-width measurements" *IEEE Trans. Nucl. Sci.* **56**: 3180–3184.

Matsuura, H., Iwata, H., Kagamihara, S., Ishihara, R., Komeda, M., Imai, H., Kikuta, M., Inoue, Y., Hisamatsu, T., Kawakita, S., Ohshima, T. and Itoh, H. 2006. "Si substrate suitable for radiation-resistant space solar cells" *Jpn. J. Appl. Phys.* **45**: 2648–2655.

McWhorter, P. J. and Winokur, P. S. 1986. "Simple technique for separating the effects of interface traps and trapped-oxide charge in metal-oxide-semiconductor transistors" *Appl. Phys. Lett.* **48**: 133–135.

Messenger, S. R., Summers, G. P., Burke, E. A., Walters, R. J. and Xapsos, M. A. 2001. "Modeling solar cell degradation in space: A comparison of the NRL displacement damage dose and the JPL equivalent fluence approaches" *Prog. Photovolt. Res. Appl.* **9**: 103–121.

Nishiuchi, M., Daito, I., Ikegami, M., Daido, H., Mori, M., Orimo, S., Ogura, K., Sagisaka, A., Yogo, A., Pirozhkov, A. S., Sugiyama, H., Kiriyama, H., Okada, H., Wakai, D., Sakaki, H., Bolton, P., Choi, I. W., Sung, J. H., Lee, J., Oishi, Y., Fujii, T., Nemoto, K., Souda, H., Noda, A., Iseki, Y. and Yoshiyuki, T. 2009. "Focusing and spectral enhancement of a repetition-rated, laser-driven, divergent multi-MeV proton beam using permanent quadrupole magnets" *Appl. Phys. Lett.* **94**: 061107.

Ohshima, T. Morita, Y., Nashiyama, I., Kawasaki, O., Hisamatsu, T., Nakao, T., Wakow, Y. and Matsuda, S. 1996. "Mechanism of anomalous degradation of silicon solar cells subjected to high-fluence irradiation" *IEEE Trans. Nucl. Sci.* **43**: 2990–2997.

Ohshima, T., Itoh, H. and Yoshikawa, M. 2001. "Effect of gamma-ray irradiation on the characteristics of 6H silicon carbide metal-oxide-semiconductor field effect transistor with hydrogen-annealed gate oxide" *J. Appl. Phys.* **90**: 3038–3041.

Oldham, T. R. 2003. "Total ionizing dose effects in MOS oxides and devices" *IEEE Trans. Nucl. Sci.* **50**: 483–499.

Orimo, S., Nishiuchi, M., Daido, H., Yogo, A., Ogura, K., Sagisaka, A., Pirozhkov, Z, Li, A., Mori, M., Kiriyama, H., Tanoue, M., Nakai, Y., Akutsu, A., Nakamura, S., Shirai, T., Iwashita, Y., Noda, A., Oishi, Y., Nemoto, K., Choi, I-W., Yu, T-J., Sung, J-H., Jeong, T-M., Kim, H-T., Hong, K H., Noh, Y-C., Ko, D-K. and Lee, J-G. 2007. "Simultaneous proton and x-ray imaging with femtosecond intense laser driven plasma source" *Jpn. J. Appl. Phys.* **46:** 5853–5858.

Reed, R. A., Marshall, P. W., Kim, H. S., McNulty, P. J., Fodness, B., Jordan, T. M., Reedy, R., Tabbert, C., Liu, M. S. T., W. Heikkila, Buchner, S. Ladbury, R. and LaBel, K. A. 2002. "Evidence for angular effects in proton-induced single-event upset", *IEEE Trans. Nucl. Sci.* **49:** 3038–3044.

Sexton, F. W. 2003. " Destructive single-event effects in semiconductor devices and ICs" *IEEE Trans. Nucl. Sci.* **50:** 603–621.

Sumita, T. Imaizumi, M., Matsuda, S., Ohshima, T., Ohi, A. and Kamiya, T. 2003. "Analysis of end-of-life performance for proton-irradiated triple-junction space solar cell" *Proc. 3rd World Conf. on Photovoltaic Energy Conversion, Osaka*: 689–692.

Summers, G. P., Burke, E. A., Shapiro, P., Messenger, S. R. and Walters, R. J. 1993. "Damage correlations in semiconductors exposed to gamma, electron and proton radiations" *IEEE Trans. Nucl. Sci.* **40:** 1372–1379.

Strobl, G. F. X., Fuhrmann, D., Guter, W., Khorenko, V., Köstler, W. and Meusel, M. 2012. "About AZUR'S "3G30-advanced" Space Solar Cell and Next Generation Product with 35% Efficiency" *Proc. 27th European Photovoltaic Solar Energy Conference and Exhibition, Frankfurt*: 104–108.

Tabbert, C., Liu, M. S. T., W. Heikkila, Buchner, S. Ladbury, R. and LaBel K. A. 2002. "Evidence for angular effects in proton-induced single-event upset", *IEEE Trans. Nucl. Sci.* **49:** 3038–3044.

Tada, H. Y., Carter, Jr., J. R., Anspaugh, B. E. and Dowing, R. G. 1982. "Solar cell radiation handbook third edition" *JPL Publications*: 82–96.

Tanaka, R., Mizuhashi, K., Sunaga, H. and Tamura, N. 1980. *Nucl. Instrum. Meth.* **174:** 201–208.

Titus, J. L., Wheatley, C. F., Gillberg, J. E. and Burton, D. I. 2001. "A study of ion energy and its effects upon and SEGR-hardened stripe-cell MOSFET technology" *IEEE Trans. Nucl. Sci.* **48:** 1879–1884.

Walters, R. J., Romero, M. J., Araujo, D., Garcia, R., Messenger, S. R. and Summers, G. P. 1999. "Detailed defect study in proton irradiated InP/Si solar cells" *J. Appl. Phys.* **86:** 3584–3589.

Wheatley, T. H., Wheatley, C. F. and Titus, J. L. 2001. "Early lethal SEGR failures of VDMOSFET's considering nonuniformity in the rad-hard device distribution" *IEEE Trans. Nucl. Sci.* **48:** 2217–2221.

Yamaguchi, M., Taylor, S. J., Matsuda, S. and Kawasaki, O. 1996. "Mechanism for the anomalous degradation of Si solar cells induced by high fluence 1MeV electron" *Appl. Phys. Lett.* **68:** 3141–3143.

Yamaguchi, M. 2001. "Radiation-resistant solar cells for space use" *Solar Energy Mater. Solar Cells* **68:** 31–53.

Yuri, Y., Miyawaki, N., Kamiya, T., Yokota, W., Arakawa, K. and Fukuda, M. 2007. "Uniformization of the transverse beam profile by means of nonlinear focusing method" *Phys. Rev. Special Topics-Accelerators and Beams* **10:** 104001.

Yuri, Y., Ishizuka, T., Yuyama, T., Ishibori, I., Okumura, S. and Yoshida, K. 2011. "Formation of a large-area uniform ion beam using multipole magnets in the TIARA cyclotron" *Nucl. Instrum. Meth. A* **642:** 10–17.

16

Analogy of Laser-Driven Acceleration with Electric Arc Discharge Materials Modification

Kai Nordlund

Flyura Djurabekova

16.1 Introduction

Ion implantation is a widely used tool in materials research and industry. For instance, doping of semiconductors is routinely carried out by ion implantation, making the field essentially a multi-billion-dollar industry. As described in Chapter 18 by S. Srivastava and D. Avasthi laser-driven acceleration could, in fact, become interesting for some aspects of semiconductor processing.

Chapter 16 focuses on comparing the differences between laser-driven and conventional ion acceleration with respect to the physics of ion irradiation of materials. Laser ion acceleration essentially opens up a new regime of peak fluxes for materials modification, one where the ion-induced collision cascades can overlap in both space and time, something that essentially never occurs during use of conventional ion accelerators. This flux regime is, however, not entirely unprecedented: during electrical arc discharges, similar fluxes are present in a limited time and space domain. The laser ion irradiation involved a very wide energy range, which also differs strongly from the conventional accelerator condition.

In Section 16.2, we briefly overview the key features of the known mechanism of conventional ion beam materials modification. In Section 16.3, we review briefly what is known from experiments and recent computer simulations about electrical arc materials modification, and use this information to conjecture in Section 16.4 on what might be the materials modification effects arising from systematic, laser-driven, high-peak flux ion irradiation.

16.2 Conventional Ion Beam Materials Modification Mechanisms

In typical ion beam usage, the aim is to use a beam of well-controlled energy to implant ions with a known, well-controlled ion depth distribution. This praxis derives mainly from accelerators being first used as nuclear physics experimental tools, which lead to a culture of very well-defined beam energies and minimal energy spreads, to carry out basic nuclear physics experiments where an accurate energy is crucial, e.g. to study narrow nuclear reaction resonances. When the use of accelerators spread from nuclear physics for ion implantation and other material physics applications in the 1960s [Mayer 1990], this culture clearly remained in the community. A narrow energy spectrum certainly has also many advantages for materials modification, the main one being the possibility of relatively accurate prediction of implantation depth profile, when the ion energy is known. The desire to achieve beams with well-defined energy and narrow angular spread also leads to the ion beam being relatively low average flux (or current), typically in the nA to mA range with beam spots of the order of a few mm.

The basic mechanisms of materials modification by implanted ions have been studied extensively by computer simulations since the 1950s [Oen 1963, Gibson 1960, Averback 1998]. They reveal that ions in the typical implantation energy range of ~1–100 keV initially affect materials by producing a collision cascade where atoms heat up to temperatures of the order 10000 K in approximately spherical or cylindrical regions of about 3–10 nm diameter [Averback 1998]. The regions do not grow larger due to so-called sub-cascade splitting [de Backer 2016]. This strongly heated region is, however, cooled back down to the ambient temperatures on a time scale of 1–10 ps. This process is schematically illustrated in Figure 16.1 and with actual atom coordinates (obtained from molecular dynamics computer simulations) for the case of 50 keV Xe irradiation of Au in Figure 16.2.

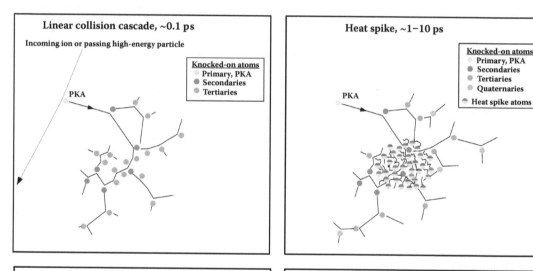

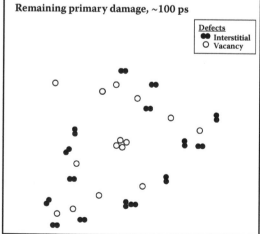 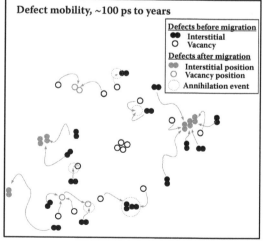

FIGURE 16.1   Schematic description of the development of a collision cascade, its time scales, and the subsequent defect mobility stage.

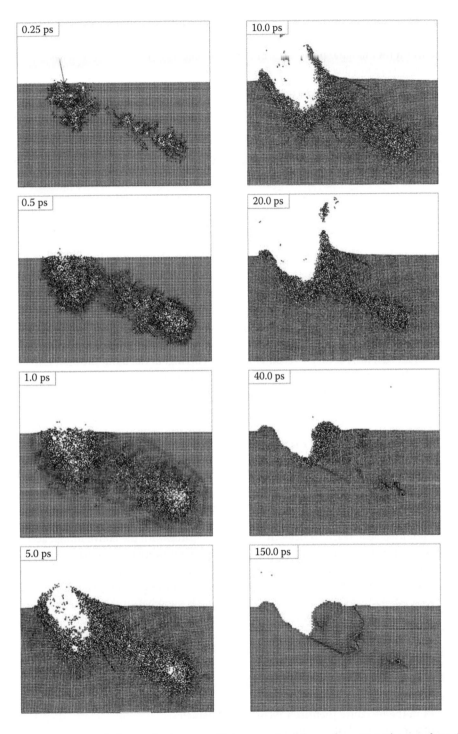

FIGURE 16.2 Formation of a heat spike due to a collision cascade and a surface crater. The snapshots show the cross section of a molecular dynamics simulation cell, modelling the effect of a 50 keV Xe ion impacting on Au. During the first 1 ps, the collision cascade formed by recoils of the high-energy ion breaks the lattice structure. The disordered zone is very hot, and forms a liquid-like zone known as a heat spike. After about 10 ps, the heat spike starts to cool down, and the lattice recrystallizes almost perfectly. However, a crater is left at the surface, and a few dislocations (1D defects surrounding the displaced lattice) are left behind inside the material.

Comparing the ion currents with the spatial and temporal scale of cascades leads to an important insight into the nature of ion beam materials modification. For a typical ion charge state of +1, a beam with 1 mA current (in the upper limit range of normal ion beam currents) that is well-focused to a 1 mm 'spot' radius translates to a flux density of

$$F = \frac{10^{-3}\,\text{A}}{e\,\pi(10^{-3}\,\text{m})^2} = 2\cdot10^{21}\,\frac{\text{ions}}{\text{m}^2 s} \tag{16.1}$$

where e is the magnitude of the electron charge. If it is assumed that a single ion affects a 10-nm cube region, the time interval between impacts of a single ion in the *same* region of space will be

$$\Delta t = \frac{1}{2\cdot10^{21}\times(10\cdot10^{-9})^2} = 5\cdot10^{-6}\,\text{s} \tag{16.2}$$

Thus, the time between impacts in the same space region is about a million times longer than the duration of a collision cascade. This gives the key insight that the ion collisional processes occur completely independent of each other.

As illustrated in Figure 16.1, the ion irradiation produces damage. The damage production mechanisms have been extensively studied [Urbassek 1991, Averback 1998, Nordlund 1998]. From these studies, it is clear that the final damage is produced when the hot region formed by the cascade (the 'heat spike') cools down. In metals, there is major recombination occurring; in the bulk, the final number of stable defects is roughly two orders of magnitude lower than the number of atoms that are part of the heat spike, while in semiconductors the molten region typically becomes a disordered (amorphous) damage pocket.

16.3 Overview of Electrical Arc Materials Modification

In Section 16.2, we showed by a simple estimate that the collision cascades that produce damage do not overlap in time. They can of course overlap in space, simply if the ion fluence (number of ions per area) is high enough that another collision cascade occurs later on in the same region of space. In other words, in the normal ion irradiation regime, the overlap in time (i.e. coincidence) is not possible, as it is not feasible to reach sufficiently high flux densities with the conventional accelerators. As mentioned in the introduction, the consequences of temporal overlap of collisional cascades can be, nevertheless, observed in the regime of plasma-wall interactions during electrical sparks and arcs. The latter are plasma discharges that carry large peak currents and can strongly modify surfaces [Jüttner 1979, Behrisch 1986, Jüttner 2001]. This surface damaging is often harmful, as in the case of linear collider components [Braun 2002], fusion reactors [McCracken 1980] or spark plugs for cars [Rager 2005], but it is also utilized for electrical discharge machining of materials [Jeong 2007]. The damage usually comes in the form of micrometre-sized craters and frozen-in liquid on the surface [Behrisch 1986], but the mechanisms by which the craters form have until recently not been clear.

Experimental images of surface damage in Cu produced by arcs obtained with a direct-current (DC) setup used recently at CERN are shown in Figure 16.3. These images, which are very similar to typical crater images in the literature [Behrisch 1986], show characteristic complex crater shapes that give an impression of a frozen-in splash of liquid.

It is well-established that single energetic ions (with energies in the range 10–100 keV) coming from an accelerator can produce craters on surfaces [Ghaly 1994, Birtcher 1999, Bringa 2002] (Figure 16.2), and also more complex features like 'fingers' that superficially resemble those of the arcing craters shown in Figure 16.3 [Nordlund 2003]. However, there are at least two reasons why the arcing craters cannot be expected to form due to the same mechanism as that for single ions. First, an overlap of several craters

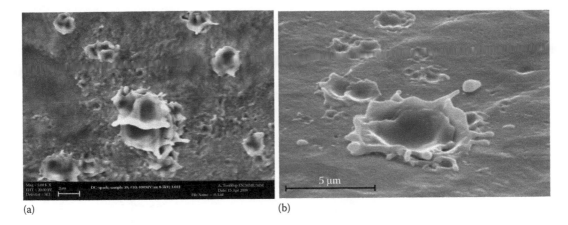

(a)   (b)

FIGURE 16.3 Craters produced during arcing at a DC arc setup at CERN. (a) Top-view scanning electron micros-copy (SEM) image showing several craters of different size. The experiments show craters of similar shape in a wide variety of sizes in the range 0.1–10 μm. (b) Tilted SEM experimental image of a few nearby craters. Reprinted with permission from Timkó, H. et al. 2010. *Phys. Rev. B*, 81:184109. Copyright (2010) by the American Physical Society.

produced by ions hitting random locations on surfaces does not produce a larger crater. Rather, simulations show that an ion hitting a pre-existing crater tends to fill it up [Henriksson 2007]. Hence, spatial overlap of craters just leads to surface roughening. Second, it is found that the arcing craters are produced on a large number of different materials, including Cu, in which single ion bombardment has not been found to produce cratering [Nordlund 1999].

The experimental investigations of arc cratering give a hint as to why craters could be formed on surfaces, however. Measurements of the electric current during an arcing event show that the peak currents I can be in the range of 1–100 A and final crater areas of the order of $A_{cr} = (10 \, \mu m)^2$ [Behrisch 1986]. Assuming a peak current of 10 A and that the electron current is 100 times higher than the ion current in the arc plasma, this gives an estimate of the ion flux density of $I/100/e/A_{cr} = 6 \times 10^{27}$ ions/m²/s (here e is the electron charge). Inserting this huge flux value into Equation 16.2 gives a time between single ion impacts on $(10 \, nm)^2$ cascade areas of 1.6 ps, i.e. a value well within the lifetime of a collision cascade. On the other hand, the lifetime of the high-current part of an arc is typically tens of nanoseconds, i.e. much longer than the time between impacts. This implies that the standard assumption of ion impact events being independent of each other does not apply during arc cratering conditions.

The other important factor for understanding arc surface modification is the properties (ion energy and incoming angle) of the ions escaping an arc plasma at a cathode when they impact on a surface. This is a standard issue of plasma physics, and can be tackled by particle-in-cell (PIC) computer simulations. Timko and Djurabekova et al. have carried out PIC simulations of arc plasmas under conditions matching experiments carried out at CERN [Timko 2010, Timko 2011, Djurabekova 2012]. The initial simulations, carried out with 1D (one spatial dimension) PIC modelling indicated that after the plasma is formed [Timko 2011] between electrodes with an initial voltage difference of 10 kV, a sheath potential forms near the cathode that can accelerate the ions to an energy of about 8 keV. However, this modelling did not reach a steady state plasma condition or account for the loss of external voltage difference after the arc current drains the electrical circuit of power. A later refined model utilizing a 2D PIC model (two spatial coordinates in cylindrical symmetry) indicated that [Timko 2015] in a steady state condition, the voltage drop near the cathode is lower, of order 100 V. In both cases, the flux density of ions hitting the cathode is very high, consistent with the experimental observations and simple analytical estimates.

The combination of high flux density and ion energies well above thermal levels offer explanations to the formation mechanisms of arc plasmas. In the reference, [Timko 2010], explicit simulations of 8 keV Cu ion impacts on Cu with flux densities corresponding to those observed in the initial PIC simulations

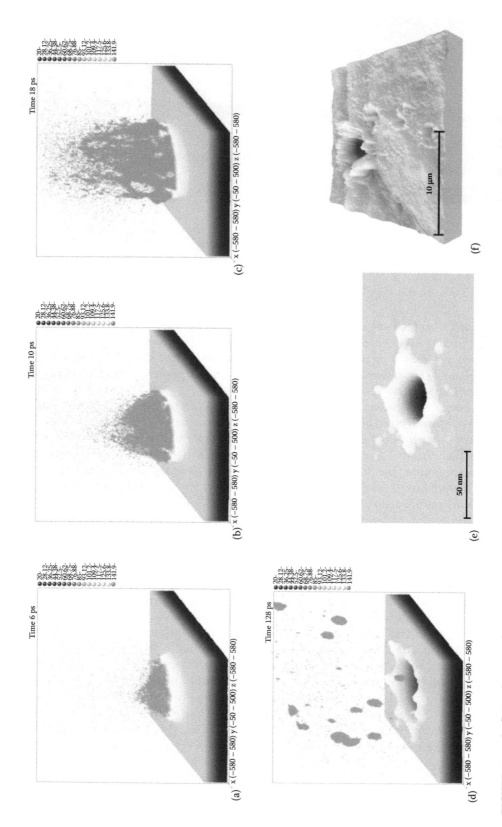

FIGURE 16.4 (a–d) Development of a crater due to multiple overlapping ion impact during the initial stages of arc plasma cratering. The development takes about 120 ps, and is caused by 200 Cu ions with an energy of 8 keV impacting on the same 30 nm diameter circular area of the surface; (e) final crater shape; (f) experimental image of an arc crater. (Reprinted with permission from Timkó, H. et al. 2010. *Phys. Rev. B*, 81:184109. Copyright (2010) by the American Physical Society.)

were carried out. The results showed that indeed the arc plasma ion bombardment leads to a condition of multiple overlapping heat spikes that can lead to formation of large craters on the surface (see Figure 16.4), similar to those observed in the experiments. Comparison of the crater depth profiles showed that the crater rim width to crater depth ratio was similar in the simulated and experimental craters [Timko 2010].

This high-energy cratering mechanism is likely to occur only during the initial stages of arcing, when the sheath potential has not yet lowered to about 100 V. It might also be relevant later when a side plasma filament from the central plasma impacts the surface on the side of the main crater evidenced by the small side craters in Figure 16.3. After the sheath potential has lowered to the 100 V level, a slightly different mechanism can explain the crater formation. Although the ion energy is lower, the very large ion flux will heat up the surface and at the same time exert a pressure on it. Recent finite element method

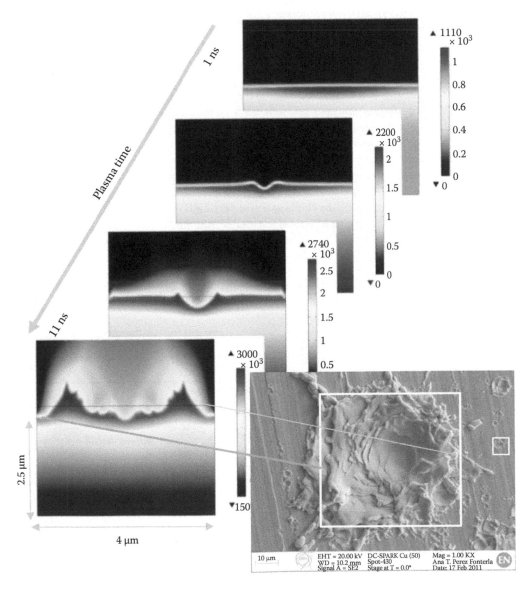

FIGURE 16.5 Illustration of the formation of a crater by plasma ions. The simulations are performed by the finite element method (FEM), where the plasma is simulated as 'ion pressure' estimated from a transferred momentum from plasma ions to the surface [Tian 2016]. The image to the bottom right shows an experimental crater image.

(FEM) simulations [Tian 2016] indicate that also this condition can lead to the formation of large craters, see Figure 16.5.

16.4 Laser-Driven Acceleration Effects on Materials

The brief overview of conventional and arc surface plasma modification given in Sections 16.2 and 16.3. indicates that laser-driven ion acceleration would open up several new interesting physics regimes for exploration. Some possibilities for these are summarized in Table 16.1.

The much wider energy spread of laser-driven ion bunches compared to conventional ion irradiation may actually offer advantages of practical value. As noted above, the strong tendency of the ion beam community to use ions with essentially a single, well-defined ion energy is to some extent just historic in origin. In fact, in modern materials processing quite often 'box-like' ion implantation depth profiles are desired (see also Chapter 18), and to achieve these, the implantation is split up into several stages that utilize different ion energies [Stichtenoth 2008]. Laser acceleration could at least in principle achieve the same goal with a single 'shot' – although this would require achieving a fairly well-defined, controllable ion energy profile with the laser-driven scheme.

Comparison with the arc plasma modification, on the other hand, indicates that laser-driven acceleration with similarly high fluxes could open up a new materials modification regime. The laser-acceleration could have the advantage over arcing that one could use any ion-material combination – arcing is limited to using the same material for both the implanted ions and the material to be implanted. Moreover, the laser acceleration would be more controllable and the ion flux densities could be better tuned. Also, a regime slightly below the 'heat spike overlap' regime could be very interesting from a basic science point of view, as it would allow studying materials modification in a regime where there is very little time for thermally driven defect migration. The combination of very high flux density and little time for defect migration could allow for making new kinds of metastable thin films.

Another potentially interesting aspect, albeit even more demanding to study theoretically, could be the combined effect of the ions and electrons simultaneously emitted during the laser-acceleration process. Such a combination of simultaneous high flux of electrons and ions for materials modification has barely been studied at all for a wide spread of ion energies. However, recent studies of swift heavy ion modification indicate that interesting synergistic effects can appear when a material is subject to

TABLE 16.1 Best Estimates of How Laser-Driven Ion Irradiation could Open Up New Avenues of Research on Ion-Solid Interactions

Ion Energy Range and Distribution	Flux (ions/m^2 s)	Type of Ion	Irradiation Area	Expected Outcome and Relation to Other Fields
keVs, but with limited energy spread	$\ll 10^{27}$	Any	Broad beam	Efficient ion implantation with box-like dopant profiles; potential use in semiconductor technology
keVs	$< 10^{27}$	Any	Any	Material modification with limited defect mobility; way to test diffusion/flux density effects on ion materials modification.
100 of eVs to keVs	$\geq 10^{27}$	Heavy	Focused	Local cratering; way to test arc plasma modification of materials
keV to MeVs	$\geq 10^{27}$	Medium or heavy	Broad beam	Ablation of material; analogy to laser ablation
> 10s of MeVs	$\geq 10^{27}$	Heavy	Focused	Track formation; analogy to swift heavy ion irradiation. Synergy effects of ions and electron material modification.

Note: (We emphasize that as most of these regimes have not been available for experimental study at all, they are deductions based on simulations and theory, and actual experiments need to be performed to verify the projections).

simultaneous ion and electron excitations [Backman 2012]. A laser-driven scheme could thus open up an entirely new realm of combined energetic charged particle–material interactions.

References for Chapter 16

Averback, R.S. and Diaz de la Rubia, T. 1998. "Displacement damage in irradiated metals and semiconductors" in H. Ehrenfest and F. Spaepen (Eds.) *Solid State Physics*, volume **51**, 281–402. Academic Press, New York.

Backman, M., Djurabekova, F., Pakarinen, O. H., Nordlund, K., Zhang, Y., Toulemonde, M. and Weber, W.J. 2012. "Cooperative effect of electronic and nuclear stopping of ion irradiation damage in silica" *J. Phys. D: Appl. Phys.* **45**: 505305.

Behrisch, R. 1986. "Surface erosion by electrical arcs" In D. E. Post and R. Behrisch (Eds.) *Physics of Plasma-Wall Interactions in Controlled Fusion*, volume **131** of NATO ASI Series, Series B: Physics, pages 495–513. Plenum Press, New York.

Birtcher, R.C. and Donnelly, S.E. 1999. "Plastic flow produced by single ion impacts on metals" *Nucl. Instr. Meth. Phys. Res. B* **148**: 194–199.

Braun, H.H., Döbert, S., Syratchev, I., Taborelli, M., Wilson, I. and Wuensch W. 2002. "CLIC high-gradient test results" *Technical Report CERN/PS* 2002–062 (RF), CERN, Cern, Geneva.

Bringa, E.M., Johnson, R.E. and Papaleo, R.M. 2002. "Crater formation by single ions in the electronic stopping regime: Comparison of molecular dynamics simulations and experiments on organic films" *Phys. Rev. B* **65**: 094113.

De Backer, A., Sand, A.E., Nordlund, K., Luneville, L., Simeone, D. and Dudarev, S.L. 2016. "Subcascade formation and defect cluster size scaling in high-energy collision events in metals" *EPL* **115**: 26001.

Djurabekova, F., Samela, J., Timko, H., Nordlund, K., Calatroni, S., Taborelli, M. and Wuensch, W. 2012. "Crater formation by single ions, cluster ions and ion "showers"" *Nucl. Instr. Meth. Phys. Res. B* **272**: 374–376.

Ghaly, M. and Averback, R.S. 1994. "Effect of viscous flow on ion damage near solid surfaces" *Phys. Rev. Lett.* **72**(3): 364–367.

Gibson, J.B., Goland, A.N., Milgram, M. and Vineyard, G.H. 1960. "Dynamics of radiation damage" *Phys. Rev.* **120**(4): 1229–1253.

Henriksson, K.O.E., Nordlund, K. and Keinonen, J. 2007. "Annihilation of craters: Molecular dynamic simulations on a silver surface" *Phys. Rev. B* **76**(24): 245428.

Jeong, Y.-H. and Min, B.K. 2007. "Geometry prediction of shapes of micro-edm process using simulation-drilled holes and tool electrode shapes of micro-edm process using simulation" *International Journal of Machine Tools and Manufacture* **47**: 1817.

Jüttner, B. 1979. "Erosion craters and arc cathode spots in vacuum" *Beiträge aus der Plasmaphysik* **19**(1): 25–48.

Jüttner, B. 2001. "Cathode spots of electric arcs" *J. Phys. D: Appl. Phys.* **34**: R103.

Mayer, J.M. and Lau, S.S. 1990. "Electronic Materials Science For Integrated Circuits in Si and GaAs" MacMillan, New York.

McCracken, G.M, 1980. "A review of the experimental evidence for arcing and sputtering in tokamaks" *J. Nucl. Mater.* **93–94**: 3–16.

Nordlund, K., Ghaly, M., Averback, R.S., Caturla, M., Diaz de la Rubia, T. and Tarus, J. 1998. "Defect production in collision cascades in elemental semiconductors and fcc metals" *Phys. Rev. B* **57**(13): 7556–7570.

Nordlund, K., Keinonen, J., Ghaly, M. and Averback, R.S. 1999. "Coherent displacement of atoms during ion irradiation" *Nature* **398**(6722): 49–51.

Nordlund, K., Tarus, J., Keinonen, J., Donnelly, S.E. and Birtcher, R.C. 2003. "Atomic fingers, bridges and slingshots: Formation of exotic surface structures during ion irradiation of heavy metals" *Nucl. Instr. Meth. Phys. Res. B* **206**: 189–193.

Oen, O.S., Holmes, D.K. and Robinson, M.T. 1963. "Ranges of energetic atoms in solids" *J. Appl. Phys.* **34**: 302.

Rager, J., Flaig, A., Schneider, G., Kaiser, T., Soldera, F. and Mücklich, F. 2005. "Oxidation damage of spark plug electrodes" *Advanced Engineering Materials* **7**: 633–640.

Stichtenoth, D., Wegener, K., Gutsche, C., Regolin, I., Tegude, T-J., Prost, W., Seibt, M. and Ronning, C. 2008. "P-type doping of GaAs nanowires" *Appl. Phys. Lett.* **92**(16): 163107.

Tian, Y., Wang, Z., Jiang, Y., Ma, H., Liu, Z., Geng, Y., Wang, J., Nordlund, K. and Djurabekova, F. 2016. "Modelling of crater formation on anode surface by high-current vacuum arcs" *J. Appl. Phys.* **120**: 126015.

Timko, H., Djurabekova, F., Costelle, L., Nordlund, K., Matyash, K., Schneider, R., Toerklep, A., Arnau-Izquierdo, G., Descoeudres, A., Calatroni, S., Taborelli, M. and Wuensch, W. 2010. "Mechanism of surface modification in the plasma-surface interaction in electric arcs" *Phys. Rev. B* **81**: 184109.

Timko, H., Matyash, K., Schneider, R., Djurabekova, F., Nordlund, K., Hansen, A., Descoeudres, A., Kovermann, J., Grudiev, A., Wuensch, W., Calatroni, A. and Taborelli, M. 2011. "A one-dimensional particle-in-cell model of plasma build-up in vacuum arcs" *Contributions to Plasma Physics* **51**(1): 5–21.

Timko, H., Sjobak, K.N., Mether, L., Calatroni, S., Djurabekova, F., Matyash, K., Nordlund, K., Schneider, R. and Wuensch, W. 2015. "From field emission to vacuum arc ignition: A new tool for simulating copper vacuum arcs" *Contributions to Plasma Physics* **55**: 299.

Urbassek, H.M. and Waldeer, K.T. 1991. "Spikes in condensed rare gases induced by keV-atom bombardment" *Phys. Rev. Lett.* **67**(1): 105.

17

Nuclear Reaction Analysis of Li-Ion Battery Electrodes by Laser-Accelerated Proton Beams

Kunioki Mima

Kazuhisa Fujita

Yoshiaki Kato

Shunsukei Inoue

Shuji Sakabe

17.1 Introduction

Li-ion batteries (LIBs) have been widely used in automobiles like hybrid (HV) and electric (EV) vehicles and in energy storage for intermittent renewable energy sources like solar batteries and wind power stations, since LIBs have characteristics of high specific energy, power availability, high efficiency and the possibility to charge/discharge many times [Armand 2008, Wakihara 2001, Whittingham 2004, Itou 2005]. The charge and discharge performance of LIBs depends on the diffusion rate of lithium (Li) ions in an electrolyte as well as in active materials contained in LIB electrodes [Ogihara 2012]. They also depend on the electrical conductivity in electrodes. When the electrochemical reactions (intercalation and deintercalation of Li ions) in LIBs are uniform and the Li-ion mobility is high enough in charge and discharge processes, the Li-ion concentrations in the cathode and anode have to be uniform. However, when the charge and discharge rate is high, and/or the electrodes are thick and the Li ion mobility is not adequately high, the electrochemical reactions may not proceed homogeneously.

Actually, some works have been devoted to simulate the Li-depth profiles during charge and discharge by computer [Smith 2006, Xiao 2012, Fuller 1994, Zhang 2007]. They show that the Li distributions in the LIB electrodes are not homogeneous. The Li density non-uniformity induces potential fluctuations and inhomogeneous electrical resistance of the electrode. These two facts give rise to a notable degradation of the energy, power density and durability of LIBs, which is undesirable from the performance point of view. Therefore, further improvements in battery development require a better understanding of the Li transport mechanisms during and after charging or discharging. However, the measurement of Li-ion motion is not a trivial task, in particular, if the spatial resolution better than a few µms are required. New *in situ* diagnostic methods like extended X-ray absorption fine structure (EXAFS), neutron small angle scattering and ion beam analysis (IBA) have been developed to diagnose Li distributions in working LIBs.

Proof of principle experiments to diagnose working LIB electrodes with IBA were recently carried out using ion accelerators in Japan and Spain. The experiment results are reported by K. Mima et al. [Mima 2012], R. Gonzalez Arrabal et al. [Arrabal 2015] and A. Yamazaki et al. [Yamazaki 2016]. They show that the Li-depth distribution in the positive electrode is non-uniform in thick electrode samples for high charge and/or discharge rates. The accelerator based IBA will be extended to IBA with laser-driven ion beams.

Several methods in the ion beam analysis (IBA) have been known [Tesmer 1995]. They are Rutherford backscattering (RBS) [Chu 1978], elastic recoil spectroscopy (ERS) [Ecuyer 1976, Deconninck 1978], particle-induced X-ray emission (PIXE) [Tadic 2000, Tadic 2001], particle-induced gamma emission (PIGE) [Tadic 2000, Tadic 2001, Raisanen 1992, Caciolli 2008], nuclear reaction particle analysis (NRA) [Amsel 1984] and nuclear particle activation analysis (NAA) [Vabdecasteele 1988]. As initially described by T. Tadic et al. [Tadic 2000, Tadic 2001], K. Mima et al. [Mima 2012], R. Gonzalez Arrabal et al. [Arrabal 2015] and A. Yamazaki et al. [Yamazaki 2016], ion beam analysis (IBA) techniques have been demonstrated to be very well-suited for fully characterizing the elemental composition of LIBs and, in particular, their Li distribution.

Usually, the electrostatic accelerators are used for IBA, but they are too large to be installed in laboratories. Recently, applications of laser-driven ion beams have become interesting not only for their relative compactness, but also due to the extraordinary high beam quality. Namely, laser-produced ion bunches are of high-energy density, high brightness, low emittance and short bunch duration [Krushelnik 2000, Bulanov 2002, Roth 2001, Borgesi 2004, Cowan 2004, Gibbon 2005, Ter-Avetisyan 2009]. These characteristics of laser-produced ion bunches could become useful in many novel applications such as ion beam cancer therapy, proton imaging and ion fast ignition in laser fusion, as well as ion beam material analysis.

Chapter 17 reviewed the applications of laser protons to IBA for imaging Li distributions in the positive LIB electrode. The samples are LIB composite positive electrodes made by two typical active materials: $LiNi_{0.80}Co_{0.15}Al_{0.05}O_2$ (LNO) and $LiFePO_4$ (LFP) which were provided by the Uchimoto group of Kyoto University and the Toyota Central Research Laboratory.

Section 17.2 describes IBA (micro-PIGE, micro-PIXE, NRA, RBA), where micro-PIGE and micro-PIXE stand for the PIGE and PIXE with micrometre-scale ion beams. In Section 17.3, examples of IBA experiments of LIBs are described. In Section 17.4, the *in situ* measurement is presented. In Section 17.5, the possibilities for applying laser-driven ion bunches to the IBA of LIB diagnostics are described.

17.2 Ion Beam Analysis

It is not easy to diagnose the depth distribution of Li-ions in the LIB electrodes, since they are embedded in the metal layers. Promising methods are NRA, particle PIGE with PIXE and RBS by irradiating ion beams. The diagnostic processes are illustrated in Figure 17.1. There are several nuclear reactions and atomic processes used for diagnosing the Li distribution together with the other elemental distributions in the electrodes.

In proton beam diagnostics, proton beam stopping in the sample should be precisely simulated. Fortunately, multi-MeV ions are mainly stopped by collisions with electrons. So, the orbit is relatively straight. Therefore, if the spatial stopping power distribution in a sample is known, the light element depth distribution can be obtained by scanning a proton beam in space or in energy. In Chapter 17, the PIGE for the reactions of $_3^7Li(p,p'\gamma)_3^7Li$, and $_3^7Li(p,n\gamma)_4^7Be$ as well as PIXE for the atomic processes: $Ni(p,p'\chi)Ni$ $E_\chi = 7.48$ and NRA for the $_3^7Li(p,\alpha)_2^4He$ are considered.

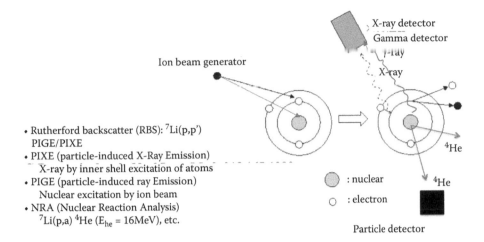

- Rutherford backscatter (RBS): ^{7}Li(p,p')
  PIGE/PIXE
- PIXE (particle-induced X-Ray Emission)
  X-ray by inner shell excitation of atoms
- PIGE (particle-induced ray Emission)
  Nuclear excitation by ion beam
- NRA (Nuclear Reaction Analysis)
  ^{7}Li(p,a) ^{4}He (E_{he} = 16MeV), etc.

FIGURE 17.1 Mechanisms of ion beam analysis [Tesmer 1995, Chu 1978, Ecuyer 1976, Tadic 2000, Tadic 2001, Deconninck 1978, Mateus 2008]; also see Chapter 5 by A. Macchi (Ion Acceleration).

σ_{ZW} (E_p)s the cross section of the nuclear reaction between a proton with energy E_p and a nucleus of charge Z to produce gamma or x-ray photons or particles. The yield of nuclear reaction products, Y_w can be evaluated as:

$$Y_w(\varepsilon,E_{p0})=N_p(E_{p0})\int_0^L dx\sigma_{ZW}\left(E_p(x,y,z;E_{p0})\right)n_Z(x)\,\delta\left[\varepsilon-E_W(E_p)\right]$$ (17.1)

Here, we address the reaction: p + Z = p' + Z' + W where p', Z', and W are scattered proton and nuclear reaction products, respectively and Y_w (ε, E_{p0}) is the yield of the reaction product, W for the energy of ε. E_{p0} and E_p (x, $y.z$; E_{p0}) for the incident proton and the proton at depth, x, respectively. $E_W(E_p$ (x, $y.z$; E_{p0})) is the energy of the product at the position (x, y, z) and $n_z(x)$ is the number density of the element, Z. The stopping power of a proton with energy E_p (x, $y.z$; E_{p0}) is supposed to be S (E_p, x, $y.z$) at the position, x, y, z and E_p (x, $y.z$; E_{p0}) is determined, for an initial proton energy, E_{p0}; by integrating the following stopping power equation from x' = 0 to x:

$$\frac{\partial E_p(x',y,z;E_{p0})}{\partial x'}=-S(E_p,x',y,z)$$ (17.2)

Then, Equation 17.1 can be rewritten as follows:

$$Y_w(\varepsilon,y,z;E_{p0})=N_p(E_{p0})\int_0^{E_{p0}} dE_p\sigma_{ZW}\left(E_p(x,y,z;E_{p0})\right)n_Z\left(x(E_p;E_{p0})\right)$$
$$\times\delta\left[\varepsilon-E_W(E_p)\right]/S(E_p,x,y,z)$$ (17.3)

The elemental density distribution can be obtained from Y_w (ε, $y.z$; E_{p0}). Then we know the energy dependent cross section, σ_{ZW} (E_p) and the local stopping power, S (E_p, x, y, z)

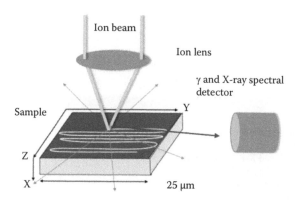

FIGURE 17.2   Beam material analysis. An ion beam is focused to a few micrometres and scans a sample. X-ray and gamma ray spectra are accumulated to obtain respective images.

The energy of a resonance line of an X-ray (PIXE) or gamma ray (PIGE) photon is basically determined by atomic and nuclear excited states and is therefore insensitive to the proton beam energy. So, the spatial distributions of X-ray or gamma ray resonance line intensities are obtained according to:

$$
P_w(y,z;E_{p0}) = \int_{E_W-\delta}^{E_W+\delta} Y_w(\varepsilon,y,z;E_{p0})d\varepsilon
$$

$$
= N_p(E_{p0}) \int_0^{E_{p0}} dE_p \sigma_{ZW}\left(E_p(x,y,z;E_{p0})\right) n_Z\left(x(E_p;E_{p0})\right)/S(E_p,x,y,z),
$$

(17.4)

which depends upon the incident proton energy, E_{p0}. When the cross section, $\sigma_{ZW}(E_p)$ features a sharp resonance (i.e. a strong proton energy dependence), the E_{p0} dependence of $P_W(y,z;E_{p0})$ can be related to the elemental depth distribution directly, since $E_{p0} - E_p$ is proportional to the depth × near the surface. If the cross section has a weak energy dependence, it is not easy to obtain the elemental depth distribution from the injected ion energy dependence of $P_{Z\gamma}(E_{p0},x,y)$. The transverse, namely y-z distribution of the nuclear element Z can be obtained if we could use a micro-ion beam. In this case, the spatial resolution of the distribution measurement is determined by the size of the ion beam, as shown in Figure 17.2.

For an example, a micro proton beam of diameter of 1.0 micron was used to obtain a y-z distribution of Li in LIB electrodes at the TIARA facility of the National Institutes for Quantum and Radiological Science and Technology (QST), Japan [Kamiya 1996]. The schematic view of the beam scanning to acquire the y-z distribution is shown in Figure 17.2. The images taken by micro-PIXE and micro-PIGE are described in the next section.

17.3 IBA with a Microproton Beam for Active Materials in the LIB Electrodes

In the Li ion battery, it is important to see Li ion behaviour in electrodes during the charge and discharge. In Figure 17.3, the Li ion motion in the Li ion battery is shown schematically. The anode contains the active material like $Li_xNi_{0.85}Co_{0.15}Al_{0.05}O_2$ and the cathode contains the active material like Li_xC_6, as shown in the Figure 17.1. In the charging processes, Li ions in the anode (the active material is $Li_xNi_{0.85}Co_{0.15}Al_{0.05}O_2$) deintercalate into the electrolyte and intercalate into the cathode (the active material is Li_xC_6). When Li ions deintercalate from the $Li_xNi_{0.85}Co_{0.15}Al_{0.05}O_2$ surface during the charge,

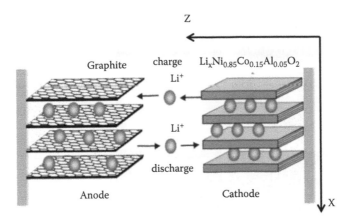

FIGURE 17.3 Li ion motion in the charge and discharge processes of a Li ion battery. Li ions move from the cathode to the anode or from the anode to the cathode. Li ions are located between two lattice meshes of graphite or $LiNi_{0.85}Co_{0.1}Al_{0.05}O_2$.

the Li ion concentration is lower near the electrolyte layer and increases near the conductor which is connected to an external power supply. The gradient of the Li concentration depends upon the charging rate, the active materials ion conductivity, the electrode structure, and so on.

We irradiated the cathode samples shown in the scanning electron microscopy (SEM) images of Figure 17.4 with a 3 MeV microproton beam from the TIARA accelerator as shown in Figure 17.2. The Figure 17.5 shows the energy-dependent cross section of the $^7_3Li(p,p'\gamma)^7_3Li$ eaction for PIGE. The threshold of this process is about 800 keV and the resonance peak exists near 1 MeV. The stopping range of 3 MeV protons is about 45 μm in $Li_xNi_{0.85}Co_{0.15}Al_{0.05}O_2$. So, the expected gamma yields induced by 3 MeV protons is about 10^8 μ/C which is enough for the required resolution. On the other hand, the cross section for $^7_3Li(p,n\gamma)^7_4Be$ has a threshold at 1.8 MeV and the resonance energy is 2.3 MeV. The gamma ray energy is 0.429 MeV. An example of the gamma ray spectrum for 3 MeV irradiation is shown in Figure 17.6. The X-ray spectrum (PIXE) was also obtained for the irradiated sample as shown in Figure 17.6. The spatial distribution of the Ni-K_β emission is shown in Figure 17.7 which indicates the Ni density distribution. Note that the PIGE image for Li (Figure 17.4b) and the PIXE image for Ni (Figure 17.7) correspond to each other. Therefore, the ratio of these PIGE and PIXE signals yields the fraction of Li relative to Ni.

For Figures 17.4 and 17.7, the size of the active particle is about 5 μm wide and 10 μm in length. So, we found that the PIGE and PIXE resolution is better than a few micrometres. Since the proton beam diameter was 1.5 μm the spatial resolution will be a few μm Figure 17.7 shows the set of data which could be used for the Li battery cathode diagnostics. Figure 17.7c shows that the distribution of the signal ratio of PIXE and PIGE is a flat-top shape. This means that the Li concentration is uniform in the 10 μm scale active particle. Similar measurements have been done for the charged cathode. The results are similar except the average signal ratio of PIGE to PIXE images is lower than that for the non-charged cathode. This means that the Li is uniformly intercalated during the discharge for $Li_xNi_{0.85}Co_{0.15}Al_{0.05}O_2$ particles.

We have demonstrated by the TIARA experiment by using the microproton, that the nuclear material analysis by ion beams is useful for diagnosing Li concentration of active materials in Li battery electrodes with a spatial resolution of 1.5 μm. This diagnostic method can be applied to developing advanced Li-ion batteries.

As an example of applying ion beam analysis to the research and development of advanced Li-ion batteries, we diagnosed the Li distribution in the z direction of Figure 17.8 (i.e. perpendicular to the anode surface from the Al electrode to separator). When the anode is charged, the Li ions are deintercalated from $LiNi_{0.8}Co_{0.15}Al_{0.05}O_2$ active particles. In this case, Li ions move out in the z direction and they are intercalated into the graphite layer.

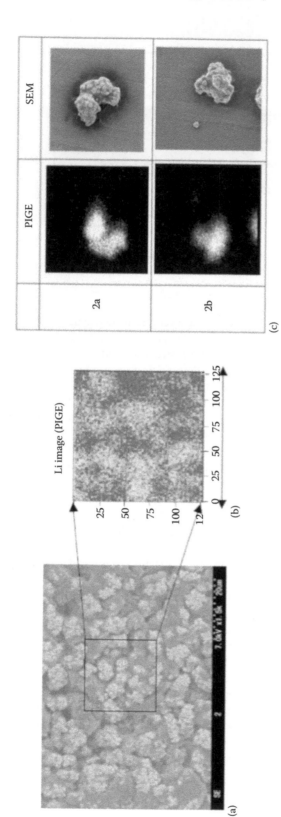

FIGURE 17.4 (a) An SEM image of the surface of a Li battery cathode and (b) a Li PIGE image taken at TIARA in Takasaki, Japan. The whole scale of the SEM is 100 μm × 100 μm and the PIGE image scale is 25 μm × 25 μm. The dark grey areas indicate low Li concentration, where secondary particles of $LiNi_{0.85}Co_{0.1}Al_{0.05}O_2$ exist. The dark grey areas are dominant and light grey areas indicate high Li concentration, where binder is provided in a CRC webfolder, where light grey areas in the black and white version correspond to light blue areas and dark grey areas to dark blue areas). The size of the secondary particle is approximately 5 μm ∼ 10 μm. (c) PIGE and SEM images for two secondary particles. SEM and PIGE images exhibit mirror symmetry relative to each other. (Reproduced with permission from Mima, K. et al., 2012. "Li distribution characterization in Li-ion batteries positive electrodes containing $Li_xNi_{0.85}Co_{0.1}Al_{0.05}O_2$ secondary particles (0.75 < x < 1.0)" *Nuclear Instruments and Methods* **B** 290: 79–84.)

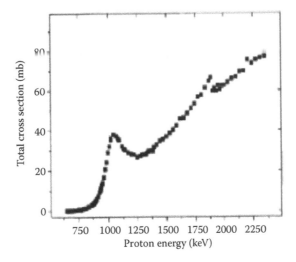

FIGURE 17.5 Energy-dependent cross section of the Li^7 (p.p'γ) Li^7 process.

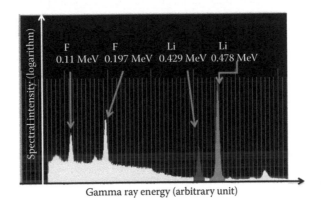

FIGURE 17.6 Gamma ray spectrum from a 3 MeV irradiated Li battery cathode surface coated with active material, LiNi$_{0.85}$Co$_{0.15}$Al$_{0.05}$O$_2$. The 0.429 MeV and 0.478 MeV peaks are from Li and the 0.11 MeV and 0.197 MeV peaks are from F contained in the binder for pasting LiNi$_{0.85}$Co$_{0.15}$Al$_{0.05}$O$_2$ particles.

If the charging process is rapid, it is expected that Li diffusion process does not smooth out the Li distribution in the anode. So, more Li ions are lost near the surface which faces the cathode. The mobility of Li depends upon how the anode is fabricated. For efficient energy storage in a rapid charge, the uniform extraction of Li in the anode is better. Therefore, it is important to measure the Li-depth distribution in the charged anode. In the Figure 17.8c, the SEM image shows the cross section of the anode, where it has been cut after a rapid charge (15% of Li contained in LiNi$_{0.8}$Co$_{0.15}$Al$_{0.05}$O$_2$ particles is extracted in 2 hours of charging). The cross section was measured by the TIARA ion beam. The white particles of the SEM image represent LiNi$_{0.8}$Co$_{0.15}$Al$_{0.05}$O$_2$. The PIGE image of the cross section and PIGE line profile along z images are shown in the Figure 17.8a and 17.8b, respectively. It is noted that the PIGE profile (b) indicates that Li density is 15 ~ 20% higher near the Al electrode surface than for the surface facing to the cathode. This means that the deintercalation of Li from the anode is not uniform in this case and the total amount of energy storage is reduced for the fast charge. This also demonstrates that the PIGE is a useful tool for Li distribution measurements.

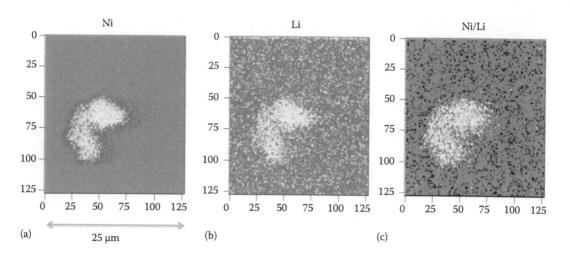

FIGURE 17.7 PIXE and PIGE images for the non-charged cathode: (a) the PIXE image for the $Ni - K_\alpha$, (b) the PIGE image for the 0.478 MeV Li gamma line, (c) the ratio of the PIXE image to the PIGE image. The full scale of the pictures is 25 µm. (Reproduced with permission from Mima, K. et al., 2012. "Li distribution characterization in Li-ion batteries positive electrodes containing $Li_x Ni_{0.8} Co_{0.15} Al_{0.05} O_{0.2}$ secondary particles (0.75 < × < 1.0)" *Nuclear Instruments and Methods B* 290: 79–84.)

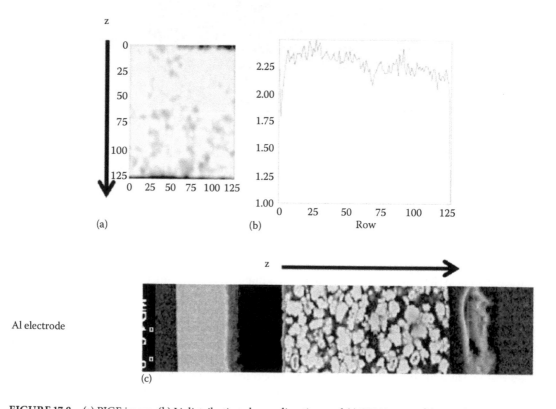

FIGURE 17.8 (a) PIGE image, (b) Li distribution along z direction, and (c) SEM image of the anode cross section.

17.4 Nuclear Reaction Analysis with Particle Emission for Li-Depth Profiling

The possible reactions available for Li characterization by using the particle emission reactions are: $^7Li(^3He,\alpha_0)$ ^6Li, $^7Li(^3He,d)^8Be$, $^7Li(^3He,p)^9Be$ and $7Li(p,\alpha)^4He$ [Paneta 2012]. For proper analysis of the results, an accurate knowledge of the reaction cross section is needed. Since accurate analytical theory of nuclear reaction cross sections does not exist at present, experimental data have to be used. However, most of the listed nuclear reaction cross sections were measured in the past for nuclear physics purposes. They are often not precise enough for NRA. When carrying out NRA experiments one has to carry out the experiments at those energies and detection angles for which the cross-section data exists and is well characterized. The cross section for Li characterization by using the $^7Li(p,\alpha)^4He$ nuclear reaction, which is the most suitable one for Li-depth profiling, has been recently reported by Paneta et al. [Paneta 2012] for proton beam energies in the range, 1.7 to 7.0 MeV and for detection angles of 140°, 150°, 160° and 170° to the beam direction.

In the case of this nuclear reaction, the ^4He particle emitted from the nuclear reaction is of energy about 7 MeV which exceeds that for incident protons (of energy near 3 MeV). We can measure the RBS protons and alpha particles simultaneously. However, because the RBS proton signal can be much higher than the alpha particle signal, special attention is needed to select the proper fluence which allows good resolution without saturating the detector.

Often more than one reaction can take place in LIB samples, emitting different particles at different energies. Therefore, advanced, detailed information on the elemental composition of the samples is very useful for implementing proper and accurate interpretation of the measured spectra. LIBs typically contain light elements like Fluorum (F) and Aluminum (Al) in the electrode and current collector which can lead to additional nuclear reactions.

We have characterized the depth distribution of Li in $LiFePO_4$ (LFP) positive electrodes, fabricated at the Graduate School of Human and Environmental Studies at the Kyoto University. The positive electrodes were fabricated from a mixture of the active material (LFP) with artificial carbon and binder (polyvinyliden fluoride: PVdF) which glues the powder of the active particles and the artificial carbon powder. SEM images (similar to that shown in Figure 17.4a) indicate that the active particles distribute randomly and homogeneously within the electrode. The average secondary active particle diameter is between 0.5 and 1.0 μm. The thickness and the areal density of the positive electrodes are 30 μm and 1.0 mg/cm2, respectively. This positive electrode is coated on an Al current collector with a thickness of 25 μm The charged samples were fabricated by attaching the electrode to the separator. Li was used as the negative electrode. The current density during the charge was 45 mA/cm^2 and the charging time was 1 min. More details about the sample preparation procedure and characterization are reported in [Orikasa 2013a, Orikasa 2013b]. It is worth noting that even though $LiFePO_4$ based systems have a reduced operating voltage in comparison to typical LiM_xO_{2x} positive electrode materials, they show a number of advantages such as non-toxicity, low metal costs and good cycle-life performance.

As reported by Sagara et al. [Sagara 1988], the $^7Li(p\alpha,)e$ is the most suitable nuclear reaction for this analysis for the following reasons: the intensity of the signal is proportional to the amount of the naturally dominant ^7Li isotope and the reaction cross section exhibits a broad maximum at a proton energy near 3 MeV which allows estimation of the depth profile directly from the energy spectrum of the α-particles emitted in the nuclear reaction.

Following Sagara's work, we have characterised the Li depth profiles by using the $^7Li(p\alpha,)e$ nuclear reaction. Measurements were carried out at the standard beam line of CNA at Seville University [Garcia 2000]. For these measurements, a H+ beam impinging onto the sample surface at normal incidence at an energy of 3 MeV was used. The beam spot and the current were selected to be 2 × 2 mm^2 and ~15 nA, respectively. Because no foil was used between the sample and the detector, the α-particles produced in the nuclear reaction and also the backscattered protons from other elements (Fe, P, O, etc.) present in the electrodes were detected by a Si-barrier detector with an active area of 50 mm², located at 10 cm from the sample and at an angle of 150° relative to the proton beam injection direction.

Figure 17.9 shows a typical NRA spectrum for a LiFePO4 sample. The two peaks observed (channel 600 to 920) correspond to the α-particles from the $^7Li(p,\alpha)^4He$ and $^{19}F(p,\alpha)^{16}O$ nuclear reactions (see inset in Figure 17.9). F is contained in the binder. The energy of the emitted α-particles in both nuclear reactions is about 7.7 MeV for ^7Li and 7.9 MeV for ^{19}F. An overlap of these two peaks cannot be avoided and the deconvolution of the two peaks can be done only by appropriate data analysis. The peaks in the low energy region of the spectra (channels 10 to 310) are the RBS protons from Fe P, O and other elements in the sample.

The depth distribution of the Li concentration in the sample can be determined by comparing the measured and calculated spectra. For the calculations the commercial computer code SIMNRA can be used [Mayer 1997]. The cross section data related to the Li and F nuclear reactions were taken from Paneta et al. [Paneta 2012] and introduced as an R33 file in the SIMNRA code. First the spectrum was calculated by assuming that the Li and F concentration were homogeneously distributed along the whole depth. As illustrated in Figure 17.10a, clear discrepancies are observed between the measured and calculated spectra. In particular, the measured yield at smaller energies, which corresponds to the signals for depths well below the sample surface, is significantly higher than the calculated one. Possible reasons for these discrepancies are that the Li concentration along depth is not homogeneous, the F concentration along depth is not homogeneous or both of them are not homogeneous. However, since the distribution of the binder in the sample is homogeneous and F is contained in the binder, there is no physical reason to assume that the F distribution along depth is inhomogeneous. Therefore, in order to properly fit the spectrum, only the Li distribution was assumed to be inhomogeneous along the electrode thickness (depth). To calculate the spectrum under this assumption, the sample was considered in several layers, each of them containing a different Li concentration. Then, calculation runs were carried out by varying the number of layers and their Li concentration until a good fit was achieved. As depicted in Figure

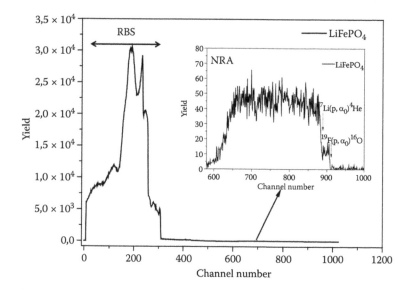

FIGURE 17.9 Particle spectra of NRA for a LiFePO$_4$ sample. The peaks in the low energy region of the spectra (channels 10 to 310) are mainly RBA protons from Fe, P, O and other elements present in the sample. The two peaks in the high-energy region of the spectra (channels 600 to 920) correspond to the α-particles from the $^7Li(p\alpha,)e$ (left) and $^{19}F(p\alpha,)O$ (right) nuclear reactions (see inset). (Reproduced with permission from Arrabal, R.G. et al., 2015. "Meso-scale characterization of lithium distribution in lithium-ion batteries using ion beam analysis techniques" *Journal of Power Sources* 299: 587–595.)

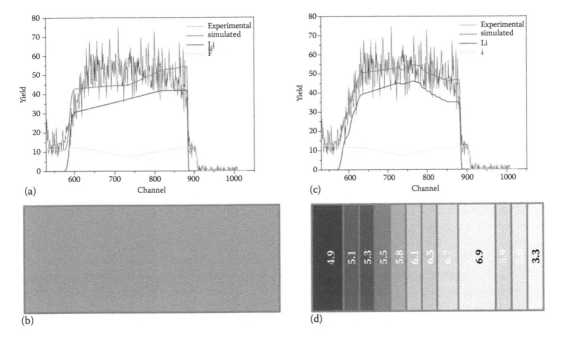

FIGURE 17.10   Measured (black line) and calculated (red line) NRA spectra for a LiFePO$_4$ positive electrode. For the calculation, the Li distribution along the depth was considered to be homogeneous (a) and inhomogeneous (b). The contributions to the calculation by Li and F are shown by the blue and green lines, respectively. The number of layers corresponding to the best fit of the spectrum together with the Li content in each of them are shown in (c), where the left-hand side corresponds to the boundary of the separator, and the right-hand side to the current collector. (The three curves from bottom to top in Figure 17.10a and 17.10c are green, blue and red in the colour version. The levels of concentration of Li in Figure 17.10b and 17.10d are indicated by various blue grey colours in the colour version located in the CRC webfolder for the ALPA book. (Reproduced with permission from Arrabal, R. G. et al., 2015. "Meso-scale characterization of lithium distribution in lithium-ion batteries using ion beam analysis techniques" *Journal of Power Sources* 299: 587–595.)

17.10c and 17.10d, the best fit was obtained by assuming that the total sample consists of 12 layers, each of them containing a certain amount of Li which increases from a 4.9 atomic percentage (at.%) for the layer close to the separator up to a 6.9 at.% for the layer close to the current collector. The layer thickness and the Li content in each layer are schematically shown in Figure 17.10d, where the left-hand side is close to the separator and the right-hand side is close to the current collector. These results indicate that the Li concentration distribution is inhomogeneous along the electrode thickness and it rises with increasing depth from the separator.

17.5   Application of Laser-Produced Proton Beams for PIGE and Concluding Remarks

We consider the possibility of using laser-produced proton beams. It has been experimentally verified that laser produced ion beams are of very low transverse emittance [Cowan 2004] and they are laminar when the beam is dispersed by energy separation in a spectrometer [Ter-Avetisyan 2009]. The reader is also referred to Chapter 5 of this book by Macchi where the state of the art for laser-driven ion acceleration is briefly reviewed. The emittance measured by Cowan et al. [Cowan 2004] is lower than 0.004 mm · mrad although the energy is not monochromatic. This means that it is possible to focus the beam to diameters of only a few micrometres with a meter-long focal length, although the laser-produced proton beam has a unique phase space shape as measured at the Max Born Institute [Ter-Avetisyan 2009] and as shown in Figure 17.11.

When the beam energy is fixed (3 MeV for example), the phase space distribution of the ion beam ressembles 'a bent thin line'. The area of the line, namely the emittance, is supposed to be less than 0.004 mm mrad [Cowan 2004, Ter-Avetisyan 2009]. Although it is necessary to develop new focusing optics for such a laser produced proton beam, it could be focused to a small spot by taking advantage of the very small natural emittance.

Because the energy spread requirement of the injected proton beam is not very strict for the PIGE imaging, the issue is how to focus a proton beam which has finite energy spread and the complex phase space shape. One possible way is that we cut out a small part of the central region (from –0.05 mm to 0.05 mm by the first slit in Figure 17.11) and then separate the proton energy by a magnet as shown in the Figure 17.11. Then the beam can be divided into a few tens of energy groups where each beam group can be focused separately with all beams subsequently recombined to a few-micron spot. The energy in each focusing device is quasi-monochromatic.

When the energy spread, $\Delta E/E$ is 5% and the beam size in the focusing quadrupole magnet is 100 μm^ϕ the focused proton beam size will be 5 μm^ϕ which meets the requirement for the application to Li battery analysis. The required proton current for the analysis was less than 500 pA which means the number of proton per pulse is 4×10^7 for 100 Hz operation with sub-picosecond duration. This will typically be 0.1 ~ 0.01% of the total yield of laser produced protons for a 1 J laser pulse energy. So, the proton flux requirement seems feasible. Further improvement can be expected, since the generation of quasi-monochromatic laser proton spectra has been achieved in recent laser ion experiments and theory [Hegelich 2006, Esirkepov 2002, Schwoerer 2006, Tonican 2006, Murakami 2009, Qiao 2009].

In conclusion, in electrostatic ion accelerator experiments, we have demonstrated that MeV micro proton beams are useful for the Li-ion battery analysis. According to the experimental results and recently published laser ion research results, a laser produced proton beam can be applicable to the material analysis. In order to test the feasibility of the application of laser proton beams to the IBA of LIBs, ion beam generation experiments have been designed and carried out at the ICR at Kyoto University.

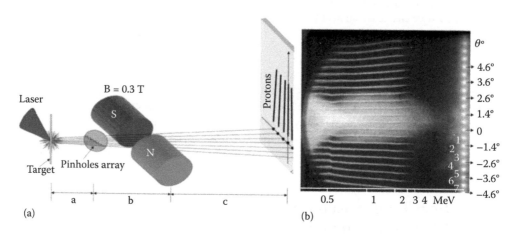

FIGURE 17.11   (a) The proton beam emittance is measured by the pin hole array and bending magnet where a = 85 mm, b = 120 mm, and c = 275 mm. (b) images of the beamlets (lines) which are steered by the magnet. The small white spots in the right edge of image (b) are the X-ray images of pinholes. The length of the lines indicates the energy spread and the width of the lines indicates the emittance of the each beamlet. The vertical scale is the divergence angle from the laser spot and indicates that the angular spread of each beamlet is less than 1.0 mrad. (Reproduced with permission from Ter-Avetisyan, S. et.al., 2009. "Characterization and control of ion sources from ultra-short high-intensity laser–foil interaction" *Plasma Phys. and Cont. Fusion: IOP Science* 51: 124046.)

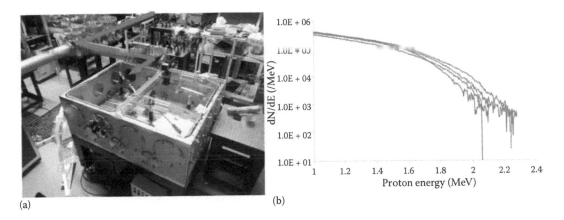

FIGURE 17.12 (a) Laser-driven proton beam generation chamber. (b) Proton energy spectra for four shots.

The laser irradiation chamber and the proton energy spectra are shown in the Figure 17.12. The laser parameters are 800 nm central wavelength and pulse energy of 0.5 J within a 100 fs pulse duration. The maximum proton energy reaches 2.2 MeV by irradiating CH foil of 10 μm thickness.

The proton beam is then injected into a LIF plate, as shown in the Figure 17.13a to detect alpha particles emitted by $^7Li(p,\alpha)^4He$ nuclear reactions. By obtaining the alpha particle energy spectrum of the $^7Li(p,\alpha)^4He$ reaction, the laser produced proton beam can be used to measure the Li depth distribution in a LIB sample which requires production of more than 10^9 protons/sec.

In Figure 17.14, $K_\alpha - X -$ rays are emitted from a sample by irradiating with laser produced MeV protons. By using this type of X-ray microscope, those K_α-X-rays can be used for imaging distributions of ions like Li. Typically, the emittance of the proton beam accelerated by the TNSA is about 1 mm mrad as indicated by the proton beam phase space distribution of the proton beam which is evaluated by the setup shown in Figure 17.11 [Ter-Avetisyan 2009]. Therefore, when the distance between the TNSA foil and the sample in Figure 17.14 is 10 cm, then the irradiation area will be a few 100 μm with appropriate ion beam optics. So, the alpha particle signal from the 100×100 μm^2 area of the sample is enough to obtain an alpha particle spectrum which can be reduced to the local Li-ion depth distribution.

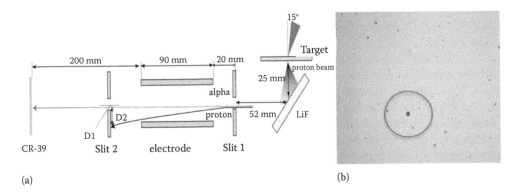

FIGURE 17.13 (a) Sample irradiation and measurement system by a laser-produced proton beam. A short-pulse laser irradiates a target to produce a proton beam which is injected onto the LIB sample. The alpha particles and scattered protons passing through the slit 1 are bent by the transverse electric field to select alpha particles to see the concentration of Li ions on the sample surface, (b) CR 39 film tracks of protons (small pits) and an alpha particle track indicated by a black circle (red in the webfolder colour version).

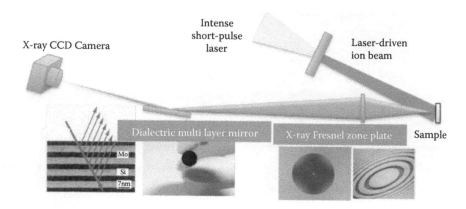

FIGURE 17.14   Example of a schematic set-up for the ion beam driven X-ray microscope.

References for Chapter 17

Amsel. G. and Lanford, W.A. 1984. "Nuclear reaction techniques of materials analysis" *Ann. Rev. Nucl. Part. Sci.* **34**: 435–460.

Armand, M. and Tarascon, J-M. 2008. "Building better batteries" *Nature* **451**: 652–657.

Arrabal, R.G. et al., 2015. "Meso-scale characterization of lithium distribution in lithium-ion batteries using ion beam analysis techniques" *J. Power Sources* **299**: 587–595.

Borghesi, M. et al., 2004. "Multi-MeV proton source investigations in ultraintense laser-foil interactions" *Phys. Rev. Lett.* **92**: 055003-1–4.

Bulanov, S.V. et al., 2002. "Oncological hadron therapy with laser ion acclerators" *Phys. Lett. A* **299**: 240–247.

Caciolli, A. et al., 2008. "Proton elastic scattering and proton induced γ-ray emission cross-sections on Na from 2 to 5 MeV" *Nuclear Instruments and Methods B* **266**: 1392–1396.

Cowan, T.E. et al., 2004. "Ultralow emittance, multi-MeV proton beams from a laser virtual-cathode plasma accelerator" *Phys. Rev. Lett.* **92**: 204801.

Chu, W.K. et al., 1978. *Back Scattering Spectroscopy*. Academic Press, New York.

Deconninck, G. 1978. *Introduction to Radioanalytical Physics, Nuclear Methods Monographs*. Elsevier Scientific Pub. Co.; distribution for the U.S.A. and Canada, Elsevier/North-Holland, Amsterdam; New York New York.

Ecuyer, J.L. et al., 1976. "An accurate and sensitive method for the determination of the depth distribution of light elements in heavy material" *J. Appl. Phys.* **47**: 381–382.

Esirkepov, T. et al., 2002. "Generation of high-quality laser-accelerated ion beams" *Phys. Rev. Lett.* **89**: 175003.

Fuller, T.F., Doyle, M. and Newman, J. 1994. "Simulation and optimization of the dual Lithium ion insertion cell" *J. Electrochem. Soc.* **141**: 1–10.

Garcia López, J., Ager, F.J., Barbadillo Rank, M. et al., 2000. "CNA: The first accelerator-based IBA facility in Spain" *Nuclear Instruments and Methods B* **161–163**: 1137–1142.

Gibbon P. 2005. *Short Pulse Laser Interactions with Matter*. 1st ed. London: Imperial College Press.

Hegelich, B.M. Albright, B.J. Cobble, J. Flippo, K., Letzring, S. et al., 2006. "Laser acceleration of quasi-monoenergetic MeV ion beams" *Nature* **439** (7075): 441–4.

Itou, Y. and Ukyo, Y. 2005. "Performance of LiNiCoO$_2$ materials for advanced lithium-ion batteries" *J. Power Sources* **146**: 39–44.

Kamiya, T., Suda, T. and Tanaka, R. 1996. "High energy single ion hit system combined with heavy ion microbeam apparatus" *Nuclear Instruments and Methods B* **118**: 423–425.

Krushelnik, K. et al., 2000. "Ultrahigh-intensity laser-produced plasmas as a compact heavy ion injection source" *IEEE Trans. Plasma Sci.* **28**: 1110–1115.

Mateus, M. et al., 2008. "PIGE analysis and profiling of aluminum" *Nuclear Instruments and Methods B* **266**: 1490–1492.

Mayer, M. 1997. "SIMNRA" Max Planck Institut für Plasmaphysik, n.d. IPP9/113.

Mima, K., Gonzalez-Arrabal, R., Azuma, H., Yamazaki, A., Okuda, C., Ukyo, Y., Sawada, H., Fujita, K., Kato, Y., Perlado, J.M. and Nakai, S. 2012. "Li distribution characterization in Li-ion batteries positive electrodes containing LixNi$_{0.8}$ Co$_{0.15}$Al$_{0.05}$ O$_{0.2}$ secondary particles (0.75 < × < 1.0)" *Nuclear Instruments and Methods B* **290**: 79–84.

Murakami, M. and Mima, K. 2009. "Efficient generation of quasi-mono energetic ions by Coulomb explosions of optimized nanostructured clusters" *Physics Plasmas* **16**: 103108.

Ogihara, N., Kawauchi, S., Okuda, C., Itou, Y., Takeuchi, Y. and Ukyo, Y. 2012. "Theoretical and experimental analysis of porous electrodes for lithium-ion batteries by electrochemical impedance spectroscopy using a symmetric cell" *J. Electrochem. Soc.* **159A**: 1034–1039.

Orikasa, Y., Maeda, T., Koyama, Y., Murayama, H., Fukuda, K., Tanida, H. et al., 2013a. "Direct observation of a metastable crystal phase of LixFePO4 under electrochemical phase transition" *J. Am. Chem. Soc.* **135**: 5497–5500.

Orikasa, Y. et al., 2013b. "Transient phase change in two phase reaction between LiFePO4 and FePO4 under battery operation" *Chem. Mater.* **25**: 1032–1039.

Paneta, V., Kafkarkou, A., Kokkoris, M. and Lagoyannis, A. 2012. "Differential cross-section measurements for the 7Li(p,p0)7Li, 7Li(p,p1)7Li, 7Li(p,α0)4He, 19F(p,p0)19F, 19F(p,α0)16O and 19F(p,α1,2)16O reactions" *Nucl. Instrum. Methods Phys. Res. Sect. B Beam Interact. Mater. At.* **288**: 53–59.

Qiao B. et al., 2009. "Stable GeV ion-beam acceleration from thin foils by circularly polarized laser pulses" *Phys. Rev. Lett.* **102**: 145002.

Raisanen, J. 1992. "Analysis of lithium by ion beam methods" *Nuclear Instruments and Methods. B* **66**: 107–117.

Roth, M. et al., 2001. "Fast ignition by intense laser-accelerated proton beams" *Phys. Rev. Lett.* **86**: 436.

Sagara, A., Kamada, K. and Yamaguchi, S. 1983. "Depth profiling of lithium by use of the nuclear reaction 7Li(p, α)4IIe" *Nucl. Instrum. Methods Phys. Res. Sect. B Beam Interact. Mater. At.* **34**: 465–469.

Schwoerer, H. et al., 2006. "Laser-plasma acceleration of quasi-monoenergetic protons from microstructured targets" *Nature* **439**: 445–448.

Smith, K. and Wang, C--Y. 2006. "Power and thermal characterization of a lithium-ion battery pack for hybrid-electric vehicles" *J. Power Sources* **160**: 662–673.

Tadic, T., Jaksic, M., Capiglia, C., Saito, Y. and Mustarelli, P. 2000. "External microbeam PIGE study of Li and F distribution in PVdF/HFP electrolyte gel polymer for lithium battery application" *Nuclear Instruments and Methods B* **161–163**: 614–618.

Tadic, T., Jaksic, M., Medunioc, Z., Quarterone, E. and Mustarelli, P. 2001. "Microbeam studies of gel–polymer interfaces with Li anode and spinel cathode for Li ion battery applications using PIGE and PIXE spectroscopy" *Nuclear Instruments and Methods. B* **181**: 404–407.

Ter-Avetisyan, S., Borgeshi, M., Nickles, P., Nakamura, T., Mima, K. et al., 2009. "Characterization and control of ion sources from ultra-short high-intensity laser–foil interaction" *Plasma Phys. and Cont. Fusion* **51**: 124046.

Tesmer, J.R. and Natasi, M. 1995. "Handbook of Modern ion beam material analysis" Material Research Society. Pittsburg, Pennsylvania, USA.

Tonican, T. et al., 2006. "Ultrafast laser-driven microlens to focus and energy-select mega-electron volt protons" *Science* **312**: 410–413.

Vandecasteele, M. et al., 1988. *Activation Analysis with Charged Particles.* Ellis Horwood Ltd., Chichester, UK.

Wakihara, M. 2001. "Recent Developments in Lithium Ion Batteries" *Materials. Science and Engineering R33:* 109–134.

Whittingham, M.S. 2004. "Lithium batteries and cathode materials" *Chem. Rev.* **104:** 4271–4301.

Xiao, M. and Choe, S.Y. 2012. "Dynamic modeling and analysis of a pouch type $LiMn_2O_4$/Carbon high power Li-polymer battery based on electrochemical-thermal principles" *J. Power Sources* **218:** 357-367.

Yamazaki, A., Orikasa, Y. et al., 2016. "In-situ measurement of the lithium distribution in Li-ion batteries using micro-IBA techniques" *Nuclear Instruments and Methods B* **371:** 298302.

Zhang, Q. and White, R.E. 2007. "Comparison of approximate solution methods for the solid phase diffusion equation in a porous electrode model" *J. Power Sources* **165:** 880–886.

18

Possible Roles of Broad Energy Distribution in Ion Implantation and Pulsed Structure in Perturbed Angular Distribution Studies

Sanjeev K. Srivastava

Devesh K. Avasthi

18.1 Introduction

Laser-accelerated ion bunches can contain multiple ion species, each with a broad energy distribution and narrow time (pulsed) structure. A detailed description of the recent developments in the field, the basic concepts and the nature of the laser-driven ions can be found in the Chapter 5, viz. 'Laser-driven Ion Acceleration' in Part I of this book. Chapter 18 brings out the applicability of the two features in materials science and microscopic study of solids, respectively, in the following two distinct ways. First, the broad energy distribution, which is generally considered as a deterrent in utilization of such ion bunches for applications, can be suitably exploited in ion implantation, with an advantage over conventional accelerator-based implantation [Nastasi 2006]. The advantage is the achievability of simultaneous and uniform distribution of a dopant in the near-surface region, i.e., from a few tens of nanometers, to about a micron in depth. The multiplicity of the laser-accelerated ions would further add to this advantage, because it enables one to do such implantations simultaneously with multiple dopants. There are a number of interesting cases in materials science, such as with requirements of low dopant concentrations or with formation of small clusters and nanoparticles buried in the near-surface region, where such ions can be suitably employed. Second, the pulsed structure, along with the achievable ion energies sufficient to produce nuclear fusion evaporation reactions [Schieck 2014], opens up another possibility to exploit these ions for perturbed angular distribution (PAD) measurements [Recknagel 1972]. PAD with laser-accelerated ions has two distinct advantages over the conventional accelerator-based technique: the availability of multiple ion species affords the generation of multiple PAD nuclear probes simultaneously, enabling one to study microscopic electronic and magnetic properties at different probe sites locally in a single experiment and the potential for control of the probe pulse duration to sub-nanosecond levels, as is further discussed in Chapter 18, would facilitate utilizing all the nuclear probes (formed in initial nuclear reactions) with half-lives as short as 1 ns, thus expanding the number of sites that could be studied simultaneously.

18.2 Broad Energy Distribution for Implantation

Ion implantation is a process of introducing or doping impurity atoms into a host or substrate matrix (i.e. doping a substrate) by bombarding it with ions of the desired species, accelerated to a predetermined energy [Nastasi 2006]. The atoms are ionized first in a conventional ion source and then accelerated typically up to 400 keV electrostatically in a low-energy accelerator, known as an ion implanter. While penetrating the substrate, these ions undergo numerous collisions with the substrate atoms. Consequently, the ions lose all their kinetic energy and are therefore stopped at a certain energy-dependent mean depth which is referred to as the ion range. The collisions being stochastic in nature, the different ions follow different paths in the substrate and undergo different number of collisions, resulting in a characteristic angular spread. As a result, they can stop at different depths characterized overall by a mean depth (i.e. the ion range) and a Gaussian depth distribution, whose full width at half maximum (FWHM) is the measure of the range straggling [Nastasi 2006]. Doping by ion implantation, as described above, has definite advantages over doping by a conventional diffusion [Schubert 2005] process. This includes the precise control of penetration depth, concentration of dopants and spatial selectivity. The penetration or implantation depth can be controlled by appropriately selecting the ion energy, whereas the desired concentration can be achieved by the ion flux density, the irradiation time interval and the range straggling. The straggling can be determined from the computer simulation program SRIM [Ziegler 2008], which is also discussed elsewhere in this chapter. The control of spatial selectivity refers to precise control of dopant distribution in the substrate not only longitudinally (in the z direction for depth) but also transversely (in the x-y plane that is transverse to the nominal ion propagation direction). Doping by diffusion, on the other hand, is governed by concentration gradients and thermodynamical considerations, and hence the achievable dopant concentrations in x, y and z directions are not as controllable [Ohring 2002]. The superiority of ion implantation over diffusion, thus, brought a revolution in the semiconductor industry in the 1960s [Rimini 1995].

Range straggling is the key parameter in determining the fluence, i.e. the number of ions per unit area, required to achieve a desired concentration of dopants in the host matrix. It can be made clear with the following example: Synthesizing buried Ag nanoparticles in a PbTe thin film by Ag implantation in order to improve its thermoelectric properties would require implanting Ag atoms in the film beyond its solubility limit, which is typically 5 atomic percent. Considering that PbTe has about 3×10^{22} atoms/cm^3, the required Ag concentration would be $\geq 1.5 \times 10^{21}$ atoms/cm^3. Further, the energy of Ag ions needs to be specified so that the ion range is well within the film thickness. So, if the PbTe film thickness is 200 nm, Ag ions of 200 keV energy would be appropriate because they will be implanted at the 52 ± 33 nm (range ± straggling) depth in the film (which can be estimated by the SRIM code). The ion fluence required to reach this specified concentration is estimated as 'required concentration × 2 × straggling', which is near 3.9×10^{16} ions/cm^2.

However, the concentrations of implanted ions typically feature a Gaussian distribution. Therefore, when a homogeneous dopant concentration is desired, several implantation steps with multiple ion energies are necessary, such that the overlap of the Gaussian implantation profiles corresponding to the various energies result in a close-to-uniform dopant concentration. Laser-accelerated ions with broad energy distributions could prove to be highly suitable for such purposes. For illustration, some interesting material science cases, wherein one requires a low dopant concentration or formation of small clusters and nanoparticles buried in the near-surface region, are addressed in what follows.

18.2.1 Deep and Uniform Fe Implantation in LiNbO$_3$

Lithium niobate (LiNbO$_3$, LN) is a well-known ferroelectric material with an excellent combination of electro-optical and non-linear optical properties, which in turn make it a photorefractive (PR) material and render it most suitable for holographic applications. It has been shown [Sada 2010] that the PR properties of LN can be enhanced further by doping it with Fe. It has also been suggested that the PR response of Fe-doped LiNbO$_3$ (Fe:LN) can be optimized by tuning the Fe content in such a way that the change in the refractive index under inhomogeneous visible light illumination is the highest, and the absorption from the Fe:LN matrix is at its minimum. Such concentrations are suggested to be around 2×10^{19} Fe/cm^3. To maintain the uniformity of the PR properties of Fe:LN throughout, it is desirable to dope LN with Fe homogeneously, i.e. with a uniform Fe concentration starting from the surface to the maximum attainable implantation depth. In an attempt to achieve such an objective, Sada et al. [Sada 2010] adopted an innovative method – implantation of LN with Fe ions of 95, 100 and 105 MeV energies in the presence of a 7.5 µm energy degrader Fe foil in front of the LN crystal. In accordance with the SRIM–simulated Fe depth profile overlaps for the three energies [Sada 2010], a very broad hump-like (quasi-homogeneous) Fe concentration profile was achieved in this experiment to a near 2 µm depth.

The above report indicates that implantation of LN with Fe ions having a broad energy distribution obtainable from laser particle acceleration could lead to a much more uniform Fe depth profile. To demonstrate this, we present results from SRIM simulations of Fe depth profiles in LN for Fe ions of various discrete energies ranging from hundreds of keV to a few MeV, and by taking equal number of ions for all energies. The simulation results are shown in Figure 18.1. It is readily observable from the figure that the lowest energy (100 keV) Fe implants are confined more toward the surface region within 20–70 nm, while those with the highest energy (2 MeV) reach a much greater depth near 2 µm. Ions of intermediate energies are, similarly, implanted in narrow zones around various depths from the surface. This way, with the laser-accelerated Fe ions of a continuous energy spectrum, one can achieve a dopant concentration which is large at the surface, decays rapidly to a depth of about 600 nm (if the energy range is similar to the example above) and becomes uniform for further depths to about 2 µm. Removal of the top 600 nm of LN (using plasma reactive ion etching, for example [Tamura 2001]) can then leave a homogeneously Fe-doped LN usable for applications.

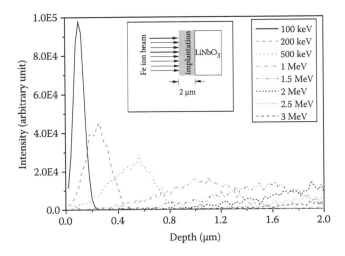

FIGURE 18.1 Depth profiles of Fe ions in LiNbO$_3$ simulated using SRIM for different ion energies ranging from 100 keV to 2 MeV.

Another important consideration in this proposed configuration is whether the required ion density (~2 × 10^{19} Fe/cm³) can be attained with laser-accelerated ions. If this concentration must be maintained from the surface down to 2 μm, the required fluence of Fe ions would be of order 10^{15} at/cm². With an achievable 10^{10} ions/pulse in the laser-acceleration case ([Daido 2012]), and an assumed 10 Hz repetition rate, this dopant concentration can be achieved in about 3 hrs of irradiation.

18.2.2 Evenly Distributed Nano-inclusions of Noble Metals in Thermoelectric Materials

Thermoelectric materials are of great interest due to their potential use as solid-state heat pumps and power generators [Sootsman 2009]. The low efficiency of devices based on conventional bulk thermo-electric materials is a key issue that limits their applications, and hence there have been continued efforts to improve their thermoelectric efficiency. Thermoelectric efficiency is measured in terms of the figure of merit (ZT), given by $ZT = S^2 T \sigma / \kappa$, where T is the temperature, S is the Seebeck coefficient, σ is the electrical conductivity and $\kappa = \kappa_E + \kappa_L$ is the sum of the electronic (κ_E) and lattice (κ_L) thermal conductivities. Thus, for a material of given S, the ZT can be enhanced by increasing σ and decreasing κ simultaneously. Further, since the Wiedemann-Franz law relates σ and κ_E together, κ_L should be independently reduced in order to achieve a better ZT [Bala 2015, Sootsman 2009]. In this direction, different research groups have recently attempted enhancing ZT of thermoelectric thin films using various approaches, one of them being the incorporation of nano-inclusions in the thin-film matrix [Bala 2015]. The nano-inclusions in this work have been synthesized by implanting the thermoelectric material (PbTe) with noble metal (Ag) atoms which form nano-aggregates due to their immiscibility with the matrix. These nano-inclusions enhance the thermoelectric power due to the low energy carrier filtering and/or act as phonon scattering centres that independently reduce the lattice part of the thermal conductivity. In addition, incorporating percolating metal nano-inclusions can be a good approach to increasing ZT by independently increasing σ without affecting κ_E.

In order to synthesize evenly distributed nano-inclusions in a thermoelectric thin film, noble metal ions with a broad energy distribution, as obtainable from laser ion acceleration, would be most suitable according to the arguments given in the last section. To demonstrate this, a SRIM simulation was performed to estimate the depth profiles of Au ions in PbTe at various discrete energies, ranging from 200 keV to 3 MeV. The results are shown in Figure 18.2. It can be easily inferred from the figure that the

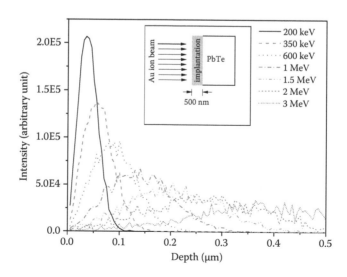

FIGURE 18.2   SRIM simulated depth profile of Au ions for different energies from 200 keV to 3 MeV.

TABLE 18.1   Ion Bunch Requirements for the Proposed Implantation Experiments

Example	Ion	Kinetic Energy Range (MeV)	Implantation Concentration (atoms/cm³)	Ion Fluence (Ions/cm²)
Au in PbTe	Au	0.2–3	1.5×10^{21}	4×10^{16}
Fe in LiNbO₃	Fe	0.1–2	2×10^{19}	10^{15}

outcome is in line with the Fe implantation profile in LN as discussed in the last section – the envelope of the Au depth profiles is sharp near the surface, decays rapidly for depths to 200 nm and beyond that, becomes uniform for depths to about 500 nm. Thus, the laser accelerated ions can help achieving the formation of evenly distributed noble metal nano-inclusions in thermoelectric material thin films in a large depth range.

The ion bunch requirements for the two examples are listed in Table 18.1.

18.3 Pulsed Structure for Perturbed Angular Distribution

In experimental condensed matter physics, it is customary to distinguish between macroscopic and microscopic measurements. The former refers to measurements of quantities such as reflectivity, resistivity, susceptibility, etc. which are properties of materials as a whole, while the latter provide information over short length and time scales. Over the years, there has been growing demand for microscopic investigation of solid state effects leading to significant development in new methodologies, such as scanning tunnelling microscopy (STM), atomic force microscopy (AFM), etc. On the other hand, nuclear techniques such as nuclear magnetic resonance (NMR), Mössbauer spectroscopy (MS), muon spin rotation/relaxation (μSR) and γ-ray angular correlation/distribution (PAC/PAD) have long proven to be extremely useful for studying properties of solids. Based on principles of hyperfine interaction, these techniques not only provide a way to investigate solid state effects on atomic length scales, but also are very effective for studying dynamic interactions.

In PAD, a bunch of ions of appropriate energy bombards a suitable target. If the ion energy is above the Coulomb barrier of nuclear reaction, the collision between the nuclei of the ion and the target can produce nuclear fusion evaporation reactions [Recknagel 1972]. The resultant nuclei possess large linear and angular momenta and are in high-energy excited states. The linear momenta help recoil implant the resultant nuclei in the solid under study, which is kept downstream and in contact with the target. The implanted nuclei in high-energy and high-angular momentum states are subsequently de-excited to lower energy and lower momentum states and may emit γ-rays. A γ-ray detector kept at an arbitrary angle from the ion-bunch direction would then record the γ-ray intensity, which decays exponentially with time according to the transition half-life. Further, the electromagnetic moments associated with the resultant nuclei are aligned perpendicular to the ion propagation direction, and hence the γ-ray intensity is anisotropic [Recknagel 1972]. If the solid in which these γ-ray emitting nuclei are implanted possesses local (microscopic) hyperfine fields (like a hyperfine magnetic field or an electric field gradient locally at the nuclear sites), then the electromagnetic moments of these nuclei undergo hyperfine interactions with these local fields. These interactions lead to the precession of the nuclear moment about the local field with the Larmor frequency determined by the moment and the field. Due to the precession, the γ-ray intensity (reaching the detector) also starts oscillating with a geometry-dependent harmonic of the Larmor frequency and temporally modulates the exponentially decaying intensity [Recknagel 1972]. If the nuclear moments are known, what remains to be determined, by the oscillation frequency of the decaying γ-ray intensity, is solely the local hyperfine field which arises basically from the electronic charge and spin distributions in the local vicinity of the nucleus, and thus gives microscopic insight into the electronic and magnetic behaviour of the solid. If the duration of the ion bunch is much smaller than the intensity oscillation period, all the ions in the bunch can be considered to produce the above effect simultaneously and to increase the counting statistics in the detector. A subsequent ion bunch, then, must arrive several half-lives later, so that the intensity–time pattern produced by this later bunch does not overlap with that of the previous bunch [Recknagel 1972].

This part of Chapter 18 demonstrates that the laser accelerated ion bunches, apart from being exploitable for PAD studies, have also the following two distinct advantages over the conventional accelerator-based technique: capability for simultaneously generating multiple nuclear probes – enabling one to perform microscopic studies at multiple sites in a single experiment – and control of the probe bunch duration to sub-nanosecond levels. This feature would facilitate utilizing all the nuclear probes with half-lives as short as one nanosecond that are formed in the initial nuclear reactions, expanding the variety of sites.

18.3.1 Essential Basics of PAD

To explore the feasibility of high power femtosecond lasers for PAD studies, it is indispensable to briefly introduce the following essential basics of PAD [Recknagel 1972].

18.3.1.1 The Magnetic Hyperfine Interaction: Example

If the nucleus of the probe atom has a spin I, it possesses a magnetic dipole moment $M_I = g_N \mu_N I$, where g_N is the nuclear g-factor and μ_N is the nuclear magneton. The Hamiltonian for the magnetic hyperfine interaction between the nuclear dipole moment and the outer electrons of the probe atom is given by:

$$\hat{H} = -M_I(B_{hf} + B_{ex}) = -g_N \mu_N I B_{eff},$$
(18.1)

where B_{ex} is the external magnetic field and B_{hf} is the hyperfine field at the nucleus arising from the rest of the atom and its environment; $B_{eff} = (B_{hf} + B_{ex})$ is the effective field acting at the nucleus. The hyperfine interaction between the nuclear moment and the effective field results in precession of the moment about the field with a frequency $\omega_{LM} = g_N \mu_N B_{eff}/\hbar$, known as the Larmor frequency. This is equivalent to a splitting of the $2I + 1$ nuclear sublevels. The energy difference ΔE between two adjacent sublevels is given by

$$\Delta E = \hbar \omega_{LM} = g_N \mu_N B_{eff}.$$
(18.2)

Thus, if one measures ω_{LM} where g_N and B_{ex} are known, one can determine the hyperfine field at the probe nucleus.

We note that, in the same manner, electric hyperfine interactions give rise to quadrupolar frequencies ω_Q of the nuclear spin precession, but this need not be discussed for the present purpose.

18.3.1.2 Angular Distribution of Gamma Rays

Consider a radiating system such as an ensemble of nuclei in an excited state with spin I. The orientation of the nuclei can be characterized by the populations a_m of different substates with magnetic quantum numbers $m = -I, ...,+I$ with respect to a quantization axis. The $(2I+1)$ independent parameters a_m satisfy the normalization condition:

$$\sum_{m=-I}^{I} a_m = 1$$
(18.3)

If all the a_m's are equal, the ensemble of nuclei is called randomly oriented. In this case, there is no preferred spatial direction and hence the angular distribution pattern of the γ-rays emitted by the nuclei is isotropic. In the case of an oriented ensemble, a_m's can have different values. These nuclei emit radiation with characteristic anisotropic angular distribution. In the case of the oriented ensemble with axial symmetry, i.e. the ensemble of aligned nuclei, the angular distribution function for emitted radiation can be written as:

$$W(\theta) = \sum_{even\ k} A_k B_k P_k(\cos\theta) \tag{18.4}$$

where θ is the angle between the axis of quantization (the direction of nuclear spin alignment) and the direction of emission of radiation, A_k's are called the angular distribution coefficients and B_k's the orientation (alignment) parameters which depend on a_m, and P_k (cos θ) is the Legendre polynomial of order k. A well-known example is the radiation pattern of a dipole, which has emission probabilities proportional to $\sin^2\theta$ and $(1+\cos^2\theta)$ for $m = 0$ and $m = \pm1$ populations, respectively.

18.3.1.3 Achieving Nuclear Orientation with Accelerated Ions

The nuclear alignment in PAD is achieved in a nuclear reaction by bombarding a spinless target with suitable spinless particles, for example in the reaction ^{45}Sc (^{12}C, p2n) ^{54}Fe*, so that the transfer of a large transverse angular momentum, from the projectile to the product nucleus, takes place. The superscript* is indicative of a nucleus emitting γ-rays of energy E_γ. The reaction is produced conventionally by short (nanosecond) ion bunches with long intervals (microseconds) between bunches. The (large) angular momenta of the compound nuclei are then all perpendicular to the ion-bunch propagation direction and the nuclei are thus completely aligned. The population parameters in this case are simply:

$$\begin{aligned} a_m &= 1, \quad &\text{for } m = 0 \\ &= 0, \quad &\text{otherwise.} \end{aligned} \tag{18.5}$$

An aligned compound nucleus, like ^{54}Fe* in the example above, then emits (evaporates) neutrons and protons and also further γ-rays successively until finally reaching the low-lying states in the resultant nucleus, from which the signature γ-rays are emitted.

The original orientation formed in the collision is retained to a considerable extent if the angular momenta transferred by the projectile are large, since the angular momenta taken away by the nucleons and early γ's are too small to induce a considerable change of orientation. The γ-rays emitted from this resultant nuclear state will then be anisotropic, with respect to the projectile direction, with an $m = 0$ dipole pattern, i.e., $W(\theta) \propto \sin^2\theta$. The three-dimensional angular distribution will be a surface of revolution of this pattern around the beam direction z. Further, the γ-ray intensity at a certain angle θ will decay exponentially in time.

18.3.1.4. Perturbation of Angular Distribution

If the nuclei in the oriented state I are subjected to external perturbations such as an external magnetic field, their orientation evolves temporally. As a result, the angular correlation or distribution pattern of the γ-rays is perturbed. For the perturbations such as randomly oriented B_{hf}'s and for B_{ext} applied perpendicular to the plane containing the ion bunch propagation direction and the detector, for example, the temporal distribution function $W(\theta, t)$ for a specified angle θ can be written as

$$W(\theta, t) \propto G(t) \sim e^{-t/\tau_N} \cos\left(2\omega_{LM}t - \phi\right) \tag{18.6}$$

where τ_N is the half-life of the nuclear level under consideration and ϕ is the phase (which is not important in this discussion). On removing the exponential decay term, the observed intensity in the detector varies periodically (sinusoidally) with time at twice the Larmor frequency ($2\omega_{LM}$). As discussed earlier, the ω_{LM} contains all the static local information, including local magnetic field effects and information about spin fluctuations in the vicinity of the probe nucleus [Srivastava 2006]. For the oscillations to be observable, there should be at least one oscillation during one half-life, which means $\dfrac{2\pi}{\omega_{LM}} \leq \tau_N$. In accordance with Equation 18.2, this condition requires $g_N\tau_N \geq \dfrac{131.2}{B_{\text{eff}}(T)}$ ns. In certain magnetic materials, B_{hf} values can reach up to 40 T. With an external field of 8 T and adequately fast electronics, one can study probe nuclei with $g_N\tau_N$ values as low as ~0.5 ns. The ion bunch duration, however, must be shorter than $g_N\tau_N$ for the oscillations to be resolvable. This methodology of PAD is known as time-differential PAD (TDPAD).

18.3.2 PAD with Laser-Accelerated Ions: Exploratory Example

In order to explore the applicability of laser accelerated ion bunches for PAD studies, we need to have, as described in Section 18.3.1 above, the accelerated ion(s) and energy(ies) provided by the laser, the nuclear reaction target and the sample under study. To illustrate, we consider in the following the H and C ions produced by a Ti-sapphire (Ti:Sa) laser using a 10 nm C target according to a compilation by Schreiber et al. [Schreiber 2016]. As evident, both the ions have a wide energy range. Thin-film ^{45}Sc (see section 18.3.1.3) is the reaction target example. It is then important to determine which useful nuclear probes (reaction products) can be generated in the H/C → ^{45}Sc reactions and in what amount. The amount $\mathcal{A}$ of a possible nuclear probe produced in the nuclear reaction would be proportional to the product of the likelihood of formation of the probe in the reaction, i.e. the nuclear reaction cross section (NRX) at a particular ion energy E, and the ion fluence of that energy reaching the reaction target. The latter is proportional to the yield $\mathcal{Y}$ of ions generated by a laser per solid angle per 1% of the ion energy, as described by Schreiber et al. [Schreiber 2016]. Thus, $\mathcal{A} \propto$ (NRX × $\mathcal{Y}$), where the product (NRX × $\mathcal{Y}$) = $\mathcal{P}$ gives a quantitative estimate of the nuclear probe and its relative amount that is available for analysis. NRXs for the generation of daughter nuclei with H/C ion projectiles are shown in Figure 18.3a using the ^{45}Sc target. The calculations are performed using the CASCADE code [Puehlhofer 1977]. The possible nuclear reactions in this case are ^{45}Sc (^{1}H, p) ^{45}Sc, ^{45}Sc (^{1}H, n) ^{45}Ti*, ^{45}Sc (^{12}C, p2n) ^{54}Fe* and ^{45}Sc (^{12}C, 3p3n) ^{51}Cr*. The I^π, g_N, $E\gamma$ and τ_N values [Raghavan 1989] for the reaction products are listed in Table 18.2.

Figure 18.3b displays the yield $\mathcal{Y}$ of H and C ions, as reproduced from one set of data in [Schreiber 2016], and the product $\mathcal{P}$, as a function of ion energy. Please note that the energies on the x-axis are those of the H and C ions and not the (recoil) energies of the daughter nuclei. The recoil energy of a particular probe is proportional to the corresponding projectile ion energy with the proportionality constant, known as the kinematic factor, determined by the conservation laws for energy and momenta. It can be inferred from the figure that the C ions produce ^{54}Fe* and ^{51}Cr* nuclei with a wide (recoil) energy range, while the H ions create ^{45}Sc and ^{45}Ti* nuclei with a smaller, yet a broad, (recoil) energy range.

Thus, it can be inferred from Figure 18.3 that this particular combination of H/C ions and ^{45}Sc target would produce high-spin nuclei in two groups: first, those produced predominantly by H ions, viz., ^{45}Ti* and ^{45}Sc, and second, those by C, viz., ^{54}Fe* and ^{51}Cr*, simultaneously. This presents an advantage over conventional accelerator-based ion beams, which are comparatively monoenergetic and mono-elemental, and can therefore produce nuclei of only one group. Further, all the states of ^{45}Ti*, ^{54}Fe* and ^{51}Cr* are useable as PAD probes because $g_N\tau_N$ is more than 1.2 ns for all these probes (this is further discussed in Section 18.3.2.1).

The nucleus ^{45}Sc, on the other hand, is stable and is available in an exceptionally large quantity. Hence, it can be used as a conventional dopant. The wide energy range for ^{45}Sc recoils obtained this way could be helpful in producing a quasi-uniform doping because of their wide ranges. The same concept is valid also for the other, radiating nuclei. Thus, the radiating nuclei themselves are implanted in a wide depth range and make the probe impurities even less interactive, whereas in the case of monoenergetic ions,

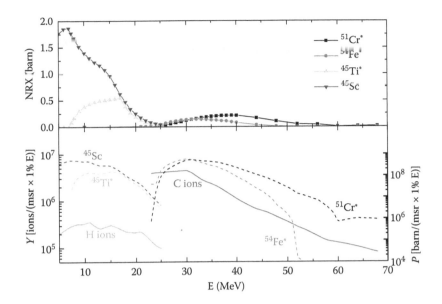

FIGURE 18.3   (a) Top, nuclear reaction cross sections (NRX) versus the laser-accelerated H/C ion energy of the daughter nuclei obtainable with bombardment of the ions on a ^{45}Sc target. (b) Bottom, laser-accelerated H and C ion yields $\mathcal{Y}$ plotted in bold continuous lines with the axis on the left. The figure also displays the product $\mathcal{P}$ (dotted lines and the axis on the right) for a qualitative estimation of the availability of the nuclear probes. (Reproduced from J. Schreiber et al. 2016. "Invited Review Article: 'Hands-on' laser-driven ion acceleration: A primer for laser driven source development and potential applications" *Review of Scientific Instruments* 87: 071101:1–10, with the permission of AIP Publishing.)

TABLE 18.2   The Nuclear Parameters Spin Parity (I^{π}), Nuclear g-Factor (g_N), the Emitted γ-Ray Energy (E_{γ}), and the Half-Life (τ_N) of the γ-Emitting State for the Nuclear Reaction Products Discussed in the Text

Nucleus	I^{π}	g_N	E_{γ} (keV)	τ_N (nanoseconds)
^{45}Ti*	3/2$^+$	+1.05	329	1.1
	5/2$^-$	-0.13	40	11.3
^{51}Cr*	3/2$^-$	-0.86	749	7.3
^{54}Fe*	10$^+$	+7.28	6527	367.0
	6$^+$	+8.22	2950	1.22
^{45}Cr	No γ-emission			

Source: Reprinted with permission from Atomic Data and Nuclear Data Tables, 42, P. Raghavan, Table of Nuclear Moments, 189 201, Copyright (1989) Elsevier.

a narrow implant depth would have a higher concentration and more localized probe distribution. For microscopic studies, ideally non-interacting probe nuclei are desirable.

18.3.2.1 Suitability of Laser Ion Parameters for PAD

Next, we examine the suitability of the ion bunch parameters obtainable from laser-acceleration for PAD studies. For a typical 1 cm^2 PAD target placed at 10 cm from the ion source (a thin foil used for ion acceleration by laser), the ion emission solid angle is about 10 msr. The integrated C ion yield in the energy range 25 MeV to 70 MeV (see Figure 18.3), the relevant energy range for generating PAD probes, is about 6.4 × 10^7 particles/msr. With the proposed geometry, this would deliver about 6.4 × 10^8 C ions for each laser pulse within 1 cm^2 area on the sample. With a typical pulse repetition rate of 10 Hz and taking the C ion charge state as +6 [Daido 2012], this is equivalent to a current of 6.4 nA. In standard ion-beam terminology, this is represented in terms of particle nanoamperes (pnA), which is equal to the current divided by the charge state, and is a measure of the ion flux. This way, a 1 cm² target will receive a flux of ~1 pnA (= 6.3 × 10⁹ particles per second) of C ions. The availability of H ions in the range 5–25 MeV for the same geometry and repetition rate is 5 × 10⁶ particles/msr, which finally would be equivalent to 0.1 pnA of H ion flux at the target. Such ion fluxes are comparable to the conventional accelerator-generated ion currents typically used for PAD studies. With these ion fluxes, one set of PAD data sufficient for analysis can be collected in approximately 3–6 hrs. Further, if one uses a glass-based laser system [Schreiber 2016], ion flux two orders of magnitude higher can be obtained, which will reduce the PAD data collection time commensurately, i.e. to about 2–4 min. This way, even with the existing 1 Hz laser sources, a PAD data acquisition time of 20–40 min would be sufficient for a PAD experiment. This would be a big advantage over PAD by conventional accelerators, where there is no scope to reduce the data acquisition time from at least two hours. The ion bunch duration at the PAD target, τ can be

expressed as $\tau = d \left(\dfrac{m}{2} \right)^{1/2} \left(\varepsilon_{min}^{-1/2} - \varepsilon_{max}^{-1/2} \right)$ [Daido 2012]. Here, m is the ion mass, d is the laser target to PAD

target distance, ε_{min} and ε_{max} are the minimum and maximum ion energies, respectively and $\varepsilon_{max} - \varepsilon_{min}$ is the energy width. By restricting the H ion energy width (using a slit) to 10 MeV (from $\varepsilon_{min} = 10$ MeV to $\varepsilon_{max} = 20$ MeV) or less, bunch durations of the order of 0.1–0.7 ns are obtainable for the proposed geometry. Similarly, for a 35 MeV (from $\varepsilon_{min} = 25$ MeV to $\varepsilon_{max} = 70$ MeV) or lower C ion energy width, bunch durations of the order of 0.3–2.0 ns can be achieved. Thus, all the γ-emitting nuclei in the reaction can be studied using TDPAD. Figure 18.4 shows a schematic comparison between the conventional

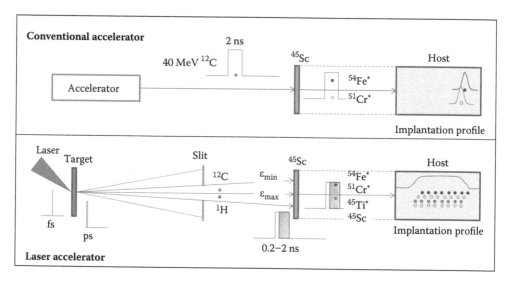

FIGURE 18.4   A schematic comparison between conventional TDPAD technique and the proposed laser-driven TDPAD with laser acceleration. Conventional TDPAD (upper panel): A pulsed ¹²C ion beam of ~2 ns bunch duration hits a ⁴⁵Sc foil. The reaction products ⁵⁴Fe* and ⁵¹Cr* with the same bunch duration are then implanted deep within the host under study with Gaussian distribution profiles. The gap between the foil and the host is just for illustration, the foil is actually kept in close contact with the host to avoid deorientation of the nuclear moments. TDPAD with a femtosecond laser (lower panel): The laser generates ¹²C and ¹H ions, with a broad energy distribution, on the rear side of the target. At the point of generation, the ion bunches have approximate picosecond duration which keeps increasing as the beam travels ahead. A slit can be used to limit the energy width $\Delta\varepsilon = \varepsilon_{max} - \varepsilon_{min}$ of the ions reaching the ⁴⁵Sc foil. By adjusting $\Delta\varepsilon$, the ion bunch duration at the foil can be varied from 0.2–2.0 ns. The two ion species would now result in four reaction products with a controllable probe durations and broad energy distributions that can be simultaneously implanted within the host with an ideally uniform implantation profile. The ion bunch requirements for the proposed PAD experiments are listed in Table 18.3.

TABLE 18.3   Ion Bunch Requirements for the Proposed H/C → ^{45}Sc PAD Experiments

Target	Ion	Energy Range (MeV)	Ions/Bunch/cm^2	Average Current with 10Hz Repetition Rate (pnA = 6.3×10^9 particles s^{-1})
^{45}Sc	H	10–20	6.4×10^8	1.0
	C	25–70	6.4×10^7	0.1

TDPAD technique and the proposed laser-driven TDPAD highlighting the advantages of the latter over the former.

18.3.2.2 A Note on Microscopic Studies with TDPAD

In most of the TDPAD studies reported in the literature, the atom containing the nuclear probe, which we call M, had been different from the atoms of the host matrix, such as an elemental or compound solid with formula unit $P_{1-x}Q_x$ ($0 \leq x \leq 1$). The reason usually is the combination of convenience of producing only certain nuclear probes with a conventional accelerator and the need for studying compound solids of different formula units. After implantation, M typically substitutes one of the host-atom sites. This way, the hyperfine fields produced by the host matrix $P_{1-x}Q_x$ at an impurity (M) site are usually studied. Basically, the microscopic properties of the host are studied via the interaction with an impurity. It would certainly be desirable to have the probe not as a lattice impurity but as a part of the lattice. In other words, having nuclear probes of either atom P or atom Q in $P_{1-x}Q_x$ would lead to a microscopic study of $P_{1-x}Q_x$ at its own lattice sites. Mohanta et al. have reported one such study [Mohanta 2013].

18.3.2.3 Feasible Case Studies with H/C→ ^{45}Sc

Based on the arguments above, three interesting candidates of condensed matter physics investigations for our H/C → ^{45}Sc example are given in what follows:

i. Dutta et al. [2014] have reported improved magnetic and ferroelectric properties of BiFeO$_3$ with Sc and Ti co-doping. In a proposed experiment with the laser-accelerated ions, the microscopic origin of this behaviour can be explored by taking a Sc and Ti doped BiFeO$_3$ and probing the hyperfine interactions at Fe and Ti simultaneously. In fact, the large amount of ^{45}Sc can be used to concomitantly dope a Ti-doped BiFeO$_3$ sample for this purpose.

ii. There is a ^{75}As-NMR study of antiferromagnetic order and superconductivity in Sr$_4$(Mg$_{0.5-x}$Ti$_{0.5+x}$)$_2$O$_6$Fe$_2$As$_2$ with electron doping [Yamamoto 2012]. One can supplement this study with Fe and Ti probes simultaneously by TDPAD using ions produced by a laser.

iii. In a study by Ma et al., correlation-induced self-doping in the iron-pnictide superconductor Ba$_2$Ti$_2$Fe$_2$As$_4$ has been observed [Ma 2014]. Studying Ti and Fe sites microscopically may give further insight into the mechanism.

18.3.2.4 Further Scope

Simultaneous production and laser-acceleration of H and C ions using lasers, as discussed above, is not the only interesting case. Laser acceleration of a multitude of ion species, ranging from H to those as heavy as Au, has already been demonstrated [Schreiber 2016]. And there is a host of PAD nuclei, including ^{28}Al, ^{37,40}K, ^{41,42}Ca, ^{44}Sc, ^{51}Cr, ^{54}Fe, ^{61}Ni, ^{62,63}Cu, ^{70}Ga, ^{69,73}As, ^{77}Se, ^{89}Zr, ^{103}Rh, ^{100,103}Pd, ^{105}Ag, ^{111,118}Sn, ^{121}Sb, ^{117,126}Te, ^{131}I, ^{138}Ba, ^{135,140}Ce, ^{141,143}Pr, ^{161}Dy, ^{152,154}Er, ^{172}Hf, ^{182}W, ^{197}Pt, ^{193}Au, ^{194}Pb and ^{209}Bi [Raghavan 1989], which can be produced using different ion–PAD target combinations and utilized to study materials containing these atoms at microscopic length and time scales. Further, the multiplicity of laser-accelerated ions is not confined to two; simultaneous production of more than two ion species has been demonstrated (e.g. simultaneous production of F, O, C and Be ions [Hegelich 2005]). The energy per nucleon for all these ions is above the respective Coulomb barrier and hence more than two PAD probes can be generated simultaneously and more complex materials can be studied microscopically.

18.4 Summary

The multiplicity, the broad energy distribution and the short time structure of laser-accelerated ions have definite viable prospects for ion implantation technology and perturbed angular distribution studies, with certain advantages over ions provided by conventional accelerators. In the case of ion implantation, the broad energy is demonstrably helpful in achieving a quasi-uniform implantation profile down to a depth of a few microns, and the multiplicity has the potential of doping multiple impurities simultaneously. In Chapter 18, it is illustrated with the help of calculations from the ion-matter interaction simulation package SRIM that the broad energy spectra of laser-accelerated ions can, for example, be used to uniformly dope $LiNbO_3$ with Fe atoms and evenly embed PbTe with Au metal nano-inclusions. The uniform doping demonstrated in the former case is crucial for synthesizing a ferroelectric thin film with improved and uniform photorefractive properties, a feature difficult to achieve using conventional accelerators. Further, an evenly-distributed, laser ion-induced embedding of thermoelectric materials with noble metal nanoclusters, as exemplified in the latter case, is vital for efficiently enhancing their thermoelectric properties. These experiments are in the scope of existing laser-acceleration capability typically within a 3–4 hr duration, provided that an enhanced laser ion pulse repetition rate (10 Hz which exceeds currently available 1 Hz) is available.

The prospective perturbed angular distribution studies with laser-accelerated ions once again have a dual advantage over the conventional accelerator-based method: while the ion multiplicity is capable of producing multiple nuclear probes simultaneously, the controllability of ion bunch duration to less than a nanosecond can expand the number of accessible probe species by including more and more nuclei with lower and lower $g_N\tau_N$ values. Using these multiple PAD probes as self-impurities in complex materials can exhibit interesting behaviour, such as multiferroic properties like Sc and Ti co-doped in $BiFeO_3$; the simultaneous presence of antiferromagnetism and superconductivity, as in $Sr_4(Mg_{0.5-x}Ti_{0.5+x})_2O_6Fe_2As_2$; and superconductivity in the presence of magnetic constituent elements, such as in $Ba_2Ti_2Fe_2As_4$. The origins of such complex behaviours can be investigated by looking at each local atomic site. This is certainly not possible with PAD that uses conventional accelerators. The feasibility of such laser ion-based experiments with currently achievable laser ion yields has been demonstrated. Such an experiment, however, needs special experimental hardware, viz., a cryostat equipped with a powerful superconducting magnet as well as a sample-holder and gamma ray detectors outside. Such arrangements are also compatible with the existing laser-accelerator geometries.

Acknowledgements

The authors acknowledge M. Bala for providing SRIM simulation data, which were used in plotting Figures 18.1 and 18.2, and S. N. Mishra for helpful discussion and guidance on PAD.

References for Chapter 18

Bala, M., Gupta, S., Tripathi, T. S. et al., 2015. "Enhancement of thermoelectric power of PbTe: Ag nanocomposite thin films" *RSC Advances* **5**: 25887–25895.

Daido, H., Nishiuchi, M. and Pirozhkov, A. S. 2012. "Review of laser-driven ion sources and their applications" *Rep. Prog. Phys.* **75**: 056401–056471.

Dutta, D. P., Mandal, B. P., Mukadam, M. D. Yusuf, S. M. and Tyagi, A. K. 2014. "Improved magnetic and ferroelectric properties of Sc and Ti codoped multiferroic nano $BiFeO_3$ prepared *via* sonochemical synthesis" *Dalton Trans.* **43**: 7838–7846.

Hegelich, B. M., Albright, B., Audebert, P. et al., 2005. "Spectral properties of laser-accelerated mid-ZMeV/u ion beams" *Phys. Plasmas* **12**: 056314 (1–6).

Ma, J-Z., van Roekeghem, A., Richard, P. et al., 2014. "Correlation-induced self-doping in the iron-pnictide superconductor $Ba_2Ti_2Fe_2As_4O$" *Phys. Rev. Lett.* **113**: 266407(1–4).

Mohanta, S. K., Mishra, S. N., Davane, S. M. et al., 2013. "Local probe studies of Fe hyperfine field in CaFe₄As₈ by time differential perturbed angular distribution (TDPAD) spectroscopy and *ab initio* methods" *Nucl. Instr. Meth. Phys. Res. B* **299**: 71–76.

Nastasi, M. and Mayer, J. W. 2006. *Ion Implantation and Synthesis of Materials*. Berlin Heidelberg: Springer-Verlag.

Ohring, M. 2002. *Materials Science of Thin Films*. San Diego: Academic Press.

Pühlhofer, F. 1977. "On the interpretation of evaporation residue mass distributions in heavy-ion induced fusion reactions" *Nucl. Phys. A* **280**: 267–284.

Raghavan, P. 1989. "Table of nuclear moments" *Atomic Data and Nuclear Data Tables* **42**: 189–291.

Recknagel, E. 1972. *Perturbed angular distribution following nuclear reactions* Berlin: Hahn-Meitner-Instistut.

Rimini, E. 1995. *Ion Implantation: Basics to Device Fabrication*. New York: Springer Science/Business Media.

Sada, C., Argiolas, N., Bazzan, M. et al., 2010. "Structural and compositional characterization of LiNbO₃ crystals implanted with high energy iron ions" *Nucl. Instr. Meth. Phys. Res. B* **268**: 2937–2941.

Schieck, gen. H. P. 2014. *Nuclear Reactions: An Introduction* Berlin Heidelberg: Springer-Verlag.

Schreiber, J., Bolton, P. R. and Parodi, K. 2016. "Invited Review Article: 'Hands-on' laser-driven ion acceleration: A primer for laser driven source development and potential applications" *Review of Scientific Instruments* **87**: 071101:1–10.

Schubert, E.F. 2005. *Doping in III-V Semiconductors*. Cambridge: Cambridge University Press.

Sootsman, J.R., Chung, D. Y. and Kanatzidis, M.G. 2009. "ChemInform Abstract: New and old concepts in thermoelectric materials" *Angewandte Chemie International Edition* **48**: 8616–8639.

Srivastava, S. K., Mishra, S. N. and Das, G.P. 2006. "Spin fluctuations of isolated Fe impurities in Pd-based dilute alloys: Effect of ferromagnetic host spin polarization" *J. Phys.: Cond. Matt.* **18**: 9463–9470.

Tamura, M. and Yoshikado, S. 2001. "Etching characteristics of LiNbO₃ crystal by fluorine gas plasma reactive ion etching" *Sci. Tech. Adv. Mater.* **2**: 563–569.

Yamamoto, K., Mukuda, H., Kinouchi, H. et al., 2012. "Antiferromagnetic Order and Superconductivity in $Sr_4(Mg_{0.5-x}Ti_{0.5+x})_2O_6Fe_2As_2$ with Electron Doping: ^{75}As-NMR Study" *J. Phys. Soc. Japan* **81**: 053702–053705.

Ziegler, J. F., Biersack, J. P. and Ziegler, M. D. 2008. SRIM, the Stopping and Range of Ions in Matter http://www.srim.org.

A Compact Proton Linac Neutron Source at RIKEN

Yoshie Otake

19.1 Introduction

Accelerator-driven compact neutron source development now attracts attention for not only neutron and X-ray fields, but also such fields as material development aimed at significant weight reduction in automobiles and other transportation, cancer therapy and nondestructive inspection technology development. A neutron beam has such unique characteristics as high penetration power for steels and metals and high sensitivity for light elements (e.g. hydrogen, lithium, boron) making the neutron beam an ideal probe for material development. Riken has been developing an accelerator-driven compact neutron source, RANS (RIKEN accelerator-driven compact neutron source) [Otake 2015] as shown in Figure 19.1, which is easy to access on site. There are two major goals of RANS's research and development. One is to establish a new, compact, low-energy neutron system of floor-standing type for industrial use. Another goal is to invent a novel, transportable compact neutron system for the preventive maintenance of large-scale construction, such as bridges and airports, using a higher energy neutron beam of above 100 keV. Such maintenance aims to find the degradation points under the floor slab of roads, bridges and airstrip surface layers. For these goals, a proton linac with 7 MeV kinetic energy is used for RANS.

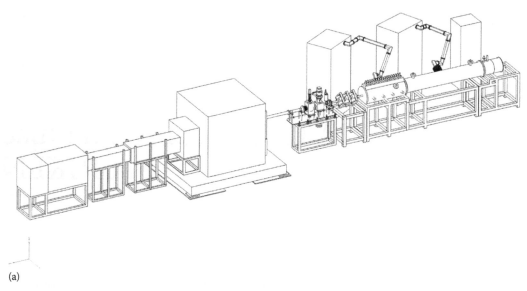

(a)

(b)

FIGURE 19.1   RIKEN accelerator-driven compact neutron source (RANS) (a) 3D graphics; (b) photograph, showing the proton linear accelerator (right), the target station (middle box) and the neutron beam line. At the 5 m downstream position, the sample and detector box with beam dump (left) is shown in 3D graphics while experimental stage (left) is shown in the picture.

In Section 19.2 the selection of ion beam and its energy for RANS, in Section 19.2.2, the long-life target development and in Section 19.2.3 neutron moderator and radiation shielding design will be explained. In Section 19.3, some recent results from experimental research at RANS are presented. This includes pioneering success in visualization of corrosion and wet–dry processes in steel under the film which was done with RANS, together with the research group of ISIJ [Yamada 2014, Taketani 2017]. We also include a description of the observation of texture change before and after deformation through the plastic process of steel materials [Ikeda 2015, Ikeda 2016] and novel demonstrations of fast neutron radiography.

19.2 RIKEN Accelerator-Driven Compact Neutron Source, RANS

19.2.1 Compact Neutron Source Based on a Proton Accelerator

To realize the compact neutron source for practical use, the RIKEN VCAD project [Makinouchi 2011] organized the Compact Neutron Source Imaging System Exploratory Committee and published a technical report in January 2010 [Makinouchi 2010]. There are two main fields where compact neutron systems are expected to be introduced on-site, one is industrial usage (as in the automobile and steel industries) and the other is nondestructive inspection of thick concrete construction, such as in bridges and highways, for construction of social infrastructure. For those future applications, the number of the neutrons for these purposes should be much more than that provided by such neutron generators as D-D or D-T neutron tubes. The requirements for the total system are clearly based on the discussions in the committee. The total size of the compact neutron system, including shielding of the target, is requested to be as small as possible, and the radioactivation should be as little as possible.

The required neutron yield at the target was determined to be 10^{12}/sec based on the actual performance of HUNS (Hokkaido University Neutron Source) [Furusaka 2014] where small and intermediate angle neutron scattering and Bragg Edge pulse neutron imaging experiments for iron and steel material samples have been performed and compared with the results at HANARO [Ohnuma 2016, Tomota 2016]

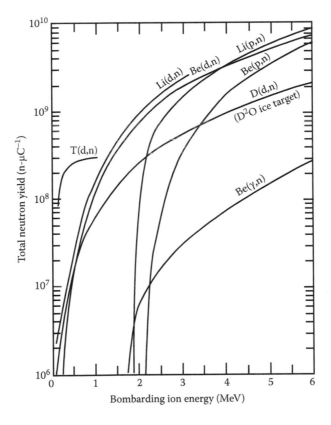

FIGURE 19.2 Total neutron yield per micro-coulomb. (From Hawkesworth, M.R. 1977, *Atomic Energy Review* **152**: 169–220.) The thick-target neutron yield, neutrons per micro-coulomb, as a function of bombarding ion energy for various accelerator neutron sources of interest for radiography.

TABLE 19.1   Specifications of the Proton Linac and the Materials Around the Target
for RANS

Parameter for RANS Source	Description
• Proton linac	
• Energy and current	7MeV (fix), from 20 to 100 μA
• RF frequency	RF frequency 425 MHz
• Ion injector energy	30 keV
• Bunch repetition range	20–180 Hz
• Pulse width range	8–200 μs
• Maximum RF power duty	1.3%
• Target moderator shielding	
• Target	Beryllium
	Diameter: φ = 50 mm, thickness z = 0.3 mm
• Backing	Vanadium; φ = 140 mm, z = 4.5 mm
• Cooling water cavity	Titanium; φ = 152 mm, z = 10 mm
• Moderator	Polyethylene; φ= 200 mm, z = 40 mm
• Reflector	Graphite; 40 cm *20 cm *40 cm
• Shielding	Lead, Borated-polyethylene, Iron; 2 × 2 × 2 m³

and other instruments at large facilities. From Figure 19.2 [Hawkesworth 1977], Li (p, n), Be (d, n), and Be (p, n) nuclear reactions are the main candidates which result in a neutron yields of 10^{12} with the current of 100 μA. Lithium has more difficult characteristics to handle with a low melting point and it is more chemically active than beryllium as a target material. The proton linear accelerator is easy to handle with low radioactivation compared with that of deuterium. RANS chose the pulsed proton linear accelerator with 7 MeV to achieve the 10^{12} neutron yield.

The proton bunch duration and repetition rate have been upgraded for shorter bunches. As a result, the proton linac can be operated with bunch durations between 8 and 200 μs at repetition rates from 10 to 180 Hz. For example, a short proton bunch duration of 8 μs can be provided at a high repetition rate of 180 Hz while keeping the proton average current below 100 μA with a 1.3% duty cycle in the ideal condition. The radio frequency of the RFQ is 425 MHz and the ion output energy from the ion source to the RFQ is 30 keV. The specifications of the proton linac are shown in Table 19.1 in detail.

19.2.2 Long-Life Be Target

On the other hand, because of major problems with targets using a low energy proton beam, the low-energy Be (p,n) reaction has not been used for stable compact neutron sources. Yamagata invented a new long-life Be target [Yamagata 2015] for which the proposed new design has a hydrogen diffusible backing. Based on finite element analysis (FEM) and Monte-Carlo ion injection simulation, the thickness of the beryllium metal and the choice and the design of the backing material has been done and, at the same time, the mechanical strength and heat removal capacity of the target was also considered. The thickness of the beryllium is sufficiently reduced to slow down the proton beam to kinetic energies as low as 2 MeV, which is slightly above the 1.85 MeV Be (p,n) reaction threshold. The beryllium thickness required for proton deceleration to 2 MeV (with respect to different incident proton energies) is calculated based on data from PSTAR (Stopping Power and Range Tables for Protons, NIST) [PSTAR 2011]. For RANS, a commercially available 7 MeV proton linear accelerator is used where the maximum average current is 100 μA. For the incident proton energy of 7 MeV, the required beryllium thickness is 368 μm. The proton beam penetrating through the beryllium is injected into the backing material and converted to hydrogen after binding an electron. Because most of the injected hydrogen will remain at

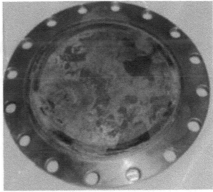

FIGURE 19.3 Target surface inspection following one year of use. No cracks or blistering can be observed on beryllium (left) while corrosion was observed on the cooling water side (right).

a certain depth due to the Bragg peak phenomenon, there is a possibility that this hydrogen will cause blistering of the backing material. To prevent this, the backing material must have a high hydrogen diffusion coefficient so that the maximum concentration of hydrogen will be reduced due to diffusion. The backing material should withstand a pressure of 1 atm (against vacuum) and cooling water pressure. The new design of the beryllium target system for RANS is a 300-μm thick beryllium plate of 50 mm diameter with a 4.5-mm thick vanadium backing of 152-mm diameter.

After two years of usage (actually 9100 μA hr usage), there were no signs of cracks or blistering on the beryllium target. Target photographs taken after one year of use are shown in Figure 19.3 [Yamagata 2015].

The proton bunch duration and repetition rate have been upgraded for shorter bunches. As a result, the proton linac can be operated with bunch durations between 8 to 180 μs at repetition rates from 10 to 200 Hz.

19.2.3 RANS Target station, Moderator, Reflector and Radiation Shielding

The slab-type design of the neutron beam line from the target and moderator is suitable for RANS use with low-energy and higher energy neutrons above 100 keV and at the same time the wing-type design is similar to that for low-energy neutron usage. For stable RANS operation, the chosen moderator material for obtaining a thermal neutron source is polyethylene. The optimal design of the moderator is determined based on Monte-Carlo simulations with the PHITS code [Sato 2013] to get the optimal neutron numbers 5 m downstream from the surface of the moderator. The diameter of the moderator is 200 mm with a 40-mm thickness. Figure 19.4 shows a cross-sectional drawing of a shield box around the target. The cross section of neutron beam from the moderator to the exit of the target station is 160 mm × 160 mm. The maximum size of the neutron beam line from the target station to the sample and detector box is 250 mm × 250 mm.

Figure 19.5 shows a neutron energy distribution per cm^2 per second per lethargy at the maximum RANS proton current of 100 μA and at the position which is 5 m straight downstream from the moderator surface (which is the sample position of some experiments). It is the simulation results of the radiation transfer calculation, GEANT4 code [GEANT4] based on the new neutron production function formation based on Be (p,n) for compact neutron source [Wakabayashi 2016]. The energy range is from meV (10^{-3} eV) to 5 MeV (10^6 eV), 9th in order of magnitude in eV. The centre energy of the left peak is 50 meV, the so-called thermal neutron beam range. Its right peak is around 1.5 MeV. The estimated number of thermal neutrons 5 m down from the target is about 10^4 s^{-1} cm^{-2}. From the radiation safety point of view, it is sufficiently low for the production of any radioactive materials to be avoided during several hours of taking measurements. Neutrons with the maximum energy of about 5 MeV are generated via the Be (p,n) reaction, and the flux-peak appears around 1 MeV. The fast neutrons are moderated in a polyethylene moderator of 40 mm thickness, and the thermal

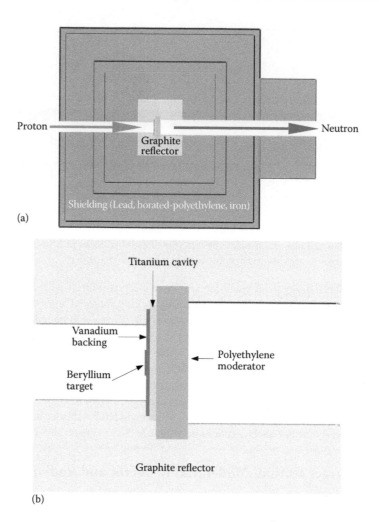

(a)

(b)

FIGURE 19.4 A schematic of the cross-sectional view of the target station of RANS in (a) and (b) an enlarged detail drawing around the beryllium target.

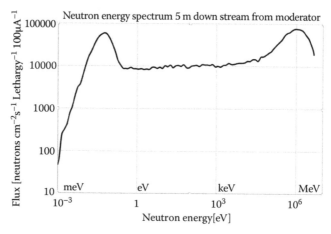

FIGURE 19.5 Energy spectrum per cm² per second per lethargy with the condition at the maximum current 100 μA of RANS at 5 m from moderator as simulated by GEANT4 code.

neutrons with approximately 0.01 eV energy (0.1 nm in wavelength), which is a suitable energy for the diffraction experiment, can be extracted from the moderator surface. Graphite blocks are placed surrounding the target and moderator as shown in Figure 19.4b, in order to increase the low energy neutron flux along the neutron beamline by gathering them into the moderator inside the neutron and gamma ray shielding layers. A shielding box around the reflector is a large cube shape with 1.8 m sides and 20 t in weight, which consists of lead (Pb) layers for gamma ray absorber and borated-polyethylene layers for neutron absorption. In this shielding box, there is a square window of area 100×100 mm^2 to pass protons to the Be target from the linac, whereas the other square window is placed in the opposite wall of the shield box through which neutrons from the polyethylene moderator transit.

The estimated thermal neutron flux 5 m downstream from the moderator is of order 10,000 neutrons per (cm^2 second Lethargy 100 μA) as shown in Figure 19.5. This sample and detector box consists of an aluminum frame including polyethylene and boron, for reducing the background noise from outside. We can build a simple optical system inside the sample and detector box, to achieve the purpose of each experiment. Table 19.1 lists the specifications of the proton linac and materials around the target for RANS.

19.3 Applications of RANS: Early Results

The development of the nondestructive inspection technique on-site for structural materials and material deterioration over extended time is one of the most important urgent issues in the world; for example, in the form of materials for automobiles, trains, airplanes and, in the form of social infrastructure, bridges and tunnels. For example, steel is one of the most widely used materials because of its price and stiffness. There are two directions of the development of various kinds of steel, one is corrosion-resistant alloy together with a film coating, and another is weight reduction development. Here, we introduce three important recent results using RANS: visualizing corrosion (rust) in the steels under the film with low-energy neutron imaging (under the wet–dry process), observing the texture evolution of steel and quantization of an austenite phase with pulsed neutron diffraction and novel fast neutron radiography and nondestructive visualization of the air gap and steel bar position difference in concrete with fast neutron imaging detector for outdoor usage.

19.3.1 Visualizing Corrosion in Steel with Low-Energy Neutron Imaging Under the Film Under the Wet–Dry Process

Because corrosion damage of steel cannot be avoided in outdoor environments [Nakayama 2005], effective countermeasures for corrosion are seriously needed. Paint coatings are the most widely used corrosion prevention method for steel infrastructures [Committee 2001]. RANS carried out the world's first nondestructive visualization of the corrosion and water movement in the different steels, normal steel and a corrosion-resistant steel alloy [Yamada 2015, Taketani 2017].

Two steel plates were prepared; one of normal steel (JIS M400) [JIS2015] and one of corrosion-resistant alloy steel (0.8Cu-0.4Ni-0.05Ti). Both were machined to plates of 6 mm thickness, 70 mm width and 140 mm height. They were coated with epoxy-modified paint of 240 μm thickness and low corrosion resistance. They were scratched by a sharp knife to fix the site of the corrosion and simulate a film defect. This scratch is considered to produce a damaged region much rougher than that in practical environments.

The plates were then subjected to a cyclic corrosion test (CCT), the procedure for which is shown in Figure 19.6, to create blisters of under-film corrosion during which the samples were placed horizontally. The salt water spray conditions were: 1.5 ± 0.5 ml/h per sample area of 80 cm^2 with 5% salt concentration by weight. These steel plates were prepared as samples for water-imaging experiments. Optical images of the samples are shown in Figure 19.7a.

Each sample was soaked in water for 2 hrs. Water saturation in the sample was confirmed by the weight change. After the sample was removed from the water, the water on the surface was removed by

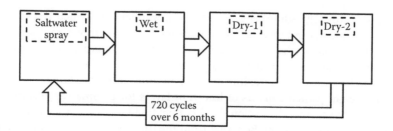

FIGURE 19.6 Procedure of the cyclic corrosion test (CCT) for the samples of 6 mm thickness, 70 mm width and 140 mm height coated with epoxy-modified paint of 240 μm thickness. (Reproduced with permission from A. Taketani et al. 2017. "Visualization of Water in Corroded Region of Painted Steels at a Compact Neutron Source" ISIJ International 57: 155.)

wiping, then the sample was mounted on a precision balance scale located in front of the image sensor and was dried in the air. Then neutron images were continuously taken by the image sensor. Neutron exposure started 5 min after the sample had been removed from the water. The average proton current on the Be target was about 15 μA. The exposure times for each image were 5 min at the beginning of the measurements and 10 min at the end. The total duration of the measurement was 150 min for each sample. The amount of water in the sample was continuously monitored via the weight variation.

The distances between the samples and the scintillator of the image sensor were 44 mm and 55 mm for the alloy sample and the normal sample, respectively. Thus, a blur of the image due to the angular divergence of the neutron beam was expected to be about 1 mm. The pixel size of the image sensor is small enough to see this effect.

Figure 19.7b shows neutron images of the normal steel (SM400) sample in dry states (right images of 19.7b). The corrosion in the painted steel is clearly observed as a shadow in the neutron imaging. This demonstrates clearly that neutron imaging is a new tool for non-destructive observation of corrosion under the paint of painted steel. The corrosion is generated by water and oxygen. The existence of water is the most important factor. The duration of water contact with bare steel, which means the water detention period in the under-film corrosion, may be considered to have a strong correlation with the corrosion. However, there are no studies that confirm this. Thus, it is necessary to measure the water detention period in the corrosion for the discussion the corrosion mechanism. Figure 19.7b left shows neutron imaging results for the fully wet sample.

These images were constructed by merging raw images to enhance the contrast for the regions of corrosion and water. The wet and dry images were obtained with equivalent exposure time of 20 min and 55 min, respectively. The contrast in the images was adjusted to enhance differences between these states.

The image of the dry sample appears to show regions of corrosion, and that of the wet sample indicates regions of corrosion and water. The difference in the neutron transmission ratio between the fully wet sample and the dry sample is a few per cent. The transmission ratio for the 6 mm–thick steel is 54%. The imaging sensor on these regions was exposed to the neutron beams directly. Thus, these regions were used for estimating the intensity of neutron exposure. The two black square boxes with a circular hole in the images are plastic markers of 10 mm thickness, which were used to superimpose photograph and blister shape information and further analysis.

Figure 19.8 presents the time dependence of the water distribution for both samples. The distribution of the water is in good agreement with the position of the blister in the under-film corrosion. The normal steel has a larger area of water than the alloy. Most of the water had disappeared after 1 hr of air-drying for the alloy steel while considerably more water remained in the normal steel.

The three-dimensional reconstructed imaging of the corrosion and water content in the normal steel under the film sample is shown in Figure 19.9. This result is believed to indicate improvement of corrosion life of painted steel structures and a usefulness of compact neutron source of the floor-standing type.

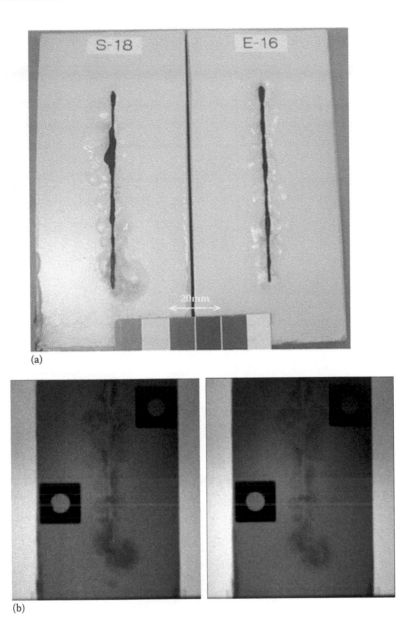

FIGURE 19.7 (a) Optical images of samples and (b) neutron images of the normal steel (SM400) sample in dry (right) and wet (left) states. (Reproduced with permission from A. Taketani et al. 2017. "Visualization of Water in Corroded Region of Painted Steels at a Compact Neutron Source" ISIJ International 57: 155.)

19.3.2 Observing Texture Evolution of Steel and the Quantization of an Austenite Phase with Neutron Engineering Diffraction Using RANS

Based on the high neutron penetrating power through metals, the matter wave characteristics, neutron diffraction and small angle neutron scattering techniques have been used to characterize metal microstructure with bulk samples [Ikeda 2015, Ikeda 2016]. Here, Sections 19.3.2 and 19.3.3 are described based on [Ikeda 2016]. Neutron diffraction is known as the only method which can measure internal strains inside crystalline materials non-destructively. Quantitative evaluation of these microscopic parameters may bring advanced understanding of the mechanical behaviour of metals by investigating the relationship with macroscopic

Normal
steel

Corrosion-
resistant alloy

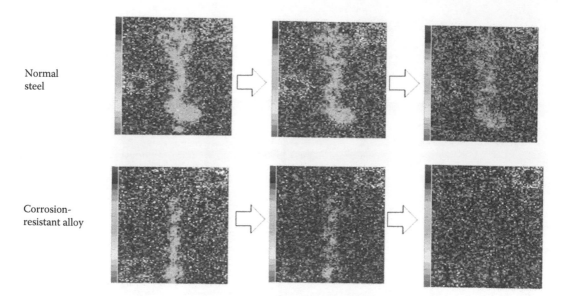

FIGURE 19.8 Time dependence of the water distributions. The colours indicate the estimated water thickness at each pixel and shows distinct reduction of water layer thickness in the corrosion-resitant alloy.

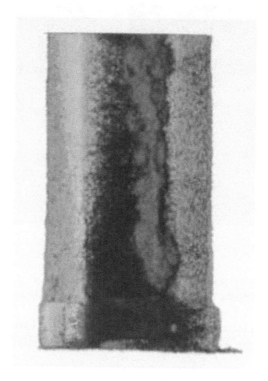

FIGURE 19.9 Reconstructed tomographic image measured at RANS, red colour part indicates the corrosion with water in the normal steel.

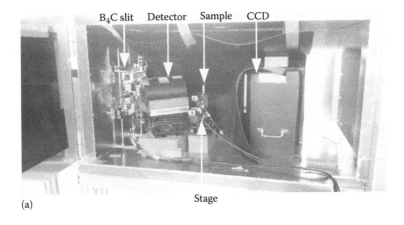

(a)

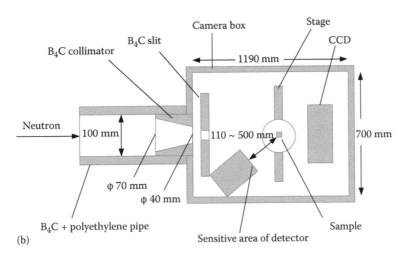

(b)

FIGURE 19.10 (a) The overview photograph and (b) the schematic illustration of the experimental setup for the diffraction experiment at the sample and detector box.

behaviour. A diffraction technique to investigate microstructural features that influence the mechanical properties of materials (for example, texture, a measure of grain orientations, lattice strain, a measure of distortion in crystals, and so on) is referred to as 'engineering diffraction'. So, the neutron diffraction technique which is carried out with RANS can be called neutron engineering diffraction. Neutron engineering diffraction is well-established and used to evaluate such parameters for design of advanced metals with outstanding properties and for development of industrial products with high reliability and low environmental impact. Development of advanced high-strength steels is one of the most critical issues for meeting societal demands in a variety of industrial fields such as the automotive industry, which urgently requires novel technologies to measure the texture evolution and the amount of retained austenite. Typically, a compact neutron source like RANS is not considered for conducting a neutron diffraction experiment or energy-resolved pulse neutron experiment, because of its low neutron flux . Neutron diffraction studies have been conducted at RANS in order to clarify its capability for neutron engineering diffraction.

The perspective of compact accelerator-based neutron sources for engineering applications is discussed based on time-of-flight (TOF) neutron diffraction, which is typically utilized in pulsed neutron sources. Figure 19.10 shows the experimental setup for measuring diffraction in steel samples. The incident neutron beam emitted from the moderator surface was formed into a square by a B_4C collimator and a B_4C slit, installed

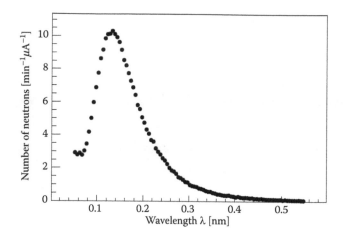

FIGURE 19.11 Neutron energy spectrum measured by incoherent scattering from a polyethylene sample, which approximately corresponds to the energy spectrum of the incident neutron beam.

before a sample at approximately 5 m (= L1) from the moderator surface. The aperture size of the slit was adjusted to completely illuminate the whole sample mounted on a sample table. The sample was carefully positioned at the centre of the incident neutron beam by using a transmission image captured by a CCD behind the specimen. A neutron detector, which consists of a ^{6}LiF/ZnS(Ag) scintillator and a position sensitive photomultiplier tube RPMT, was installed inside the sample and detector box at diffraction angles from 90° to 150°.

The energy spectrum of the incident neutron beam is determined by measuring the energy spectrum scattered from a polyethylene sample which is set at the sample place in Figure 19.10b, which approximately corresponds to it, as shown in Figure 19.11. An energy range from 0.32 eV to 0.003 eV, corresponding to 0.05 to 0.5 nm in wavelength is utilized to measure the diffraction pattern from the steel in this study.

Figure 19.12 shows the diffraction pattern of the ferritic steel specimen, taken for 10 min, which is given as a function of wavelength. These diffraction patterns were calibrated by time-focusing to the centre of the detector. Furthermore, the intensity was normalized by the approximate energy spectrum of the incident

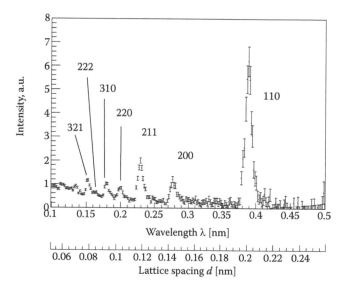

FIGURE 19.12 Diffraction pattern of the ferritic steel measured at a 150° diffraction angle. The diffraction peaks can be identified using a body centred crystal structure.

beam obtained by a polyethylene sample shown in Figure 19.11. As shown in Figure 19.12, several distinct diffraction peaks can be found in the wavelength range from 0.1 to 0.5 nm, and they can be indexed by the body centred cubic (BCC) structure. This result indicates that it is possible to take a recognizable diffraction pattern with a large sampling volume of 10 mm³ for an acceptable measurement time, i.e. 10 minutes. Actually, the measurement efficiency was improved by a factor of 10 by using the position sensitive detector of ³He with the size of 100 mm × 600 mm and by installing additional shielding to reduce background in between sample and the detector; as a result, a diffraction pattern can now be taken within 1–2 min.

To know texture and the amount of retained austenite in steel is important for assessing deformability as well as the material strength. Neutron diffraction is an important probe to assess these parameters in bulk average. Therefore, the compact neutron source can be a useful technology because we can evaluate these parameters easily at our own laboratory, not only at the large neutron experimental facilities.

Texture is an important parameter related to the deformability of material, which is commonly measured by X-ray diffraction or neutron diffraction [Wenk 2004]. We investigated the possibility of the texture measurement using a compact neutron source. The sample used was commercial ferritic steel (JIS JSC440W), which was plastically deformed by approximately 20% in tension along the rolling direction. A rectangular-shaped specimen of volume approximately 10 × 10 × 30 mm³ was prepared by the following procedure. The small pieces of the plate with a size of 10 × 10 × 1.2 mm³ were, at first, taken from a tensile-deformed specimen by shear-cutting, and then assembled together into the rectangular shape, preserving the orientation of the pieces. The diffraction patterns were taken for 90 min to get sufficient intensity. Figure 19.13 shows the diffraction pattern in the loading direction of the deformed specimen, in comparison with that of the non-deformed specimen. It can be seen in Figure 19.13 that the as-received specimen may originally have strong texture even before tensile deformation because the intensity of the 110 reflection is much higher than that of other peaks. After applying tensile deformation by 20%, the 110 reflection is preferentially oriented more to the loading axis. The intensity of the 110 reflection is obviously increased by 2.2 times, while that of the 200 and 211 reflections that are perpendicular to the 110 reflection is decreased by 0.5–0.7 time. This is a typical texture evolution for the BCC structure, caused by tensile plastic deformation [Morooka 2006]. This result suggests that the texture evolution due to the plastic deformation can be observed by RANS. TOF neutron diffraction is an advantage for efficient texture measurement by measuring multiple neutron TOF histograms in various directions of the pole figure, simultaneously measured by a wide area detector [Wenk 2003]. For RANS based on TOF neutron diffraction, therefore, it is also expected to measure the texture of material for relatively short time by using a similar technique.

Retained austenite in steel is an important parameter related to the toughness of material, which is commonly measured on the material surface by X-ray diffraction [Magner 2002]. However, it is actually

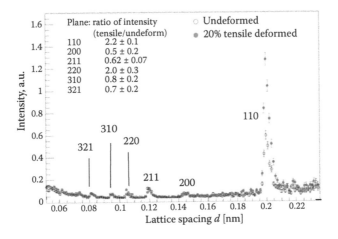

FIGURE 19.13 Comparison of the diffraction patterns of the ferritic steel before and after 20% tensile deformation.

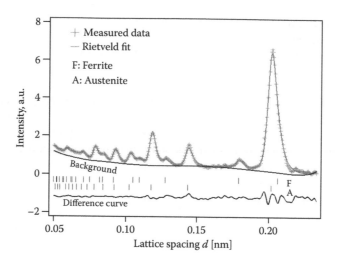

FIGURE 19.14 Diffraction pattern of the dual phase mock specimen. The + markers show the measured data smoothed with binominal smoothing and the solid line under or near the + marks through them is the Rietveld fit. The lower graph is the difference curve between the measured data and the Rietveld fit.

necessary to measure this parameter as a bulk average since it may have through-thickness variation. Therefore, the neutron diffraction technique is useful to assess the amount of retained austenite for the development of the advanced steels. We investigated here the possibility of measuring the amount of retained austenite by using a compact neutron source. The dual-phase mock specimen was prepared by stacking small cubes of the ferritic steel (JIS SM400A) with the cylindrical austenitic stainless-steel plates (JIS SUS316), to control the volume fraction of the austenite to be 19.1%. The specimen was continuously rotated during measurement in order to reduce the influence of texture. Figure 19.14 shows the measured diffraction pattern, taken for 60 min, which is smoothed by binominal smoothing. Diffraction peaks derived from the ferritic and austenitic steels are recognizable, and the volume fraction of the austenitic phase is estimated to be 17.0 ± 3.5 [17 ± 03.5]% by fitting the diffraction pattern using Z-Rietveld code [Oishi 2009]. This value agrees with the actual value 19.1% within the error bar, and it is expected to improve the measurement accuracy more by optimizing the measurement condition.

19.3.3 Effect of Moderation Time on Instrument Resolution

As shown in Figure 19.12, the peak width of each reflection associated with the resolution seems to be larger compared with that of typical neutron engineering diffractometers. In the present study, the resolution was determined by normalizing the full width half maximum (FWHM) of the reflection by its peak position. The peak position and FWHM were obtained by Gaussian fitting. Typical resolution for the engineering diffractometer in the large neutron facility, i.e. TAKUMI in J-PARC/MLF [Harjo 2010], is optimized to be 0.2–0.4%, in order to perform accurate strain measurements by neutron diffraction. In contrast, the resolution of the 110 reflection in Figure 19.12 obtained by RANS is calculated to be approximately 3.2%, which is of lower resolution than the typical engineering diffractometer. This result indicates that much higher resolution is required to perform the strain measurement using RANS. Therefore, we discuss here the optimum optical system, especially in terms of the moderation time, to realize the high resolution of TOF neutron diffraction in a compact environment.

The dispersion of the wavelength is related to the dispersion of the flight time of neutrons, which is determined by the neutron pulse width. The neutron pulse width depends on the proton pulse width, t_p and the moderation time, t_{mod}. Furthermore, the dispersion of the diffraction angle is affected by a difference between the estimated neutron path and the real neutron path. We define z as an effective sample width, which can be seen from the detector.

At first, a change in the resolution was experimentally measured as a function of t_p. The proton pulse width, t_p, was manually changed from 190 to 8 μs, and the resolution of the 110 reflection was calculated for each t_p. Note that the sample width, z was constant at 10 mm in this measurement. It can be found in Figure 19.15 that the resolution increased almost linearly with an increase of t_p. The reason why the resolution at t_p = 20 μs is higher than that in Figure 19.12 is a predominant effect of smaller solid angle of the detector due to longer distance from the sample to the detector. Secondly, the effective sample width, z was changed by changing the distance between two samples as shown in an insertion of Figure 19.16, and a change in the resolution was experimentally measured as a function of z. Note that the proton pulse was constant at 20 μs in this measurement. As shown in Figure 19.16, the resolution increased almost linearly with the increase of z. According to the results in Figures 19.15 and 19.12, the minimum resolution is predicted to be about 2.5%, which is still insufficient for the strain measurement by neutron diffraction. Consequently, a way to improve the resolution is investigated below by focusing on the moderation time.

The moderation time for RANS is estimated to be 30 μs in the standard deviation by fitting the plots in Figures 19.15 and 19.16 with following equations:

$$\sigma = \sqrt{\sigma_{mod}^2 + \sigma_p^2 + \sigma_z^2} \tag{19.1}$$

$$\sigma_p \propto t_p \tag{19.2}$$

$$\sigma_z \propto z \tag{19.3}$$

where σ is a dispersion of the diffraction peak, determined by three factors, i.e. a moderation time width, σ_{mod}, a proton pulse width, t_p, and a sample width, z, which are established by the experiments. The standard deviation of a square wave is given by 0.3 times its width. According to the radiation transport Monte-Carlo simulation, the moderation time necessary to obtain 10–20 meV was calculated to be 30 μs, which is equivalent to that of the experimental estimation. Note that the energy range utilized for this simulation was 0.2–0.3 nm in wavelength, corresponding to the 110 reflection. This moderation time width is much longer than the shortest proton pulse duration of 8 μs, and it is equivalent to approximately 1.8% of the resolution at the 110 reflection. This result indicates that it is difficult to achieve a 1.8% resolution (or less) even by optimizing

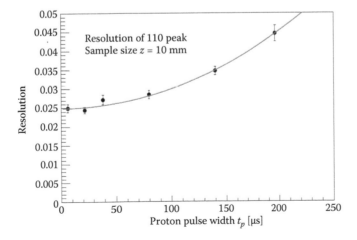

FIGURE 19.15 Change in the time resolution of the 110 reflection as a function of the proton pulse width, t_p for RANS. The solid line shows the fitted curve using Equation 19.1.

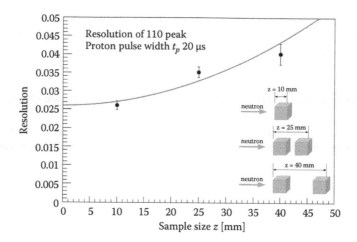

FIGURE 19.16 Change in the time resolution of the 110 reflection as a function of the sample size z for RANS. The solid line shows the fitted curve using Equation 19.1.

the optical layout, since the moderation time predominantly determines the resolution for the diffraction measurement using RANS. Therefore, it is necessary to optimize the shape and the material of the moderator to obtain shorter moderation time in an attempt to improve the resolution achievable at RANS.

According to Figure 19.17, that shows the simulation results of timing distribution for neutron emission at the moderator surface, by decreasing the thickness of the moderator from current 40 to 20 mm we can expect to improve the resolution by a factor of 2. In addition, the application of the decoupler system might improve the resolution as well. Although these refinements can work better for the strain measurement using the compact neutron source, it is conversely a disadvantage for the texture measurement because of lower counting statistics.

In this study, we discussed engineering applications of a compact accelerator-driven neutron source based on TOF neutron diffraction. However, discussion of the perspective for compact neutron sources for engineering applications should not be limited to the TOF method, which inherently requires long flight paths for achieving a reasonable resolution level. In contrast, angular dispersive neutron

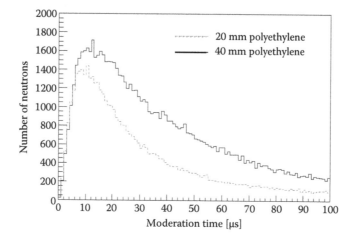

FIGURE 19.17 Simulated timing distribution of neutron emission at the moderator surface in the energy range of 10–20 meV, for RANS by Geant4 code. The moderator thicknesses were assumed to be 20 mm and 40 mm for discussing the effect of thickness on the resolution.

diffraction is the other notable technique for improving resolution in a compact environment. A monochromatic beam obtained by a monochromator is expected to bring a reasonable resolution and lower background noise for the diffraction experiment, which may realize the strain measurement by neutron diffraction with compact accelerator-driven neutron sources.

19.3.4 Nondestructive Visualization of the Air Gap and Steel Bar Position Difference in Concrete with Fast Neutron Imaging

Aging of large scale infrastructures such as bridges is another serious global issue that we have to address as soon as possible. In Japan, there are 720,000 bridges, and beginning in 2014, all the bridges in Japan should be examined every 5 years. But until now there has been no technique for observing inside a thick concrete slab, or finding corrosion inside the junction part of large steel bridges in order to make an accurate assessment or diagnosis of which bridges seriously need rapid maintenance and when.

Figure 19.18 shows the neutron transmission probability for different concrete slab thicknesses for neutrons of energy from 1 meV to 2 MeV. For example, for the 50-cm concrete slab the transmission probability of fast neutrons with energy above about 100 keV grows more than ten million times bigger than that for lower energy 1 eV neutrons. For the 30-cm concrete thickness it becomes a hundred thousand times bigger.

The fast neutron imaging detector development and transmission experiments are now explained in what follows [Otake JDR 2017, Seki NIMA 2017].

Figure 19.19 shows the macroscopic scattering cross section of ordinary concrete for neutrons and X-rays based on the data of Table 19.2. As seen in the figure, the penetration power of neutrons above a few MeV is almost the same as that of 500 keV X-rays which can be produced by portable electron linacs [Uesaka 2014].

There are two advantages to using fast neutrons with the energy above 100 keV over high-energy X-rays with the energy above 500 keV. One is that with fast neutron imaging, we can get high signal-to-noise imaging with TOF measurement to avoid scattered neutrons while X-ray imaging cannot use TOF. Another is that neutrons have high sensitivity for the water, so that inside the concrete slab, fast neutron can distinguish an air hole, void or water clearly.

We have developed a fast neutron imaging detector. Its design can be characterized with the following features: (1) the detector is considered for use outdoors on a mobile truck, (2) the detector provides observation in real time, (3) high efficiency is necessary for low dose and rapid inspection, (4) a spatial resolution of 3 cm is required to identify inner objects in concrete and (5) the detector is ready for time-of-flight measurement. We adopted a simple combination of plastic scintillators (BICRON BC-408) and

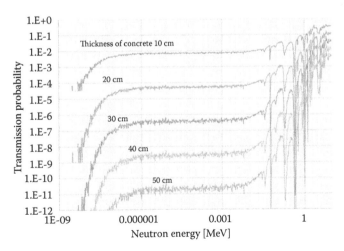

FIGURE 19.18 Transmission probability for neutrons through thick concrete slabs of 10 cm, 20 cm, 30 cm, 40 cm and 50 cm thicknesses. The neutron energy range is from 1 meV to 2 MeV.

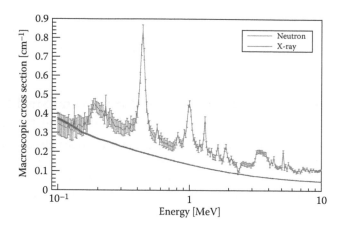

FIGURE 19.19 Macroscopic cross sections for neutrons and X-rays in concrete. The error bars represent the variation of hydrogen content indicated in Table 19.2.

TABLE 19.2 Relative Elemental Composition of
Concrete in Mole Ratio (the Density of Concrete is
Assumed to be 2.2 g/cm³)

Elements of Concrete	Relative Composition Ratio
• H	0.16 ± 0.05
• O	0.55(+0.02/−0.03)
• Si	0.16(+0.02/−0.03)
• Ca	0.05 ± 0.01
• Al	0.04 ± 0.01
• Na	0.02 ± 0.01
• Others	0.03 ± 0.01

Note: Other elements include iron, magnesium, sulphur, potassium, titanium, phosphorus and manganese.

a semiconductor photon sensor (multi-pixel photon counters (MPPCs); HAMAMATSU S10931-050P) as the configuration of the unit pixel [Seki NIMA 2017]. The scintillator is 30-mm high, 30-mm wide and 50-mm deep. The MPPC has an effective area of 3 mm × 3 mm. The prototype has 16 (4 × 4) pixel arrays with the total area of 120 mm × 120 mm. These elements are easy to handle outdoors because the plastic scintillators do not have inflammability (unlike liquid scintillators) and MPPCs are compact with low power consumption in comparison with photomultiplier tubes.

The detector and samples were installed in the sample and detector box. A 300-mm-thick set of three ordinary concrete blocks was located just behind the 250 mm × 250 mm beam entrance of the sample and detector box. The beam incidence plane of the concrete was 5 m away from the moderator surface. Each concrete block was 300 mm in height, 300 mm in width, and 100 mm in thickness with a density of 2.2 g/cm³. The detector was located 220 mm behind the most upstream concrete block, and was fixed during the measurement.

The first object was a steel bar with the length of 300 mm and the square cross section of 100 mm². The second object was water in a quartz cell with the height of 50 mm, width of 10 mm, thickness of 20 mm (inner dimensions) and wall thickness of 0.625 mm. The quartz cell was attached on a supporting steel bar as shown in Figure 19.20a. Each of these two samples was mounted on a motor-driven stage between concrete blocks and the detector, and moved in steps of 5 mm in order to observe smaller objects than the pixel width (30 mm) of the detector. As for the second object, image scanning was performed with water and without water, respectively. The third object was an 18-mm-diameter void hole in a concrete block

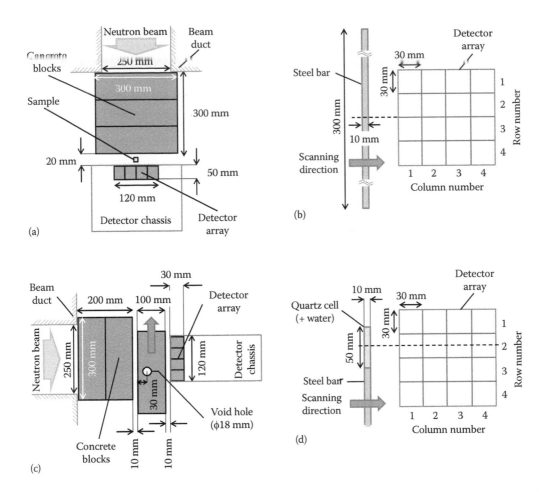

FIGURE 19.20 Experimental setup for (a) steel bar and water (top-view); (b) steel bar (front-view); (c) void hole (side-view); (d) water (front-view).

which was located behind the totally 200-mm-thick ones shown as in Figure 19.20c. The hole penetrated through the 100-mm-thick concrete block horizontally at the position of 30 mm from the beam incident plane. In the measurement for this sample, the concrete block in the most downstream position had the horizontal hole, and was moved vertically in 5-mm steps in front of the detector [see Figure 19.20c].

The accelerator was adjusted to provide beam with an average proton current of 11 µA, a pulse width of 100 µs and a 20 Hz repetition frequency. The time gate of 150 µs for the detector was synchronized after 30 µs delay with proton pulse timing in order to remove slow neutrons. The exposure time was 10 min at each sample position. The detector counts were between about 30,000 and 100,000 neutrons/min/pixel which differed depending on the position behind the concrete blocks and also the strength of optical coupling between scintillators and MPPCs. The steel bar and water were scanned at 37 and 11 points from right to left, when viewed downstream, with a total measurement time of about 6 hrs and 4 hrs, respectively. The concrete block with the void hole was scanned at 20 points in the upward direction with a total measurement time of about 5 hrs.

The incident flux fluctuated by about 4% due to the instability of the proton beam current. This fluctuation was calibrated by using the neutron counts of the pixels which was thought to not be affected by the sample object; that is, when the object was in front for the right- and left-half side of the detector, the neutron counts at each scanning position were normalized by using the total counts of the leftmost/rightmost columns of pixel array, respectively.

By drawing the relative intensity data, obtained on a certain column (row) of the detector array, in the order of the sample position, we can display two-dimensional transmission images of samples with the 5 mm (scanning step) × 30 mm (scintillator width) pixels. Figure 19.21a shows a transmission image of the steel bar obtained on the second column. The smaller object than the 30-mm-wide scintillator could be clearly imaged by the scanning method. Since the relative intensity was attenuated by the steel bar, the image of steel bar is represented as a dark shadow with the grayscale. Figure 19.21b shows a transmission image of the void hole obtained on the second row and its projection, respectively. In contrast to the steel bar, the presence of the void hole increased the relative intensity as shown in Figure 19.21c and gave rise to a bright image.

The second row in Figure 19.22a and 19.22b shows transmission images of the quartz cell with and without water, obtained on the second column, respectively. The image of water itself was conveniently produced from the ratio of Figure 19.22a and 19.22b, as shown in Figure 19.22c, and successfully observed. The projection of each image is shown in Figure 19.22d.

The measurement time for fast neutron transmission imaging can become much shorter (such as a few minutes) by increasing the sensitivity of the fast neutron detector and the number of the fast neutrons from the source.

Our final goal is to realize a transportable neutron system for nondestructive inspections and health diagnostics for social infrastructures such as bridges and highways by using fast neutron imaging. The results presented here are the first demonstration of fast neutron imaging of steel bars, air holes and water though 30-cm thick concrete using the plastic fast neutron imaging detection which can be safely used outdoors. For the transportable case, the Japanese radiation safety regulation allows usage of linear accelerators with energy less than 4 MeV for nondestructive inspection of bridges. At RIKEN, we are now developing a more compact neutron source with a lower energy (2.49 MeV) proton linac. In Section 19.4, our future plan for this compact neutron source is explained briefly.

We have shown that fast neutrons are useful for nondestructive inspection of social infrastructural. This is also quite important for the laser-driven neutron source, since the generated neutron energy is higher than a few MeV. Although the low energy neutron intensity in the laser-driven case will likely be much less than 10^{12} neutrons per second, which is appropriate for low energy neutron scattering applications, laser-driven hot neutron production would be useful for fast neutron imaging which can include TOF methods where the potential for short neutron bunch generation is attractive. The reader is encouraged to view Chapter 6 by M. Roth which presents neutron production yields using irradiation by laser-accelerated protons and deuterons. Also, a relevant review of laser-driven ion acceleration in general can be found in Chapter 5 by A. Macchi.

19.4 Future Plans

We are confronting challenges in three general subject areas: more compact neutron system development; optimized neutron beam design and production; and detection and data-analysis system development and optimization according to each application. To realize a more compact neutron source which can be easily used at many places, the most important issue is to find a suitable shielding method for the whole compact neutron system which is to be as small as possible while keeping the instrument background as low as possible. The present RANS operating condition with 7 MeV protons with a pulsed average maximum current of 100 μA delivers 10^{12} neutrons per second at the target position. We have also shown here that neutron radiography and neutron diffraction for the steel samples can be implemented with the measurement times of several to about 10 min. A neutron source with more neutrons per second requires shielding more than 10^{12} neutrons per second at the target and the neutron beam line becomes bigger, so that it would not be as compact as what industry can use on-site.

We are planning to construct RANS 2, a more compact lighter neutron source with a 2.49 MeV proton linac whose length is about 2.5 m and whose target station weight will be less than 1 t (note that the RANS proton linac is 5-m long and the weight of the target station is more than 20 t).

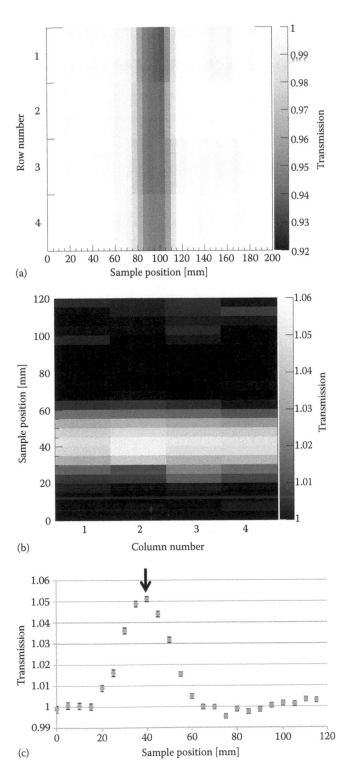

FIGURE 19.21 Result of observation through 30 cm of concrete: (a) results of steel bar behind concrete, transmission image on the second column of detector array and; (b) results of the void hole in concrete transmission image on the second row of detector array; (c) projection of transmission image.

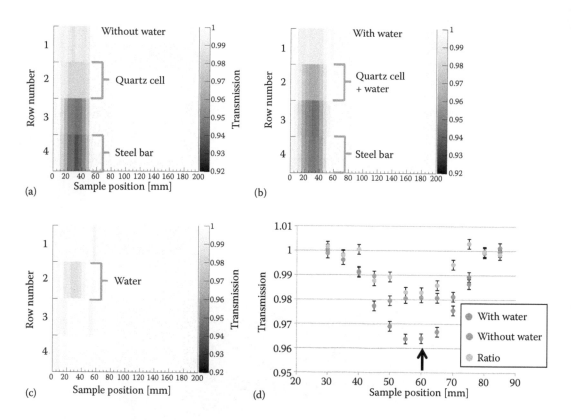

FIGURE 19.22 Results from observation of water: (a) transmission image of the empty quartz cell; (b) transmission image of the quartz cell with water on the second column of detector array; (c) transmission image of water (ratio between [a] and [b]); (d) projection of transmission images.

In addition to compact neutron source development, as we have explained here, we are developing fast neutron imaging techniques for social safety of large-scale construction. With acceleration of protons to several MeV driven by high-powered lasers, more useful neutron imaging techniques could be developed due to the unique capability for production of intrinsically shorter neutron bunches in the laser case.

Acknowledgements

This work has been partially supported by the Council for Science, Technology and Innovation (CSTI), Cross-Ministerial Strategic Innovation Promotion Program (SIP), Infrastructure maintenance, renovation and management (Funding agency: JST), and the Photon and Quantum Basic Research Coordinated Development Program from the Ministry of Education, Culture, Sport, Science and Technology, Japan. Furthermore, this work was supported by JSPS KAKENHI Grant Numbers 25289265 and 25420078. The author would like to thank the Iron and Steel Institute of Japan (ISIJ) Research Group for their beneficial assistance.

References for Chapter 19

Committee on Cost of Corrosion in Japan 2001. "Cost of Corrosion in Japan" *Zairyo-to-Kankyo*, **50**: 490–512.

Furusaka, M., Sato, H., Kamiyama, T., Ohnuma, M. and Kiyanagi, Y. 2014. "Activity of Hokkaido University Neutron Source, HUNS" *Physics Procedia* **60**: 167–174; doi:10.1016/j.phpro.2014.11.02.

[GEANT4] http://geant4.cern.ch [20 May 2017].

Harjo, S. et al., 2010. "Aspire to Become TAKUMI – TAKUMI Present Status and Research Topics" *Mater. Sci. Forum* **652**: 99.

Hawkesworth, M.R. 1977. "Neutron Radiography: Equipment and methods" *Atomic Energy Review* **152**: 169–220.

Ikeda, Y., Takamura, M., Taketani, A., Sunaga, H., Kumagai, M., M., Oba, Y., Otake, Y. and Suzuki, H. 2016. "Prospect for application of compact accelerator-based neutron source to neutron engineering diffraction" *Nuclear Instruments and Methods in Physics Research Section* A**833**: 61–67; doi:10.1016/j.nima.2016.06.127.

Ikeda, Y., Takamura, M., Taketani, A., Sunaga, Otake, H., Suzuki, H., Kumagai, M., M., Oba, Y. and Hama, T. 2015. "Measurement of neutron diffraction with compact neutron source RANS" IL NUOVO CIMENTO, **38C**: 177, DOI: 10.1393/ncc/i2015-15177-y.

Japanese Industrial Standards JIS, JIS G 3106: 2015 "Rolled steels for welded structure".

Magner, S.H., De Angelis, R.J., Weins, W.N. and Makinson, J.D. 2002. "A Historical review of retained austenite and its measurement by x-ray diffraction" *Advances in X-ray Analysis* **45**: 92.

Makinouchi, A. et al., 2010. "Report of Compact neutron source imaging system exploratory committee".

Makinouchi, A. et al., 2011. "VCAD System Research Program2010~Future of VCAD system", March, 2011, RIKEN Symposium, ISBN978-4-9904431-2-2.

Morooka, S., Tomota, Y., Adachi, S. and Kamiyama, T. 2006. "Influence of <110> Fiber Texture on Internal Stresses Evolved by Plastic Deformation for a Pearlite Steel" *Advanced Materials Research* **15–17** (2006) 912.

Nakayama, T., Fukumoto, H. and Fujii, Y. 2005. "Ferrous Materials for Construction" *Bull. of ISIJ* **10**: 932.

Ohnuma, M., Furusaka, M., Nakamichi, H. and Murakami, T. 2016. CAMP, ISIJ, **29**: 88.

Oishi, R. et al., 2009. "Rietveld analysis software for J-PARC" *Nucl. Instrum. Methods A* **600**: 94–96.

Otake, Y., Seki, Y., Wakabayashi, Y., Ikeda, Y., Hashiguchi, T., Yoshimura, U., Sunaga, H., Taketani, A., Mizuta, M., Oshima, Y. and Ishida, M. 2017. "Research and Development of a Non-destructive Inspection Technique with a Compact Neutron Source" *Journal of Disaster Research* **12**[No.3]: 585–592; doi: 10.20965/jdr.2017.p0585.

Otake, Y., Uesaka M., Kobayashi H. 2015. "Compact Neutron Sources for Energy and Security" Reviews of Accelerator-Science and Technology 'Accelerator Applications in Energy and Security' **Vol. 08**: 181–207, World Scientific DOI: 10.1142·S1793626815020014.

PSTAR 2011. Program is provided by the National Institute of Standards and Technology; http://physics.nist.gov/PhysRefData/Star/Text/PSTAR.html.

Sato, T. et al., 2013. "Particle and Heavy Ion Transport code System, PHITS, version 2.52" *J. Nucl. Sci. Technol.* **50** [9]: 913–923.

Seki, Y., Taketani, A., Hashiguchi, T., Wang, S., Mizuta, M., Wakabayashi, Y., Otake, Y., Yamagata, Y., Baba, H., Kino, K., Hirota, K. and Tanaka, S. 2017. "Fast neutron transmission imaging of interior of large-scale concrete structures using a newly developed pixel-type detector" *Nucl. Instrum. Methods A* **870**: 148–156.

Taketani, A., Yamada, M., Ikeda, Y., Hashiguchi, T., Sunaga, H., Wakabayashi, Y., Ashigai, S., Takamura, M., Mihara, S., Yanagimachi, S., Otake, Y., Wabayashi, T., Kono, K., and Nakayama, T. 2017. *"Visualization of water in corroded region of painted steels at a compact neutron source"* ISIJ International **Vol. 57** [No. 1]: 155–161; DOI: 10.2355/isijinternational.ISIJINT-2016-448.

Tomota, Y. and Sekido, N. 2016. CAMP ISIJI, **29**: 87.

Uesaka, M. et. al., 2014. "On-site nondestructive inspection by upgraded portable 950 keV/3.95 MeV X-band linac x-ray sources" *J. Phys. B: At. Mol. Opt. Phys.* **47**: 23400.

Wakabayashi, Y., Taketani, A., Ikeda, Y., Hashiguchi, T., Kobayahsi, T., Wang, S., Yan, M., Harada, M., Ikeda, Y. and Otake, Y. 2016. "A function formation of source neutron production by the 9Be + p reaction at 7MeV at RANS" JAEA-Conf 004 (2016) 135–140.

Wenk, H.-R. and Van Houtte, P. 2004. "Texture and anisotropy" *Reports on Progress in Physics* **67**: 1367.

Wenk, H.-R., Lutterotti, L. and Vogel, S. 2003. "Texture analysis with the new HIPPO TOF diffractometer" *Nucl. Instrum. Methods Phys. Res. A* **515**: 575.

Yamada, M., Otake, Y., Taketani, A., Sunaga, H., Yamagata, Y., Wakabayashi, T., Kono, K. and Nakayama, T. 2014. "Non-Destructive Inspection of Under-Film Corroded Steels Using A Compact Neutron Source" **Testu-to-Hagane 100**: 429–431.

Yamagata, Y., Hirota, K., Ju, J., Wang, S., Morita, S.-Y., Kato, J.-I., Otake, Y., Taketani, A., Seki, Y., Yamada, M., Ota, H., Bautista, U. and Jia, Q. 2015. "Development of a neutron generating target for compact neutron sources using low energy proton beams" *J. Radioanal Nucl. Chem.* **Vol. 305**: 1–8.

20

Neutron Science with Highly Brilliant Beams

Peter Böni

Winfried Petry

Illuminating suitable targets with intense γ-rays may yield highly brilliant neutron beams that are suitable for the investigation of the static and dynamic properties of very small samples and samples under extreme conditions. Conducting such experiments are presently not an easy task at existing neutron sources due to their limitations in brilliance. To make the neutrons from reactors and spallation sources useful for experiments, they have to be brought into thermal equilibrium in a moderating material, such as water. Only the neutrons diffusing in the direction of the experiment can be extracted thus reducing the available flux dramatically. Moreover, using spallation, the diffusion process leads to a temporal broadening of the width of the pulsed neutron beams, thus reducing the energy and momentum resolution of the spectrometers. In contrast, if directed neutron beams with the proper energy can be produced, the dynamic range for spectroscopy with neutrons can be significantly widened. Future applications of highly brilliant neutron beams will be the *in situ* and real-time investigation of materials under extreme conditions and the characterization of interfaces and surfaces. Foreseeable applications are in materials science, soft and condensed matter, as well as for industrial and medical needs. If neutrons can be extracted as envisaged using modern neutron optics, the brilliance of neutron beams may be increased rather dramatically with the development of new accelerator and laser technologies. The prospect of laser-driven particle accelerators contrasts the present stagnation of the neutron flux at existing sources due to limitations in the power densities in fission as well as spallation processes. As the development of the technology for these highly brilliant beams is not yet assured, a step in between might be attractive, for example the usage of laser-driven particle acceleration technology combined with innovative focusing moderators that may lead to a high brilliance. For more details on accelerator-driven neutron generation see also Chapter 6 'Neutron Generation' by M. Roth and Chapter 19 'A Compact Proton Linac Neutron Source at RIKEN' by Y. Otake.

20.1 Introduction

Neutron scattering and scattering by synchrotron radiation are ideal probes for the investigation of condensed matter on an atomic scale. The microscopic information on the structure of materials and excitations are the basis for any kind of research in physics, biology, chemistry and materials science, as well as for device development. Although, at the existing neutron sources the average brilliance

$$\Psi = d^4N/dtdAd\lambda d\Omega,\qquad(20.1)$$

i.e. the number of neutrons that are produced per second, unit area, wavelength interval and solid angle is very small when compared with synchrotron light sources; neutron scattering is indispensable because slow neutrons have properties that can hardly be matched by any other experimental technique.

We summarize the salient features in the eight points that follow [Furrer 2009]. Neutrons interact with the nuclei and are therefore sensitive to light and heavy atoms. Moreover, isotopic substitution enables adjustment of the scattering contrast, thus allowing for example to highlight the functionality of parts of large macromolecules by deuteration. The wavelength of thermalized neutrons is of order atomic lengths, thereby ideally suited to make visible these dimensions. The energy of thermal neutrons is of the order of the energy of elementary excitations in condensed matter. Therefore, the dynamic properties of materials which are relevant at room temperature can be measured with high accuracy. As neutrons have no charge, their penetration depth is large for most materials. Therefore, the volumetric properties of materials can be identified. Moreover, studies under extreme conditions, such as high pressure and high magnetic fields, can be performed. The dominant interaction of neutrons with matter is a nuclear process, i.e. it depends on the particular isotope thereby allowing a distinction of isotopically labelled molecules in an environment of chemically identical molecules. Because of the weak interaction strength, neutrons do not cause significant radiation damage and the cross sections can be interpreted in terms of linear response theory. In addition, the data can be put on an absolute scale, facilitating their interpretation. Neutrons carry a magnetic moment and are thus very sensitive to the static and dynamic magnetic properties of materials. Finally, thermal neutrons might be absorbed by a nucleus and induce a nuclear reaction, thereby creating radio isotopes.

Unfortunately, the number of brilliant neutron sources is rather limited due to the huge capital investments required. Therefore, sources with a high flux density, i.e. a high number of neutrons per cm²s can usually only be realized by large international collaborations. As the beam time is expensive and scarce, the neutron beams have to be used as efficiently as possible. Hence, it is difficult to establish neutron scattering as a laboratory technique. Therefore, the availability of compact laser-driven neutron sources would ease conducting experiments on a small scale and educate young scientists in the field. This is in contrast to scattering of X-rays, where the education can be performed in small laboratories using conventional X-ray equipment. With the planned shutdown of the medium flux neutron sources such as the BER-II in Berlin (Germany) and the Orphée reactor in Saclay (France) the availability of neutrons in Europe will decrease significantly. To cope with this situation on a long term, affordable sources based on accelerator technologies should be developed.

In Chapter 20, we will discuss the various means for producing neutrons, followed by an introduction into neutron transport and focusing systems that allow taking advantage of brilliant neutron sources. Finally, a few applications of neutron scattering on small samples are shown.

20.2 Neutron Sources

Neutron scattering is a flux-limited technique. Therefore, the quality of the experiments is essentially limited by the brilliance of the neutron source. Assuming that the neutron losses during the transport of the neutrons from the source to the sample are kept small, it is straightforward to calculate the maximum possible intensity I at the sample position of a beam line using the expression [Böni 2014]:

$$I(\text{s}^{-1}) = A \cdot \Delta\lambda \cdot \Omega \cdot \Psi \qquad (20.2)$$

Here $A(\text{cm}^2)$, $\Delta\lambda(\text{Å})$, $\Omega(\text{sterad})$, and $\Psi(\text{cm}^{-2}\text{s}^{-1}\text{Å}^{-1}\text{sr}^{-1})$ designate beam size at the sample position, wavelength band, solid angle and brilliance of the neutron source, respectively.

As an example, we consider the beamline MIRA at MLZ [Georgii 2015], Munich (Germany), that provides cold neutrons from the medium flux reactor FRM II. For a typical experiment [Kugler 2015] the following configuration is used: $\lambda = 4.5$ Å, $\Delta\lambda = 0.065$ Å (i.e. a bandwidth $\Delta\lambda/\lambda = 1.4\%$) and a horizontal and vertical divergence of 0.5° and 1°, respectively, yielding $\Omega = 1.5 \cdot 10^{-4}$ rad. In addition, we have assumed a transport efficiency of 50% due to the finite reflectivity of the supermirror coatings and the monochromator. Reading the brilliance from the 'Blue Book' of the FRM II [Bluebook 2011], i.e. $\Psi(4.5$ Å$) = d^2\Phi/(d\Omega d\lambda) = 2.2 \cdot 10^{12}$ $\text{cm}^{-2}\text{s}^{-1}\text{Å}^{-1}\text{sr}^{-1}$, one obtains for a beam size of 1 cm² an intensity $I = 1.1 \cdot 10^7$ neutrons per second, which represents a typical value also achieved at the most powerful neutron sources worldwide. According to Liouville's theorem, there is no means to improve this intensity, not even with fancy focusing techniques.

It is difficult to compare the time averaged intensity as obtained at a continuous neutron source, such as the FRM II with the neutron fluence obtained by means of laser-driven ion bunches [Roth 2017] because the energy range of the produced neutrons, as well as the emission characteristics, are very different. We quote that Higginson et al. obtained an unmoderated neutron fluence of up to $8 \cdot 10^8$ neutrons per steradian per laser pulse (360 J in 9 ps) in the forward direction [Higginson 2011], which seems to be rather low.

Realistically, one should consider beam losses of about 50% between the source and the sample area. Simple calculations based on Equation 20.2 benchmark the feasibility of experiments at various neutron sources. To a good approximation, the brilliance $\Psi = d^2\Phi/(d\Omega d\lambda)$ of thermal beams scales with the neutron flux density Φ of the respective neutron source, because the moderation processes are similar, i.e. the moderation takes place in a large volume of D_2O (reactors) or H_2O (spallation sources). In contrast, the brilliance of cold moderators may differ strongly due to the different and still improving geometrical designs of the cold sources.

Until now, all neutron sources use nuclear reactions for the release of neutrons. Figure 20.1 shows the rapid development of the peak neutron flux density Φ between 1930 and 2020. Neutrons were produced for the first time by Chadwick in 1932 who irradiated beryllium with α-particles from the decay of polonium. The following years witnessed a rapid increase of Φ when fission reactors came into operation. However, due to limitations in removing the heat, as produced during the fission reaction, the maximum thermal neutron flux density is limited to approximately 10^{15} $\text{cm}^{-2}\text{s}^{-1}$ and was achieved by means of the high-flux reactor (HFR) at the Institut Laue-Langevin (ILL) in Grenoble (France). The proposal to construct an even more powerful source, the Advanced Neutron Source (ANS) in the US was cancelled in 1995 [Rush 2015] and the ambitious project of the future Russian high-flux neutron source PIK at Gatchina (Russia) still has to prove its capabilities.

A more prospective route to achieve a high-flux density is the use of the spallation process. Here, a heavy metal target is bombarded by protons or electrons. The energy released per neutron is approximately 10 times smaller than during the course of a fission process, where approximately 210 MeV are produced. Moreover, because the neutrons are released in short pulses, the peak flux density Φ can be further increased. The future European Spallation Source ESS will deliver neutron bunches with a repetition rate of 14 Hz and a width of 2.86 ms yielding $\Phi \approx 6 \cdot 10^{16}$ $\text{cm}^{-2}\text{s}^{-1}$ (Figure 20.1). Similarly, as for fission sources, the maximum flux density seems also to approach a limit, which may be of order of 10^{17} $\text{cm}^{-2}\text{s}^{-1}$ that cannot easily be overcome due to the problems of removing the heat. In contrast to sources for synchrotron radiation, where the FELs promise many orders of increase in brilliance, neutron sources will

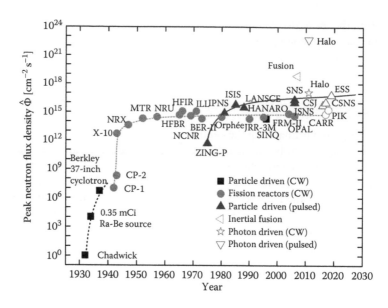

FIGURE 20.1 Peak neutron flux density $\hat{\Phi}$ of major neutron sources. Inserted are also the performances of the recently proposed accelerator-based sources namely the source based on halo neutrons, i.e. neutrons in highly excited states orbiting the nucleus at large radii [Habs 2011] and the compact source CSJ [Rücker 2016]. Open symbols indicate neutron sources that are not yet in operation or in a conceptual phase.

remain limited in their magnitude in the near future. This statement will even be true if neutrons may become available by means of fusion [Taylor 2007], where an effective neutron flux density of order of $6 \cdot 10^{18}$ cm^{-2}s^{-1} is anticipated. Such a facility will be extremely expensive, technologically complex, and will produce large amounts of nuclear waste. Therefore, considering the moderate gains in flux density this concept may not justify the tremendous investments and environmental impact.

The major problem with conventional neutron sources is twofold. Firstly, the sources and the moderators are much larger than typical sample volumes and therefore not adapted. Secondly, the moderators emit neutrons into a solid angle of $\simeq 1$ sr, i.e. only a fraction of ~10^{-5} of the emitted neutrons reaches the sample. In order to circumvent the process of moderation, Habs et al. proposed producing neutrons by means of the emission of low energy polarized neutron beams from neutron halo isomers of stable nuclei [Habs 2011]. These isomers could provide beams with a high brilliance. The concept is based on the (γ,n) reaction, i.e. energy levels directly below the threshold for the release of neutrons are excited by means of γ-beams of high intensity and small bandwidth. Low-energy neutrons from these long-lived states (100 ps–μs) can be released in pulses or quasi-continuously by photons in the visible or X-ray regime, for example, by a laser (Figure 20.2). The γ-beams can be produced by Compton backscattering of laser beams off brilliant electron beams as produced by energy recovery linacs (ERL) [Hajima 2009].

As shown by Habs et al. [Habs 2011], using present day technology, an up to a 10^2- to 10^6-fold increase in brilliance may be achieved when compared with fission and spallation sources, respectively, using accelerator technology that will become available in the near future. Because the targets for neutron production have only a diameter d of typically a few tens of mm, the total number of produced neutrons per second (or pulse) will, of course, be much smaller when compared with conventional sources, however, much more confined in space. As has been shown for synchrotron radiation, the further development of accelerators for example by laser-driven technology will help to increase the brilliance even further. Clearly, to make this concept work, suitable neutron halo isomers, i.e. nuclei that allow neutron excitation far out into free space much beyond the nuclear core radius, have to be identified.

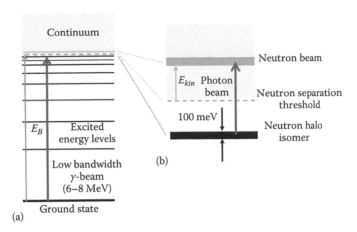

FIGURE 20.2 Schematic picture of the production scheme using neutron halo isomers. (a) With an intense narrow-bandwidth γ-beam neutrons are excited to energy levels close to the binding energy E_B of the neutrons (bold vertical arrow). The level scheme shows the increasing number of compound nuclear resonances with increasing excitation energy. The halo isomer is admixed to several high-lying resonances, resulting in halo isomers with different binding energies within a window of approximately 100 meV. (b) In a second step, a photon beam of much lower energy shown as bold vertical arrow, generates the neutron beam with a kinetic energy E_{kin} by dissociating the neutron halo state.

Recently, an alternative concept for a highly brilliant compact neutron source was proposed by Rücker et al. [Rücker 2016]. Singly charged deuterons D^+ are accelerated by a linear accelerator to an energy of approximately 25–50 MeV providing a current of 50–100 mA. The pulsed beam will have a frequency between 10–300 Hz that can be selected to optimize the experiment. The D^+ beam hits a Be-target. By means of the reaction

$$^9\text{Be} + d \rightarrow n + ^{10}\text{Be} \tag{20.3}$$

neutrons will be released and afterwards thermalized in small moderators. These moderators have a thickness $t \simeq 1$ cm and can be individually designed for the forseen application (diffraction, spectroscopy, SANS, imaging, etc.). It is expected that the performance of these instruments may become competitive with beamlines at the spallation neutron source ISIS at Rutherford Appleton Laboratory, Didcot (UK) [Gutberlet 2016] because the flux densities will be similar (Figure 20.1).

Summarizing, the performance of the present-day neutron sources based on fission and spallation seems to converge towards a peak flux density $\Phi \approx 10^{15}$ and $\Phi \sim 10^{18}$ cm^{-2}s^{-1}, respectively, mostly due to limitations in dissipating heat during the production process of the neutrons. As they are thermalized in volumes, which are much larger than typical sample sizes, most produced neutrons cannot be made available for the experiments. A major increase in brilliance of neutron sources seems only to be possible by building compact, accelerator-based sources which do not use moderators and emit neutrons into a small solid angle. This way, the heat loads and the nuclear inventory can be significantly reduced thus improving the safety and security of neutron sources.

20.3 Neutron Transport and Focusing

As discussed in Section 20.2, the moderators of the existing neutron sources are large when compared with the typical sample size of a few cm^3. For example, in research reactors the moderators D_2O for thermal neutrons and liquid D_2 for cold neutrons occupy typically a volume of 2000 l and 20 l, respectively

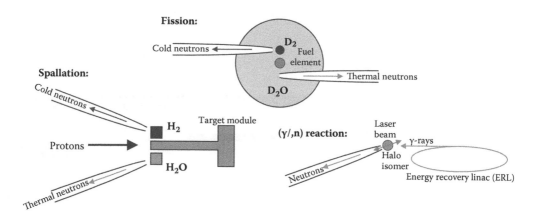

FIGURE 20.3 Moderator configurations for neutron sources using fission, spallation and the (γ,n) reaction. The moderators for fission sources are large (many litres) to bring the neutrons in thermal equilibrium. To keep the pulse width short, the moderators at spallation sources have a volume of only about 1 litre yielding under-moderated beams. The volumes of the targets anticipated for the (γ,n) reaction are of the order of 1 mm³ leading to a highly brilliant beam. The entrance of the neutron guides at future neutron sources will be moved as close as possible to the neutron source in order to guarantee a full illumination of the guides and to minimize the background by fast neutrons and γs from the neutron production process.

(Figure 20.3). The large size is required for achieving a proper thermalization of the neutrons. At spallation sources, the moderators occupy typically a volume of 0.5 l to 1.0 l in order to keep the pulse length as short as reasonably possible. The neutrons are under-moderated and contain a significant fraction of epithermal neutrons (200 meV < E < 100 eV). The extracted beams have typically a size of order of 60 mm (width) × 170 mm (height) [Zeitelhack 2006]. If the (γ,n) reaction can be realized, no moderator is required because the neutrons are directly released with the proper energy from the small target containing the halo isomer.

To adjust the large neutron beams to the sample, the beam size is either reduced by means of collimators and/or slits or by focusing devices. Focusing monochromators reduces the beam size to typically 20 mm × 20 mm. This size is given by the reflecting area of the individual monochromator crystals comprising the focusing device [Bührer 1994, Komarek 2011, Skoulatos 2011, Link 2000, Eckold 2014]. Using focusing parabolic or elliptic guides, the beam size can be reduced to approximately 1 mm × 1 mm [Hils 2004]. A further reduction to 33 μm × 55 μm was achieved by means of Montel mirrors at the SNAP beamline at the Spallation Neutron Source (SNS) at Oak Ridge National Laboratorly (USA) [Ice 2010]. Of course, other focusing techniques such as magnetic lenses [Iwashita 2008], bulk lenses [Eskildsen 1998], Fresnel zone plates [Altissimo 2008], capillary lenses [Kumakov 1992], etc. are available.

Despite all these available focusing techniques, the brilliance of neutron beams cannot be increased using passive focusing elements due to the theorem of Liouville, which states that the phase space is conserved. Let us consider an extraction scheme at a typical reactor or spallation neutron source. The phase space at the entrance of the guide is given by the product of the cross section of the beam multiplied with its solid angle. To increase the solid angle for the neutrons to reasonable values, supermirror coatings with large angles of reflection θ_C are used (Figure 20.4). They are measured in units of the angle of total reflection of nickel, which is given by

$$\theta_c^{Ni} = (N_{Ni} b_{Ni} / \pi)^{1/2} \lambda, \tag{20.4}$$

where N_{Ni}, b_{Ni}, and λ are the density of the nuclei, the scattering length, and the wavelength of the neutrons, respectively. Inserting the parameters for Ni yields θ_c^{Ni} (rad) = 0.0173λ (nm) (or in degrees:

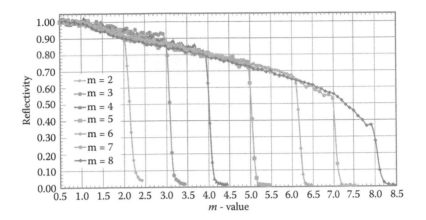

FIGURE 20.4 Reflectivity profiles of supermirrors with m-values $2 < m < 8$. A value $m = 1$ corresponds to a coating made from natural nickel. The slope of R is largely independent of the m-value indicating a stable interface roughness, which is essentially constant for all layers. (SwissNeutronics Website 2017, available at http://www.swissneutronics.ch/index.php?id=24; Schanzer, C. et al. 2016. "Neutron Optics: Towards Applications for Hot Neutrons" *J. Phys.: Conf. Ser.* 746: 012024.)

θ_c^{Ni} $(^0) = 0.99\lambda$ (nm)). For example, for $m = 3$, the divergence of the neutrons that are accepted by the guide is $\alpha = 2\theta_c = 2m\,\theta_c^{Ni} = 2 \cdot 3 \cdot 0.99 \cdot 0.2 = 1.2^0$ for a typical wavelength $\lambda = 0.2$ nm. For a beam size of 10 cm × 10 cm one may define as a measure for the phase space the number B_{in} by

$$B_{in} = (10\ \text{cm})^2 \times (1.2^0)^2 = 144\ \text{cm}^2 (^0)^2. \tag{20.5}$$

Here we have evaluated the solid angle as square of the accepted divergence.

A typical experiment is performed using a sample with a size of 1 cm × 1 cm and a horizontal and vertical divergence of 2^0. Hence the phase space number at the sample, B_{out}, becomes

$$B_{out} = 1\ \text{cm}^2 \times (2^0)^2 = 4\ \text{cm}^2 (^0)^2. \tag{20.6}$$

Therefore, $B_{in}/B_{out} = 36$ times more neutrons are transported by the neutron guide to the instrument than are actually used. If the sample size was reduced to 1 mm^2, 3600 times more neutrons are extracted than actually used, leading to a higher background and requiring costly shielding. Moreover, the neutron guides may degrade more quickly due to radiation damage. We note that in an inelastic experiment another two or more orders of magnitude of unused neutrons are transported by the guide system relative to the number of useful neutrons. Clearly, in design work for future neutron sources it should be considered extracting neutrons from an area of the moderator that is comparable to the sample size by moving the entrance of the neutron guides as close as possible to the moderator (Figure 20.3). This goal can be achieved by means of supermirror coated metallic neutron guides [Schanzer 2010].

We conclude that in the design phase of future neutron sources a major effort should be directed towards the implementation of small moderators and appropriately designed beam extraction systems. Here, compact and affordable neutron sources using accelerator-based technologies such as proposed recently [Habs 2011, Rücker 2016] seem to be very attractive. When these sources are combined with non-linearly tapered guides [Schanzer 2004, Böni 2008] using the recently developed new generation of supermirrors with $m = 8$ times the critical angle of Ni (Figure 20.4) [SwissNeutronics Website 2017, Schanzer 2016] the extraction of neutron beams with a large divergence exceeding $\eta = 1.6°$ and $\eta = 3.2°$ for $\lambda = 1$ Å and $\lambda = 2$Å, respectively, will become feasible. Such a large divergence is sufficient for most experiments.

20.4 Applications

In the following sections, we expose a few examples where a high brilliance neutron source combined with advanced neutron optics is applied to conduct experiments.

20.4.1 Determination of the Order Parameter in NiS_2

Suppressing phase transitions to zero temperature by the application of pressure is an elegant technique for discovering unconventional metallic and superconducting phases in strongly correlated electron systems [v. Löhneysen 2015]. For this application, neutron diffraction is particularly well-suited because of the large penetration depth of the neutrons. However, the availability of usually only small quantities of sample and the limitation of the application of high pressure to small samples severely limits the application of neutron diffraction due to the small intensities available. Therefore, highly brilliant beams on tiny samples are required.

Using two elliptic neutron guides for focusing and defocusing the neutron beam before and after the sample, respectively, we investigated the temperature dependence of the antiferromagnetic order parameter in NiS_2 in a single crystal with the dimensions $1 \times 1 \times 0.1$ mm^3 [Niklowitz 2009] using the triple axis spectrometer PANDA at MLZ.

Rocking curves of the antiferromagnetic Bragg peak with Miller's indices (100) are shown in Figure 20.5a for various temperatures T. The Neel temperature $T_N \simeq 40$ K where the peak vanishes as well as the determined magnitude of the magnetic moment are in good agreement with values reported in the literature [Sudo 1985]. The major advantage of using focusing guides is that by concentrating the beam only onto the sample, the neutron scattering from the bulky high-pressure cell will be reduced, thus increasing the signal to noise ratio. In addition, the intensity of the diffracted neutrons is increased, of course on the expense of divergence.

The experiment shows that the magnetic signal from the sample can be measured within reasonable counting times. Based on the brilliance of the cold source of the FRM II [Bluebook 2011], we estimate that approximately $1.1 \cdot 10^7 \times 0.1$ mm^2/100 mm^2 = $1.1 \cdot 10^4$ neutrons hit the sample per second (the intensity $I = 1.1 \cdot 10^7$ s^{-1} was taken from Section 20.2). Recently, the suppression of magnetic domains in antiferromagnetic Cr under uniaxial pressure was studied using a similar set-up [Adams 2014].

20.4.2 Measurements of Transverse Acoustic Phonons in Lead

The electron phonon interaction is the basis for the formation of Cooper pairs in the theory for superconductivity by Bardeen, Cooper and Schrieffer (BCS theory) [Bardeen 1957]. Modern *ab initio* calculations are able to predict the dispersion of the phonon branches throughout the entire Brillouin zone. Lead is one of the most prominent examples to test the theoretical predictions [Furrer 1970]. Therefore, we have benchmarked the feasibility of conducting inelastic neutron scattering using the focusing set-up at MIRA; a cold, triple-axis spectrometer at the MLZ on a single crystal with the dimensions $2 \times 2 \times 2$ mm^3 [Brandl 2015].

Transverse (TA) and longitudinal (LA) acoustic phonons were measured near the (200) and the (111) zone centres, respectively. A TA phonon is shown in Figure 20.5b. The data witnesses an impressive increase of the intensity of the scattered neutrons (red circles) by approximately a factor of 35 when compared with the data without focusing (green squares). The gain in intensity is much larger than the gain obtained in elastic scattering from Bragg peaks (see Section 20.4.1). The reason is that during the process of diffraction only a very small divergence (depending on the number of coherently reflecting planes of the sample) of the incident beam is accepted because momentum (Q) and energy (E) conservation involve δ-functions. An increase of the divergence does not necessarily lead to an increase in intensity. Therefore, focusing is not as effective for diffraction as might be expected [Adams 2014]. In contrast, in inelastic scattering, the dispersion of the excitations, $\omega(Q)$, allows a

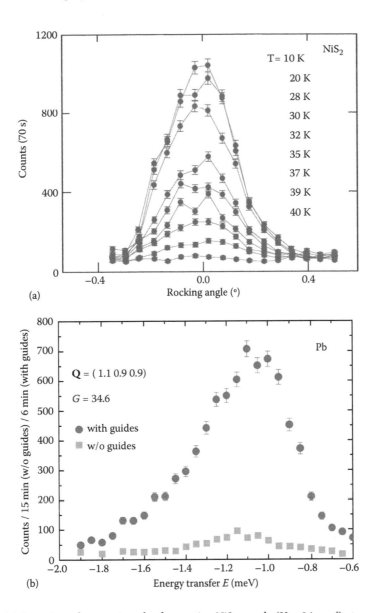

(a)

(b)

FIGURE 20.5 (a) Detection of magnetic order from a tiny NiS_2 sample ($V = 0.1$ mm^3) at zero pressure, using elliptic neutron guides at PANDA. (Niklowitz, P. G. et al. 2009. "New Angles on the Border of Antiferromagnetism in NiS_2 and URu_2Si_2," *Physica B* 404: 2955–2960.) The vertical axis provides the number of detected neutrons per 70 s when the sample is rotated around the Bragg position. The intensity decreases with increasing temperature and vanishes around $T = 40$ K. (b) Measurement of the inelastic scattering from a transverse acoustic phonon in lead with (red circles) and without (green squares) guides as measured using the focusing set-up at MIRA. Focusing yields an intensity gain $G = 34.6$.

collection of neutrons from an extended volume in (Q, E)-space. Therefore, increasing the divergence leads to a significant increase of the intensity. The inelastic data agrees very well with the literature [Furrer 1970, Stedman 1967].

In contrast to TA phonons, the intensity gains for LA phonons are much smaller. For details see Brandl et al. [Brandl 2015]. It turns out, that the spectra are strongly distorted. The reason is that the large divergence of the incident and scattered neutrons leads to a very coarse, longitudinal Q-resolution.

Therefore, an extended range of LA phonons is sampled whose energy changes significantly due to the dispersion relation $\omega = cq$, where ω and q are the frequency and the wavenumber of the phonon, respectively. The TA phonons are barely affected as long as the reduced momentum q = Q − G_{hkl} is not too small. Here, G_{hkl} corresponds to the nearest Bragg peak of the reciprocal space. Recently it was shown that phonons can be measured in single crystals of SiO_2 which had a volume of only 1 mm³ [Böni 2011].

We conclude that the focusing set-up works excellently if the dispersion depends only weakly on Q along those Q-directions where the focusing leads to a coarse Q-resolution. Therefore, focusing techniques are most appropriate for the measurement of phonons with transverse polarisation, optic phonons or crystal electric field excitations. The latter two are usually only weakly dispersing. Of course, if compact and highly brilliant neutron sources become available, it may not be required to apply very strong focusing.

20.4.3 Protein Crystallography

Thanks to the large cross section and the possibility of contrast variation of thermal neutrons with hydrogen and deuterium, neutron crystallography is best suited to probe the position and reaction pathways of hydrogen in protein structures, provided single crystalline samples with volumes in the range of 0.1–1 mm³ are available. Two competing diffraction methods are in use; single crystal diffraction typically with $\Delta\lambda/\lambda < 2.8\%$ as available at the instrument BIODIFF at MLZ [Ostermann 2015] and quasi-Laue diffraction typically with $\Delta\lambda/\lambda = 20\%$–30% like the instrument LADI III at the ILL [Blakely 2010]. The complex biological structures require a divergence $\eta < 0.5°$.

A typical example highlighting the advantages of neutron cryo-crystallography in probing reaction mechanisms and visualizing the role of hydrogen binding in enzyme reactions might be the study of the ferric derivative of cytochrome peroxidase and its ferryl intermediates [Casadei 2014]. It turns out that the ferryl heme is an Fe(IV) = 0 species and is not protonated, i.e. it does not contain a proton (H⁺). The distal histidine becomes protonated on formation of transient iron-oxo-(ferryl) intermediates of the heme iron, which has implications for the understanding of O-O bond cleavage in heme enzymes. For this study, a single crystal with a volume of 0.7 mm³ was soaked in D_2O to exchange H by D.

Neutron crystallography on proteins is severely limited by the availability of sufficiently large crystals, and if possible completely deuterated. Figure 20.6 gives an idea of what is feasible today at the best neutron diffractometers available for that purpose. Today hydrogen positions in protein structures just become solvable at deuterated crystals as small as 0.07 mm³. More brilliant beams for use with samples < 0.01 mm³ are the ultimate goal.

FIGURE 20.6 Diffraction pattern of fatty acid binding protein taken at BIODIFF at MLZ using a crystal size of 0.07 mm³, a beam divergence $\eta = 0.5°$, $\Delta\lambda/\lambda = 1.5\%$, and a measuring time of 4 hours. The complete structure would need about 400 different orientations consuming beam time at BIODIFF of 2–3 weeks. The structure has been determined by quasi-Laue diffraction at LADI III with a similar amount of beam time. (Howard, E.I. et al. 2016, *IUCrJ* 3: 115–126.)

20.4.4 Diffraction Simulating the Inner Earth

There is high demand in the earth and material science communities for the ability to perform neutron diffraction studies at high pressure and high temperature conditions. In general, there is an inverse relationship between the maximum attainable pressure and the sample volume. For example, conventional multi-anvil pressure cells that generate pressures up to about 30 GPa have sample volumes in the 1–10 mm^3 range, while diamond anvil pressure cells that can achieve pressures higher than 200 GPa commonly use sample volumes $\ll$ 0.001 mm^3. Modern X-ray and synchrotron facilities have the necessary optics and intensity for measuring even these small sample volumes, which allowed great progress to be made in recent time to better understand the interior of our planet.

For neutrons which are much richer in contrast, however, today's instruments are limited to pressures up to about 20 GPa at room temperatures and below [Klotz 2013]. A new generation of cubic multi-anvil devices allowing stable experiments under high pressure and high temperature conditions has recently started operation at J-PARC [Sano 2014, Hattori 2015]. Another device is currently setup at MLZ [SAPHiR 2016] that is designed for pressures of 10–15 GPa at temperatures $T > 2000$ K [Sano 2014]. A further increase of the pressure to 20–25 GPa at reduced volumes of 5–10 mm^3 maintaining temperatures $T > 2000$ K is foreseeable. Hence, these instruments can only simulate the first 300–500 km depth of the earth, which means that more than 80% of the volume of the Earth cannot currently be investigated with neutrons. Due to the size and cost of these multi-anvil presses and the necessary infrastructure for maintenance and preparation of experiments, they will also in the future require access to the traditional large neutron sources such as J-PARC, ILL, FRM II, SNS or ESS.

However, a new approach to use diamond anvil cells (DAC) is currently under development. Successful tests with pressures up to about 94 GPa at room temperature have been reported [Boehler 2013]. Here, the challenge is to obtain meaningful results with sample volumes of only 0.02 mm^3. For further development of this technique, submillimetre neutron focusing at moderate divergence is needed, which requires new neutron optics and, if possible, higher neutron flux densities. The high potential of this technique for a new generation of specialized beamlines is the small size and good portability of DAC.

20.4.5 Nondestructive 3D Imaging of Elemental Composition

Today neutron tomography reaches a standard resolution of 50 μm in all three dimensions. Provided a large enough L/D ratio (L = distance pin hole to sample, D = diameter of the pin hole) is feasible, the resolution is limited from the spatial extent of the light flash that appears after the absorption process of the neutron in the scintillator that generates a photon which in turn produces the light flash [Trtik 2015]. Ultra-thin scintillators might bring this limit down to 10 μm at the expense of intensity [Hess 2011]. Imaging of elemental composition to a concentration sensitivity $\ll 10^{-6}$ becomes possible by a neutron tomography setup for prompt gamma activation analysis (PGAA). Such a prompt gamma activation imaging (PGAI) is realized at the MLZ [Söllradl 2015]. Here the spatial information obtained with neutron tomography is combined with the elemental information determined with PGAA.

At the PGAA instrument at the MLZ, a cold neutron flux density of up to 6×10^{10} cm^{-2}s^{-1} is available in the focus of an elliptically tapered neutron guide with horizontal and vertical divergence of $\pm 3.46^0$ and $\pm 4.41^0$. With an L/D ratio of 200, a spatial resolution of 250 μm is achieved, but now combined with elemental analysis with a sensitivity largely superior to 10^{-6}. On account of the very intense beam further collimation up to 1 mm^2 is realistic, thereby also decreasing the divergence. Smaller beams with less divergence would allow further increase in the spatial resolution eventually to 15 μm.

Figure 20.7 shows a high-resolution reconstruction of an electronic amplifier tube on which tomography was performed at the PGAA instrument at the MLZ [Söllradl 2014]. The results demonstrate that a resolution of approximately 100 μm can be achieved with the used setup. Using the tomographic reconstruction, various sections of the tube can be visualized. Moreover, the different materials comprising the tube can be visualized. Highly divergent and intense beams as possibly provided by accelerator-based

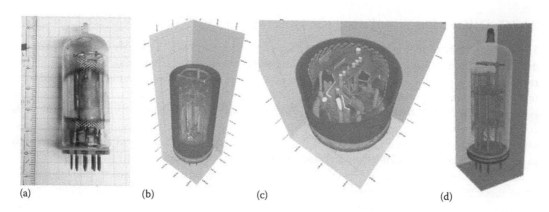

(a)   (b)   (c)   (d)

FIGURE 20.7   Image and tomographic reconstruction of an amplifier tube with different sections. The distance between the marks in (b) and (c) is equal to a distance of 10 mm. (a) Photograph of the amplifier tube (70 mm × 23 mm). (b) Cut along the centre axis of the tomographic reconstruction of the amplifier tube. (c) Magnified section of the bottom of the tomographic reconstruction of the amplifier tube. (d) Trial to segment the different materials based on their greyscale values. (Adapted with permission from S. Söllradl et al. 2014. "Developments in prompt gamma-ray neutron activation analysis and cold neutron tomography and their application in non-destructive testing", PhD Thesis, University of Bern, Switzerland.)

neutron sources are ideally suited for neutron imaging because these beams allow a magnification of the sample, thus reducing the effects of blurring in the detector [Kardjilov 2005].

20.4.6 Compact Neutron Sources for Nuclear Medicine

Radioisotopes for nuclear medicine are produced either by proton and neutron beams yielding proton- and neutron-rich nuclei, respectively. These radioisotopes are labelled with biomarkers to serve as radiopharmaceuticals for molecular imaging or therapeutic purposes. Depending on the particular need, these radioisotopes emit α-, β-, or γ-radiation of different energy. In some cases, γ- and particle-irradiation are combined enabling therapy and at the same time imaging of the therapeutic result by the same application. The lifetime of the radioisotope is of prime importance to reduce radiation doses of healthy tissue. In ideal cases the life time of the radioisotope is in the order of several hours to a maximum of one day, again to minimize radiation dose for healthy tissue. The latter ultimately asks for fast radiochemistry to produce the radiopharmaceutical. Nowadays, this is achieved for proton-induced radiopharmaceuticals by compact proton accelerators including the necessary processing at dedicated labs inside the hospitals, thereby minimizing the transport chain. ^{18}F for positron emission tomography and with a half-life time of 110 min is the most prominent example.

For neutron-based radioisotopes, large-scale neutron sources are used, often thousands of kilometres away from the physicians and patients. ^{99m}Tc, i.e. its mother isotope ^{99}Mo, with about 30 million applications every year worldwide, the by far most used radioisotope is produced by about 10 research reactors distributed over the continents. ^{99m}Tc serves for molecular imaging and functional studies of the brain, thyroid, lungs, kidneys, blood or tumours. The mother isotope ^{99}Mo has a half-life time of 66 hrs. Its complex production chain from irradiation, chemical processing, generator production, to worldwide distribution to hospitals and physicians makes it that the isotope comes to application typically only 10–14 days after production, i.e. with considerable losses. Production of short living neutron rich radio-isotopes with compact neutron sources located at the hospital therefore might become attractive. This the more as the isotope production asks for neutron intensity at a target near by the source and does not need brilliance. Under such circumstances a flux density of 10^{12} slow neutrons per cm²s might be sufficient. Furthermore, the now drastically reduced time from production to application gives access to the class of short living, neutron-rich isotopes opening the door for new and eventually advanced radiopharmaceuticals.

Fast neutrons are suited for hadron therapy [Wagner 2017]. Direct irradiation of tumours by neutrons achieves typically a biological efficiency BE $\simeq$ 3 [ICRP 1990] which, for instance, is considerably higher than BE $\simeq$ 1.1 for proton treatment. At MLZ, patients are treated at the clinical neutron irradiation therapy station MEDAPP [Wagner 2008, Genreith 2015] by means of a rather monoenergetic beam around 2 MeV fission neutrons with a flux of 10^8 fast neutrons per cm²s. Typically a total dose of around 6 Gy is applied in about 5 sessions each of which requires 2–3 min. Despite of the well proven efficiency of the treatment in particular for palliative applications [Specht 2015] only a maximum of 40 patients are treated per year. This is mainly caused by the irregular availability of the fast neutron beam. Research reactors are operated according to nuclear regulations and not according to the immediate need of the patient. Compact and accelerator driven sources of fast neutrons installed at the clinic for radiation oncology could be a way out of this dilemma. This might become possible with a neutron flux in the order of that of MEDAPP because the fast neutrons from the accelerator driven system need not be moderated. However, filtering of the neutron spectrum is mandatory in order to achieve well-defined irradiation volumes. Accelerator driven fast neutron hadron therapy is not new [Wagner 2012]. However, the existing facilities use rather large accelerator systems with heavy gantries and a neutron spectrum with long tails towards high energy E_n > 10 MeV. Progress can be expected from small compact sources with well-defined filtered beams. It might be that such systems will become available in the near future [Otake 2017, Kumada 2014]. Tuning the energy of the neutrons towards the epithermal spectrum, these compact systems might become an option for boron neutron capture therapy (BNCT) [Bavarnegin 2017].

20.4.7 Application of Montel Optics for Thin Film Research

Magnetic layers and heterostructures are the basic building blocks for a large number of magneto-electronic devices such as read-heads of hard disk drives or magnetic random-access memories (MRAMs). The performance of these devices relies heavily on the morphology and microstructure of the individual layers and on the coupling between them. All these parameters can change during the growth process. It is therefore important for the understanding and optimization of magnetic layers and heterostructures to accurately monitor the magnetic and structural properties of the multilayers during the deposition process. The specular reflection of polarized neutron beams is ideally suited for measuring the evolution of magnetism *in situ* because of its sensitivity to the nuclear and magnetic structure of thin films.

To overcome the problem of low intensity with samples consisting of only a few monolayers of material, highly brilliant and well-focused beams are required in order to avoid long data accumulation times which may involve a contamination of the surface of the layers during growth. Recently, a new concept for conducting experiments on small sample volumes has been developed. Here, a converging beam geometry within the scattering plane using Montel mirrors [Montel 1957] is used (Figure 20.8). Because the set-up had to be installed at the existing beamline AMOR at the spallation source SINQ of the Paul Scherrer Institut (PSI) in Villigen (Switzerland), the neutrons could not be directly extracted from the moderator by means of Montel mirrors which resulted in an intensity penalty. Still, the simultaneous measurement of neutrons reflected over a wide angular range and wavelength band can be accomplished, thus increasing the signal by more than an order of magnitude when compared with a conventional set-up [Stahn 2011, Stahn 2012].

As an example, we show the results of a recent experiment that was performed at the beamline AMOR [Kreuzpaintner 2017]. The epitaxial growth of monolayers of Fe on Cu(111) was monitored during growth (Figure 20.9). The splitting of the spin-up and spin-down reflectivity as measured from a film consisting of nominally two atomic layers of Fe demonstrates directly the evolution of an in-plane magnetization in the layers. The measurement time of 15 min per spin channel is sufficiently short to avoid any contamination of the film during the measurement. Using highly brilliant beams like those that will become

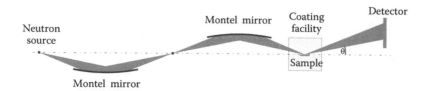

FIGURE 20.8 A pulsed, polarized neutron beam is focused by means of two Montel mirrors onto the sample that is located in a deposition chamber. The reflected neutrons are recorded with a two-dimensional detector. Ideally, the neutrons are directly extracted by the first Montel mirror from the neutron source.

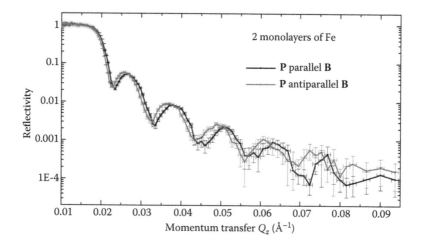

FIGURE 20.9 The data shows that in nominally two monolayers of Fe that were grown epitaxially on Cu(111) magnetism can be recorded *in situ* within a measuring time of 15 min per spin channel.

available at ESS or at future laser-driven accelerator sources will enable this kind of reflectivity measurement with polarized neutrons *in situ* and in real time because the measurement times will be below 1 sec.

20.5 Conclusions

We have shown that the combination of a brilliant neutron source with modern neutron optics using parabolic or elliptic mirrors allows the investigation of static and dynamic properties of materials, which are only available in small quantities either due to the difficulty of producing large samples or as imposed by extreme conditions (like high pressure). Using Montel optics [Montel 1957], it has recently become possible to even investigate the emergence of magnetism in single atomic layers during the *in situ* growth [Kreuzpaintner 2017].

It is for the investigation of small samples in particular that compact accelerator-based sources might offer distinct advantages when compared with conventional neutron sources, such as nuclear reactors or spallation sources, because of their compact design and their high brilliance. In addition to being more affordable, such compact sources might overcome technical limitations of heat transport more easily and facilitate the installation of advanced neutron optics [Böni 2008]. While the brilliance of conventional sources approaches their technical limits, laser-driven sources might not be restricted yet by any obvious limitations. Moreover, the significant reduction of the nuclear inventory will mitigate problems associated with safety and security measures. Of course, in order to become competitive with conventional neutron sources, the brilliance, as well as the neutron fluence of the compact sources, has to be strongly increased.

The major obstacle in producing highly brilliant beams with a large fluence is the need of a moderator that is still required for the presently proposed accelerator-based neutron sources [Roth 2017, Rücker 2016]. Therefore, future research should concentrate on the production of highly-brilliant beams using laser-driven compact accelerators and the circumvention of the moderation process for the thermalization of the neutrons, which distributes the flight direction of the neutrons over 4π. Here, using the concept of releasing neutrons from neutron halo isomers of stable nuclei [Habs 2011] along the direction of the electric field of a laser may bring a solution.

Acknowledgements

We are grateful for our discussions with D. Habs, T. Gutberlet and S. Söllradl. The experiments were performed in collaboration with G. Brandl, U. Filges, R. Georgii, K. Hradil, W. Kreuzpaintner, S. Mühlbauer, P. Niklowitz, C. Pfleiderer, B. Roessli, C. Schanzer, M. Schneider, J. Stahn, and B. Wiedemann. This work was supported by the Deutsche Forschungsgemeinschaft via the Transregional Research Center TRR 80 and by the European Union within the Sixth Framework Program FP6 under Contract no. 505925.

References for Chapter 20

Adams, T. et al. 2014. "Versatile module for experiments with focussing neutron guides" *Appl. Phys. Lett.* 105: 123505.

Altissimo, M. et al. 2008. "Neutron diffraction from macroscopic objects and transverse coherence of the wavefunction: The Fresnel zone plates" *Nucl. Instr. and Meth. A* 586: 68–72.

Bardeen, J. et al. 1957. "Theory of Superconductivity" *Phys. Rev.* 108: 1175–1204.

Bavarnegin, E. et al. 2017. "Neutron beams implemented at nuclear research reactors for BNCT" *Journal of Instrumentation (JINST)* 12: 05005.

Blakely, M. P. et al. 2010. "Neutron macromolecular crystallography with LADI-III" *Acta. Cryst. D* 66, 1198–1205.

BlueBook of the FRM II 2011. http://cdn.frm2.tum.de/fileadmin/stuff/instruments/BlueBook/efaccs4 Januar2011verlinkt2.pdf.

Boehler, R. et al. 2013. "Large-volume diamond cells for neutron diffraction above 90 GPa" *High Pressure Research* 33: 546–554.

Böni, P. 2008. "New concepts for neutron instrumentation" *Nucl. Instr. and Meth. A* 586: 1–8.

Böni, P. et al. 2011. "Inelastic neutron and x-ray scattering from incommensurate magnetic systems" *J. Phys.: Condens. Matter* 23: 254209.

Böni, P. 2014. "High Intensity Neutron Beams for Small Samples" *J. Phys.: Conf. Ser.* 502: 012047.

Brandl, G. et al. 2015. "Compact turnkey focussing neutron guide system for inelastic scattering investigations" *Appl. Phys. Lett.* 107: 253505.

Bührer, W. 1994. "Triple-axis instrument with doubly focusing ('zoom') monochromator and horizontally focusing analyser: Seven years experience" *Nucl. Instr. and Meth. A* 338: 44–52.

Casadei, C. M. et al. 2014. "Neutron cryo-crystallography captures the protonation state of ferryl heme in a peroxidase" *Science* 345: 193–197.

Eckold, G. and Sobelov, O. 2014. "Analytical approach to the 4D-resolution function of three axes neutron spectrometers with focussing monochromators and analysers" *Nucl. Instr. and Meth. A* 752: 54–64.

Eskildsen, M. et al. 1998. "Compound refractive optics for the imaging and focusing of low-energy neutrons" *Nature* 391: 563–566.

Furrer, A. and Hälg, W. 1970. "Experimental Phonon Frequencies and Widths of Lead at 5, 80, and 290 K" *Phys. Status Solidi* 42: 821–833.

Furrer, A. et al. 2009. *Neutron Scattering in Condensed Matter Physics*. World Scientific Series on Neutron Techniques and Applications Vol. 4, Singapore.

Genreith, C. 2015. "MEDAPP: Fission neutron beam for science, medicine, and industry" *Journal of Large-Scale Research Facilities* 1: A18.

Georgii, R. and Seemann, K. 2015. "MIRA: Dual wavelength band instrument" *Journal of Large-Scale Research Facilities* 1: A3.

Gutberlet, T. 2016. Private communication.

Habs, D. et al. 2011. "Neutron halo isomers in stable nuclei and their possible application for the production of low energy, pulsed, polarized neutron beams of high intensity and high brilliance" *Appl. Phys. B* 103: 485–499.

Hajima, R. et al. 2009. "Detection of radioactive isotopes by using laser Compton scattered γ-ray beams" *Nucl. Instr. and Meth. A* 608: S57–S61.

Hattori, T. et al. 2015. "Design and performance of high-pressure PLANET beamline at pulsed neutron source at J-PARC" *Nucl. Instr. and Meth. A* 780: 55–67.

Hess, K.-U. et al. 2011. "Advances in high-resolution neutron computed tomography: Adapted to the earth sciences" *Geosphere* 7: 1294–1302.

Higginson, D. P. et al. 2011. "Production of neutrons up to 18 MeV in high-intensity, short-pulse laser matter interactions" *Physics of Plasmas* 18: 100703.

Hils, T. et al. 2004. "Focusing parabolic guide for very small samples" *Physica B* 350: 166–168.

Howard, E. I. et al. 2016. "High-resolution neutron and X-ray diffraction room-temperature studies of an H-FABP-oleic acid complex: Study of the internal water cluster and ligand binding by a transferred multipolar electron-density distribution" *IUCrJ* 3: 115–126.

Ice, G. E. et al. 2010. "Nested neutron microfocusing optics on SNAP" *Appl. Phys. A* 99: 635–639.

ICRP 1990. International Commission on Radiological Protection. "The 1990 Recommendations on Radiological Protection" *Annals of the ICRP* 60: ICRP Publication.

Iwashita, Y. et al. 2008. "Variable permanent magnet sextupole lens for focusing of pulsed cold neutrons" *Nucl. Instr. and Meth. A* 586: 73–76.

Kardjilov, N. et al. 2005. "Characterization of a focusing parabolic guide using neutron radiography method" *Nucl. Instr. and Meth. A* 542: 248–252.

Klotz, S. 2013. *Techniques in High Pressure Neutron Scattering*. CRC Press, Taylor & Francis Group.

Komarek, A. C. et al. 2011. "Parabolic versus elliptic focusing – Optimization of the focusing design of a cold triple-axis neutron spectrometer by Monte-Carlo simulations" *Nucl. Instr. and Meth. A* 647: 63–72.

Kreuzpaintner, W. et al. 2017. "In situ polarized neutron reflectometry: Epitaxial thin-film growth of Fe on Cu(001) by dc magnetron sputtering" *Physical Review Applied* 7: 054004.

Kugler, M. et al. 2015. "Band Structure of helimagnons in MnSi resolved by inelastic neutron scattering" *Phys. Rev. Lett.* 115: 097203.

Kumada, H. et al. 2014. "Project for the development of the linac based NCT facility in University of Tsukuba" *Applied Radiation and Isotopes* 88: 211–215.

Kumakhov, M. A. and Sharov, V. A. 1992. "A neutron lens" *Nature* 357: 390–391.

Link, P. et al. 2000. "The new thermal triple-axis spectrometer PUMA at the Munich neutron-source" *Physica B* 276–278: 122–123.

Montel, M. 1957. *X-Ray Microscopy and Microradiography*. Academic Press, New York.

Niklowitz, P. G. et al. 2009. "New Angles on the Border of Antiferromagnetism in NiS_2 and URu_2Si_2" *Physica B* 404: 2955–2960.

Ostermann, A. and Schrader, T. E. 2015. "BIODIFF: Diffractometer for large unit cells" *Journal of Large-Scale Research Facilities* 1, A2.

Otake, Y. 2017. "A Compact Proton Linac Neutron Source at RIKEN" in *Applications of Laser-Driven Particle Acceleration*, ed. P. Bolton, K. Parodi and J. Schreiber, Chapter 19. CRC Press, Boca Raton, FL.

Roth, M. 2017. "Neutron Generation", in *Applications of Laser-Driven Particle Acceleration*, ed. P. Bolton, K. Parodi, and J. Schreiber, Chapter 6. CRC Press, Boca Raton, FL.

Rücker, U. et al. 2016. "The Jülich high-brilliance neutron source project" *Eur. Phys. J. Plus* 131: 19.

Rush, J. J. 2015. "US Neutron Facility Development in the Last Half-Century: A Cautionary Tale" *Phys. Perspect.* 17: 135–155.

Sano-Furukawa, A. et al. 2014. "Six-axis multi-anvil press for high-pressure, high-temperature neutron diffraction experiments" *Rev. Scientific Instruments* 85: 113905.

SAPHiR 2016. http://www.mlz-garching.de/instrumente-und-labore/struktur/saphir.html.

Schanzer, C. et al. 2004. "Advanced geometries for ballistic neutron guides" *Nucl. Instr. and Meth. A* 529: 63–68.

Schanzer, C. et al. 2010. "High Performance Supermirrors on Metallic Substrates" *J. Phys.: Conf. Ser.* 251: 012082.

Schanzer, C. et al. 2016. "Neutron Optics: Towards Applications for Hot Neutrons" *J. Phys.: Conf. Ser.* 746: 012024.

Skoulatos, M. and Habicht, K. 2011. "Upgrade of the primary spectrometer of the cold triple-axis spectrometer FLEX at the BER II reactor" *Nucl. Instr. and Meth. A* 647 (2011): 100–106.

Söllradl, S. 2014. "Developments in prompt gamma-ray neutron activation analysis and cold neutron tomography and their application in non-destructive testing", PhD Thesis, University of Bern, Switzerland.

Söllradl, S. et al. 2015. "Development and Test of a Neutron Imaging Setup at the PGAA Instrument at FRM II" *Physics Procedia* 69: 130–137.

Specht, H. M. et al. 2015. "Paving the Road for Modern Particle Therapy – What Can We Learn from the Experience Gained with Fast Neutron Therapy in Munich?" *Frontiers in Oncology* 5: 262.

Stahn, J. et al. 2011. "Study on a focusing, low-background neutron delivery system" *Nucl. Instr. and Meth. A* 634, S12–S16.

Stahn, J. et al. 2012. "Focusing specular neutron reflectometry for small samples" *Eur. Phys. J. Appl. Phys.* 58, 11001.

Stedman, R. et al. 1967. "Phonon-frequency distributions and heat capacities of aluminum and lead" *Phys. Rev.* 162: 549–557.

Sudo, S. and Miyadai, T. 1985. "Weak ferromagnetism and antiferromagnetism in the insulator phase of $NiS_{2-x}Se_x$ with pyrite structure" *J. Phys. Soc. Jpn.* 54: 3934–3941.

SwissNeutronics Website 2017. http://www.swissneutronics.ch/index.php?id=24.

Taylor, A. et al. 2007. "A route to the brightest possible neutron source?" *Science* 315: 1092–1095.

Trtik, P. et al. 2015. "Improving the spatial resolution of neutron imaging at Paul Scherrer Institut The Neutron Microscope Project" *Physics Procedia* 69: 169–176.

v. Löhneysen, H. and Vojta, M. (Eds.) 2015. "Quantum phase transitions in correlated electron systems" *European Physical Journal Special Topics* 224: 969.

Wagner, F. M. et al. 2008. "The Munich fission neutron therapy facility MEDAPP at the research reactor FRM II" *Strahlenther. Onkol.* 184: 643–646.

Wagner, F. M. et al. 2012. "Neutron medical treatment of tumors – A survey of facilities" *Journal of Instrumentation (JINST)* 7: C03041.

Wagner, F. M. et al. 2017. "Fast Neutron Therapy: A Status Report" *Sibirsky onkologich-eskiy zhurnal* 6: 5–12.

Zeitelhack, K. et al. 2006. "Measurement of neutron flux and beam divergence at the cold neutron guide system of the new Munich research reactor FRM-II" *Nucl. Instr. and Meth. A* 560: 444–453.

21

'Fission–Fusion': Novel Laser-Driven Nuclear Reaction Mechanism

Peter G. Thirolf

21.1 Introduction

High-power, short pulse lasers during the last decade have proven their capacity for the production of high energy γ-rays, charged particles and neutrons, and for inducing a variety of nuclear reactions. With laser and optical technology continuously improving and next-generation (multi-) PW laser facilities like ELI (Extreme Light Infrastructure) on the horizon, nuclear reaction studies based on high-power lasers will open new perspectives for various applications in nuclear physics research, in particular for nuclear astrophysics. Laser-driven ion acceleration with adequately tailored subsequent interactions will complement the portfolio of nuclear reactions so far available from conventionally accelerated particle beams. This chapter outlines a unique perspective for laser-driven nuclear reactions, targeting the astrophysically relevant nucleosynthesis of heavy elements beyond the regime of thermonuclear fusion occurring in stars. A prototypical scenario is studied as achievable with laser systems like the upcoming 10 PW-class lasers at ELI-Nuclear Physics (ELI-NP) in Bucharest. Following a short introduction to the astrophysical motivation, the perspectives originating from the expected uniquely high-density ion bunches accelerated via the radiation pressure acceleration mechanism will be presented. This will allow for a novel nuclear reaction mechanism ('fission–fusion'), capable of generating extremely neutron-rich isotopes in the vicinity of the astrophysically important 'waiting point' of the r-process heavy-element nucleosynthesis path at the waiting point of the magic nucleon number $N = 126$.

21.2 Prospects for Laser-Driven Ultra-Dense Ion Bunches: 'Fission–Fusion' Mechanism

Heavy elements beyond iron are produced via the rapid neutron capture process (r-process) at astrophysical sites such as merging neutron star binaries or (core collapse) supernova type II explosions. When aiming to improve our understanding of these nuclear processes by measuring the properties of heavy nuclei on (or near) the r-process path, we note that the lower-mass path of the r-process for the production of heavy elements is well-explored. Thereby the nuclei around the $N = 126$ waiting point, which comprise the pronounced third peak in the elemental abundance distribution near mass numbers A = 180–200 (corresponding to the group of elements near gold, platinum and osmium), critically determine this element production mechanism. However, at present, basically nothing is known about these nuclei, since the relevant N = 126 r-process nuclei with Z near 70 are about 15 neutrons away from the last known isotope and will stay out of reach even for next-generation conventional accelerator facilities like FAIR [FAIR 2006] or FRIB [FRIB 2009]. Nuclear properties to be studied in this area are nuclear masses, lifetimes, beta-delayed neutron emission probabilities and the underlying nuclear structure.

In the context of high-power, short-pulse, laser-driven particle acceleration, a novel nuclear reaction mechanism is proposed, that draws on the uniquely high density of laser-accelerated ion bunches. It is envisaged to exploit the (so far more theoretically investigated) efficient radiation pressure acceleration (RPA) mechanism for ion acceleration. This mechanism requires very thin targets and ultra-high contrast laser pulses to avoid the preheating and expansion of the target before the interaction with the main laser pulse. For details on laser-driven ion acceleration, the reader is referred to Chapter 5 of this book. In the RPA 'hole-boring' scenario the laser pulses interact with targets thick enough to allow driving target ions ahead of it in a piston-like manner, but without interacting with the target rear surface [Robinson 2009]. The RPA mechanism promises generating energetic ion bunches with potentially solid-state density (10^{22}–10^{23}/cm³), which would improve ion bunch densities from classical (i.e. not laser-driven) accelerators by about 14 orders of magnitude. The basic concept of the fission–fusion reaction scenario draws on this ultra-high density of laser accelerated ion bunches. Choosing fissile isotopes as target material for a first target foil ('production target') to be accelerated by an intense laser driver will provide a dense bunch of fission fragments which in turn interacts with a second target foil ('reaction target'), also consisting of fissile isotopes. So, finally, in a second step of the reaction process, the fusion between (neutron-rich) beam-like and target-like (light) fission products will become possible, generating extremely neutron-rich ion species. This scheme is depicted in Figure 21.1. Here ^{232}Th was chosen as fissile target material, primarily because of its long half-life of $1.4 \cdot 10^{10}$ years, which avoids extensive radioprotection precautions during handling and operation. As shown in Figure 21.1, the target arrangement, optimized for maximum reaction yield, will make use of multi-layered targets, where low-Z layers are combined with the heavy fissile target material. The accelerated thorium ions will undergo fission in the CH_2 layer of the reaction target, whereas the accelerated carbon ions and deuterons from the production target will generate thorium fragments in the thick thorium layer of the reaction target. This scenario makes efficient use of the laser pulse energy that is naturally partitioned amongst multiple ion species. In view of the available energy in the accelerating driver laser pulse, the optimized production target should have a thickness of about 0.5 μm for the thorium as well as for the CD_2 layers. The thorium layer of the reaction target would have a thickness of about 50 μm.

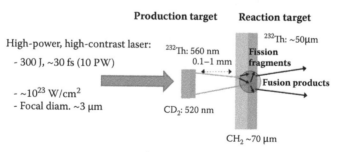

FIGURE 21.1　Sketch of the target arrangement envisaged for the fission–fusion reaction process based on laser ion acceleration, consisting of a production and a reaction target from a fissile material (here ^{232}Th), each of them covered by a layer of low-Z materials (CD_2 and CH_2, respectively). The thickness of the CH_2 layer as well as the second thorium reaction target have to be limited to 70 μm and 50 μm, respectively, in order to enable fission of beam and target nuclei. This will allow for fusion between their light fragments, as well as enable the fusion products to leave the second thorium reaction target.

In order to allow for an optimized fission of the accelerated Th bunch, the thicker Th layer of the reaction target, which is closely positioned behind the production target, is covered by about 70 μm of polyethylene. This layer serves a twofold purpose: primarily it is used to induce fission of the impinging Th ions, generating the beam like fission fragments. Here, polyethylene is advantageous when compared to a pure carbon layer, because of the increased number of atoms that can induce fission of the impinging Th ions. Secondly, the thickness of this CH_2 layer has been chosen such that the produced fission fragments will be decelerated to a kinetic energy which is suitable for optimized fusion with the target-like fission fragments generated by the light accelerated ions in the Th layer of the reaction target, minimizing the amount of evaporated neutrons. For practical reasons, we propose to place the reaction target about 0.1 mm behind the production target, as indicated in Figure 21.1. After each laser shot, a new double-target has to be inserted. In the Th layer of the reaction target, target-like fission fragments will be generated primarily by fission induced via the accelerated low-Z species (d and C). Assuming a 300 J laser pulse energy with an efficiency near 10% for conversion from photon energy to ion kinetic energy and about $2.8 \cdot 10^{11}$ laser-accelerated deuterons (plus $1.4 \cdot 10^{11}$ carbon ions), we conclude the corresponding fission probability in the Th layer of the reaction target to be about $2.3 \cdot 10^{-5}$ which corresponds to about $3.2 \cdot 10^6$ target-like fission fragments per laser pulse. A thorium thickness of 50 microns in the reaction target could be exploited, where the proton kinetic energy would exceed the fission Coulomb barrier over the full target depth. In the second step of the fission–fusion scenario, we consider the fusion between the light fission fragments of beam and target to a compound nucleus with a central value of A ~ 182 and Z ~ 75. Again, we employ geometrical arguments for an order of magnitude estimate of the corresponding fusion cross section, resulting in a fusion probability of about $1.8 \cdot 10^{-4}$. Very neutron-rich nuclei still have comparably small production cross sections, because weakly bound neutrons ($S_n \sim 3$ MeV) will be evaporated easily. A detailed discussion of the achievable fission–fusion reaction yield is given in Ref. [Habs 2011].

In addition to the scenario discussed so far, the unprecedented ion bunch density may lead to collective effects that do not occur with conventional ion beams: when sending the energetic, solid density ion bunch into a solid carbon or thorium target, the plasma wave driven by the ion bunch with a phase velocity corresponding to the thorium ion velocity has a wavelength near 5 nm (i.e. much smaller than the ion bunch length ≈ 560 nm) making collective acceleration and deceleration effects cancel and only the binary collisions remain and contribute to the stopping power. In this case, the first layers of the impinging ion bunch will attract the electrons from the target and, like a snow plough, will take up the decelerating electron momenta. Hence the predominant part of the ion bunch is screened from electrons and we expect a significant reduction of the stopping power. This potential effect requires detailed experimental investigations, aimed at verifying the perspective to use a significantly thicker reaction target, which, in turn, would significantly boost the achievable fusion yield.

Table 21.1 indicates the achievable yield per laser pulse of neutron-rich fusion products in the mass range of A ~ 180–190 (i.e. around N = 126) for two scenarios of projectile-target interaction: based on the assumption of a laser pulse with an energy of 300 J and 10% conversion efficiency into laser-accelerated ions, the middle column lists the expected rates for 'normal stopping', i.e. projectile energy loss according to the conventional Bethe-Bloch equation (middle column). In the right column, the scenario of a reduced stopping behaviour in the target foil by collective effects induced by the ultra-high density of the impinging ion bunch is listed, exemplarily assuming a reduction factor of 100. Numbers of accelerated heavy and light projectiles and resulting beam- and target-like fission fragments are given. Based on the fusion probability, extremely neutron-rich fusion product yields between 1.5 and about $4 \cdot 10^4$ per laser pulse are estimated. It is obvious that two laser parameters are desirable to be optimized in the future: the reaction yield is primarily governed by the energy in the laser pulse, determining the number of accelerated ions. The second parameter is the laser repetition rate. Presently, next-generation multi-PW systems like the one foreseen at ELI-NP in Bucharest will exhibit a rate of about 1 pulse per minute. Here, ongoing laser development towards higher repetition-rated systems based on diode-pumped systems for example, may lead to improvements of several orders of magnitude. Along with required mass production and exchange strategies for sophisticated targets, this would greatly improve the perspectives for the outlined fission–fusion reaction scheme.

TABLE 21.1 Expected Yield of Neutron-Rich Fusion Products in the Relevant Mass Range A ~ 180–190 per Laser Pulse (Last Row) for Normal Ion Stopping (Middle Column) and Assumed Reduction by Collective Effects by a Factor of 100 (Right Column)

Laser Pulse: 300 J, $\varepsilon = 10\%$	Normal Stopping	Reduced Stopping
Accelerated: ^{232}Th	$1.2 \cdot 10^{11}$	$1.2 \cdot 10^{11}$
C	$1.4 \cdot 10^{11}$	$1.4 \cdot 10^{11}$
protons	$2.8 \cdot 10^{11}$	$2.8 \cdot 10^{11}$
Fission: beam-like fragments	$3.7 \cdot 10^{8}$	$1.2 \cdot 10^{11}$
target-like fragments	$3.2 \cdot 10^{6}$	$1.2 \cdot 10^{11}$
Fusion probability (F_L(beam) + F_L(target))	$1.8 \cdot 10^{-4}$	$1.8 \cdot 10^{-4}$
Neutron-rich fusion products (A ≈ 180–190)	1.5	4.10^{4}

Source: Habs, D. et al. 2011. "Introducing the Fission-Fusion reaction process: Using a laser-accelerated Th beam to produce neutron-rich nuclei towards the N = 126 waiting point of the r process." *Appl. Phys. B* 103: 471–484.

Note: Numbers of accelerated projectiles, resulting beam- and target-like fission fragments and expected fusion probabilities are given as well for the two scenarios.

Figure 21.2 displays a closer view into the region of nuclides around the N = 126 waiting point of the r-process, where nuclei on the r-process path are indicated by the green colour, with dark green highlighting the key bottleneck r-process isotopes [NRC 2007] at N = 126 between Z = 66 (Dy) and Z = 70 (Yb). For example, one should note that for Yb the presently last known isotope is 15 neutrons away from the r-process path at N = 126. The elliptical contour lines indicate the range of nuclei envisaged to become accessible with our new fission–fusion scenario on a level of 50%, 10% and 10^{-3} of the maximum fusion cross section between two neutron-rich light fission fragments in the energy range near 2.8 MeV/u.

Exploring this 'terra incognita' of yet unknown isotopes towards the r-process waiting point at N = 126 calls for a staged experimental approach, starting with the theory-assisted development of laser acceleration of heavy ions and an investigation of potential collective effects on the stopping power of

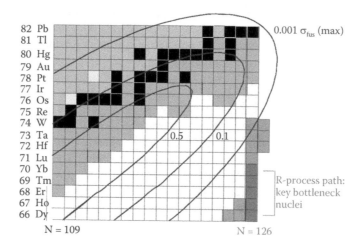

FIGURE 21.2 Chart of nuclides around the N = 126 waiting point of the r-process path. The blue ellipses denote the expected range of isotopes accessible via the novel fission–fusion process. The indicated lines represent 0.5, 0.1 and 0.001 of the maximum fusion cross section after neutron evaporation. In green, the N = 126 nuclides relevant for the r-process are marked, with the dark green colour indicating the key bottleneck nuclei for the astrophysical r-process. (With kind permission from **Springer Science+Business Media**: *Appl. Phys. B*, Introducing the Fission-Fusion reaction process: Using a laser-accelerated Th beam to produce neutron-rich nuclei towards the N = 126 waiting point of the r process, 103, 2011, 471–484, Habs, D. et al.)

the dense, laser-accelerated ion bunches when interacting with matter. A critical prerequisite of identifying potentially generated exotic ion species is the efficient separation of the multitude of ion species resulting from the nuclear interaction, comprising non-reacting projectiles and fission fragments, as well as fusion products from different reaction channels. This separator has to accept a much broader momentum and charge state range than typically requested from existing comparable devices operated at conventional accelerator facilities. For example, the ions of interest could be selected in a two-stage velocity filter, preparing them for identification and further investigation, e.g. after thermalization in a buffer-gas stopping cell [Neumayr 2006], by cooling and bunching in a radiofrequency quadrupole ion guide, to finally transfer them to perform either selective spectroscopic studies or precision mass measurements.

With the described novel nuclear 'fission–fusion' reaction scheme, exclusively available for laser-accelerated particle bunches, a region of the terra incognita in the chart of nuclei could become accessible that can shed light onto the fundamental processes of heavy element formation in the universe.

21.3 Conclusion

Laser-driven ion acceleration bears great potential to complement conventional accelerator facilities, in particular regarding the production of exotic isotope species far-off stability, which are crucial for our understanding of the nucleosynthesis of heavy elements in the universe beyond the iron peak. The key property of laser-driven ion bunches in this context is their unprecedented high density, approaching solid-state density and, as such, many orders of magnitude denser than conventionally accelerated ion beams. This may allow for exploiting new nuclear reaction schemes, like the described 'fission–fusion' mechanism. However, one has to state that at present the performance of such a reaction scheme is still limited by two laser beam parameters: the maximum achievable focused intensity with present lasers allows observing the onset of the RPA mechanism, expected from simulations to dominate beyond 10^{23} W/cm^2. Nuclear reaction schemes, conventionally drawing from well-defined beam energies, would greatly benefit from the expected (quasi-) monoenergetic energy characteristics of laser-driven ion bunches in this regime. Moreover, the reaction yield, at present strongly limited by laser pulse repetition rates of about 1/min (as for the ELI-NP 10 PW laser), would considerably benefit from an increase of the repetition rate into the (multi-) Hz regime. Obviously, also the development of suitable targetry will play a decisive role in this context in the future as well [Lindner 2017].

Acknowledgements

This work was supported by the German BMBF (Verbundforschung, 05P15WMEN9).

References for Chapter 21

http:www.fair-center.de/fileadmin/fair/publications_FAIR/FAIR_BTR_1.pdf.

http://www.frib.msu.edu/_files/pdfs/frib_scientific_and_technical_merit_lite_0.pdf.

FAIR 2006, Baseline Technical Report, FAIR/GSI.

Habs, D. et al. 2011. "Introducing the Fission-Fusion reaction process: Using a laser-accelerated Th beam to produce neutron-rich nuclei towards the N=126 waiting point of the r process" *Appl. Phys. B* **103**: 471–484.

Lindner, F.H. et al. 2017. "Swift Ion Bunch Acceleration by High Power Laser Pulses at the Centre for Advanced Laser Applications (CALA)" *Nucl. Instr. Meth. B* **402**: 354–357.

Neumayr, J.B. et al. 2006. "Performance of the MLL-IonCatcher" *Rev. Sci. Instr.* **77**: 065109.

NRC Rare Isotope Science Assessment Committee (RISAC) Report, National Academies Press, Washington/DC, USA (2007).

Robinson, A.P.L. et al. 2009. "Relativistically correct hole-boring and ion acceleration by circularly polarized laser pulses" *Plasma Phys. Control. Fusion* **51**: 024004.

22

Nuclear Reactions in a Laser-Driven Plasma Environment

David Denis-Petit

Ken W.D. Ledingham

Paul McKenna

David Mascali

Salvatore Tudisco

Klaus M. Spohr

Medhi Tarisien

22.1 Introduction

The matter surrounding us usually occurs in the three different states: solid, liquid and gaseous. Under extreme conditions of temperature, pressure and density, matter can assume a fourth state: plasma. This state is reached when we apply such high energies that individual electrons are torn from the electron shell of the atoms that make up matter. A system of free, negatively charged electrons and positive ions is thus created. Although, in our daily lives, we are not familiar with plasmas, it is the dominant state of matter in the universe. More than 99% of matter in the universe exists as plasma. This electron–ion plasma can exist across a large range of temperatures and densities, and can change into various phases. In a hot plasma, various mechanisms of nuclear excitation and de-excitation may appear. Besides direct interaction with free electrons and X-rays/γ-rays through mechanisms such as photoexcitation, electron inelastic scattering and stimulated gamma-ray emission, other excitation/de-excitation mechanisms can also occur that involve the bound states of the electron shell: internal conversion (IC), i.e. nuclear de-excitation resulting in the emission of an orbital electron to the continuum; bound internal conversion (BIC), which is similar to IC except the electron is promoted to a bound state; NEEC (nuclear excitation following electron capture from continuum) which is the inverse of IC; and NEET (nuclear excitation following an electronic de-excitation transition), being the inverse of BIC. These processes will occur with different rates compared to isolated atoms or materials in normal conditions. Moreover, significant changes in nuclear lifetimes are predicted in hot and dense plasmas. Laser–matter interactions of high-powered, short laser pulses (like the ones projected for the ELI laser facilities) with solid targets are capable of generating plasma conditions that open a new world of possibilities to study their behaviour, for the first time, under the extreme temperature and pressure conditions present in the inner cores of planets and stars.

22.2 Nuclear Reactions in Laser Plasmas

Plasma is a common form of matter known in the universe. One of the fundamental questions related to the nuclear–plasma science is what happens to the nuclear structure, nuclear properties and/or to the nuclear reactions inside plasmas. For example, the discovery that in nuclear reactions at very low energies the electrons of the target atoms partially screen the Coulomb barrier between the projectile and the target nuclei, with a resulting enhancement of the fusion cross section, opens new and important perspectives in astrophysics and cosmology. For example, obtaining information on 'electron screening' is extremely valuable for improving our knowledge of nuclear processes in stellar plasmas. A further important issue concerns the influence of nuclear excited states on the fusion cross section. Nuclear reactions, relevant for astrophysical studies, have been performed in the past with both target and projectile in their ground state. However, in high temperature plasmas ($>10^8$ K), excited nuclear states can play an important role, as is thoroughly discussed in the pioneering theoretical work of Bahcall and Fowler [Bahcall 1969]. In that case, the authors studied the influence of low-lying excited ^{19}F states on the final ^{19}F(p,γ) reaction, predicting an increase of the reaction rate by a factor of about 3 at temperatures near 10^9 K. Thus, determining the appropriate experimental conditions that allow the evaluation of the role of excited states in the stellar environment could strongly contribute to the development of nuclear astrophysics.

Direct measurements of reaction rates in a plasma offer this opportunity. In addition, other new topics can be conveniently explored by measuring nuclear reactions in plasmas, such as three- body fusion reactions (such as those predicted by Hoyle [1946]), lifetime changes of unstable elements [Limata 2006] or nuclear and atomic levels [Hannachi 2007] in different plasma environments and fundamental physics aspects, such as non-extensive statistical thermodynamics [Tsallis 2009]. This last item can be investigated in order to validate or refute the general assumption of local thermal equilibrium that is traditionally implied for plasmas [Semerok 1999].

22.2.1 Perspectives of Studies at the Extreme Light Infrastructure

In this context, several experiments will be conducted at the future multi-petawatt laser facility ELI-NP (extreme light infrastructure nuclear physics) in Bucharest, Romania. Among the various nuclear reactions – which have attracted attention for astrophysical or cosmological reasons – to be investigated as a first case study, the reactions ^{13}C(^{4}He, n)^{16}O and ^{7}Li(d, n)^{4}He-^{4}He have been selected. The first one for its relevance in the framework of stellar nucleosynthesis, while the second one was chosen for its role played in the Big Bang primordial nucleosynthesis. In these studies, the aim is to produce plasmas, through the laser–target interaction, containing mixtures of ^{13}C+^{4}He and ^{7}Li+d at temperatures of order 10^8 K in order to investigate inner-plasma thermonuclear reactions.

Further investigations are also planned concerning weakly bound nuclear states, for which an interesting case is represented by the ^{11}B(^{3}He, d)^{12}C* reaction. Nucleonic matter displays a quantum–liquid structure, but in some cases finite nuclei behave like molecules composed of clusters of protons and neutrons. Clustering is a recurrent feature in light nuclei, from beryllium to nickel, drawing on the tight binding of α-particle substructures in these nuclei. Cluster structures are typically observed as excited states close to the corresponding decay threshold; the origin of this phenomenon lies in the effective nuclear interaction, but the detailed mechanism of clustering in nuclei has not yet been fully understood. It is extremely interesting to study this aspect also in a plasma. It is important to note that all the elements involved in the reactions previously mentioned are stable nuclei.

To conduct such activities, an innovative experimental set-up will be built, where two laser beams generate two colliding plasmas. The use of colliding plasma plumes suitable for nuclear physics studies was proposed few years ago [Mascali 2010] and recently adopted in some studies [Labaune 2013].

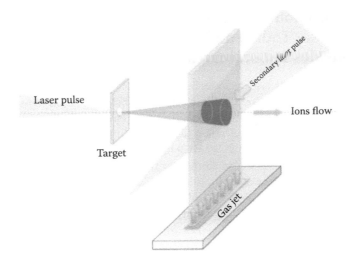

FIGURE 22.1 Experimental layout; a laser pulse impinging on a thin foil generates a primary plasma plume (red) which impacts on a plasma slab produced through the interaction of a second laser pulse with a gas jet.

The idea adopted for investigations in preparation at ELI-NP is illustrated in Figure 22.1. A laser pulse impinges on a thin solid target and produces (depending on the target material) boron, carbon or lithium plasmas [Tudisco 2016]. The rapidly streaming plasma impacts on a secondary plasma target, prepared through the interaction of a second laser pulse in a gas jet target (with densities ranging at about 10^{18}–10^{20} atoms/cm³). The charged particles, neutrons and gamma rays generated in nuclear reactions will be detected by using a highly segmented detection system. The segmentation is required for the reconstruction of the reaction kinematics. A prototype detection system is currently under study at the Laboratori Nazionali del Sud – INFN, Catania and will constitute detection of neutrons and charged particles.

An ideal neutron detection module must provide high efficiency, good discrimination of gammas from neutrons and good timing performance for ToF (time-of-flight) neutron velocity reconstruction. In addition, it must be able to work under harsh experimental conditions imposed by the high-intensity laser–matter interaction. All these aspects may be met by a configuration based on $50 \times 50 \times 50$ mm³ PPO-plastic scintillators [Zaitseva 2012] plus a SiPM [Privitera 2008] read-out and a totally digital acquisition of the multi-hit signals.

Concerning the charged particle detection and identification, research and development activities are focused on the silicon carbide (SiC) material. Recently, SiC has been proven to have excellent properties [Moll 2003, Chaudhuri 2013] in terms of energy and time resolution, resistance to radiation damage, insensibility to visible light and, probably in the near future, decreasing cost. The blindness to the electromagnetic visible radiation is the key aspect for the charged particle detection and identification in a plasma environment, and subsequently for the activities at ELI-NP and other petawatt class laser systems.

In conclusion, new and extremely exciting perspectives for nuclear reaction studies in laser plasmas will open up in the near future at the ELI-NP facility [Negoita 2016]. From these studies, we are sure to obtain information on some fundamental questions associated with the nuclear–plasma science. Measurements of reaction rates or cross sections and astrophysical factors are possible also with conventional accelerators, but not under plasma conditions. Laser–matter interactions will provide this opportunity. What needs to be improved in the laser-driven scheme is the repetition rate with which (single-shot) experiments can be conducted, since reaction yields directly scale with this parameter.

22.3 Production and Decay Studies of Cosmogenic ^{26}Al in Laser-Induced Plasma: Towards a Nuclear-Astrophysics Laboratory with PW Laser Systems

The development of high-power laser facilities in the PW class, such as what is under development at ELI-NP, can potentially initiate a paradigm shift regarding the way nuclear experiments will be conducted in the 21st century. With the anticipated high-power levels, particle acceleration will likely be governed by the radiation pressure acceleration (RPA) domain [Ur 2015, Dover 2012] (see Chapter 5), which will allow the realization of intense beams of highly energetic ions that will induce warm dense matter (WDM) and eventually even stellar plasma conditions in laboratory settings [Ping 2010]. Hence, nuclear phenomena can be experimentally investigated in entropy conditions, resembling astrophysical environments in the laboratory [Takabe 2001, Ledingham 2003, Ledingham 2010]. A prominent feature in this respect will be the ability to produce ground and even short-lived excited states of exotic isotopes simultaneously with considerable yields. High-intensity laser–plasma-driven accelerators will differ in this respect from all existing accelerator systems, which typically provide low peak fluxes.

The Scottish Universities Physics Alliance (SUPA) collaboration aims to study nuclear cross sections in plasma which are crucial to understanding the isotopic abundances in the universe, starting with the radioactive, cosmogenic ^{26}Al because of its outstanding astrophysical significance.

In a recent overview article, Iliadis et al. even describe ^{26}Al as the most important nucleus for stellar astrophysics [Iliadis 2011], as its high intergalactic abundance gives clear evidence that stellar nucleosynthesis is still ongoing [Lee 1976, Mahoney 1984]. It is thought that ^{26}Al is created in explosive carbon, oxygen and silicone burning cycles [Lee 1979] or explosive helium burning [Arnould 1980] which take place in novae or supernovae explosions, Wolf-Rayet stars, red giants, massive stars and supermassive stars [Diehl 2006]. The total galactic abundance of ^{26}Al is estimated to be around 2–3 solar masses and since current theory only accounts for a minute fraction of this value, the related production cross section in plasma needs to be studied. The theoretical evaluation of the ^{26}Al abundance in the universe is challenged by the isotope's rather complex decay pattern [Dunford 1995]. The ß$^+$ decay from the 5$^+$ ground state decays with a long half-life of $t_{1/2} = 0.72$ Megayears predominantly to the first excited state in ^{26}Mg at 1809 keV by a second order forbidden ß$^+$ decay. Its first excited state is an isomer at 228 keV (^{26m}Al), which decays via a super-allowed 0$^+$→0$^+$ ß$^+$ transition directly into the ground state of the daughter nuclei ^{26m}Mg$_{g.s.}$ with $t_{1/2} = 6.35$ sec. The direct M5 electromagnetic transition from the isomer to the ground state 0$^+$→5$^+$ has an extremely low transition probability. Therefore, ^{26}Al$_{g.s.}$ and ^{26m}Al are treated as different species in the cold environments $kT \ll 1$ eV, which are characteristic for experiments with continuous particle beams provided by conventional electrostatic accelerators. However, theory predicts that a thermal equilibrium between the long-lived 5$^+$ ground state and the first isomeric 0$^+$ state at 228 keV will occur at high temperatures (10^6–10^9 K) via a manifold of interlinking transitions with short-lived (fs–ns) high lying energy levels, resulting in a much shorter effective lifetime for the ^{26}Al species [Ward 1980, Coc 1999, Gupta 2001].

Experiments which focus on the creation and destruction processes of ^{26}Al are already very challenging with established low intensity accelerator systems that provide particle beams with current of order μA for which high spatial and temporal densities of short-lived excited states are elusive. Moreover, the isotope is very rare and precious and as such, high-powered laser experimentalists are best advised to focus initially on the inverse reaction ^{26}Mg(p, n)^{26}Al to produce states in ^{26}Al *in situ*. Fortunately, the individual cross sections for ^{26}Al$_{g.s.}$, ^{26m}Al and the 3$^+$ level at 417 keV, ^{26}Al$_{417}$ keV, have already been studied with extremely high accuracy by Skelton et al. [Skelton 1987] for energies up to 1 MeV above the reaction threshold of $E_p \sim 5.0$ MeV. A crucial benefit and unique feature of laser-acceleration is the high peak flux density of the proton bunch. This allows one to create sufficient yields of excited states faster than they can decay. The resulting ratios of the population distribution mimic those of a thermodynamic equilibrium at very high temperatures for as long as the high flux density bunch can be sustained.

It is also hoped for the distant future that an exawatt (EW) class of optical accelerators can be developed that will ultimately provide stellar-like temperatures for electrons and ions *in situ*.

At present laser–plasma accelerators can already provide WDM conditions for fleeting time spans in the ps time regime. As for the concrete case of ^{26}Al depicted here, the target temperatures are eventually 10^6–10^9 K, which are currently still far out of reach.

The SUPA collaboration performed a *prima facie* experiment to demonstrate the feasibility of ^{26}Al studies with the VULCAN PW laser system at the Rutherford Appleton Laboratory in Didcot, UK. This campaign followed our original proposal in the original ELI-NP White Book (2009) which was also cited in the NuPECC (Nuclear Physics European Collaboration Committee) publication 'Light to Reveal the Heart of Matter' in 2015.

Current detector technology is unable to measure the prompt radiation spectra during a high-energy laser pulse impact. This restriction will apply for any future experiment with high-powered laser systems such as those at ELI-NP, as the strong electromagnetic pulse (EMP), which is induced at laser impact, will instantaneously saturate any conventional detection system placed in the close proximity to the irradiation zone. In addition, the decay of ^{26}Al$_{g.s.}$ eludes direct detection due to its extremely long lifetime. Both shortfalls are crucial and demand the use of, low-intensity accelerator sites, for example, and isotope separation techniques to support any laser nuclear experiment in the future.

In the VULCAN experiment (see Figure 22.2), we used two synchronized optical beam lines for proton (laser 1–300 J, 10 ps) and X-ray production (laser 2–ca. 100 J, 2 ps). The light pulse for proton production was focussed onto primary plastic or gold foils. Protons were accelerated via the target normal sheath acceleration (TSNA) mechanism and energies above the reaction threshold of $E_p \sim 5.0$ MeV were reached with sufficient yield (10^{10-11} protons/bunch). Their energy distribution was measured and

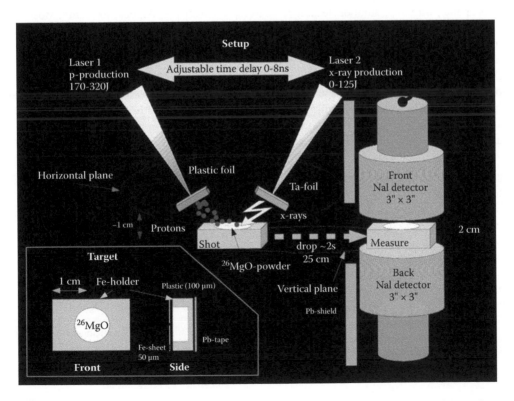

FIGURE 22.2 Schematic layout of the experimental set-up, the inset in the lower left corner shows a front and cross-sectional view of the target.

approximated by a Maxwell-Boltzmann distribution function with $k_B T_p \sim 2$ MeV. As we chose a close geometry between the primary production foil and the secondary target ($d \sim 1$ cm) in which the nuclear reaction takes place (see Figure 22.2), we achieved reaction-driving proton currents with peak values >1 A for a short time span of 100 ps–200 ps. The X-ray driving laser pulse was directed onto a 1–mm tantalum radiator to create *bremsstrahlung*. Isotopically-enriched ^{26}MgO powder was chosen as the target material because of its high melting point of 2900 °C. A thickness of 1.0 mg/cm^2 was deemed necessary to safeguard the expensive material from overheating. The lose magnesium powder was placed inside an iron canvas and then sealed by a thin PET (polyethylene terephthalate)/iron layer. After irradiation, the target was quickly dropped within (1–2 seconds) into a small gap between two NaI detectors facing each other at 180° in close proximity which were also placed inside the large vacuum vessel. During a series of 48 shots in the first campaign, we measured a single Gaussian peak at 511(9) keV in the delayed gamma energy spectrum after irradiation with reaction-driving proton bunches from initially ~300 J laser pulse energy. The dominant component, accounting for 80(10)% of the signal intensity, showed a half-life of $t_{1/2} = 6.6(3)$ sec. We also identified, as expected, a much less intense, delayed component where 7 min < $t_{1/2} < 11$ min, assigned to the ß$^+$ decay of ^{13}N from ^{16}O(p,α)^{13}N with $t_{1/2} = 9.96$ min.

As the VULCAN experiment did not allow measurement of the prompt radiation of the target, we performed a complementary experiment with four highly efficient germanium detectors at the ALTO Tandem accelerator of the Institute Physique Nucléaire Orsay (IPNO), where we used the same two thick targets from the VULCAN campaign and a dc proton beam, providing an average current of 1 nA. This measurement unambiguously confirmed all our results obtained with the VULCAN laser. Crucially, we also identified the 417 keV photopeak corresponding to the E2 transition between the 3$^+$ ^{26}Al$_{417}$ keV and 5$^+$ ^{26}Al$_{g.s.}$ in the prompt spectra. We therefore concluded successful production of ^{26m}Al *in situ* with the VULCAN array. The measured activity was about 300 kBq–400 kBq per shot for a proton-driving laser pulse near 300 J with 10 ps duration, leading to an isotopic density $\rho(^{26m}$Al) near 10$^{6-7}$ cm^{-3}. Most importantly, by identifying the decay of ^{26m}Al, we can also deduce that for times in the range of the duration of the production-driving proton bunch (100 ps–200 ps), ^{26}Al$_{g.s.}$ and ^{26}Al$_{417\,keV}$ were created simultaneously with comparable yield.

The first test experiment was successful and particularly showed the specific challenges and technological arrangements which need to be applied to make nuclear physics experiments comply with high-powered laser accelerator systems. We identified the influence of the EMP as the core problem in all *in situ* measurements. Moreover, the presence of a third, temperature-probing laser beamline, would have been a tremendous asset. It needs to be pointed out that targets will differ substantially from

TABLE 22.1 Experimental Possibilities of High-Powered Laser-Induced Accelerators in Comparison with Tandem/Cyclotron Systems

Accelerator System	I_{beam}	Experimental Possibilities & Limits
Tandem & cyclotron	~1 μA (average)	Precision measurements of cross sections at low temperatures
		No temperature conditions resembling stellar-like conditions possible
		No *in situ* exposure of created isotopes to *bremsstrahlung* possible
Laser–plasma system >100 TW	>1 A in bunch (~10$^{10-11}$ protons per bunch in ~100 ps)	Direct measurements of astrophysical cross sections possible with next generation laser accelerators
		Creation of astrophysical isotopes in excited states with high yields
		Non-equilibrium distribution between short-lived, low excited states can be achieved in cold environments, which mimic high-temperature entropy conditions for fleeting time spans (ps-regime)
		In situ exposure of the created isotopes in ground and excited state to *bremsstrahlung* is possible via second, synchronized, laser pulse, hence cross section measurements on excited states are possible
		On the pathway to establish stellar-like conditions in laboratories, with ELI-NP, the creation of WDM conditions via isochoric heating could be envisaged

conventional thin targets currently used in experiments with continuous ion beams, which can survive even long measurement campaigns without being destroyed. Supplementary investigations using low intensity accelerators and isotope separation techniques are required to characterise the targets, and to understand the yields and hence the nuclear processes involved. The short duration of the reaction-driving proton bunches and the associated population distributions are the most interesting features of high-powered laser experiments. Our initial investigation suggests that comparable yields of the first three states in ^{26}Al can be present within the short time span of the reaction-driving proton bunch. With predicted ELI-NP repetition rates in the Hz regime, the VULCAN yields will be enhanced by a factor of 10^4 or even more with RPA. As such we are optimistic that with the unique features of the PW class of laser accelerators, such as will be available at ELI-NP, one can produce very high yields of isotopes in excited states thus mimicking astrophysical conditions. We conclude that the dawn of a new era in experimental nuclear physics is looming, promising exciting times and groundbreaking experiments (Table 22.1).

22.4 Laser-Driven De-excitation of ^{84m}Rb

Nuclear properties have been investigated worldwide in solid, liquid and gaseous material states. Nevertheless, the plasma state is the most abundant one in the visible universe in which it is important to investigate the nuclear properties. An example of a property that can change in the plasma environment is the apparent lifetime of isomeric states [Gosselin 2007]. In hot and dense plasmas, nuclear excitation processes can more rapidly depopulate isomeric states, reducing their apparent lifetimes. This change has been suggested for ^{176}Lu [Klay 1991], ^{93}Mo, ^{201}Hg [Gosselin 2007], etc. One nucleus kept our attention: ^{84}Rb.

Figure 22.3 presents the dependence of the apparent lifetime of the 463 keV isomeric state ^{84m}Rb on plasma temperature and for a plasma density of $\rho = 10^{-2}$ g/cm^3, calculated with the ISOMEX code [Comet et al. 2015]. The plasma temperature is defined as the average energy of the free electrons in the plasma. In the local thermodynamic equilibrium (LTE) hypothesis, this temperature is also equal to the photon and ion temperatures. The 20.26 min lifetime of the isomeric state is not affected at plasma temperatures below 150 eV, but dramatically reduces to 200 μs for temperatures between 150 and 400 eV. Temperatures up to 270 eV have been obtained at the PHELIX facility (GSI) with a laser intensity of about 10^{15} W/cm^2 in a plasma of density 10^{-2} g/cm^3 [Denis-Petit et al. 2014], corresponding to the critical density for a laser wavelength of 1.06 μm. Lower temperatures can be obtained by defocusing the laser.

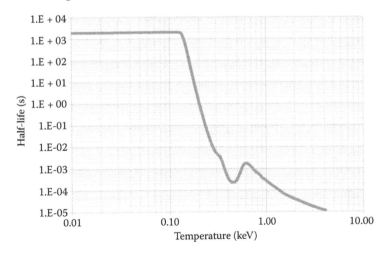

FIGURE 22.3 Dependence of the half-life of the ^{84m}Rb 6$^-$ isomer on the temperature of a plasma of density $\rho = 10^{-2}$ g/cm^3. These calculations have been made with the ISOMEX code.

For higher temperatures, some special cup-shaped targets or heating set-up (using electrons or X-rays) needs to be developed. The advantage of ^{84m}Rb is that it de-excites by the emission of few hundred keV photons, energies that are much higher than the few hundred eV surrounding plasma temperatures. This facilitates the de-excitation detection in plasma environments. In the following, we present the nuclear excitation mechanisms responsible for changes of the apparent lifetime in the plasma and an experiment to evidence this phenomenon.

22.4.1 Excitation of ^{84m}Rb in a Plasma

A partial level scheme of the ^{84}Rb nucleus is shown in Figure 22.4. The $J^\pi = 6^-$, $T_{1/2} = 20.26(3)$ minute, isomeric state of the unstable ^{84}Rb nucleus lies at an excitation energy of 463.618(3) keV, which is 3.498(6) keV lower than a $J^\pi = 5^-$, $T_{1/2} = 9$ ns state [Denis-Petit et al. 2014, Denis-Petit et al. 2017]. In a plasma of T = 400 eV (LTE), four different excitation processes [Harston 1999] can be involved in the excitation of the 6^- state to the 5^- one via an M1 transition, thus reducing the apparent lifetime of the 6^- state. These processes are illustrated in Figure 22.5. The photo-excitation process is a resonant process, where a photon from the plasma directly transfers its energy to the nucleus. The inelastic electron scattering is a process, where an energetic free electron from the hot plasma transfers part or all its kinetic energy to the nucleus. NEEC occurs, when a free electron from the plasma is captured by the ion to fill a vacancy within an atomic orbital. The energy released in the process is transferred to the nucleus via a virtual photon. Both the atomic capture and the nuclear excitation take place within the same atom. Finally, NEET [Kishimoto 2000] occurs when a vacancy in a lower-lying orbital of the atom is filled by an electron from a higher level. In this process, the nucleus can also be excited by the energy released in the

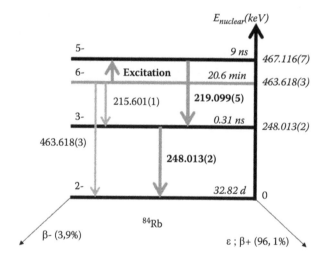

FIGURE 22.4   Partial level scheme of the ^{84}Rb nucleus.

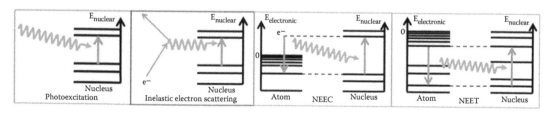

FIGURE 22.5   Nuclear excitation processes occurring in a hot and dense plasma.

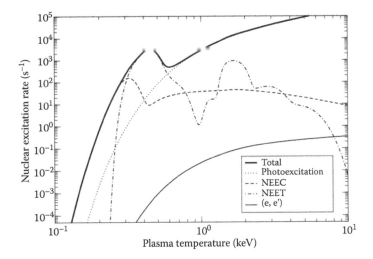

FIGURE 22.6 ISOMEX calculations of excitation rates of ^{84m}Rb in a 10^{-2}g/cm^3 plasma, as functions of its temperature (in the LTE approximation).

atomic transition. The NEET process requires both the electronic and nuclear transitions to carry the same amounts of energy and angular momentum; it is a resonant process.

The resonance conditions are usually not fulfilled in neutral atoms, but can be reached in ionized atoms for which the electron binding energies are modified, due to reduced screening of the nuclear Coulomb potential. In plasmas charge state distributions can be broad and electron–atom collisions also significantly modify the atomic state widths [Baranger 1958]. This helps fulfilling the nucleus–atom resonance conditions.

The contributions of the four processes mentioned above to the excitation of the 6$^-$ nuclear state of ^{84}Rb towards the 5$^-$ state are presented in Figure 22.5 as a function of the plasma temperature. At low temperatures, the NEEC effect is dominant up to 150 eV, where the NEET effect presents a maximum at 400 eV, corresponding to an excitation rate per nucleus of 4000 excitations per second. At higher temperatures (above 700 eV), the photoexcitation becomes dominant. The estimations presented in Figure 22.3 and Figure 22.6 were obtained using the ISOMEX code based on the relativistic average atom model (RAAM) [Comet 2015]. This statistical approach to describe a resonant process such as NEET in a light nucleus and at this low-plasma density has been reinvestigated.

A refined NEET rate estimation has been performed using a multi-configuration Dirac-Fock (MCDF) code, taking into account the electronic configurations of each ion present in the plasma. Due to the extremely resonant condition and the measurement uncertainty of 6 eV on the nuclear transition energy, a range of excitation rates is obtained. The expected excitation rate lies between 2.2/sec and 7.4 × 10^6/sec in an LTE plasma of T = 400 eV and ρ = 10^{-2} g/cm^3 [Denis-Petit et al. 2017]. The large value span is directly related to the highly resonant character of the NEET process with respect to the energy difference between the atomic and nuclear transitions.

22.4.2 Experiment Design

Because of its 20 min half-life, the ^{84m}Rb has to be produced *in situ* before being excited in a plasma. Two lasers are thus required: one ultra-short, high-intensity laser to accelerate a beam of particles used to produce the radio isotopic target (production laser pulse) and a long-duration, high energy laser to heat the target and form the plasma (plasma–laser pulse). One possible facility offering these two powerful synchronized laser pulses will be ELI-NP at Bucharest. Table 22.2 presents the laser characteristics

TABLE 22.2 Laser Characteristics Required to Perform Isomer Excitation Experiments in a Plasma: Those accessible in the Next Few Years, and The Ideal Ones to Tend Toward

	Future Laser:		Ideal Laser:	
	Ion Production Laser	Plasma Laser	Ion Production Laser	Plasma Laser
Energy (J)	3	250	250	1000[a]
Pulse duration	30 fs	0.5 ns	30 fs	1 ns
Focal spot diameter (μm)	5	250	5	500 μm[a]
Intensity (W/cm^2)	5×10^{20}	5×10^{15}	5×10^{22}	5×10^{15}
Repetition rate	10 shot/sec	1 shot/min	1/sec	1/sec

[a] Higher laser pulse energy associated with larger focal spots can produce larger plasmas, allowing de-excitation of many more nuclei.

required to perform isomer-excitation experiments in plasmas in the near future and the characteristics of an ideal laser facility to perform these experiments.

We plan to use the ^{76}Ge(^{12}C,p+3n)^{84m}Rb nuclear reaction in order to produce the isomers. This reaction requires as many carbon ions as possible with energies ranging from 30 MeV to 90 MeV. Currently, we do not have reliable estimations of the number of ions one can accelerate with a 10-PW laser at ELI-NP. So, we based our calculations on an experiment performed at a focused peak laser intensity of 7×10^{20} W/cm^2, where carbon ions have been accelerated [Carroll 2010]. The facility used is an analogue of the planned ELI-NP 100 TW, which, however, will operate at a much higher repetition rate (10 Hz instead of 0.05 Hz). Using ELI-NP at a 10^{22} W/cm^2 peak laser intensity will obviously change our estimates significantly. About 3×10^{11} carbon ions with energies adequate for creating 5×10^5 ^{84m}Rb nuclei would be obtained per shot and after one hour of irradiation 7×10^9 ^{84m}Rb nuclei in a 3-μm thick layer are expected. This layer thickness is about the one which can be heated by the plasma laser. Some other ways to produce the ^{84m}Rb have been also investigated, such as (γ,n) reactions requiring photons between 13 and 21 MeV [Plaisir 2012], obtained via the *bremsstrahlung* process accompanying the deceleration of laser-produced electrons in matter. The electron source terms for the very high intensities of ELI-NP are not yet known, but if the ability to create more than 10^{17} electrons between 13 and 21 MeV is demonstrated in the future, this way to produce the isomer would become feasible.

The plasma will be created and heated to 400 eV (LTE), on account of the 250 J pulse of the high-powered laser system (HPLS) with a 500 ps duration, focused to a spot of 250 μm diameter. To optimize the set-up, the ^{84m}Rb nuclei must be produced in this focal spot area. For this purpose, some particle focusing device (coils, structured targets, etc.) must be investigated. If a 100-TW configuration system can be used as the production laser and if all the produced ^{84m}Rb nuclei can be immersed within a 400-eV plasma, the number of triggered de-excitations during one laser shot will be between 12 and 4.2×10^6. If the number of de-excitations is close to the lower estimate, the phenomenon may only be observable with a 10-PW HPLS such as the one planned for ELI-NP.

22.5 Conclusion

Laser–matter interaction provides an incredible tool to mimic astrophysical plasmas otherwise inaccessible in the laboratory. The innovative experimental set-up based on colliding plasmas, actually in preparation phase for the installation at the ELI-NP facility in Bucharest, will be useful to investigate the fundamental properties of nuclei (interactions and decays) also under astrophysical conditions. Hot and dense plasmas with astrophysical properties can be formed in laboratories via the interaction of high-powered lasers with matter. Such plasmas can be used to understand the nuclear reactions and excitations occurring in stars better, modifying for example the apparent lifetime of the nuclei. Facilities delivering inside the same reaction chamber a powerful laser pulse to create a plasma and a particle beam from an accelerator able to create some instable isotopes are limited. When the produced radioisotopes have lifetimes of some nanoseconds, accelerators do not provide sufficient particle flux to produce and adequately place nuclei inside a plasma to study their properties in such an extreme environment. Laser-accelerated bunches of particles give the opportunity to produce in few nanoseconds short-lived nuclei or isomers that can be studied in a plasma.

However, there are major problems to be solved before performing measurements for this new topic of nuclear physics: the detection of nuclear observables necessary to do spectroscopy. Indeed, during a powerful laser shot, many X-rays are emitted from the formed plasmas. The intense X-ray pulse blinds the detectors conventionally used in nuclear physics. In addition to the electromagnetic perturbation, this X-ray burst can destroy fragile detection systems such as Germanium detectors. Therefore, more robust detectors, such as scintillators coupled to photomultipliers, must be considered. Besides, conventional acquisition chains used in nuclear physics (shaping amplifiers + peak-sensing analogue to digital converters) cannot be used anymore. Instead, the detector signals have to be digitized in order to extract the spectroscopic information from them. Nowadays, gamma spectroscopy can only be done a few milliseconds after the creation of isomers formed by a laser pulse [Negoita 2015]. In a very recent experiment, it was possible to decrease the detector blindness to only few tens of microseconds. The difficulty detecting nuclear observables is still a barrier to laser-driven nuclear-physics experiments, but we expect significant improvements within the next three years. A new generation of detectors will be ready for use at new high-intensity laser facilities, which offer a 10-PW pulse every minute. A high repetition rate is essential also for the rise of laser-driven nuclear physics, which is governed by statistical rules and needs the accumulation of thousands of nuclear reactions to bring to light the underlying physics. The *prima facie* studies at ^{26}Al show that isotopes of pronounced astrophysical interest can be produced in high yields employing TW and PW laser systems and can be exposed to *bremsstrahlung* radiation *in situ*. Interestingly, it was found that the short duration of the reaction-driving proton bunch allows for the simultaneous preparation of all three low-lying excited states in ^{26}Al with comparable yields, which includes the short-lived (1.25 ns) state at 417 keV. Consequently, reaction cross section measurements on excited states can be envisaged. Moreover, the complex interplay between of population and depopulation of excited states and their influence on astrophysical lifetimes can eventually be studied, thus allowing verification of astrophysical theories. This strain of research will be emphasized dramatically once PW laser systems such as ELI-NP will be able to provide warm dense matter conditions in coincidence with the nuclear reactions taking place.

Acknowledgements

Valuable contributions to the content of Section 22.3 of this chapter are acknowledged and provided by R. Clarke, D.C. Carroll, R.J. Gray, M. Hassan, D. Johnson, M. Labiche and T. McCanny. We are deeply obliged to our colleagues from the CENBG laboratory, F. Hannachi, F. Gobet, M. Versteegen and to G. Gosselin, V. Méot and P. Morel from CEA/DIF for their help in writing Section 22.4.

References for Chapter 22

Arnould et al., 1980. "Synthesis of Al-26 in explosive hydrogen burning" *Astrophys. J.* **237**: 931–950.

Bahcall, N.A. and Fowler, W.A. 1969. "The effect of excited nuclear states on stellar reaction rates" *Astrophysical Journal* **157**: 645–657.

Baranger, M. 1958. "General impact theory of pressure broadening" *Phys. Rev.* **112**: 855–865.

Carroll, D.C. et al., 2010. "Carbon ion acceleration from thin foil targets irradiated by ultrahigh-contrast, ultraintense laser pulses" *New Journal of Physics* **12**: 045020.

Chaudhuri, S. et al., 2013. "High resolution alpha particle detection using 4H-SiC epitaxial layers: Fabrication, characterization, and noise analysis" *Nucl. Instr. Meth. A* **728**: 97–101.

Coc, A., Porquet, M.-G. and Nowacki, F. 1999. "Lifetimes of 26Al and 34Cl in an astrophysical plasma" *Phys. Rev. C* **61**: 015801.

Comet, M. et al., 2015. "NEET rate confidence interval in a [201]Hg LTE plasma" *Phys. Rev. C* **92**: 054609.

Denis-Petit et al., 2014. "Identification of X-ray spectra in the Na-like to O-like rubidium ions in the range of 3.8–7.3 Å" *J. Quant. Spectrosc. Radiat. Transf.* **148**: 70–89.

Denis-Petit, D. et al., 2017. "Calculation of the rate of nuclear excitation by electron transition in an [84m]Rb plasma under the hypothesis of local thermodynamic equilibrium using a multiconfiguration Dirac-Fock approach" *Phys. Rev. C* **96**: 024604.

Diehl, R. et al., 2006. "Radioactive 26Al from massive stars in the galaxy" *Nature* **439**: 45–47.

Dover, N.P. and Najmudin, Z. 2012. "Ion acceleration in the radiation pressure regime with ultrashort pulse lasers" *High Energy Density Physics* **8**: 170–174.

Dunford, C. and Burrows, T. 1995. "Online nuclear data service" *Rep. IAEA-NDS-150, Int. At. Energy Agency, Vienna, Austria.*

Gosselin, G. Méot, V. and Morel, P. 2007. "Modified nuclear level lifetime in hot dense plasmas" *Phys. Rev. C* **76**: 044611.

Gupta, S. and Meyer, B. 2001. "Internal equilibration of a nucleus with metastable states: 26Al as an example" *Phys. Rev. C* **64**: 025805.

Hannachi, F. et al., 2007. "Prospects for nuclear physics with lasers" *Plasma Phys. Control. Fusion* **49**: B79–B86.

Harston, M.R. and Chemin, J.F. 1999. "Mechanisms of nuclear excitation in plasmas" *Phys. Rev. C* **59**: 2462–2473.

Hoyle, F. 1946. "The synthesis of the elements from hydrogen" *Monthly Notices of the Royal Astronomical Society* **106**: 343–383.

Iliadis, C. et al., 2011. "The effects of thermonuclear reaction rate variations on 26Al production in massive stars: A sensitivity study" *Astrophys. J. Suppl. Ser.* **193**: 16 (23pp).

Kishimoto, S. et al., 2000. "Observation of nuclear excitation by electron transition in [197]Au with synchrotron X-rays and an avalanche photodiode" *Phys. Rev. Lett.* **85**: 1831–1834.

Klay, N. et al., 1991. "Nuclear structure of Lu-176 and its astrophysical consequences. 2. Lu-176, a thermometer for stellar helium burning" *Phys. Rev. C* **44**: 2839–2849.

Labaune, C. et al., 2013. "Fusion reactions initiated by laser-accelerated particle beams in a laser-produced plasma" *Nature Comm.* **4**: 42506.

Ledingham, K., McKenna, P. and Singhal, R. 2003. "Applications for nuclear phenomena generated by ultra-intense lasers" *Science* **300**: 1107–1111.

Ledingham, K. and Galster, W. 2010. "Laser-driven particle and photon beams and some applications" *New Journal of Physics* **12**: 045005.

Lee, T., Papanastassiou, D. and Wasserburg, G. 1976. "Demonstration of 26Mg excess in Allende and evidence for 26Al" *Geophys. Res. Lett.* **3**: 41–44.

Lee, T. et al., 1979. "60Fe in the early solar system and its nucleosynthetic implications" *Astrophys. J.* **232**: 854–862.

Limata, B.N. 2006. "New measurement of 7Be half-life in different metallic environments" *Eur. Phys. J. A* **27**, s01:193–196.

Mahoney, W. et al., 1984. "Heao 3 discovery of 26Al in the interstellar medium" *Astrophys. J.* **286**: 578–585.

Mascali, D. et al., 2010. "Colliding laser-produced plasmas: A new tool for nuclear astrophysics studies" *Rad. Eff. and Def. in Sol.* **165**: 730–736.

Moll, M. 2003. "Development of radiation hard sensors for very high luminosity colliders – CERN-RD50 project" *Nucl. Instr. Meth. A* **511**: 97–105.

Negoita, F. et al., 2015. "Perspectives for neutron and gamma spectroscopy in high-power laser-driven experiments at ELI-NP" *AIP Conference Proceedings* **1645**: 228–236.

Negoita, F. et al., 2016. "Laser-driven nuclear physics at ELI-NP" *Romanian Report in Physics* **68**: s37–s144.

Ping, Y. et al., 2010. "Warm dense matter created by isochoric laser heating" *High Energy Density Physics* **6**: 246–257.

Plaisir, C., Hannachi, F. and Gobet, F. 2012. "Measurement of the ^{85}Rb(γ,n)^{84m}Rb cross section in the energy range 10–19 MeV with Bremsstrahlung photons" *Eur. Phys. J. A* **48**: 68 (5).

Privitera, S. et al., 2000. "Single photon avalanche diodes: Towards the large bidimenional arrays" *Sensors* **8**: 4636–4655.

Semerok, A. et al., 1999. "Experimental investigations of laser ablation efficiency of pure metals with femto, pico and nanosecond pulses" *Applied Surface Science* **138–139**: 311–314.

Skelton, R., Kavanagh, R. and Sargood, D. 1987. "^{26}Mg(p,n)^{26}Al and ^{23}(γ,n)^{26}Al reactions" *Phys. Rev. C* **35**: 45–54.

Takabe, H. 2001. "Astrophysics with Intense and Ultra-Intense Lasers "Laser Astrophysics" *Progress of Theoretical Physics, Suppl.* **143**: 202–265.

Tsallis, C. 2009. "Introduction to Nonextensive Statistical Mechanics: Approaching a Complex World" Springer.

Tudisco, S. et al., 2016. "Investigation on target normal sheath acceleration through measurements of ions energy distribution" *Rev. Sci. Instrum.* **87**: 02A909.

Ur, C. et al., 2015. "The ELI-NP facility for nuclear physics" *Nucl. Instr. Meth. B* **355**: 198–202.

Ward, R. and Fowler, W. 1980. "Thermalization of long-lived nuclear isomeric states under stellar conditions" *Astrophys. J.* **238**: 266–286.

Zaitseva, N. et al., 2012. "Plastic scintillators with efficient neutron/gamma pulse shape discrimination" *Nucl. Instr. Meth. A* **668**: 88–93.

23

Advances in Nondestructive Elemental Assaying Technologies

Robert J. Ledoux

23.1 Introduction to Photon Sources in Nonintrusive Inspection

There has been significant progress recently in nonintrusive inspection (NII) technologies that measure the elemental composition of cargo. These new NII technologies will greatly benefit from innovations in laser-driven particle accelerators for the production of multi-MeV photon beams. This chapter discusses some of the newest technologies being applied to NII, as well as the state of the art of current photon beam sources. The final section is 'aspirational' to the field of laser-driven acceleration and puts forward specifications for photon sources that would fully exploit new NII technologies and usher in low-dose, high spatial resolution materials identification. Yes, that is a challenge!

23.2 Conventional High Energy X-ray Imaging NII

Nonintrusive Inspection (NII) of commercial cargo is limited to a maximum of 10 MeV photons (imposed by the World Health Organization). Therefore, this discussion is limited to photon beams less than 10 MeV. Of course, with new monochromatic photon sources, this requirement could be relaxed since the total dose and production of radioactive by-products could be significantly reduced.

All existing high-energy photon beam sources used in commercial NII are produced by electron *bremsstrahlung*. The most ubiquitous NII technology is the oldest; 2D projective X-ray imaging. Selection of the optimal beam energy (<10 MeV) and intensity for conventional X-ray imaging is determined by factors such as the maximum total attenuation (maximal penetration) and size of the cargo, as well as allowed inspection time and shielding footprint or standoff. The vast majority of existing high energy X-ray scanners use RF electron LINACs with very low duty cycle, typically a few 0.1%. The use of electron LINAC's is driven by their relatively low cost and commercial maturity. X-ray inspection is limited by threat or contraband determination based on the appearance of the projected image without any independent material discrimination or identification.

One improvement upon this basic transmission X-ray technique is to compare two images of the cargo obtained with *bremsstrahlung* beams with end point energies that emphasize the differential mass absorption for materials of different atomic number Z, e.g. 4 and 9 MeV. This dual energy method provides some knowledge of the average Z of the projective image. A shortcoming of this materials discrimination method is that it is integral and, therefore, the differential image is complicated because it combines the different Z materials along the path of the transmitted photons. This technique ideally requires a rapid (<100 μs) interlacing (e.g. alternating pulses between 4 MeV and 9 MeV) of two beam energies to maintain equivalent scan times.

23.3 Overview of New Material Discrimination Technologies for NII

Passport Systems, Inc. (Passport) has pioneered new NII technologies that utilize physical processes beyond attenuation of a transmitted beam to extract materials information from backscattered high-energy photons [Betrozzi 2005, Bertozzi 2007] and photo-fission neutrons [Danagoulian 2010]. These new technologies for NII require a high duty cycle photon beam since the extraction of materials properties depends on the measurement of properties of individual photons and neutrons. Given the bandwidth limits of existing photon and neutron sensors, it is not practical to use low duty cycle LINACS for these advanced materials determination technologies. Further, 3D imaging of high energy backscattered photons is possible using an accurate and rapidly scanned photon beam which dictates energy and emittance requirements on the electron beam.

23.3.1 Physical Principles

The next three sections provide an overview of the physical processes involved in three new NII technologies: nuclear resonance fluorescence (NRF) [1–3], effective Z in three-dimensions (EZ-3D™) [Bertozzi 2007, Bertozzi 2011a] and prompt neutrons from photo-fission (PNPF) [Danagoulian 2010].

23.3.1.1 Material Identification via Nuclear Resonance Fluorescence

The process of NRF corresponds to the excitation of a nuclear state by photons and having that state decay to the ground state or an excited state by the emission of a photon. A schematic of the physical processes which are important to NRF are shown in Figure 23.1. NRF cross sections typically have very large peak values that correspond to hundreds of barns (1 barn is 10^{-24} cm^2), but with energy widths of order 1 eV. These narrow states are broadened by the zero-point motion of the atom and thermal motion. For light nuclei, such as nitrogen and oxygen, the broadening can be approximately 20 eV resulting in a reduction in peak cross sections but wider energy width. However, cross sections for NRF states that are used in NII are still considerably larger than those of the usual electromagnetic processes: photoelectric effect, Compton scattering and pair production in that energy region. The elements carbon, nitrogen and oxygen have strong NRF states as shown in Figure 23.1. Oxygen has NRF states at 6.917 and 7.117 MeV, nitrogen at 7.029, 5.691, 4.915 and 2.313 MeV, carbon ^{12}C at 4.439 MeV and ^{13}C at 3.685 and 3.089 MeV. Measurement of NRF states enables a direct determination of the composition of cargo. For best signal-to-noise ratio (SNR), the NRF photon emission associated with state de-excitation is measured at back angles to the photon beam direction (i.e. scattering angles greater than 90° relative to the incident photon beam direction).

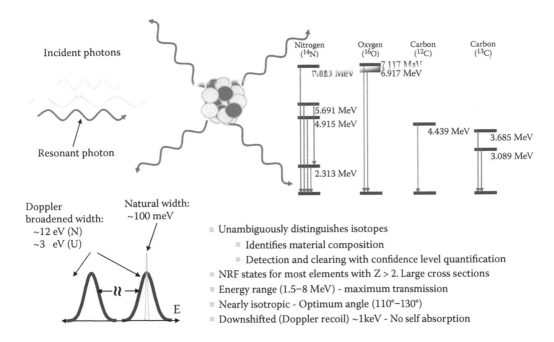

Incident photons

Resonant photon

Nitrogen (^{14}N) Oxygen (^{16}O) Carbon (^{12}C) Carbon (^{13}C)

7.117 MeV
7.083 MeV 6.917 MeV

5.691 MeV
4.915 MeV

4.439 MeV
3.685 MeV
3.089 MeV

2.313 MeV

Doppler broadened width:
~12 eV (N)
~3 eV (U)

Natural width: ~100 meV

E

- Unambiguously distinguishes isotopes
  - Identifies material composition
  - Detection and clearing with confidence level quantification
- NRF states for most elements with Z > 2. Large cross sections
- Energy range (1.5–8 MeV) - maximum transmission
- Nearly isotropic - Optimum angle (110°–130°)
- Downshifted (Doppler recoil) ~1keV - No self absorption

FIGURE 23.1 Overview of NRF physical processes.

A *bremsstrahlung* beam can sample many NRF states simultaneously, but obviously most of the beam is producing dose (i.e. depositing energy through ionizing radiation) and not signal! The requirements of the 'ideal' photon beam for NRF NII are the most stringent of any technology discussed herein, given their narrow energy width and wide range of energies. They are:

- Rapidly tunable in energy – within seconds
- Narrow energy width of order 10 eV
- Very high spectral flux – 10^7 γ/s/eV
- Practical photon beam directional scan mechanism

23.3.1.2 Fissionable Materials Detection via Prompt Neutrons from Photo-Fission (PNPF)

Photo-fission is the process by which a nucleus fissions after absorbing a photon. Photo-fission near energy threshold in actinides is very similar to spontaneous fission in terms of the number of emitted decay neutrons and their energy distribution. A summary of photo-fission attributes is given in Figure 23.2. Prompt neutrons from photo-fission (PNPF) is an extremely sensitive tool for the detection of actinides [Danagoulian 2010]. The threshold photon energy for photo-fission for almost all actinides is approximately 5.5 MeV. Therefore, a source of photons with energy significantly above the photo-fission threshold is needed for fissionable material detection. A photon energy of approximately 9 MeV is a good choice since it is well above threshold for photo-fission but near or below (γ,n) thresholds for most isotopes. Neutrons from (γ,n) reactions are a low probability source of false photo-fission alarms. A measurement of the neutron energy distribution can differentiate photo-fission from (γ,n) neutrons and hence remove these false alarms.

In NII for fissionable materials, the spectrum of high energy (greater than a few hundred keV) neutrons is compared with the known neutron backgrounds and possible (γ,n) sources. PNPF requires a high duty cycle beam since the energy of individual photo-fission neutrons must be measured. Further, all photons below photo-fission threshold only contribute to unnecessary dose. The ideal photon beam energy for PNPF would be centred at approximately 9 MeV with a width of order 1 MeV.

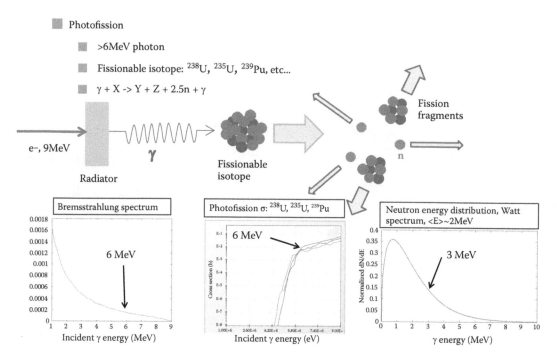

FIGURE 23.2   Overview of prompt neutrons from photo-fission.

23.3.1.3 Material Discrimination via EZ-3D™

The EZ-3D™ technology developed by Passport [Bertozzi 2007, Bertozzi 2011a] depends on the analysis of the entire spectrum of photons scattered from a target when irradiated by energetic photons. As in NRF, the best signal-to-noise is obtained at back angles to the incident photon beam. The nature of the spectrum depends on the complex interaction of multiple processes and depends sensitively on the Z of the target. An example of the multiple processes that contribute to the strong dependence of high energy photons from targets of different Z is shown in Figure 23.3 which displays the backscattered energy spectrum from a 2.8 MeV endpoint *bremsstrahlung* photon beam from a variety of targets where the different spectra are normalized to the 511 keV annihilation peak. Note the dramatic increase of the energy distribution above the 511 keV peak with increasing Z of the target. This technique can be further exploited to generate a 3D map of the density and 'effective Z' of cargo. In order to generate this 3D map, we will see in the next section that the beam must be scanned across a slice of the cargo and the backscattered detectors must be collimated to view only a portion of the beam at any scan position. This new imaging technique is called EZ-3D™. It can be used alone as a powerful NII technology or combined with the other NII technologies discussed. For example, EZ-3D™, X-ray imaging and PNPF can be used together to more effectively identify contraband. If any ambiguity exists, then NRF can be used to confirm or clear the anomaly with very high probability of detection and very low false alarms.

The EZ-3D™ method requires a high duty cycle photon beam since the energy of individual backscattered photons must be measured. The portion of the *bremsstrahlung* energy spectrum below 2 MeV is not useful for NII and it would be desirable to eliminate it.

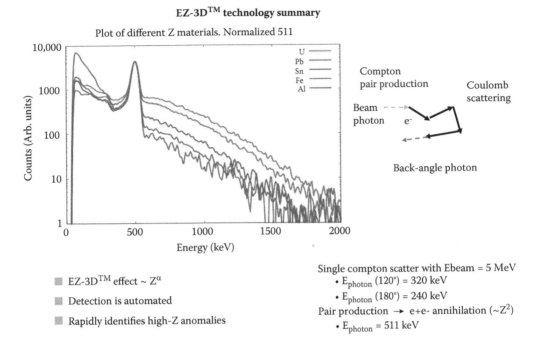

FIGURE 23.3 Processes involved in EZ-3D™ 3D density and average Z imaging.

23.3.2 Putting it All Together: The Passport Systems SmartScan3D™ NII

Unlike current X-ray technologies which use a pulsed X-ray LINAC beam, SmartScan 3D™ uses a continuous wave (CW) 9 MV X-ray beam. This enables materials discrimination capabilities via the incorporation of EZ-3D™ and PNPF in all primary scans and material identification via NRF in secondary scans. The SmartScan 3D™ system makes use of the IBA Rhodotron TT100 CW electron accelerator capable of delivering a 9 MeV electron beam with power up to 40 kW. This ebeam is rapidly (order 10 ms across 3 m) magnetically scanned across a *bremsstrahlung* radiator with highly collimated photon beam ports. The primary scan time is approximately 1 min with secondary NRF scans requiring a few min per anomaly. Figure 23.4 highlights the primary scan capabilities of the system. 'Slices' of the container cross section are obtained as the cargo is transported through the scanner and used to reconstruct a three-dimensional image of the cargo (and in the case shown the cab and wheels). An example of an NRF secondary inspection of an anomaly detected via EZ-3D™ is shown Figure 23.5. Using NRF the anomaly is quickly identified as consistent with C4 high-energy explosives. The incorporation of these new NII technologies makes possible the automated detection of a wide range of threats (nuclear, explosives, etc.) and contraband. The first deployed SmartScan 3D™ system has been installed in the Port of Boston, MA, USA in 2016 under a contract from the US Department of Homeland Security, Domestic Nuclear Detection Office (DHS/DNDO).

Before proceeding to discuss the 'ideal' photon source for these new materials identification techniques, a brief overview of the properties of existing electron/photon sources is now given.

Passport full capability scanner

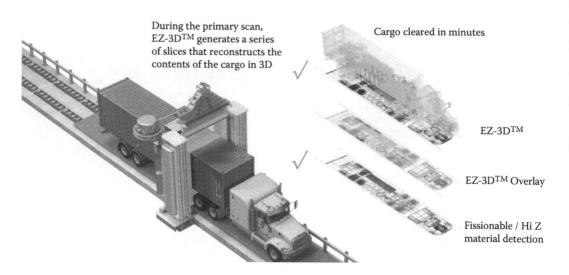

During the primary scan, EZ-3D™ generates a series of slices that reconstructs the contents of the cargo in 3D

Cargo cleared in minutes

EZ-3D™

EZ-3D™ Overlay

Fissionable / Hi Z material detection

This full configuration identifies anomalies and resolves potential threats

FIGURE 23.4 SmartScan 3D™ System including: EZ-3D™ 3D image and PNPF and EZ-3D™ overlays on high resolution transmission image. See text for details.

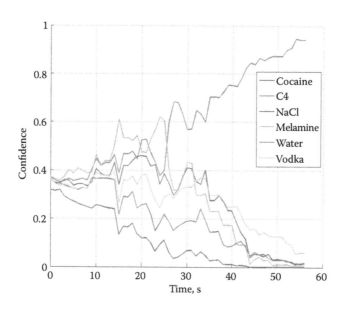

FIGURE 23.5 NRF secondary scan results for region of interest identified in primary scan by EZ-3D™ as a potential explosive.

TABLE 23.1   Electron Beam Requirements for Different NII Technologies Using *Bremsstrahlung* Photon Sources

Modality	Energy	Average Spectral Flux	Required Duty Cycle
Radiography	2–10 MeV	10^4–10^5 γ/s/eV 10–100 μA e-	30.1%
NRF	1–8 MeV	10^7 γ/s/eV–1 mA e-	10–100%
EZ-3D™	4–10 MeV	10^7 γ/s/eV–1 mA e-	10–100%
Prompt photo-fission	5.5–10 MeV	10^6 γ/s/eV 100 μA e-	10–100%

23.4 Summary of Existing Bremsstrahlung Photo Source Used in NII

As already mentioned, all commercial cargo scanners use *bremsstrahlung* beams for nonintrusive inspection. The beam requirements using a *bremsstrahlung* high-energy photon source for NII for different imaging modalities are summarized in Table 23.1. All currents represent average values over one second. The variation in duty cycle requires that the average current be fixed but the duty cycle can vary, depending on the geometry and density of cargo and ultimately the bandwidth of the detectors. Typically, radiography utilizes integrating detectors that measure integrated transmitted flux from each beam pulse with average beam currents between 10–100 μA yielding spectral fluxes (few MeV photons) of order 10^4–10^5 γ/s/eV. The other techniques require the measurement of the energy of individual photons or neutrons.

Typical NRF or EZ-3D™ measurement for large cargo screening requires of order 10^7 photons/sec/eV in 10^{-4} steradians over time intervals of ms to a minute (for shielded NRF targets). To achieve such high photon spectral flux values using bremsstrahlung requires e- beam currents of order 1 mA CW yielding a delivered power of order 10 kW. For each of the new materials identification techniques, only a portion of the *bremsstrahlung* spectrum is useful as discussed below. This unnecessarily increases dose and background to the cargo without any inspection benefit. The next section discusses how laser-driven particle acceleration could lead to major improvements in NII.

23.5 Future of NII: Sources of Nearly Monochromatic High-Energy Photons

Existing *bremsstrahlung* photon sources with high duty cycle have enabled, for the first time, the practical implementation of NII with elemental determination. However, in varying degrees, these new technologies utilize only a portion of the *bremsstrahlung* spectrum thus increasing delivered dose and higher backgrounds. Monochromatic or quasi-monochromatic sources could improve some of these shortcomings. Table 23.2 shows a comparison of various properties of *bremsstrahlung* and monochromatic

TABLE 23.2   Comparison of *Bremsstrahlung* and Monochromatic Photon Sources for NII

Property	Bremsstrahlung	Monochromatic
Photon flux	High	Low (at present)
Beam divergence	High	Low
Dose	Acceptable	Low
Background	High	Low
Multiple NRF States simultaneously excited	Many	1 (or few)
Inspection time	Acceptable	Not yet acceptable
Cost	Medium	High

photon sources. Of course, the details of the monochromaticity (e.g. energy width) matter in this analysis but in broad terms, monochromatic sources offer the possibility of much lower dose, more sensitive elemental differentiation (via better SNR), and greater source-to-cargo separation due to lower beam divergence. Such improvements would make NII safer, more sensitive and possibly faster. Of course, cost, footprint and reliability must be factored into the cost/benefit assessment. Nonetheless, given the great improvement that monochromatic sources would provide to NII, the potential commercial market for practical monochromatic photon sources is very large.

23.5.1 Examples of Existing Monochromatic MeV Photon Sources

There are a number of demonstrated sources of MeV monochromatic photons. There are three broad classes: 'active' electron sources: which includes coherent production (e.g. XFEL) and laser Compton backscatter, nuclear reactions and laser-driven particle accelerators (which are discussed in more detail in Chapter 4 of Part I). These different monochromatic photon sources are at different levels of maturity and have, at present, different 'sweet spots'. Since the focus of the present discussion is to derive requirements for the use of monochromatic photon sources in NII, the example of laser backscatter off energetic electrons (i.e. Compton backscatter of laser light from electrons) suffices to illustrate the properties required to make a useful system.

The LINAC, laser and photon production capabilities of the Lawrence Livermore National Laboratory T-REX laser-Compton 2 MeV facility serve as a good example of a monochromatic source applied to NRF measurements. This test facility [Bertozzi 2011a] demonstrated the feasibility of using the T-REX photon beam to perform NRF transmission measurements through dense cargo. The 235 MeV electron beam is produced by a high brightness photoinjector and accelerated by an rf LINAC. The 2.33 eV photons were produced by a frequency-doubled Nd:YAG laser. The energy of the upscattered photons is 2 MeV. The product of the number of photons per pulse (4.6×10^8), and the repetition rate (1 kHz) divided by the energy spread (0.5% at 2 MeV) yields approximately few $\times 10^7$ photons/sec/eV in the beam. This is a practical requirement for NRF measurements and is equivalent to that discussed above using *bremsstrahlung* beams. With the T-REX beam, dose and background are reduced by approximately a factor of 100 for detection of a single isotope. However, an important consideration is the setup time for a new energy. Multi-element determination is required for many threats such as high energy explosives where the ratios of carbon, oxygen and nitrogen are required to differentiate them from other plastics. Further, cost, footprint and reliability will need to be improved if such a technology can be applied broadly for NII.

23.5.2 Requirements for Monochromatic Sources in NII Applications

We are now in the position to summarize the requirements of monochromatic sources used in the next generation NII technologies. We will summarize our analysis for each NII technology with explicit recommendations for what the ideal photon source requirements are. In all cases, the photon source needs to have enough output power to satisfy NII requirements as previously discussed so this requirement is implicit for all photon sources.

23.5.2.1 X-ray Transmission Imaging

Dual-energy *bremsstrahlung* photon sources provide some degree of material discrimination in X-ray transmission images. Multiple monochromatic sources would allow a much better degree of elemental separation than that achieved with dual-energy *bremsstrahlung* sources with much lower background. However, for a practical multi-energy NII, the multiple energies (e.g. 4 and 9 MeV) must be interleaved (i.e. alternated every other pulse) with a frequency of at least 1 kHz frequency otherwise the cargo will have to be independently transported through the scanner at each energy thus increasing the scan time. Besides better material separation, the dose delivered in this scheme may be low enough that the driver of the truck could be safely irradiated making scan times shorter.

23.5.2.2 EZ-3D™

The EZ-3D™ technology most effectively utilizes photon energies above 2.0 MeV. Therefore, a quasi-monochromatic photon source with a low energy cut off would significantly reduce the dose to cargo without a reduction in material discrimination. EZ-3D™ does require that the beam be scanned across the cargo rapidly since this is essential in producing a 3D image of the cargo composition.

23.5.2.3 PNPF

PNPF has essentially the same requirements as EZ-3D™ but, in this case, the energy threshold is approximately 6 MeV. Moreover, comparing the production of high-energy neutrons at 8 and 12 MeV for example (if allowed by the World Health Organization for cargo screening) would increase sensitivity and reduce false alarms due to neutron photo-production from nonfissionable isotopes. Polarized photon beams may also make possible the differentiation of fissile versus fissionable materials.

23.5.2.4 NRF

NRF has the most stringent NII beam requirements including: CW e-beam, rapid beam energy changes since NRF states exist over a very wide range, 1–9 MeV photon energy. In many NII applications, the list of isotopes of interest is extensive. For example, the detection of high explosives requires the measurement of nitrogen, carbon and oxygen NRF states that span the energy region from 2.3 to 7.1 MeV (see Figure 23.1). For this reason, a *bremsstrahlung* spectrum is very useful since it can essentially sample all NRF states simultaneously although the SNR may not be ideal at a single endpoint energy. There are situations where the delivered dose may be prohibitive and a monochromatic source that could rapidly change energy would be highly advantageous both in dose reduction and increased SNR. This is the Holy Grail of photon sources!

In the case of NRF, the energy width is also decisive in determining the properties of the photon detector. Figure 23.6 illustrates some of the essential points in this regard. If the energy width is smaller than the Doppler broadened NRF state from zero-point motion, then a very sensitive differential measurement can be made by performing on- and off-resonance measurements [Pruet 2006]. The off-resonance measurement should be performed 'just' off- resonance (i.e. near resonance) so that the 'normal' absorption properties of the resonantly absorbed and subsequently emitted photons are identical. Not only is this a very accurate differential measurement which minimizes systematic measurement error, but in transmission measurements (if practical) it can be accomplished with detectors with limited energy resolution. In fact, under proper geometry and shielding conditions, an integrating detector with no

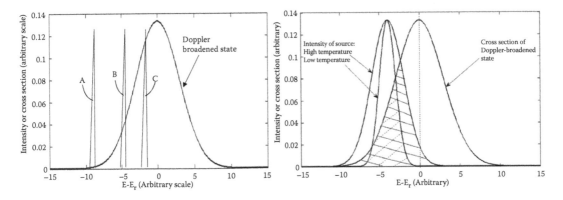

FIGURE 23.6 Relationship between photon beam energy width and detector specification used in NRF measurements. (From Bertozzi, W. and Ledoux, R.J. 2011. "Methods and Apparatus for the Identification of Molecular and Crystalline Materials by the Doppler Broadening of Nuclear States Bound in Molecules, Crystals and Mixtures Using Nuclear Resonance Fluorescence" US patent 8,023,618 issued 9/2011.)

energy resolution would be adequate. This could significantly reduce the cost of the NII system. In 3D mode, many benefits are still accrued by this differential process using a detector with medium energy resolution which is necessary since scatter and pair production photons must be separated from the NRF photons for high SNR. The efficacy of this differential technique is good for beam energy widths of less than a few 10's of eV with a reduction in usefulness as the width increases [Bertozzi 2011b].

There is also another intriguing use of NRF made possible by beam widths of order 10 eV or less. It has been observed via NRF that, as predicted, the vibrational state of $^{15}N_2$ adhering to a surface selects one specific vibrational mode [Mureh 2000]. This measurement used a neutron capture gamma emission from ^{54}Cr that was coincidentally within 30 eV of a ^{15}N NRF state at 6324 keV. The degree of overlap between this monochromatic beam and the two $^{15}N_2$ zero-point vibration modes which broaden the intrinsic NRF state width determined how the molecule adhered to a graphite surface. Thus, there is a real possibility that energy-selective monochromatic sources of narrow width may further enhance NRF's NII capabilities by differentiating isotopes in different molecular configurations. This would be a major breakthrough in NII.

23.6 In Conclusion: Laser-Driven NII Requirements

Given all of these considerations, the wish list of photon source properties in a coarsely prioritized order is:

- CW or high duty cycle greater than approximately 10%
- Rapidly tunable from 1 to 10 MeV (higher if WHO allows)
- Narrow energy width $\approx$ 10 eV
- Flexible beam scan mechanism
- Controlled beam divergence of order 1–10 mrad
- Reliable
- Small footprint
- Low cost

It is important to note that some of these requirements are dependent on the current state of the art of photon and neutron detectors. For example, a laser-driven photon source with narrow energy width (order 1 eV) enables the use of integrating detectors in many scenarios thereby lessening the CW and high duty cycle requirements. Thus, although many of the requirements of Table 23.1 will need to be met in a practical NII system, the unique features of laser-driven photon sources may drive other innovations in detectors to exploit its advantages. Ultimately the overall acceptance of these new photon sources will depend on the commercial features of reliability, cost and footprint. However, it is the author's belief that once the full power of the tunable monochromatic sources is demonstrated, the resources will be found to make them practical.

References for Chapter 23

Bertozzi, W. and Ledoux, R.J. 2005. "Nuclear Resonance Fluorescence imaging in non-intrusive cargo inspection" *Nucl. Inst. Meth. B.* **241**: 820–825.

Bertozzi, W., Korbly, S.E., Ledoux, R.J. and Park, W. 2007. "Nuclear resonance fluorescence and effective Z determination applied to detection and imaging of special nuclear material, explosives, toxic substances and contraband" *Nucl. Inst. Meth. B.* **261**: 331–336.

Bertozzi, W., Franklin, W., Korbly, S.E., Ledoux, R.J., Niyazov, R. and Swenson, D.R. 2011a. "Accelerators for Homeland Security" *Int. Journal of Mod. Phys. A* **26** [10 & 11]: 1713–1735.

Bertozzi, W. and Ledoux, R.J. 2011b. "Methods and Apparatus for the Identification of Molecular and Crystalline Materials by the Doppler Broadening of Nuclear States Bound in Molecules, Crystals and Mixtures Using Nuclear Resonance Fluorescence" US patent 8,023,618 issued 9/2011.

Danagoulian, A., Bertozzi, W., Hicks Jr. C.L., Klimenko. A.V., Korbly. S.E., Ledoux, R.J. and Wilson, C.M. November 2010. "Prompt Neutrons from Photofission and its use in Homeland Security Applications" presented at the 2010 IEEE Conference on Technologies for Homeland Security, Waltham, Massachusetts.

Moreh, R., Finkelstein, Y., Nemirovsky, D. 2000. "Nuclear Resonance Photon Scattering Studies of N_2 adsorbed on Grafoil and $NaNO_2$ Single Crystal" *J. Res. Natl. Inst. Stand. Technol.* **105**: 159–166.

Pruet, J., McNabb, D.P., Hagmann, C.A., Hartemann, F.V., Barty, C.P.J. 2006. "Detecting clandestine materials with nuclear resonance fluorescence" *Appl. Phys.* **99**: 123102.

24

ALPA Conclusion

Paul R. Bolton

Katia Parodi

Jörg Schreiber

Much more needs to be done as we seek to better understand the science and develop technologies relevant to laser-driven particle acceleration. Incremental progress can be further motivated and marked by its association with meaningful technologies and applied investigations. The Chapters of Part II (as well as final sections of Chapter 6 and final application comments in Chapters 2, 3, and 4 in Part I) have clearly indicated a broad range of potential uses for initial consideration and each with unique beam requirements. Most of the described applications in this book relate to laser-driven protons (and other ions). But, as stated already, we regard these examples as an initial subset of what is possible.

24.1 Comments on Presented Application Requirements

Commensurate with our expectations for accelerators in general and particularly for laser-driven sources, a great variety of beam requirements (for particles and photons) have been explicitly given in some of the Part II chapters. Without repeating all these application topics, we make some comments in what follows about important aspects of laser-driven sources.

The applications include the use of particles and photons as a diagnostic probe and/or pump (in this case meaning a driver bunch meant to affect some result without necessarily implying a pump-probe configuration) in an assortment of experimental methodologies. Good examples of the probe application can be found in Chapters 6 ('Neutron Generation'), 12 ('Charged Particle Radiography and Tomographic Imaging'), 17 ('Nuclear Reaction Analysis of Li-ion Battery Electrodes by Laser-Accelerated Proton Beams'), 18 ('Possible Roles of Broad Energy Distribution in Ion Implantation and Pulsed Structure in Perturbed Angular Distribution Studies') and 23 ('Advances in Nondestructive Elemental Assaying Technologies').

This broad range of requirements is anticipated. For example, applications presented in Chapters 11 ('Laser-driven Ion Beam Radiotherapy (LIBRT)'), 12, 13 ('Radioisotope Production and Application'), 15 ('Space Irradiation Effects on Solar Cells'), 17, 19 ('A Compact Proton Linac Neutron Source at RIKEN'), 20 ('Neutron Science with Highly Brilliant Beams') and 23 will likely require controlled repetition-rated bunches delivered by beam lines of greater technical sophistication than others which might need relatively little more than a filtered, collimated portion of the emergent spray from the plasma source region. Note that Chapters 13 and 23 call for relatively high repetition rates at the KHz level.

The high repetition rate is typically needed in cases where the final integration over many (possibly thousands or more) bunches (and therefore laser pulses) is required in order to accumulate a desired net fluence (number of particles/cm^2) or dose. This is especially the case in work described in Chapters 11, 13, 15, 17 and 23, where integration times can be several minutes or hours. More specifically, laser-driven ion beam radiotherapy (LIBRT) discussed in Chapter 11 can require, in a single fraction, thousands of bunches (each containing about 10^9 protons) at a moderate repetition rate (for example near 10 Hz) for a few minutes summing to few Gy dose levels. We note that, for a given delivered dose level, accuracy can be improved at higher repetition rates with fewer protons per bunch, where the dose in any region is then the integrated result of multiple bunches. For reaching desired activity levels with radioisotope production discussed in Chapter 13, integrating of order 10^5 proton bunches (each containing about 10^{10} protons) at a high repetition rate (of order 1 KHz) over several minutes can be necessary. For ground-based solar cell irradiation described in Chapter 15, integrating of order 10^7 bunches (each containing about 10^{10-11} particles) at a high repetition rate (of order 1 KHz) for a few hours yields appropriate net fluence levels near 10^{15-16} particles/cm^2 over typically large solar cell areas (for convenience in this example, we assume a 100 cm^2 area). For these important cases, stable, controlled bunch delivery over extended time intervals can require adequate bunch-to-bunch stability in addition to long-term reliability. This is the ILDIAS (ILDEAS) requirement for ions (electrons).

Spectral requirements for particles and photons given in Part II also vary significantly. By 'spectral' we refer generally to the energy dependence of a differential spectral amplitude. It can be typically specified in different ways (from first to fourth order) as in the following: the spatially integrated 'number of particles (or photons) per (energy slice)', $\dfrac{dN}{dE}$ for which the energy integration is just the total number of particles (or photons) in a single bunch; the 'number of particles (or photons) per (energy slice x unit solid angle)', $\dfrac{d^2N}{dEd\Omega}$ for which the energy integration sums to the single bunch differential amplitude in units of 'number of particles (or photons) per (unit solid angle)'; the 'number of particles (or photons) per (energy slice x unit solid angle x unit area)', $\dfrac{d^3N}{dEd\Omega dA}$; the 'number of particles (or photons) per (energy slice x unit solid angle x unit time)', $\dfrac{d^3N}{dEd\Omega dt}$ or the 'number of particles (or photons) per (energy slice x unit solid angle x unit area x unit time)', $\dfrac{d^4N}{dEd\Omega dAdt}$, which is also photon beam brilliance. The energy requirements provided in some Part II chapters give us an approximate idea about the spectral needs of an application. Although most of the proton (ion) applications included in this work require a few to 10's of MeV (MeV/u) kinetic energies, the applications of Chapters 11 and 12 are more challenging, where proton energies must at least be 250 to 300 MeV, and Chapters 14 ('Space Radiation and its Biological Effects') and 23, where electron energy requirements can be at the 100's of MeV to GeV level.

Perhaps the most valuable application enabler (which is not explicitly included in Part II) is the dedicated laser-driven facility itself that can serve as an application resource as well as a hub for relevant fundamental science and new technology (for example, radiobiology, radiochemistry, material studies, intense laser pulse diagnostics based on laser–plasma yields, etc.). It would also be a natural test bed for developing integrated system components such as instrumentation, repetition-rated targetry and beam line optics for delivery (therefore the basis for ILDIAS (ions) or ILDEAS (electrons) development for example). Such a facility could be further developed to also feature capability for delivering synchronized multiple particle and photon sources (i.e. multiple radiation fields, modalities or qualities) in beams that offer precise control of relative timing on account of a shared laser-driver. This would clearly require sophisticated multiple parallel targets, each in vacuum chambers that would be irradiated by an energy sampled 'beamlet' taken from the main pulse of a high-power laser-driver. In this way, the targetry design (and even repetition rate) in each chamber could be independently optimized according to the desired particle or photon yield. Optical delay lines upstream of each chamber could then control with high precision the relative timing between these multiple beam sources. The material presented in this book makes clear that such a synchronized, multi-source facility is uniquely possible with a high-power laser driver. This is a unique concept for new accelerator design.

24.2 Prospects for the Laser-Driven Case

One can expect that initial (near-term) applications might be those that take advantage of some of the following four features: (i) use the emergent spray from the laser–plasma with minimal (or no) beam collection optics, (ii) pose low kinetic energy requirements for particles (for example, few MeV protons or 100 MeV electrons), (iii) require low (~ few Hz) repetition rates where single-shot is included as the low limit and (iv) be realistically doable in the near term. Because less energetic laser systems (i.e. lower pulse energy) are more likely capable of higher repetition rate operation, we can expect criteria (ii) and (iii) to be coupled. Note that laser-driven neutron production (in Chapter 6) appropriately exemplifies some of these attractive features. Although near-term possibilities might presently be underdeveloped, we anticipate that many more new ideas will soon emerge for realistic executable programmes. Near-term (and even intermediate-term) applications can be suitably considered as *en route* to the more challenging longer-term ones which can mandate more advanced development. In this sense, they are effectively 'keep-alive' efforts that help sustain and promote the essential ongoing development necessary for realizing more advanced integrated laser-driven systems for distant aspirations such as laser-driven ion beam radiotherapy.

Where higher particle energy or higher bunch charge (at a given energy) is needed, higher laser pulse energy (and therefore single pulse power) is also needed (and typically at the cost of limiting repetition-rate capability). All-optical solutions ('all-optical' here means laser-driven) for the integrated system rely completely on the high-power laser-driven source(s) with or without added laser-driven post-acceleration sections. Choices for this kind of laser-driven injection, acceleration stages and related beamline architecture will ultimately be made based on particle energy requirements and net acceleration efficiency. Laser-driven injection might also be used with post-acceleration sections that are not laser-driven (i.e. the hybrid integrated systems such as HILDIAS for ions or HILDEAS for electrons). A possible configuration is the use of a conventional RF accelerating section downstream of a laser–plasma source. In general, some kind of post-acceleration is warranted if the laser-driven source alone cannot produce enough particles of adequate application-specific energy. Although not the direct subject matter of this book, the multi-staged integrated accelerator system (for example, ILDIAS or its hybrid, HILDIAS) is more complicated, and should be treated as the general laser-driven case.

Candidate source studies for highest achievable particle energy yields demand lasers with extremely high, single-pulse powers (of order 100's TW to several PW) operating at very low repetition-rates; typically in the 'single-shot' mode. Note that it is experimentally well-established that high laser power alone is not enough; specialized single shot technologies (for example, for pulse shaping and diagnostics) are needed to obtain optimum particle yields with these high power systems. Understandably, 'tailoring' both laser pulses and targetry will remain critical as we 'finesse' laser–plasma interactions to better control and understand the underlying physics and related design issues for producing optimum sources, especially at extreme laser intensities. However, an inclusive development strategy also calls for parallel development of repetition-rated beams with lower particle energies generated at relatively reduced laser intensities. Parallel advancement means that, in accordance with incremental successes, key aspects of these two parallel strategic paths can merge with increased energy over time.

We require integrated accelerator systems like ILDIAS to deliver stabilized, controlled, well-directed repetition-rated particle beams. In the near-term, we understandably anticipate ILDIAS (ILDEAS) prototypes to be large and costly based on footprints for current extreme power lasers and anticipated particle beam lines (so far, it is only the laser–plasma source at the target site that is relatively quite small). For the laser driver, in particular, there is a communal hope for industrially lead inflective behaviour that will establish a market-driven trend toward cost and size reduction. Although the present status of recently developed repetition-rated, solid-state pump laser systems is encouraging, neither the likelihood nor the time scale for significant cost and size reduction has been adequately addressed. Nonetheless, one must be careful to appropriately use cost and size projections and comparisons when assessing the relative merit of laser-driven and conventional accelerator systems for a given application (a typical example being laser-driven ion beam radiotherapy, LIBRT). If cost and size are prime motivators for using a laser-driver then it is critical to independently verify that, with the desired reductions, all state-of-the-art scientific and technical requirements can be met without impactful compromise. Furthermore, the relevant state of the art must not be the current one, but instead one that is projected to a future timescale compatible with the time estimated to realize the desired cost and size improvements. Otherwise, beam parameter and performance 'trade-offs' or compromises must be considered, altering merit assessments that will likely be more complicated. It is therefore strongly advisable to develop laser-driven sources with the motivation to enable new modalities and effects that simply cannot be achieved by any other means. This is a uniqueness strategy.

Mindful of near and distant horizons for laser-driven acceleration, it is useful in this closing summary to review in compact listed form the important aims of this book that are addressed throughout the Introduction (*italic font* highlights some key terms):

Inform experts and non-experts about laser-driven particle acceleration, associated energetic photon and neutron generation and the large assortment of research opportunities uniquely afforded by applying such novel source capability

Appreciate the present early stage development of laser-driven candidate sources and the *distinction of the integrated laser-driven particle accelerator system* as the integrated 'machine' apart from the application

Inspire growth of pioneering and engaging communities representing a broad spectrum of scientific and technical inquiry and *motivate potential users to envision and develop meaningful and doable applications* that can realistically advance the science and technology of the laser-driven case

Motivate the reader to *adopt a practical comprehensive view* of achieved progress with 'source' studies, usher a *heightened awareness* of trends in evolving capability and *consider expansive future prospects*

Appreciate dual, reciprocal and symbiotic character of multiple laser uses in pioneering investigations of high power lasers and enabled extreme laser–plasma studies of the petawatt era

Bring *practical added value* to laser-driven acceleration schemes sustaining its development for the longer term

Emphasizing together these applications and aims highlights the unique applied potential for laser-driven particle acceleration. From an accelerator viewpoint, the more constructive and innovative strategy is the complementary one that pursues research, development and applications, which are clearly not possible with conventional accelerators. As already stated in the introduction, intrinsic generation of ultra-short and short (fsec to psec) particle bunches (at and very near the source) and the generation of synchronous multiple beams (particle and photon) are accessible unique features. For example, space solar cell irradiation well exemplifies an important need for simultaneous energetic electron and proton beams that cannot be met by any conventional machine. We must pursue such a visionary strategy that exploits uniqueness (that is, a uniqueness strategy as opposed to a replicative one), adding new or improved capability and augmenting what accelerators can do overall as the laser-driven contribution to accelerator technology matures. This can include even those features that seem adverse to accelerator development and therefore pose a great challenge (such as the significant energy spread and angular divergence of typical at-source particle yields). The ensuing specialized niche applications will afford greater options for research and development.

Further, as the uniqueness strategy is not replicative with respect to any application, it promotes uniquely additive contribution to accelerator concepts, design and capability. So, it does not aim specifically to compete with or replace conventional machines which, in general, are demonstrable evidence of a history of outstanding accelerator accomplishments. As implied in the introduction, it is in this spirit that lasers have already been contributing to accelerator advancement for decades.

At the basic level, the natural exploration of laser-driven acceleration is an instinctive progression from our accelerator history in which our interests can be consuming with passionate motivations. However, in more practical terms, an assortment of meaningful doable applications must be the key part of the justification for the laser-driven case and serve as lodestar. In this sense, the practical foundation for laser-driven particle acceleration and associated photon and neutron source generation must be achieving levels of scientific and technical maturity that inspire and foster continued design and implementation of realistic meaningful applications with an enlightened creative vision.

Index

Page numbers followed by f and t indicate figures and tables, respectively.